T0281030

Lecture Notes in Computer Science 14499

Advanced Research in Computing and Software Science

Subline of Lecture Notes in Computer Science

More information about this series at https://link.springer.com/bookseries/558

Rayna Dimitrova · Ori Lahav · Sebastian Wolff
Editors

Verification, Model Checking, and Abstract Interpretation

25th International Conference, VMCAI 2024
London, United Kingdom, January 15–16, 2024
Proceedings, Part I

 Springer

Editors
Rayna Dimitrova 🔟
CISPA Helmholtz Center for Information
Security
Saarbrücken, Germany

Ori Lahav 🔟
Tel Aviv University
Tel Aviv, Israel

Sebastian Wolff 🔟
New York University
New York, NY, USA

ISSN 0302-9743   ISSN 1611-3349 (electronic)
Lecture Notes in Computer Science
ISBN 978-3-031-50523-2   ISBN 978-3-031-50524-9 (eBook)
https://doi.org/10.1007/978-3-031-50524-9

This Springer imprint is published by the registered company Springer Nature Switzerland AG
The registered company address is: Gewerbestrasse 11, 6330 Cham, Switzerland

Paper in this product is recyclable.

Preface

This volume contains the proceedings of VMCAI 2024, the 25th International Conference on Verification, Model Checking, and Abstract Interpretation. The VMCAI 2024 conference was co-located with the 51st ACM SIGPLAN Symposium on Principles of Programming Languages (POPL 2024), held at the Institution of Engineering and Technology in London, UK, during January 15–16, 2024.

VMCAI is a forum for researchers working in verification, model checking, and abstract interpretation. It attempts to facilitate interaction, cross-fertilization, and advancement of methods that combine these and related areas. The topics of the conference include program verification, model checking, abstract interpretation, program synthesis, static analysis, type systems, deductive methods, decision procedures, theorem proving, program certification, debugging techniques, program transformation, optimization, and hybrid and cyber-physical systems.

VMCAI 2024 received a total of 88 submissions, of which 74 went through the peer review process (14 were desk-rejected). After a rigorous single-blind review process, with each paper reviewed by at least three Program Committee (PC) members, followed by an extensive online discussion, the PC accepted 30 papers for publication in the proceedings and presentation at the conference. The main selection criteria were quality, relevance, and originality. Out of the 30 accepted papers, four are tool papers and one is a case study, while the rest are regular papers. In addition to the contributed papers, the conference program included two keynote talks: David Harel (Weizmann Institute of Science, Israel) and Hiroshi Unno (University of Tsukuba, Japan).

By now, artifact evaluation is a standard part of VMCAI. The artifact evaluation process complements the scientific impact of the conference by encouraging and rewarding the development of tools that allow for replication of scientific results as well as for shared infrastructure across the community. Authors of submitted papers were encouraged to submit an artifact to the VMCAI 2024 artifact evaluation committee (AEC). We also encouraged the authors to make their artifacts publicly and permanently available. All submitted artifacts were evaluated in parallel with the papers. We assigned two members of the AEC to each artifact and assessed it in two phases. First, the reviewers tested whether the artifacts were working, e.g., there were no corrupted or missing files and the evaluation did not crash on simple examples. For those artifacts that did not work, we sent the issues to the authors, for clarifications. In the second phase, the assessment phase, the reviewers aimed at reproducing any experiments or activities and evaluated the artifact based on the following questions:

1. Is the artifact consistent with the paper and are the results of the paper replicable through the artifact?
2. Is the artifact well documented and easy to use?
3. Is the artifact available?

We awarded a badge for each of these question to each artifact that answered it in a positive way. Of the 30 accepted papers, there were 14 submitted artifacts, all of which were awarded two or all three Artifact Evaluation Badges.

The VMCAI program would not have been possible without the efforts of many people. We thank the research community for submitting their results to VMCAI and for their participation in the conference. The members of the Program Committee, the Artifact Evaluation Committee, and the external reviewers worked tirelessly to select a strong program, offering constructive and helpful feedback to the authors in their reviews. The PC and the external reviewers contributed a total of 233 high-quality reviews to the review process. The VMCAI steering committee provided continued encouragement and advice. We warmly thank the keynote speakers for their participation and contributions to the program of VMCAI 2024. We also thank the general chair of POPL 2024, Philippa Gardner, and the organization team for their support. We thank the publication team at Springer for their support, and EasyChair for providing an excellent conference management system.

November 2023   Rayna Dimitrova
                                                                   Ori Lahav
                                                              Sebastian Wolff

Organization

Program Committee Chairs

Rayna Dimitrova	CISPA Helmholtz Center for Information Security, Germany
Ori Lahav	Tel Aviv University, Israel

Artifact Evaluation Committee Chair

Sebastian Wolff	New York University, USA

Program Committee

Ezio Bartocci	TU Wien, Austria
Nathalie Bertrand	Inria, France
Emanuele De Angelis	IASI-CNR, Italy
Coen De Roover	Vrije Universiteit Brussel, Belgium
Jyotirmoy Deshmukh	University of Southern California, USA
Bruno Dutertre	Amazon Web Services, USA
Michael Emmi	Amazon Web Services, USA
Grigory Fedyukovich	Florida State University, USA
Nathanaël Fijalkow	CNRS, LaBRI, University of Bordeaux, France
Hadar Frenkel	CISPA Helmholtz Center for Information Security, Germany
Liana Hadarean	Amazon Web Services, USA
Jochen Hoenicke	Certora, Germany
Hossein Hojjat	Tehran Institute for Advanced Studies, Iran
Qinheping Hu	Amazon Web Services, USA
Marieke Huisman	University of Twente, The Netherlands
Amir Kafshdar Goharshady	Hong Kong University of Science and Technology, Hong Kong
Joost-Pieter Katoen	RWTH Aachen University, Germany
Daniela Kaufmann	TU Wien, Austria
Bettina Koenighofer	Graz University of Technology, Austria
Burcu Kulahcioglu Ozkan	Delft University of Technology, The Netherlands
Anna Lukina	Delft University of Technology, The Netherlands
Roland Meyer	TU Braunschweig, Germany
David Monniaux	CNRS / VERIMAG, France
Kedar Namjoshi	Nokia Bell Labs, USA
Jens Palsberg	University of California, Los Angeles, USA
Elizabeth Polgreen	University of Edinburgh, UK
Arjun Radhakrishna	Microsoft, USA

Robert Rand   University of Chicago, USA
Francesco Ranzato   University of Padova, Italy
Xavier Rival   INRIA / ENS Paris, France
Philipp Rümmer   University of Regensburg, Germany
Anne-Kathrin Schmuck   Max-Planck-Institute for Software Systems, Germany
Mihaela Sighireanu   ENS Paris-Saclay, France
Gagandeep Singh   VMware Research and UIUC, USA
Hazem Torfah   Chalmers University of Technology, Sweden
Zhen Zhang   Utah State University, USA
Lenore Zuck   University of Illinois Chicago, USA

Additional Reviewers

Armborst, Lukas
Balakrishnan, Anand
Ballarini, Paolo
Bardin, Sebastien
Beutner, Raven
Biere, Armin
Biktairov, Yuriy
Blicha, Martin
Boker, Udi
Boulanger, Frédéric
Cailler, Julie
Cano Córdoba, Filip
Chakraborty, Soham
Cheang, Kevin
Chen, Mingshuai
Chen, Yixuan
Chiari, Michele
Daniel, Lesly-Ann
Eberhart, Clovis
Esen, Zafer
Fleury, Mathias
Fluet, Matthew
Golia, Priyanka
Gruenke, Jan
Gupta, Priyanshu
Hamza, Ameer
Helouet, Loic
Huang, Wei
Klinkenberg, Lutz
Lammich, Peter
Larrauri, Alberto
Liang, Chencheng
Lopez-Miguel, Ignacio D.
Mainhardt, Ana

Maseli, René
Matricon, Théo
Meggendorfer, Tobias
Milanese, Marco
Mora, Federico
Morvan, Rémi
Mousavi, Mohammadreza
Mutluergil, Suha Orhun
Nayak, Satya Prakash
Parker, Dave
Paul, Sheryl
Piribauer, Jakob
Pranger, Stefan
Quatmann, Tim
Rappoport, Omer
Rath, Jakob
Refaeli, Idan
Riley, Daniel
Rubbens, Robert
Saglam, Irmak
Schafaschek, Germano
Shah, Ameesh
van der Wall, Sören
Vandenhove, Pierre
Viswanathan, Mahesh
Waldburger, Nicolas
Williams, Sam
Wolff, Sebastian
Xia, Yuan
Ying, Mingsheng
Zanella, Marco
Zavalia, Lucas
Zhao, Yiqi

Invited Keynote Talks

Two Projects on Human Interaction with AI

David Harel[1]

The Weizmann Institute of Science

A significant transition is under way, regarding the role computers will be playing in a wide spectrum of application areas. I will present two work-in-progress projects that attempt to shed new light on this transition.

The first we term *"The Human-or-Machine Issue"*. Turing's imitation game addresses the question of whether a machine can be labeled intelligent. We explore a related, yet quite different, challenge: in everyday interactions with an agent, how will knowing whether the agent is human or machine affect that interaction? In contrast to Turing's test, this is not a thought experiment, but is directly relevant to human behavior, human-machine interaction and also system development. Exploring the issue now is useful even if machines will end up not attempting to disguise themselves as humans, which might become the case because they cannot do so well enough, because doing so will not be that helpful, because machines exceed human capabilities, or because regulation forbids it.

In the second project, we propose a systematic programming methodology that consists of three main components: (1) a modular incremental specification approach (specifically, scenario-based programming); (2) a powerful, albeit error-prone, AI-based software development assistant; and (3) systematic iterative articulation of requirements and system properties, amid testing and verification. The preliminary results we have obtained show that one can indeed use an AI chatbot as an integral part of an interactive development method, during which one constantly verifies each new artifact contributed by the chatbot in the context of the evolving system.

While seemingly quite diverse, both projects have human-machine interaction at their heart, and bring in their wake common meta-concepts, such as trust and validity. What, for example, are the effects of the presence or absence of trust in such inter-actions, and how does one ensure that the interaction contributes to achieving one's goals, even when the motives of the other party or the quality of its contributions are unclear?

In addition, the kinds of interactions we discuss have a strong dynamic nature: multi-step interaction on requirements elicitation, for example, is not just a search for something that should have been specified in full earlier. Rather, it is often a constructive process of building new knowledge, acquiring new understanding, and demarcating explicit boundaries and exceptions that are often absent from even the most rigorous definitions.

---

[1] Research joint with Assaf Marron, Guy Katz and Smadar Szekely.

Automating Relational Verification of Infinite-State Programs

Hiroshi Unno

University of Tsukuba

Hyperproperties are properties that relate multiple execution traces of one or more programs. A notable subclass, known as k-safety, is capable of expressing practically important properties like program equivalence and non-interference. Furthermore, hyperproperties beyond k-safety, including generalized non-interference (GNI) and co-termination, have significant applications in security.

Automating verification of hyperproperties is challenging, as it involves finding an appropriate alignment of multiple execution traces for successful verification. Merely reasoning about each program copy's executions separately, or analyzing states from running multiple copies in lock-step, can sometimes lead to failure. Therefore, it necessitates synthesizing a scheduler that dictates when and which program copies move, to ensure an appropriate alignment of multiple traces. With this alignment, synthesizing relational invariants maintained by the aligned states enables us to prove k-safety. Additionally, verifying GNI and co-termination requires synthesis of Skolem functions and ranking functions, respectively.

In this talk, I will explain and compare two approaches for automating relational verification. The first approach is constraint-based, reducing the synthesis problem into a constraint solving problem within a class that extends Constrained Horn Clauses (CHCs). The second approach is based on proof search within an inductive proof system for a first-order fixpoint logic modulo theories.

Contents – Part I

SAT, SMT, and Automated Reasoning

Contents – Part II

Program and System Verification

Runtime Verification

Security and Privacy

Abstract Interpretation

Formal Runtime Error Detection During Development in the Automotive Industry

Jesko Hecking-Harbusch$^{(\boxtimes)}$, Jochen Quante, and Maximilian Schlund

Bosch Research, Renningen, Germany
{jesko.hecking-harbusch,jochen.quante,maximilian.schlund}@bosch.com

Abstract. Modern automotive software is highly complex and consists of millions lines of code. For safety-relevant automotive software, it is recommended to use sound static program analysis to prove the absence of runtime errors. However, the analysis is often perceived as burdensome by developers because it runs for a long time and produces many false alarms. If the analysis is performed on the integrated software system, there is a scalability problem, and the analysis is only possible at a late stage of development. If the analysis is performed on individual modules instead, this is possible at an early stage of development, but the usage context of modules is missing, which leads to too many false alarms.

In this case study, we present how automatically inferred contracts add context to module-level analysis. Leveraging these contracts with an off-the-shelf tool for abstract interpretation makes module-level analysis more precise and more scalable. We evaluate this framework quantitatively on industrial case studies from different automotive domains. Additionally, we report on our qualitative experience for the verification of large-scale embedded software projects.

1 Introduction

A vehicle comprises many software modules that are relevant for its safety. Think about the electronic stability control, which improves the stability of a vehicle by automatically applying the brakes at a wheel to avoid skidding. A malfunction of such a system blocking a single wheel could result in danger for the vehicle's occupants and other traffic participants. To prevent such incidents, the quality of the deployed embedded software is of great importance. The standard ISO 26262 [38] comprises the state-of-the-art of technologies to ensure the *functional safety* of electronic systems including their software in road vehicles. Functional safety requires the "absence of unreasonable risk due to hazards caused by malfunctioning behavior of electrical/electronic systems" [38,48].

A possible reason for software malfunctions is the occurrence of runtime errors, e.g., null pointer dereferences or array-out-of-bounds accesses. In languages like C and C++, runtime errors lead to undefined behavior, which is disastrous in terms of functional safety. ISO 26262 recommends *abstract interpretation* [23,50] to prove the absence of runtime errors because it is sound and can be scalable. Abstract interpretation provides a general framework based

R. Dimitrova et al. (Eds.): VMCAI 2024, LNCS 14499, pp. 3–26, 2024.
https://doi.org/10.1007/978-3-031-50524-9_1

on the theory of semantic abstraction to analyze software. Concrete states of the software are overapproximated by finitely many abstract states such that a fixpoint can be calculated in a finite number of steps. This fixpoint overapproximates all behavior of the software and is checked for runtime errors.

Soundness is an essential property of abstract interpretation, which requires that all runtime errors are found. So, if none are found, then the absence of runtime errors is proven. However, soundness comes at the price of possibly getting more runtime errors reported than there actually are. This means that in addition to the *true alarms* reported by abstract interpretation, which represent runtime errors, there can be *false alarms* that do not represent runtime errors but are caused by the overapproximation in abstract interpretation.

Commercial tools like Astrée [2], Polyspace Code Prover [4], and TrustInSoft Analyzer [6] enable the usage of abstract interpretation in industrial contexts [25, 31,40,51,52]. They are used for verifying safety-critical software in the aerospace industry, the automotive industry, and other industrial domains.

There are two ways to apply abstract interpretation in an industrial context: It can be applied to each isolated module separately, or on the integrated system consisting of all implemented modules and their interactions. The first approach can be performed in early development stages, but the usage context of modules is missing because other modules are not implemented yet. This leads to many false alarms. For the second approach, the analysis is only possible at a late stage of development when fixing errors causes enormous costs [14]. Furthermore, its runtime might not scale well with the size of the software. For such systems, the analysis can only be performed with low precision settings of the analyzer leading again to many false alarms. In general, static program analysis is perceived as burdensome when taking too long, producing too many false alarms, or causing considerable additional manual effort [20,29,39].

In this case study, we show how to make abstract interpretation applicable to individual modules of automotive software [28] early during development without the downsides mentioned above. Applying abstract interpretation early during development is the most efficient way to handle alarms as developers get feedback quickly and still know the code when evaluating and possibly fixing alarms [55].

We develop a general framework around an underlying tool for abstract interpretation. Our framework prepares the abstract interpretation task such that the tool for abstract interpretation can analyze it with higher precision. To this end, our framework automatically generates a *verification harness* for a module under analysis. The verification harness contains the module under analysis and additional code to check all functions and their interplay with meaningful inputs. We use Astrée [12,13,41] as the underlying tool for abstract interpretation. The core concepts of our framework can be easily transferred to other tools for abstract interpretation, but their technical realization might differ.

When generating the verification harness early during development, we rely on *contracts* at the interfaces of modules. Contracts enrich the syntactic interface between modules by semantic annotations. They can define that pointers represent arrays of certain lengths and can constrain the domain of variables.

They can also state preconditions and postconditions of functions and whether these functions are called for initialization or cyclically in an embedded system.

A further core component of our framework is the *automatic inference* of contracts. This is the only viable option to obtain contracts for large legacy code bases that cannot be manually annotated as this would cause excessive effort. For automatic inference of contracts, we use different sources of information. First, we translate information from interface descriptions of modules into contracts. Second, we utilize information from abstract interpretation results of previously analyzed modules. We propagate information over all modules and iterate until a fixpoint is reached. During this iterative refinement, the contracts become more and more precise, and thus the number of found alarms can decrease significantly.

We evaluate our framework on real-world projects from the automotive domains of driver assistance, braking, and cruise control and show that it reduces the number of found alarms significantly. For example, for the domain of braking, our framework decreases the number of alarms by approx. 50% while increasing code coverage by approx. 50%. By using contracts in module-level abstract interpretation, our framework makes it also possible for developers to use abstract interpretation early during development and benefit from low analysis runtimes for individual modules. Due to the automatic inference of contracts, they rarely need to add contracts manually. Thus, our framework enables developers to analyze modules in projects that were not analyzable in reasonable time before.

When deploying our framework in different projects from different business units, we learned that automation is key in successfully using abstract interpretation. Developers often lack time and need to justify how to spend it. To overcome this, our framework provides automation. People can get first results after a few days with our framework in comparison to needing up to a month to being able to use Astrée properly. We designed our framework to be adaptable to new projects with custom extensions to account for the specific architecture models, processes, and conventions used in the projects. The first results are often already helpful for developers and make them want to continue to use our framework. Developers appreciate the short runtimes of module-level analysis because they do not block their usual workflows. With our framework, developers can quickly benefit from years of experience in how to deploy abstract interpretation in the automotive domain.

Overall, this paper shows how to use abstract interpretation to detect runtime errors in real-world automotive software. Our key contributions are

- bringing together abstract interpretation and contracts to enable the verification of automotive software early during development,
- techniques for automatically inferring contracts of automotive software to minimize manual effort for developers,
- a quantitative evaluation of our framework of module-level abstract interpretation with automatically inferred contracts on three case studies from the automotive domain, and
- a qualitative report on our experiences of using our framework on large-scale production code early during development.

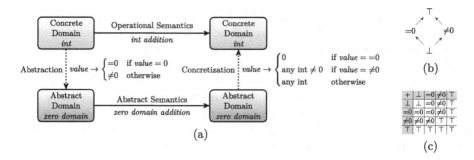

Fig. 1. Example of abstract interpretation for addition of two integers: The general idea is illustrated in 1a. The *zero*-lattice shown in 1b tracks whether a value is zero or not. The abstract semantics of the addition operator in the zero domain is given in 1c.

2   Background

Abstract interpretation [23,50] is a technique for static program analysis, i.e., the code is analyzed without executing it. The idea is to interpret the given code with *abstract values*. Abstract values can come for example from the *abstract domains* of intervals, linear equalities, or congruence relations on integers. Abstract values are elements of complete lattices. A *complete lattice* is a partially ordered set in which all subsets have both a *greatest lower bound* ⊓ and a *least upper bound* ⊔. The latter are generalizations of intersection and union on sets. The maximum and minimum elements ⊤ and ⊥ of the entire lattice are called top and bottom, respectively. *Abstract operations* on abstract values are monotonic functions. By repeatedly calculating the abstract operations for all instructions of the code, a fixpoint is reached after some iterations due to the monotonicity property. This ensures sound overapproximation of all possible concrete values.

Example 1. In Fig. 1, we perform an addition of two integer variables that both are abstracted by the zero domain, which tracks whether an integer variable is zero ($=0$) or nonzero ($\neq 0$). Figure 1a shows the abstraction and concretization functions. The lattice for the zero domain is given in Fig. 1b where the maximum element ⊤ encodes that the variable can be zero and nonzero, and the minimum element ⊥ encodes that the variable is neither zero nor nonzero. The minimum element can be helpful to encode that a certain variable is not reachable or not initialized. Abstraction of a concrete integer value is straightforward by using the appropriate member of the zero domain. Addition in the zero domain is defined by the table in Fig. 1c. For example, the sum of two variables equal to zero is also zero, whereas the sum of two variables different from zero can either be zero or nonzero and hence is ⊤. Notice that calculating the sum of two variables both abstracted by nonzero in the abstract semantics represents the addition of almost all concrete values for the two variables.

Using an adequate abstract domain, abstract interpretation can prove certain properties of software. For example, an interval domain may be used to prove that arithmetic overflows cannot happen for given range limitations of the input values. Current abstract interpretation tools like Astrée [12,13,41] can scale up to millions of lines of code. They guarantee soundness at the price of potentially many false alarms but offer the possibility to add annotations in order to feed the analysis with additional information that can greatly improve precision. However, even with a thorough understanding of abstract interpretation, intensive manual work by developers is required to come up with annotations that reduce the number of false alarms. To enable module-level analysis, further manual work is required to create *stubs* for emulating missing dependent functions and *drivers* for supplying subject functions with adequate input.

Next, we introduce annotations specific to Astrée that will be used throughout the paper. If they are encountered in code by Astrée, these annotations alter abstract values. The directive __ASTREE_modify((x;full_range)) changes the abstract value of the given variable x to $\top$, i.e., previous restrictions on the abstract value do not hold anymore. Assuming x is of type **unsigned char** and the interval abstract domain is used, Astrée continues calculations after the directive with the interval $[0, 255]$ for x. The directive __ASTREE_assert((x > 0)) makes Astrée check the given condition on the abstract value. Executing the previous two annotations in the given order without other instructions in between will result in the assertion failing as the abstract value will contain zero due to being set to $\top$ before. The directive __ASTREE_known_fact((x > 0)) allows defining conditions that are assumed as true to alter the abstract value. This makes it possible to make abstract values more precise. For the previous example of setting x to $\top$, this directive can exclude zero from the interval afterwards.

Contracts as introduced by Meyer refine interface specifications for software modules [49]. For a given function, they define that certain *preconditions* must be fulfilled when the function is called and that the function then guarantees certain *postconditions* when it returns. Below is an example for a *function contract*:

```
/// [[ requires: x >= 0.0 ]]
/// [[ ensures: return >= 0.0 ]]
float sqrt(float x);
```

Traditionally, such contracts are checked dynamically by adding assertions at the beginning and end of a function. In Sect. 3.2, we will see how contracts can be translated into Astrée annotations. Another class of contracts are *invariants*. They are two types of invariants: A class invariant specifies that a condition must always hold at interface borders, i.e., it is implicitly added to all preconditions and postconditions of the class. A global invariant is globally valid and thus must be checked whenever a relevant variable is modified. In this paper, global invariants on single variables are called *variable contracts*. Contract specifications for C++ are being discussed for inclusion in an upcoming C++2x version [30].

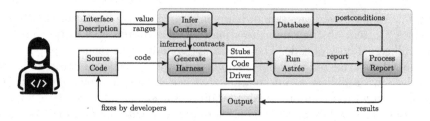

Fig. 2. The inner workings of our framework (highlighted in blue) and how developers interact with it are depicted. Source code and interface description for software modules are given to our framework, which internally uses Astrée for abstract interpretation. The results are aggregated over subsequent runs in a database to infer more contracts and output to the developers who then can fix potential runtime errors in their code. (Color figure online)

3 Module-Level Abstract Interpretation with Contracts

In *module-level analysis*, we apply abstract interpretation to a given module. There is some information missing in comparison with *integration analysis* where the entire software project is analyzed in late stages of development. Module-level analysis requires a starting point (i.e., a main function that invokes the module's functions), initialization of variables with abstract values, and meaningful stubs for missing functions. In this section, we describe how to automatically generate the *verification harness* [37] for module-level analysis based on results of code analysis and how contracts [49] can be used to enhance the harness.

Figure 2 shows how the first step called "Generate Harness" fits into our framework. The code to be checked along with the generated stubs and verification driver are given to Astrée for analysis. The report from the Astrée run is processed and interpreted with the knowledge of how the harness was generated. This is done to report potential runtime errors and contract violations to the developer and to infer additional contracts for subsequent runs. For inferring contracts, relevant abstract values are stored in a database shared across the complete software project.

3.1 Generating a Verification Harness

To generate a verification harness [37] for a given module, we perform the following three steps, which are shown as pseudocode in Algorithm 1:

(i) Dependency analysis (line 1): We identify the complete interface of the module under analysis, i.e., provided and required types, functions, and global variables, which is used in all subsequent steps.
(ii) Stub generation (lines 2 to 5): For each function the module under analysis depends upon, we generate a *stub* that returns ⊤ for all return values. Note that Astrée automatically generates such a stub for missing functions, but we will build upon this when using contracts in Sect. 3.2.

Algorithm 1: Verification of a Module.

Orange parts are added when using contracts.

Cyan parts are added for inferring contracts.

Input   : Module m, function contracts FC, variable contracts VC

Output: Potential runtime errors in m, return and variable values

1 perform dependency analysis of m;
2 include the code of m;
3 **foreach** *undefined but used function f* **do**
4 | generate a stub for f that considers $FC(f)$;
5 **end**
6 generate a main function header;
   /* start main function body   */
7 **foreach** *used global variable v* **do**
8 | generate initialization according to $VC(v)$;
9 **end**
10 create required C++ instances;
11 generate an endless loop with random switch;
12 **foreach** *function f to be called from driver* **do**
13 | generate a distinct case; // containing:
14 | initialization of each argument according to preconditions from $FC(f)$;
15 | call of f with these arguments;
16 | postcondition check for $FC(f)$;
17 | directives to output return and variable values;
18 **end**
   /* end main function body   */
19 $(alarms, values)$ = run Astrée analysis on this harness;
20 **return** $(alarms, values)$;

(iii) Verification driver generation (lines 6 to 18): We generate an entry point from where the analysis should commence following the basic structure of an embedded system:

```
initialization();
while(1) { cyclic_tasks(); }
```

During `initialization()`, global variables are set to $\top$ and required C++ instances are created. For `cyclic_tasks()`, all public functions of the module are called in all possible orders and supplied with all possible inputs.

Example 2. In (ii), the function `sqrt()` from Sect. 2 yields the following stub:

```
float sqrt(float x) {
  float res;
  __ASTREE_modify((res;full_range));
  return res;
}
```

The stub accepts all possible float values and returns all possible float values.

Example 3. Consider a module with a specific initialization function void g(), two public functions void f1() and int f2(int x), and a global integer variable glob_v. In (iii), our framework then generates the following code with comments added for explanatory purposes:

```
int glob_v;
// initialize global variables
__ASTREE_modify((glob_v;full_range));
// call the module's init-function
g();
unsigned char decision;
while(1) {
  // non-deterministic choice
  __ASTREE_modify((decision;full_range));
  // decision -> [0, 255]
  switch(decision) {
    case 0: {
      f1();
      break;
    }
    case 1: {
      int p1;
      // set f2's input to full range
      __ASTREE_modify((p1;full_range));
      int res = f2(p1);
      break;
    }
  }
}
```

This analysis driver calls the public functions f1 and f2 after the initialization function g in all possible call orderings such as g, f1, f1, ... and g, f2, f1, f2, ... as decision is set to full range.

3.2 Contracts and Their Effects

To increase the precision of Astrée when analyzing the verification harness, we have to substantiate stubs and driver. We use *contracts* [49] to do so. The contracts in our framework are geared towards proving the absence of *runtime errors* using abstract interpretation. They are not aimed at proving functional properties and are usually not expressive enough to do so. However, this lack in expressiveness makes them checkable with low overhead using tools like Astrée. Contracts can be provided to our framework by writing them as comments into header files. We illustrate the syntax in the following examples.

Pre- and Postconditions: Via requires, we can express conditions the inputs of a function need to fulfill, and via ensures, we can express guarantees on its outputs. This is used for two purposes: If the function is stubbed, then the contract refines the stub; if the function is under check, then the contract refines the driver.

Remember the two contracts for function sqrt() from Sect. 2. Using these contracts, the stub is refined to:

```
float sqrt(float x) {
    float res;
    // precondition:
    __ASTREE_assert((x >= 0.0));
    // postcondition:
    __ASTREE_modify((res;full_range));
    // res -> [-3.4028e38, 3.4028e38]
    __ASTREE_known_fact((res >= 0.0));
    // res -> [0.0, 3.4028e38]
    return res;
}
```

This refined stub leads to more precise results for modules that call function sqrt() without knowing its implementation. The reason for this improvement is that the sign of the return value is now known to the callers.

When checking the function, our verification driver calls it with the input [0.0, FLOAT_MAX], and the *ensures* condition is translated into assertions that check the implementation for compliance with the contract. The case calling sqrt() in the verification harness would be:

```
case 0: {
    float p;
    __ASTREE_modify((p;full_range));
    __ASTREE_known_fact((p >= 0.0));
    float res = sqrt(p);
    __ASTREE_assert((res >= 0.0));
    break;
}
```

Through this dual use of the contract, our framework a) ensures that an implementation fulfills the contract and b) allows analyzing function invocations with higher precision. The extensions for considering contracts in Algorithm 1 are highlighted in orange. Note that contracts for postconditions can also specify side effects on other variables than the return value, e.g., global variables or arguments passed by reference.

Array Specifications: In C/C++, pointers given to functions are often arrays. Our framework offers the possibility to specify that such a pointer is an array of a certain length. The right-hand side of this specification can refer to another argument, a global variable or constant, or a simple expression like var + const.

Consider the function memcmp from the C standard library, which compares the first n bytes of the block of memory pointed to by ptr1 to the first n bytes pointed to by ptr2. A possible contract for this function is:

```
/// [[ arrayspec: length(ptr1) >= n ]]
```

```
/// [[ arrayspec: length(ptr2) >= n ]]
int memcmp(const void *ptr1, const void *ptr2, size_t n);
```

In the generated stub, the length check will yield an invalid-dereference alarm in Astrée if it does not succeed:

```
int memcmp(const void *ptr1, const void *ptr2, size_t n) {
  ((char *)ptr1)[n-1]; // check length
  ((char *)ptr2)[n-1]; // check length
  int ret;
  __ASTREE_modify((ret; full_range));
  return ret;
}
```

The `arrayspec` contracts are also considered when creating input data for the corresponding function in the driver.

Invariants: Our invariants are expressions on global variables that should always hold. The example below shows that the counter of the current cylinder is always at most the maximum number of cylinders defined in the project.

```
/// [[ invariant: System_ctCyl <= NUM_CYL ]]
uint8 System_ctCyl;
```

We can also specify constraints on members of structs:

```
typedef struct {
  /// [[ invariant: Id <= LED_NUMBER ]]
  uint8 Id;
  [...]
} LED;
```

Our framework translates all such invariants into `__ASTREE_global_assert` statements, which makes Astrée check the condition each time the respective variable is modified.

Sequence Specifications: A developer can specify when functions from the public interface of the module under analysis are called. This can either happen initially once or cyclically afterwards (as introduced in (iii) in Sect. 3.1):

```
/// [[ sequence: init ]]
void initialization();
/// [[ sequence: cyclic ]]
void run();
```

In the generated driver, init functions are called before the while loop and cyclic functions in the cases within it. Cyclic functions often depend on prior initialization, so the initialization step included in the sequence specification prevents many false alarms in module-level analysis.

4   Automatically Inferring Contracts

Large code bases and legacy code often lack contract annotations, and adding them manually to the source code at scale is infeasible. To make our framework applicable to both large code bases and legacy code, automatic inference of contracts during module-level analysis is an essential part of it. Two ways to infer contracts are supported: First, we use interface specifications as contracts. Second, we derive additional contracts by leveraging results of previous Astrée runs. Both techniques correspond to the upper part of Fig. 2. Although inferring contracts from abstract interpretation is not a new idea (see related work in Sect. 8), we did not find prior work doing so in module-level analysis.

4.1   Contracts from Interface Specifications

There are several standards for software interfaces between automotive electronic control units. Examples are AUTOSAR [3] or ASAM MDX [1]. These interface descriptions ensure that modules in the automotive industry can be reused in different car models from different carmakers.

When value ranges for the incoming and outgoing values of a given module are defined in the interface descriptions in AUTOSAR, they are automatically parsed into preconditions, postconditions, and array specifications. For arrays, this is possible because they are explicitly represented as types with corresponding lengths. ASAM MDX focuses on messages being passed between modules. Here, value ranges can be automatically parsed into invariants.

Example 4. An AUTOSAR interface can look as follows:

```
<DATA-CONSTR>
  <SHORT-NAME NAME-PATTERN="{anyName}">
    RangeX
  </SHORT-NAME>
  [...]
  <PHYS-CONSTRS>
    <LOWER-LIMIT INTERVAL-TYPE="CLOSED">
      0.0
    </LOWER-LIMIT>
    <UPPER-LIMIT INTERVAL-TYPE="CLOSED">
      32000.0
    </UPPER-LIMIT>
  </PHYS-CONSTRS>
  [...]
</DATA-CONSTR>
```

The above AUTOSAR interface defines a closed interval $[0.0, 32000.0]$ that can be referred to as RangeX in the definition of a variable of a module and yields a function contract like: [[requires: x >= 0.0 && x <= 32000.0]]

Similar specifications exist for curves and maps that are used to adapt the software to the specifics of a given car model. Leveraging all this information turns out to be a great resource for contracts which drastically reduce the number of false alarms in module-level analysis.

4.2   Contracts from Abstract Interpretation

The second way our framework infers contracts is to automatically utilize results from previous runs of Astrée. This relies on the precision of Astrée.

Example 5. Consider the following function:

```
float foo(float x) {return 1.0f / bar(x);}
```

Knowing nothing about the function `bar()`, Astrée reports a potential division by zero in this function as `bar()` might return zero. However, we might have analyzed the implementation of `bar()` as part of some other module:

```
float bar(float x) {return max(x, 1.0f);}
```

For that function, Astrée detects that its return value is always greater than or equal to one yielding the inferred contract [[ensures: return >= 1.0]]. Storing this knowledge and reusing it for the analysis of function `foo()` then helps to prevent the false division by zero alarm in `foo()`.

Generally, after letting Astrée run on a module, we read out abstract values for all output and return messages from functions in the module and store them in a database. This corresponds to step "Process Report" in Fig. 2. The same is done for function parameters of stubbed functions to find out how these functions are used. Our framework extracts the resulting abstract values via additional directives in our generated verification harness (line 17 in Algorithm 1). In case of multiple outputs for the same variable or function, e.g., due to multiple writers or dynamic binding, these abstract values are combined via least upper bound over all possible results. They can then be used as preconditions in the next iteration and lead to more precise analysis of the functions that use these values.

More formally, let G be the set of global variables, L the abstract domain (a lattice), M the set of modules, F_m the set of functions in module $m \in M$, and $R_m : F_m \times G \to L$ the function that returns which value a global variable has after running a function from F_m. The function R_m returns $\bot$ when the function from F_m does not change the variable. Then, the possible values of a variable $v \in G$ after analysis of module m is calculated as follows:

$$\text{Value}_m(v) = \bigsqcup_{f \in F_m} R_m(f, v) \tag{1}$$

$\bigsqcup$ is the least upper bound function. To aggregate results from multiple modules, the same least upper bound function is used:

$$\text{Value}(v) = \bigsqcup_{m \in M} \text{Value}_m(v) \tag{2}$$

Algorithm 2: Runtime Error Analysis with Contract Inference.

    Input : Modules M, function contracts FC, variable contracts VC
    Output: Potential runtime errors in M, inferred function and variable
             contracts FC', VC'

1 FC'=empty, VC'=empty;
2 **repeat**
3   **forall** $m \in M$ **do**
4   ($alarms$, $values$) = run **Algorithm 1**(m, $FC \oplus FC'$, $VC \oplus VC'$);
5   update Value_m, Param_m, and Return_m based on $values$;
6   update Value, Param, and Return;
7   update FC' and VC' based on that;
8   **end**
9 **until** no more changes in FC' and VC';
10 **return** ($alarms$, FC', VC')

Analogous calculations are done for function parameters $\text{Param}(f, p)$ and return values $\text{Return}(f)$.

To create contracts based on these abstract values, knowledge about the abstraction is required. For an interval domain, the abstraction contains the lower and upper limit of possible values. Those can easily be translated to a contract: `[[ invariant: x >= lower && x <= upper ]]` For a set domain, the contract is a disjunction of possible values: `[[ invariant: x == 0 || x == 10 ]]` In the following, the sets of resulting contracts are denoted by FC' and VC' for function and variable contracts, respectively.

Algorithm 2 shows the overall algorithm for performing module-level runtime error analysis with contract inference. As input, we start with function contracts FC and variable contracts VC which both can be manually written or derived from standardized interfaces. Inferred function contracts FC' and inferred variable contracts VC' are empty initially (line 1). Variables not concerned by the contracts are set to the full range of abstract values. On each module, Algorithm 1 is run using the current set of contracts (line 4). The operator $\oplus$ denotes the merging of contracts, where only elements not already present on the left-hand side are added. In line 5, the results of abstract interpretation are used to update Value_m, Param_m, and Return_m according to Eq. 1. In line 6, the aggregation functions Value, Param, and Return are calculated based on all modules' results according to Eq. 2. Finally, the contracts FC' and VC' are derived from the aggregation functions using knowledge about the underlying abstract domains (line 7). The updated contracts are used to analyze the next module. This process is repeated, as new contracts may be identified that lead to changed results in a subsequent analysis run. Eventually, this process reaches a fixpoint, as contracts can only be refined by this process (additional and increasingly refined restrictions added via $\oplus$). The number of needed iterations depends on the number of modules with effects on each other and on the analysis order. Contract inference mostly finds pre- and postconditions.

Table 1. Characteristics of subject systems.

System	Driver Assistance (A)	Braking (B)	Cruise Control (C)
Language	C	C++	C++
KLOC	17	6	7
#functions	66	124	95
#ifs	70	102	36
#loops	49	1	28

A database is used to store results for individual modules to enable analysis in a distributed setting. Every analysis recalculates contracts based on module-level results for all modules (similar to lines 6 and 7) that are retrieved from the database to take changes on the module level into account.

Note that during development of a module, it is not required to run this entire process after every change. It is sufficient to run the analysis only on the changed module, which leads to an update of the database. This way, the derived contracts may not always be up-to-date, but it is sufficient for addressing immediate issues at this stage. The full check up to the fixpoint will still be done regularly (e.g., in a nightly build), and the database is updated accordingly.

5   Case Study

We implemented the described framework in Scala [5] and applied it to three automotive embedded subsystems under development[1]. Table 1 shows an overview of their characteristics. The first one is the hardware abstraction layer of a driver assistant subsystem (A), the second one part of the application software of a braking subsystem (B), and the third one a cruise control subsystem with a focus on control engineering functionality (C). System A is written in C, whereas systems B and C are written in C++. System B contains the highest number of functions and ifs as well as the lowest number of loops.

We apply integration analysis, which analyzes the complete software project where contracts and stubs are not necessary as all implementations are available. This baseline is called stage 0. We then perform module-level analysis as described in Sect. 3 with incremental stages of contracts. In stage 1, we use module-level analysis only with implicit contracts based on type information such as enum ranges and floats excluding invalid values like NaN. In stage 2, we also use automatically inferred contracts as described in Sect. 4: Only for system B, interface specifications are available and used, but for all three systems, contracts resulting from abstract interpretation up to a fixpoint are used. In stage 3, we additionally use contracts that were manually written by experts. Overall, 6, 0, and 13 contracts were written manually for systems A, B, and

---

[1] The tool is a company-internal development, which we cannot share. The subject systems consist of production code, which we cannot share, either.

Table 2. Case study results. Alarm classes: IPA = invalid usage of pointers or arrays; ISA = invalid shift argument; IRO = invalid ranges and overflows; DMZ = division or modulo by zero; UIV = uninitialized variables; UFC = unknown function called; DCF = data and control flow alarms; CPP = C++ specific alarms; ASR = failed asserts.

System/Stage	Total Alarms	IPA	ISA	IRO	DMZ	UIV	UFC	DCF	CPP	ASR	Definite Alarms	Coverage [%]
A0	131	21	11	17			51	33			8	83.7
A1	131	24	11	20			37	39			8	**90.4**
A2	108	20	1	11			35	41			7	89.4
A3	**102**	14	1	11			35	41			6	89.4
B0	21			21							0	64.8
B1	92	3	3	61	7					18	2	93.8
B2	**12**	1		8	1					2	2	**99.3**
C0	190	22	4	129	6	10	17			2	2	75.8
C1	304	24		167	6	61	5	8	19	14	20	**99.8**
C2	301	22		166	6	61	5	8	19	14	18	**99.8**
C3	**173**	3		139	6	12	5			8	3	**99.8**

C, respectively, whereas 28, 147, and 127 contracts were inferred automatically for those systems. For all these cases, we use the same set of Astrée settings, count the number of alarms of different classes, and calculate the code coverage. By code coverage, we mean the share of statements that are reached and executed by abstract interpretation. Code coverage is a valuable metric for abstract interpretation as all reached statements are verified if no alarms are reported.

Table 2 shows the results. The first column indicates system and stage, and the second column shows the total number of alarms. The columns in the middle denote the individual alarm classes as a decomposition of the total number of alarms. These alarm classes originate directly from the report of Astrée and are resolved in the caption of the table. The last two columns are discussed later.

In all three systems, the total number of alarms increases or stays the same when using module-level analysis (stage 1) in comparison to integration analysis (stage 0). It decreases below the values from stage 0 when additionally using contracts. The initial increase in alarms is caused by an increase in code coverage. In systems A and B, automatic contract inference in stage 2 suffices to largely eliminate additional false alarms. In system A, the alarms from classes UFC (unknown-function-called) and DCF (data-and-control-flow-alarms) originate from the use of compiler specific functions that Astrée does not know and from intentional infinite loops that are used while waiting for a reset to handle internal errors. Without these, the total number of alarms is reduced from 49 for A0 to 26 for A3. In system B, the results of stage 2 were so good that no manual contracts had to be added, therefore B3 is not required. In system C, manually written contracts are required for satisfactory results. This is due to complex dependencies between modules in this system.

B2 shows a significant reduction in the number of alarms in the class IRO (invalid-ranges-and-overflows) demonstrating how well automatically inferring contracts works when contracts can express the respective property. In B1 and B2, we get some alarms from classes that do not occur in B0. These alarms stem from limitations of the expressiveness of our contracts that are necessary to keep them checkable by Astrée. An example for such a limitation is that we cannot express the relationship between subsequent elements in an array. When an element is subtracted from the previous element in the array and the result is used as a divisor, then a division by zero occurs if both elements are the same. Although we observe that subsequent elements are always different, we cannot express this property in a contract and thus cannot avoid the false alarms.

Alarms from class ASR (failed-asserts) often represent violations of contracts as preconditions and postconditions are encoded by assertions. This is an additional benefit of our framework as it facilitates easier evaluation of alarms to be true or false alarms. If a precondition is not fulfilled, the corresponding assertion alarm is close to the calculation of the responsible value. Without the assertion, another alarm would happen deeper in the function, which is harder to evaluate.

Table 2 also shows the number of definite alarms. A *definite alarm* occurs when all abstract values of a variable lead to an alarm. An example for a definite alarm is dereferencing a pointer that is always null at that point as in this example: `int *p = NULL; int x = *p;`. This is in contrast to a *possible alarm* where at least one but not all abstract values of a variable lead to an alarm. Here, Astrée continues the analysis with the remaining values, excluding the values causing the error. In case of a definite alarm, Astrée cannot continue with any remaining value and thus aborts the execution path. Therefore, developers should fix definite alarms that are followed by other code to increase code coverage. Note that the number of definite alarms can decrease from stage 0 to stage 3 because the increasing number of contracts can restrict the value range of input variables, which can remove possible and definite alarms.

Coverage almost always increases for all considered systems. In most cases, some functions are not used in the integrated system, but module-level analysis checks all of them. In other cases, as analysis aborts analysis paths on definite alarms, more subsequent code may not be reached in integration analysis compared to module-level analysis, where each method is called individually from the main loop. We also manually compared alarms from integration analysis with alarms found in stages 2 and 3 and observed that no relevant alarms were missing, but additional alarms were found in code that was not reached before.

Table 3 shows the runtimes of integration versus module-level analysis on an Intel Xeon 4 GHz with 4 cores and 32 GB RAM as client and an Intel Xeon 3.2 GHz with 32 cores and 512 GB RAM running the Astrée server. There are big differences between C and C++ systems. The C system (A) takes 710 s for integration analysis, but only 4 s (median) per module for module-level analysis. For C++, the relatively high overhead of C++ preprocessing by Astrée leads to module-level analysis times of 20 s (B) and 48 s (C) (median). In integration analysis, this overhead is only due once and the total runtime is only 41 s (B) and 157 s (C), respectively. This means the savings when checking an individual module are rather small in the C++ cases, but the advantage of being able to

Table 3. Runtime comparison. Runtimes for all modules of respective system are in T.

System	Module-level analysis						Integration analysis
	median(T)	avg(T)	max(T)	#outliers(T)	$|T|$	sum(T)	total
A	4 s	13 s	89 s	2	18	241 s	710 s
B	20 s	21 s	32 s	3	45	950 s	41 s
C	48 s	116 s	789 s	2	13	1510 s	157 s

perform analysis early during development when integration is not yet possible still remains – along with increased coverage.

Analysis times of individual modules vary a lot. For example, in system C, one of the modules takes 789 s, which is more than half of the total module-level analysis runtime for this system. This module contains nested loops, which are expensive to analyze when loop unrolling is active in Astrée. The analysis time for this module is several times higher than the total integration analysis time, which indicates that a more concrete scenario is checked in the integration case, and thus loops have to be analyzed less often. However, the majority of modules is analyzed within at most two minutes. The number of outliers (according to Tukey's fences [53] with k = 1.5) is low as shown in the table and gives an indication about the share of modules requiring more processing time. Conducting this case study also showed that contract inference causes only a low runtime overhead (e.g., 10% in case of system B).

In summary, the evaluation of the case study shows that our framework can help developers in their daily work to detect runtime errors early during development. This is due to low analysis runtime for individual modules, almost no manual effort for writing contracts, and high precision.

6   Practical Experiences with Large Systems

We describe our experiences from letting developers of different projects use our framework on two large systems: a control software and a software library.

6.1   Embedded Control Software

For a large embedded control software system of approx. two MLOC in about 3,000 C files, the integration analysis of the system ran for more than three days with low precision settings. A run with high precision settings for Astrée using loop unrollings, partitionings, and relational domains was aborted after several weeks. By contrast, using module-level analysis via our framework reduces the runtime to 12 h. This could be reduced even further by massive parallelization, which is only possible due to module-level analysis.

Additional to the runtime improvements, our framework made it possible to run Astrée with high precision settings. The disadvantage of losing context information could be compensated by inferring contracts from abstract interpretation as described in Sect. 4.2: The number of alarms in module-level analysis was reduced by 28% when using contracts from abstract interpretation. The reduced number is still higher than in integration analysis mainly because all code from all modules is covered – and not only the code that is actually used. The main benefit of our framework in this case is that the analysis of modules can be executed at an early stage when the system is not yet integrated.

6.2   Embedded Software Library

Another team applied our framework to a foundational library used in multiple automotive software systems. The software library comprises approx. one MLOC in 2,000 C files and is organized in modules. Each module consists of 10 to 50 compilation units that may share common data, but its external interface only consists of C functions. Our framework analyzes each module individually. The analysis reaches a high code coverage of more than 90% in most modules, which is important for a library. As each module consists of only a few thousand lines of code, Astrée can be utilized with high precision settings, e.g., most loops can be fully unrolled. This leads to 10 to 100 reported alarms per module. The alarms have a short call-stack of at most three calls and hence are easy to review by a human. This is an advantage compared to integration analysis where call stacks are often large and hard to follow. The share of alarms that should lead to code changes is approx. 10%, which is a remarkably high rate for abstract interpretation in industrial use according to our experience.

Notice that the alarms that should lead to code changes include so-called *justified* alarms. An alarm is justified when it may occur by use of a function respecting its contracts. For library verification, a justified alarm is different from a bug in production code. As example, consider the following library function:

```
int Curve[10];
int getEntry(int i) {return Curve[i];}
```

In the code using the library, this function might always be called correctly, following implicit assumptions. Thus, if the integrated system is analyzed, then the statement `Curve[i]` never yields an array-out-of-bounds alarm. However, from a library point-of-view, the above implementation of the function is buggy if used incorrectly. Our framework discovers such issues: Our driver calls the function with arbitrary input for i by setting the contents of the parameter variable to full range when no contracts exist.

Here, contracts are useful in the interaction of verification engineers and core developers when these roles are distributed over different teams, time zones, or countries: Verification engineers can directly write contracts for the implicit assumptions of the library. This benefits both verification engineers and core developers: Verification engineers do not encounter alarms with the same root cause again and again. Also, contracts and the resulting Astrée directives are both more convenient and safer than commenting alarms to dismiss them. Core developers benefit from implicit assumptions being made explicit by the contracts in case they do not change their code to check them. Both verification engineers and core developers benefit from the automatic detection of future violations of the implicit assumptions being made explicit via contracts.

7   Lessons Learned

One issue preventing successful use of abstract interpretation in the automotive industry is that developers perceive it as a burdensome activity due to many false alarms. Oftentimes, alarms are only manually annotated if identified as false alarms, and this process is repeated when parts of the software are reused. A contributing factor to this issue is that abstract interpretation is only applied to the integrated system in late stages of development.

We identify module-level abstract interpretation with contracts as solution to these problems, because developers get feedback early during development and can fix issues while they still know the code. To deploy module-level analysis, we learn that central availability and automation is a prerequisite. Also, contracts are needed to resolve missing contexts of modules. We identify the contract types of preconditions, postconditions, array specification, sequence specifications, and invariants as a good middle ground in terms of difficulty of automatic inference, provided expressiveness, and needed overhead for analysis in the automotive domain. These types of contracts are also intuitive for developers such that they can write them manually on newly produced code. With the described contract inference, they can be automatically obtained for large legacy code bases.

Developers get first results with our framework after a few days because we designed it to be adaptable to new projects with custom extensions to account for the specific architecture models, processes, and conventions used in the projects. The first results are often already helpful for developers and convince them of the benefits of using abstract interpretation. Developers appreciate the short runtimes of module-level analysis as they do not block their usual workflows.

When developing and deploying our framework, we also found possibilities for improvements when applying Astrée to automotive software. Analyzing C++ has a higher overhead compared to C, but every speedup in processing of C++ would be appreciated. Also, templates are difficult to statically analyze. For container classes, it is often unclear how to instantiate them with representative instances.

Our framework focuses on Astrée as the underlying engine for abstract interpretation, although the general ideas behind our framework can also be used with abstract interpretation tools like Frama-C EVA plugin [15,16], Polyspace Code Prover [4], and TrustInSoft Analyzer [6]. Astrée is highly customizable, and we benefit from a large knowledge base on settings for different projects.

8   Related Work

Contract Specification: The concept of contracts was introduced by Meyer for Eiffel and used for dynamic checks [49]. SPARK extended Ada with contracts [18] that can be checked statically or at runtime. Leavens adapted contracts for Larch/C++ [43,44] and then co-authored JML, a behavioral interface specification language for Java [19,45], which later evolved into OpenJML [21]. The latter is still under active development. Microsoft Research developed Spec# as a C# variant with contracts [11,46]. They then changed to an API based approach called CodeContracts [33,34,47]. Frama-C supports a language called ACSL for specifying contracts, which is embedded into comments and supports predicate logic [27]. Hatcliff et al. give an overview of behavioral specification languages and their commonalities and differences [36]. Our contract specification syntax uses similar concepts as all of these publications – there seems to be a broad consensus of how contracts should look like.

Contract Verification: The first static contract checking tool for JML was called ESC/Java [35]. It translated source code and specifications into verification conditions and passed them to a theorem prover. Its successor ESC/Java2 offered support for more JML specifications [19,22] and Java 1.4. OpenJML is the current JML verification tool [21]. It is based on OpenJDK, which enables easy adaptation to new language versions. The static checker for Spec# is Boogie, which has an own intermediate language making it usable for multiple languages [10]. A checker for CodeContracts exists in the form of cccheck/Clousot [34,47]. Similar to us, they transform pre- and postconditions into assume/assert calls and then apply their abstract interpretation engine. Frama-C is capable of checking certain contracts written in ACSL [27,42]. It is also based on abstract interpretation [15,16]. Despite the availability of these approaches and tools, formal methods are rarely used when developing automotive software [7]. A contributing factor for that is the increased effort for developers [11], which we address by a high grade of automation.

Contract Inference: Abstract interpretation [23] can also be seen as a technique to infer invariants, as it iterates up to a fixpoint. We use this technique to find postconditions. Daikon uses dynamic analysis and expression patterns to infer likely invariants from Java code [32]. Ammons et al. use dynamic analysis to infer API usage protocols from code [8]. Arnout et al. investigate .NET code for implicit contracts, which they find in form of exceptions and documentation [9]. Fähndrich and Logozzo infer preconditions in simple cases – when the respective variables are not modified [34]. Wei et al. use dynamic analysis to infer contracts from Eiffel code [54]. Cousot et al. describe how to use abstract interpretation to infer necessary preconditions [24,26] (implemented in cccheck/Clousot [17]). Our framework can use contracts obtained by any of these approaches.

9   Conclusion

We presented a framework to detect runtime errors in automotive software with low manual effort. Our framework enables analysis early during development which has the following benefits: First, developers get feedback quickly and can fix problems when they still deeply understand the software. Second, the analysis runs faster on an individual module in comparison to the entire software system.

Our framework combines abstract interpretation with contracts. Contracts provide detailed information about the behavior of other modules that might not yet be implemented. They can be written manually or derived fully automatically. We use the abstract interpretation tool Astrée under the hood and presented how contracts can be utilized by Astrée to improve precision.

We evaluated our framework on several automotive software systems of different size. Our case study showed that the number of found alarms, the runtime, and the code coverage can be greatly improved. For the domain of braking, our framework decreased the number of alarms by approx. 50% while increasing code coverage by approx. 50%. Our framework is used by developers in their daily work on projects with over a million lines of code. The feedback so far has been very positive, and the rollout to other projects is planned.

References

1. ASAM Metadata Exchange Format for Software Module Sharing. https://www.asam.net/standards/detail/mdx/. Accessed 24 Aug 2023
2. Astrée. https://www.absint.com/astree/index.htm. Accessed 24 Aug 2023
3. Automotive Open System Architecture (AUTOSAR). https://www.autosar.org/. Accessed 24 Aug 2023
4. Polyspace Code Prover. https://www.mathworks.com/products/polyspace-code-prover.html. Accessed 24 Aug 2023
5. Scala. https://www.scala-lang.org/. Accessed 24 Aug 2023
6. TrustInSoft Analyzer. https://trust-in-soft.com/product/trustinsoft-analyzer/. Accessed 24 Aug 2023
7. Altinger, H., Wotawa, F., Schurius, M.: Testing methods used in the automotive industry: results from a survey. In: Proceedings of the 2014 Workshop on Joining AcadeMiA and Industry Contributions to Test Automation and Model-Based Testing, JAMAICA@ISSTA, pp. 1–6. ACM (2014). https://doi.org/10.1145/2631890.2631891
8. Ammons, G., Bodík, R., Larus, J.R.: Mining specifications. In: Conference Record of POPL 2002: The 29th SIGPLAN-SIGACT Symposium on Principles of Programming Languages, pp. 4–16. ACM (2002). https://doi.org/10.1145/503272.503275
9. Arnout, K., Meyer, B.: Finding implicit contracts in .NET components. In: de Boer, F.S., Bonsangue, M.M., Graf, S., de Roever, W.-P. (eds.) FMCO 2002. LNCS, vol. 2852, pp. 285–318. Springer, Heidelberg (2003). https://doi.org/10.1007/978-3-540-39656-7_12
10. Barnett, M., Chang, B.E., DeLine, R., Jacobs, B., Leino, K.R.M.: Boogie: a modular reusable verifier for object-oriented programs. In: FMCO 2005. LNCS, vol. 4111, pp. 364–387. Springer, Heidelberg (2005). https://doi.org/10.1007/11804192_17
11. Barnett, M., Fähndrich, M., Leino, K.R.M., Müller, P., Schulte, W., Venter, H.: Specification and verification: the Spec# experience. Commun. ACM 54(6), 81–91 (2011). https://doi.org/10.1145/1953122.1953145

12. Blanchet, B., et al.: Design and implementation of a special-purpose static program analyzer for safety-critical real-time embedded software. In: Mogensen, T., Schmidt, D.A., Sudborough, I.H. (eds.) The Essence of Computation. LNCS, vol. 2566, pp. 85–108. Springer, Heidelberg (2002). https://doi.org/10.1007/3-540-36377-7_5
13. Blanchet, B., et al.: A static analyzer for large safety-critical software. In: Proceedings of the ACM SIGPLAN 2003 Conference on Programming Language Design and Implementation, pp. 196–207. ACM (2003). https://doi.org/10.1145/781131.781153
14. Boehm, B.W.: Software Engineering Economics. Prentice Hall (1981)
15. Bühler, D.: Structuring an Abstract Interpreter through Value and State Abstractions: EVA, an Evolved Value Analysis for Frama-C. Ph.D. thesis, University of Rennes 1, France (2017). https://tel.archives-ouvertes.fr/tel-01664726. Accessed 24 Aug 2023
16. Bühler, D., et al.: Frama-C: the EVA plug-in (2023). https://frama-c.com/download/frama-c-eva-manual.pdf. Accessed 24 Aug 2023
17. Carr, S.A., Logozzo, F., Payer, M.: Automatic contract insertion with CCBot. IEEE Trans. Softw. Eng. **43**(8), 701–714 (2017). https://doi.org/10.1109/TSE.2016.2625248
18. Carré, B., Garnsworthy, J.R.: SPARK - an annotated ADA subset for safety-critical programming. In: Proceedings of the Conference on TRI-ADA 1990, pp. 392–402. ACM (1990). https://doi.org/10.1145/255471.255563
19. Chalin, P., Kiniry, J.R., Leavens, G.T., Poll, E.: Beyond assertions: advanced specification and verification with JML and ESC/Java2. In: de Boer, F.S., Bonsangue, M.M., Graf, S., de Roever, W.-P. (eds.) FMCO 2005. LNCS, vol. 4111, pp. 342–363. Springer, Heidelberg (2006). https://doi.org/10.1007/11804192_16
20. Christakis, M., Bird, C.: What developers want and need from program analysis: an empirical study. In: Proceedings of 31st IEEE/ACM International Conference on Automated Software Engineering (ASE 2016), pp. 332–343. ACM (2016). https://doi.org/10.1145/2970276.2970347
21. Cok, D.R.: OpenJML: JML for Java 7 by extending OpenJDK. In: Bobaru, M., Havelund, K., Holzmann, G.J., Joshi, R. (eds.) NFM 2011. LNCS, vol. 6617, pp. 472–479. Springer, Heidelberg (2011). https://doi.org/10.1007/978-3-642-20398-5_35
22. Cok, D.R., Kiniry, J.R.: ESC/Java2: uniting ESC/Java and JML. In: Barthe, G., Burdy, L., Huisman, M., Lanet, J.-L., Muntean, T. (eds.) CASSIS 2004. LNCS, vol. 3362, pp. 108–128. Springer, Heidelberg (2005). https://doi.org/10.1007/978-3-540-30569-9_6
23. Cousot, P., Cousot, R.: Abstract interpretation: a unified lattice model for static analysis of programs by construction or approximation of fixpoints. In: Conference Record of the Fourth ACM Symposium on Principles of Programming Languages, pp. 238–252. ACM (1977). https://doi.org/10.1145/512950.512973
24. Cousot, P., Cousot, R., Fähndrich, M., Logozzo, F.: Automatic inference of necessary preconditions. In: Giacobazzi, R., Berdine, J., Mastroeni, I. (eds.) VMCAI 2013. LNCS, vol. 7737, pp. 128–148. Springer, Heidelberg (2013). https://doi.org/10.1007/978-3-642-35873-9_10
25. Cousot, P., Cousot, R., Feret, J., Mauborgne, L., Miné, A., Rival, X.: Why does Astrée scale up? Formal Methods Syst. Des. **35**(3), 229–264 (2009). https://doi.org/10.1007/s10703-009-0089-6

26. Cousot, P., Cousot, R., Logozzo, F.: Precondition inference from intermittent assertions and application to contracts on collections. In: Jhala, R., Schmidt, D. (eds.) VMCAI 2011. LNCS, vol. 6538, pp. 150–168. Springer, Heidelberg (2011). https://doi.org/10.1007/978-3-642-18275-4_12
27. Cuoq, P., Kirchner, F., Kosmatov, N., Prevosto, V., Signoles, J., Yakobowski, B.: Frama-C. In: Eleftherakis, G., Hinchey, M., Holcombe, M. (eds.) SEFM 2012. LNCS, vol. 7504, pp. 233–247. Springer, Heidelberg (2012). https://doi.org/10.1007/978-3-642-33826-7_16
28. Dajsuren, Y., van den Brand, M. (eds.): Automotive Systems and Software Engineering - State of the Art and Future Trends. Springer, Cham (2019). https://doi.org/10.1007/978-3-030-12157-0
29. Do, L.N.Q., Wright, J.R., Ali, K.: Why do software developers use static analysis tools? A user-centered study of developer needs and motivations. IEEE Trans. Software Eng. **48**(3), 835–847 (2022). https://doi.org/10.1109/TSE.2020.3004525
30. Doumler, T., Spicer, J.: A proposed plan for Contracts in C++. Tech. Rep. P2695R0, C++ SG21 (Nov 2022). https://www.open-std.org/jtc1/sc22/wg21/docs/papers/2022/p2695r0.pdf. Accessed 24 Aug 2023
31. Duprat, S., Lamiel, V.M., Kirchner, F., Correnson, L., Delmas, D.: Spreading static analysis with Frama-C in industrial contexts. In: 8th European Congress on Embedded Real Time Software and Systems (ERTS 2016) (2016)
32. Ernst, M.D., Cockrell, J., Griswold, W.G., Notkin, D.: Dynamically discovering likely program invariants to support program evolution. IEEE Trans. Softw. Eng. **27**(2), 99–123 (2001). https://doi.org/10.1109/32.908957
33. Fähndrich, M., Barnett, M., Logozzo, F.: Embedded contract languages. In: Proceedings of the 2010 ACM Symposium on Applied Computing (SAC), pp. 2103–2110. ACM (2010), https://doi.org/10.1145/1774088.1774531
34. Fähndrich, M., Logozzo, F.: Static contract checking with abstract interpretation. In: Beckert, B., Marché, C. (eds.) FoVeOOS 2010. LNCS, vol. 6528, pp. 10–30. Springer, Heidelberg (2011). https://doi.org/10.1007/978-3-642-18070-5_2
35. Flanagan, C., Leino, K.R.M., Lillibridge, M., Nelson, G., Saxe, J.B., Stata, R.: Extended static checking for Java. In: Proceedings of the 2002 ACM SIGPLAN Conference on Programming Language Design and Implementation (PLDI), pp. 234–245. ACM (2002). https://doi.org/10.1145/512529.512558
36. Hatcliff, J., Leavens, G.T., Leino, K.R.M., Müller, P., Parkinson, M.J.: Behavioral interface specification languages. ACM Comput. Surv. **44**(3), 16:1–16:58 (2012). https://doi.org/10.1145/2187671.2187678
37. Holzmann, G.J., Joshi, R.: Model-driven software verification. In: Graf, S., Mounier, L. (eds.) SPIN 2004. LNCS, vol. 2989, pp. 76–91. Springer, Heidelberg (2004). https://doi.org/10.1007/978-3-540-24732-6_6
38. ISO 26262:2018. Road vehicles - Functional safety (all parts). International Organization for Standardization (2018)
39. Johnson, B., Song, Y., Murphy-Hill, E.R., Bowdidge, R.W.: Why don't software developers use static analysis tools to find bugs? In: 35th International Conference on Software Engineering (ICSE 2013), pp. 672–681. IEEE Computer Society (2013). https://doi.org/10.1109/ICSE.2013.6606613
40. Kästner, D., et al.: Finding all potential run-time errors and data races in automotive software. Tech. rep, SAE Technical Paper (2017)
41. Kästner, D., et al.: Astrée: proving the absence of runtime errors. In: Proceedings of Embedded Real Time Software and Systems (ERTS2 2010), vol. 9 (2010)

42. Kirchner, F., Kosmatov, N., Prevosto, V., Signoles, J., Yakobowski, B.: Frama-C: a software analysis perspective. Formal Aspects Comput. **27**(3), 573–609 (2015). https://doi.org/10.1007/s00165-014-0326-7
43. Leavens, G.T.: An overview of larch/C++: behavioral specifications for C++ modules. In: Object-Oriented Behavioral Specifications, pp. 121–142. Springer, Boston (1996). https://doi.org/10.1007/978-0-585-27524-6_8
44. Leavens, G.T., Baker, A.L.: Enhancing the pre- and postcondition technique for more expressive specifications. In: Wing, J.M., Woodcock, J., Davies, J. (eds.) FM 1999. LNCS, vol. 1709, pp. 1087–1106. Springer, Heidelberg (1999). https://doi.org/10.1007/3-540-48118-4_8
45. Leavens, G.T., Baker, A.L., Ruby, C.: JML: a notation for detailed design. In: Kilov, H., Rumpe, B., Simmonds, I. (eds.) Behavioral Specifications of Businesses and Systems, pp. 175–188. Springer, Boston (1999). https://doi.org/10.1007/978-1-4615-5229-1_12
46. Leino, K.R.M., Müller, P.: Using the spec# language, methodology, and tools to write bug-free programs. In: Müller, P. (ed.) LASER 2007–2008. LNCS, vol. 6029, pp. 91–139. Springer, Heidelberg (2010). https://doi.org/10.1007/978-3-642-13010-6_4
47. Logozzo, F.: Practical verification for the working programmer with CodeContracts and abstract interpretation. In: Jhala, R., Schmidt, D. (eds.) VMCAI 2011. LNCS, vol. 6538, pp. 19–22. Springer, Heidelberg (2011). https://doi.org/10.1007/978-3-642-18275-4_3
48. Luo, Y., Saberi, A.K., van den Brand, M.: Safety-driven development and ISO 26262. In: Automotive Systems and Software Engineering, pp. 225–254. Springer, Cham (2019). https://doi.org/10.1007/978-3-030-12157-0_10
49. Meyer, B.: Object-Oriented Software Construction, 1st edn. Prentice-Hall (1988)
50. Rival, X., Yi, K.: Introduction to Static Analysis: An Abstract Interpretation Perspective. MIT Press (2020)
51. Souyris, J., Delmas, D.: Experimental assessment of Astrée on safety-critical avionics software. In: Saglietti, F., Oster, N. (eds.) SAFECOMP 2007. LNCS, vol. 4680, pp. 479–490. Springer, Heidelberg (2007). https://doi.org/10.1007/978-3-540-75101-4_45
52. Todorov, V., Boulanger, F., Taha, S.: Formal verification of automotive embedded software. In: Proceedings of the 6th Conference on Formal Methods in Software Engineering, FormaliSE 2018, collocated with ICSE 2018, pp. 84–87. ACM (2018). https://doi.org/10.1145/3193992.3194003
53. Tukey, J.W.: Exploratory data analysis. In: Addison-Wesley Series in Behavioral Science: Quantitative Methods. Addison-Wesley (1977)
54. Wei, Y., Furia, C.A., Kazmin, N., Meyer, B.: Inferring better contracts. In: Proceedings of the 33rd International Conference on Software Engineering (ICSE 2011), pp. 191–200. ACM (2011). https://doi.org/10.1145/1985793.1985820
55. Yamaguchi, T., Brain, M., Ryder, C., Imai, Y., Kawamura, Y.: Application of abstract interpretation to the automotive electronic control system. In: Enea, C., Piskac, R. (eds.) VMCAI 2019. LNCS, vol. 11388, pp. 425–445. Springer, Cham (2019). https://doi.org/10.1007/978-3-030-11245-5_20

Abstract Interpretation-Based Feature Importance for Support Vector Machines

Abhinandan Pal[1] , Francesco Ranzato[2(✉)] , Caterina Urban[3] ,
and Marco Zanella[2]

[1] School of Computer Science, University of Birmingham, Birmingham, UK
a.pal@bham.ac.uk
[2] Dipartimento di Matematica, University of Padova, Padova, Italy
{francesco.ranzato,marco.zanella}@unipd.it
[3] INRIA and Ecole Normale Supérieure,
Université PSL, Paris, France
caterina.urban@inria.fr

Abstract. We study how a symbolic representation for support vector machines (SVMs) specified by means of abstract interpretation can be exploited for: (1) *enhancing the interpretability* of SVMs through a novel feature importance measure, called abstract feature importance (AFI), that does not depend in any way on a given dataset or the accuracy of the SVM and is very fast to compute; and (2) *certifying individual fairness* of SVMs and producing concrete counterexamples when this verification fails. We implemented our methodology and we empirically showed its effectiveness on SVMs based on linear and nonlinear (polynomial and radial basis function) kernels. Our experimental results prove that, independently of the accuracy of the SVM, our AFI measure correlates much strongly with stability of the SVM to feature perturbations than major feature importance measures available in machine learning software such as permutation feature importance, therefore providing better insight into the trustworthiness of SVMs.

1 Introduction

Machine learning (ML) software is increasingly being employed in high-stakes or sensitive applications. [11, 26, etc.]. As a consequence, research in ML verification rapidly gained popularity [28, 45], and the quest for interpretable ML models is becoming more and more pressing [43].

A fundamental and popular interpretability methodology is *feature importance*, that is, techniques for measuring the contribution of each input feature to a model prediction [5]. Nowadays, the most influential and used feature importance measures are Permutation Feature Importance (PFI) [6, 18], Local Interpretable Model-agnostic Explanations (LIME) [38], and SHapley Additive exPlanations (SHAP) [29]. PFI observes the decrease in predictive performance when a feature value is randomly shuffled: an increased loss is indicative of how much that feature is important for the predictive model. LIME approximates the prediction model locally by training an interpretable surrogate model on points in a meaningful neighborhood around a given input. SHAP

R. Dimitrova et al. (Eds.): VMCAI 2024, LNCS 14499, pp. 27–49, 2024.
https://doi.org/10.1007/978-3-031-50524-9_2

is a framework based on locally estimating so-called Shapley values [42]. The downsides of PFI and SHAP are that their outcome may greatly vary depending on the dataset and may be misleading when features are correlated [25]. Furthermore, they also have a high computational cost when the number of features is large, even for small models. Moreover, the quality of the output of PFI strongly depends on the accuracy of the model. Notably, model variance to feature perturbations [23] and PFI are strongly correlated only when the model generalizes well. Similarly, it is unclear how to define a meaningful optimal neighborhood for LIME and this may lead to explanations that are unstable and manipulable. More importantly, LIME assumes that the decision boundary of the underlying model is locally linear, but there is no guarantee that this actually happens.

Contributions. In this work, we focus on the interpretability of Support Vector Machines (SVMs), which nowadays are used in extensive repertoire of critical and high-stakes applications such as credit card fraud detection, facial recognition, and melanoma classification [9]. In particular, we propose a novel feature importance measure for SVMs, called *Abstract Feature Importance* (AFI), that: (a) leverages an approximation of the underlying predictive model which is *formally correct-by-construction*; (b) does *not* depend on a given dataset or on model accuracy; and (c) is extremely *fast to compute*, independently of the number of features. We support both linear and nonlinear kernels, in particular the polynomial and radial basis function (RBF) kernels.

We derive our importance measure from a symbolic representation of a SVM based on *abstract interpretation* [12]. Specifically, the concrete vectors being manipulated by model computations are symbolically represented through an abstract domain, which defines their abstract counterparts and their data structure representations, as well as algorithms to process them by approximating the behaviour of operations, such as additions and dot products, used in model computations.

We leverage existing numerical abstract domains such as intervals (i.e., geometric hyperrectangles) [13] and affine forms (i.e., geometric zonotopes) [31] that we combine with a novel abstract domain tailored for precisely representing computations with one-hot encoded categorical input features. We show the effectiveness of this new combined abstract domain for certifying model stability against feature perturbations. In particular, we focus on verifying *individual fairness* [16] of SVMs, which, to the best of our knowledge, has not been investigated before. We evaluate our approach by certifying SVMs trained on the reference datasets in the literature on ML fairness [30] and by considering different similarity relations. Our approach is *sound by design*, meaning that an individually fair abstract representation of a SVM implies that this SVM is actually fair. Thus, the fraction of successful fairness verifications over a test dataset turns out to be a *lower bound* on the real individual fairness of a SVM. On the other hand, our approach is *incomplete*, as there are cases in which the SVM is fair whereas the verification of its abstract representation fails, due to imprecisions introduced by the abstraction process. Our third contribution is a method to leverage this abstract representation of SVMs to generate concrete counterexamples when the abstraction is unable to prove individual fairness, i.e., we deliver concrete similar inputs to a SVM that result in different classifications. The fraction of successful counterexample searches over a test dataset yields a lower bound on how biased an SVM is, and therefore, by complement, an *upper bound* on the real individual fairness of a SVM.

Finally, we conduct an extensive experimental comparison between our new feature importance measure AFI and popular interpretability methods like PFI and LIME. The experimental results show that AFI is better correlated with stability of a SVM model to feature perturbations, independently of the accuracy of the model.

Related Work. Feature importance measures can be *local*, i.e., measuring feature importance for a specific prediction, or *global*, i.e., measuring importance over the entire input space of the predictive model. We also distinguish *model-agnostic* measures, which can be applied to any model, and *model-specific* measures. Finally, we classify importance measures in *performance-based*, i.e., measuring importance w.r.t. the predictive performance of the model (thus requiring knowledge of the ground truth values), and *effect-based*, measuring importance based on the magnitude of change in the predicted outcome due to changes in the feature value (requiring no knowledge of the ground truth values).

PFI [6,18] is a global, model-agnostic, performance-based measure. LIME [38] and SHAP [29] are both local, model-agnostic, effect-based measures. Our novel feature importance measure AFI is *specific for SVMs* but can be used *both* as *global and local* measure, is *effect-based* and is *dataset independent*, thus removing bias to the dataset. LIME can also be extended to a global measure by considering several local instances [38, Section 4]. However, the choice of how many and which instances to select remains not obvious, thus not providing correctness guarantees. By contrast, our AFI measure is based on a formally correct-by-construction approximation of the underlying predictive model.

Several other model-agnostic importance measures have been proposed in the literature. Prominent effect-based measures are visual tools such as partial dependence (PD) [19], individual conditional expectation (ICE) [22], and accumulated local effects (ALE) [4] plots. Visual tools, such as individual conditional importance (ICI) and partial importance (PI) curves [8], are also proposed for local performance-based measures. [8] proposes a Shapley feature importance measure, called SFIMP, that allows comparing feature importance across different models. Input gradient [24] is a local measure that can be both effect-based and performance-based. Feature importance measures specific for SVMs are typically limited to linear SVMs or face scalability issues w.r.t. the number of features, e.g. [10,32]. By contrast, our AFI measure supports nonlinear kernels and has no scalability issues.

Our work generally contributes to the research ecosystem around the verification of ML models using formal methods [2,28,45]. Most approaches consider (deep) neural networks [40,41,44,48, etc.], while here we focus on SVMs. Our work leverages a SVM verifier called SAVer [36]. In addition, we introduce here a more precise abstraction for one-hot encoded features that we integrated within SAVer. Our fairness analysis is in line with the approach investigated in [35], that evaluated the individual fairness of decision tree ensembles trained by a new fairness-aware learning technique. Similar works either consider an orthogonal notion of fairness or a different "threat model", in most cases both. [47] evaluates security against flipping a few labels to maximize classification error. [21] considers group and causal fairness metrics. [27] deals with robustness of SVMs against adversarial attacks. [17] proposes a new fairness metric where it adds a new feature with random values and bias individuals on this feature: the model is fair when it recovers the original labels. [34] puts forward a protocol to protect sensitive information and trains a fair model using homomorphic encryption. [30,46] discuss several fairness metrics used to verify a variety of ML models.

2   Background

Support Vector Machines. SVMs [14] are supervised machine learning models based on separation curves that partition the input vector space into regions that best fit binary classification labels $L = \{-1, +1\}$. Separation curves are computed by maximizing their distance (margin) from the closest vectors in the training dataset. The simplest SVM is linear, which in its primal form boils down to a hyperplane $\mathbf{w} \cdot \mathbf{z} - b = 0$, where $\mathbf{w} \in \mathbb{R}^n$ and $b \in \mathbb{R}$ are learned parameters, that determines whether an input $\mathbf{z} \in \mathbb{R}^n$ falls above/below (i.e., $\text{sgn}(\mathbf{w} \cdot \mathbf{z} - b) = \pm 1$) w.r.t. the hyperplane. This approach is extended to nonlinear SVMs through a projection to a high-dimensional space via a kernel function $k : \mathbb{R}^n \times \mathbb{R}^n \to \mathbb{R}$. More precisely, given an input space $X \subseteq \mathbb{R}^n$, a training dataset $T = \{(\mathbf{x}_1, y_1), ..., (\mathbf{x}_N, y_N)\} \subseteq X \times \{-1, +1\}$, a kernel function k and parameters $w_i, b \in \mathbb{R}$, with $i \in [1, N]$, a SVM classifier $C : \mathbb{R}^n \to \{-1, +1\}$ is represented in its dual form by the function $C(\mathbf{z}) \triangleq \text{sgn}\left(\sum_{i=1}^{N} (w_i y_i k(\mathbf{x}_i, \mathbf{z})) - b\right)$. Most common kernels are: (i) *linear*, where $k(\mathbf{x}, \mathbf{z}) = \mathbf{x} \cdot \mathbf{z}$; (ii) *polynomial*, where $k(\mathbf{x}, \mathbf{z}) = (\mathbf{x} \cdot \mathbf{z} + c)^p$, for some hyperparameters $c \in \mathbb{R}$ and $p \in \mathbb{N}$; (iii) *radial basis function* (RBF), where $k(\mathbf{x}, \mathbf{z}) = e^{-\gamma \|\mathbf{x} - \mathbf{z}\|_2^2}$, for some hyperparameter $\gamma > 0$. In multi-classification for labels $L = \{y_1, ..., y_m\}$, with $m > 2$, the standard approach is a reduction into multiple binary classification problems combined by leveraging a voting over different labels.

Example 2.1. Let us consider a space $X \subseteq \mathbb{R}^2$ of values normalized in the range $[-1, 1]$, thus $X = \{\mathbf{x} \in \mathbb{R}^2 \mid -1 \leq \mathbf{x}_1, \mathbf{x}_2 \leq 1\}$. A toy linear SVM $C : X \to \{-1, +1\}$ with two support vectors $\mathbf{v}_1 = (-0.5, 1), \mathbf{v}_2 = (0.5, -1) \in X$, respectively labeled as $-1, +1$, with weights $w_1 = w_2 = 0.5$ and bias $b = 0$, is represented in its dual form by the function $C(\mathbf{z}) = \text{sgn}(-1 * 0.5(\mathbf{v}_1 \cdot \mathbf{z}) + 1 * 0.5(\mathbf{v}_2 \cdot \mathbf{z}))$.   □

Abstract Interpretation. A tuple $\langle A, \sqsubseteq^A, \gamma^A \rangle$ is a *numerical abstract domain* (or abstraction) when $\langle A, \sqsubseteq^A \rangle$ is a partially ordered set of abstract values and $\gamma^A : A \to \wp(\mathbb{R}^n)$ is a concretization function which maps an abstract value a to the set $\gamma^A(a)$ of real vectors represented by a, and monotonically preserves the ordering relation, i.e., $a_1 \sqsubseteq^A a_2$ implies $\gamma^A(a_1) \subseteq \gamma^A(a_2)$. Intuitively, an abstract domain defines a symbolic representation of some sets of vectors in $\wp(\mathbb{R}^n)$: given $X \in \wp(\mathbb{R}^n)$ and $a \in A$, if $X \subseteq \gamma^A(a)$ then a is a *sound* representation, or approximation, of the set X, while if $X = \gamma^A(a)$ holds then a is called a *precise* (or *exact*) representation of X. We sometimes use the notation A_n for highlighting that A_n is an abstraction of sets of n dimensional vectors ranging in $\wp(\mathbb{R}^n)$.

Given a k-ary operation $f : (\mathbb{R}^n)^k \to \mathbb{R}^n$, for some $k > 0$, a corresponding abstract function $f^A : A^k \to A$ is a *sound* approximation of f on $(a_1, ..., a_k) \in A^k$ when $\{f(\mathbf{x}_1, ..., \mathbf{x}_k) \mid \mathbf{x}_i \in \gamma^A(a_i)\} \subseteq \gamma^A(f^A(a_1, ..., a_k))$ holds. Moreover, f^A is a *complete* approximation of f on its input $(a_1, ..., a_k)$ when equality holds. In words, soundness means that $f^A(a_1, ..., a_k)$ never misses a concrete computation of f on some input $(\mathbf{x}_1, ..., \mathbf{x}_k)$ abstractly represented by $(a_1, ..., a_k)$, while completeness entails that

$f^A(a_1, ..., a_k)$ is precisely a symbolic abstract representation of these concrete computations of f. When soundness/completeness of f^A holds for any abstract input then f^A is called a sound/complete approximation of f.

Abstract Domains. We consider the well-known abstract domain of *hyperrectangles* (or *intervals*) [12,13]. The hyperrectangle abstract domain HR_n consists of n-dimensional vectors of real intervals $h = ([l_1, u_1], ..., [l_n, u_n]) \in \mathrm{HR}_n$, with lower and upper bounds $l_i, u_i \in \mathbb{R} \cup \{-\infty, +\infty\}$ such that $l_i \leq u_i$. Hence, the concretization function $\gamma^{\mathrm{HR}} : \mathrm{HR}_n \to \wp(\mathbb{R}^n)$ is defined by $\gamma^{\mathrm{HR}}(h) \triangleq \{\mathbf{x} \in \mathbb{R}^n \mid \forall i. l_i \leq \mathbf{x}_i \leq u_i\}$. Conversely, hyperrectangles also have an abstraction map $\alpha^{\mathrm{HR}} : \wp(\mathbb{R}^n) \to \mathrm{HR}_n$ that provides the smallest hyperrectangle approximating a set of vectors: $\alpha^{\mathrm{HR}}(X) \triangleq ([\inf\{\mathbf{x}_i \in \mathbb{R} \mid \mathbf{x} \in X\}, \sup\{\mathbf{x}_i \in \mathbb{R} \mid \mathbf{x} \in X\}])_{i=1}^n$. Abstract operations are defined by extending componentwise the following additions and multiplications of intervals:

$$[l_1, u_1] +^{\mathrm{HR}} [l_2, u_2] \triangleq [l_1 + l_2, u_1 + u_2],$$
$$[l_1, u_1] *^{\mathrm{HR}} [l_2, u_2] \triangleq [\min(l_1 l_2, l_1 u_2, l_2 u_1, l_2 u_2), \max(l_1 l_2, l_1 u_2, l_2 u_1, l_2 u_2)].$$

It is known that a compositional abstract evaluation on HR of an expression can be imprecise, e.g., the evaluations of the simple expressions $x - x$ and $x \cdot x$ on an input interval $[-c, c]$, with $c > 0$, yield, resp., $[-2c, 2c]$ and $[-c^2, c^2]$, rather than the exact intervals $[0, 0]$ and $[0, c^2]$. This *dependency problem* can trigger a significant source of imprecision for the hyperrectangle abstraction of a nonlinear (i.e., polynomial/RBF) SVM classifier. Thus, following [36], for our SVM abstract representations, we leverage the *reduced affine form* (RAF) abstraction, which is a domain of zonotopes representing dependencies between components of input vectors. A RAF for vectors in $\mathbb{R}^n$ is given by an expression $a_0 + \sum_{i=1}^n a_i \epsilon_i + a_r \epsilon_r$, where the ϵ_i's are symbolic variables ranging in the real interval $[-1, 1]$ representing a dependence from the i-th component of the vector, while ϵ_r is a further symbolic variable in $[-1, 1]$ that accumulates all the approximations introduced by nonlinear operations such as multiplications and exponentials. Thus, $\mathrm{RAF}_n \triangleq \{a_0 + \sum_{i=1}^n a_i \epsilon_i + a_r \epsilon_r \mid a_0, a_1, ..., a_n \in \mathbb{R}, a_r \in \mathbb{R}_{\geq 0}\}$, where $a_r \in \mathbb{R}_{\geq 0}$ is the radius of the accumulative error of approximating all nonlinear operations during abstract computations. A RAF represents a real interval through the concretization map $\gamma^{\mathrm{RAF}} : \mathrm{RAF}_n \to \wp(\mathbb{R})$ defined by $\gamma^{\mathrm{RAF}}(a_0 + \sum_{i=1}^n a_i \epsilon_i + a_r \epsilon_a) \triangleq \{x \in \mathbb{R} \mid a_0 - \sum_{i=1}^n |a_i| - |a_r| \leq x \leq a_0 + \sum_{i=1}^n |a_i| + |a_r|\}$. RAFs are a restriction to a given length—in our case to the dimension n of $\mathbb{R}^n$—of the zonotope domain used in numerical program analysis [20]. Linear operations, namely additions and scalar multiplications, admit a complete approximation on the RAF abstraction, so that RAFs resolve the aforementioned dependency problem for linear expressions. Instead, nonlinear abstract operations, such as multiplications and exponentials, must still necessarily be approximated.

Example 2.2. Let us consider again the toy linear SVM C from Example 2.1. Its input space $X = [-1, 1]^2$ is abstracted as the RAF $a = (0 + \epsilon_1, 0 + \epsilon_2) \in (\mathrm{RAF}_2)^2$. □

Robustness. We consider an input space $X \subseteq \mathbb{R}^n$, a set of labels $L = \{y_1, ..., y_m\}$, and a dataset $T = \{(\mathbf{x}_1, y_1), ..., (\mathbf{x}_N, y_N)\} \subseteq X \times L$. A classifier trained on the dataset T is modeled as a map $C_T : X \to L$.

An adversarial region for an input sample $\mathbf{x} \in X$ is designated by a perturbation $P(\mathbf{x}) \subseteq X$ such that $\mathbf{x} \in P(\mathbf{x})$. Usually, a perturbation function $P : X \to \wp(X)$ is defined through a metric μ to measure similarity between inputs as their distance w.r.t. μ. The most common metric in ML [7] is induced by the ℓ_∞ maximum norm defined as $\|\mathbf{x}\|_\infty \triangleq \max\{|\mathbf{x}_1|, ..., |\mathbf{x}_n|\}$, so that the corresponding perturbation $P_\infty^\epsilon(\mathbf{x})$ includes all the vectors $\mathbf{z} \in X$ whose ℓ_∞ distance from $\mathbf{x}$ is bounded by a magnitude $\epsilon \in \mathbb{R}^+$, that is, $P_\infty^\epsilon(\mathbf{x}) \triangleq \{\mathbf{z} \in X \mid \|\mathbf{x} - \mathbf{z}\|_\infty \leq \epsilon\}$. Given a perturbation function P, a classifier C is *robust* (or *stable*) on $P(\mathbf{x})$, denoted by $\mathrm{robust}(C, P(\mathbf{x}))$, when for all $\mathbf{z} \in P(\mathbf{x})$, $C(\mathbf{z}) = C(\mathbf{x})$ holds. Robustness to a perturbation P is used as a major metric [23] to assess a classifier C on a test set $T \subseteq X \times L$ as follows: $\mathrm{rob}_T(C, P) \triangleq |\{(\mathbf{x}, y) \in T \mid \mathrm{robust}(C, P(\mathbf{x}))\}|/|T|$.

SAVer. Our work leverages SAVer (**SVM Abstract Verifier**), an automatic tool for robustness certification of SVMs [36,37]. Given a SVM $C : X \to L$, SAVer leverages an n-dimensional abstraction A_n of $\wp(\mathbb{R}^n)$ to first achieve a sound abstraction $P^\sharp(\mathbf{x}) \in A_n$ of a perturbation $P(\mathbf{x})$, i.e., $P(\mathbf{x}) \subseteq \gamma^A(P^\sharp(\mathbf{x}))$ must hold, and then applies sound abstract versions of the numerical functions used in C—notably, vector additions and dot products, scalar multiplications, exponentials—to design an abstract SVM classifier $C^\sharp : A \to \wp(L)$ that computes an over-approximation of the labels assigned to inputs in $P(\mathbf{x})$, i.e., $\{C(\mathbf{z}) \in L \mid \mathbf{z} \in P(\mathbf{x})\} \subseteq C^\sharp(P^\sharp(\mathbf{x}))$. If $C^\sharp(P^\sharp(\mathbf{x})) = \{y_i\}$, then every input in $P(\mathbf{x})$ is classified as y_i, so C is proved robust on the perturbation $P(\mathbf{x})$. In the binary classification case $L = \{-1, +1\}$, the abstract SVM $C^\sharp$ consists of an abstract function $\mathcal{A}_C^\sharp : A_n \to A_1$ that computes an over-approximation $\mathcal{A}_C^\sharp(P^\sharp(\mathbf{x}))$ of the distances between samples in $P^\sharp(\mathbf{x})$ and the SVM separation curve, namely, it computes an abstract value $a = \mathcal{A}_C^\sharp(P^\sharp(\mathbf{x})) \in A_1$ such that $\{\sum_{i=1}^{N}(w_i y_i k(\mathbf{x}_i, \mathbf{z})) - b \mid \mathbf{z} \in \gamma^{A_n}(P^\sharp(\mathbf{x}))\} \subseteq \gamma^{A_1}(a)$ holds. Afterwards, an over-approximation of the set of labels is inferred as follows:

$$C^\sharp(P^\sharp(\mathbf{x})) \triangleq \text{ if } \gamma^{A_1}(\mathcal{A}_C^\sharp(P^\sharp(\mathbf{x}))) \subseteq \mathbb{R}_{<0} \text{ then } \{-1\}$$
$$\text{elseif } \gamma^{A_1}(\mathcal{A}_C^\sharp(P^\sharp(\mathbf{x}))) \subseteq \mathbb{R}_{>0} \text{ then } \{+1\}$$
$$\text{else } \{-1, +1\}.$$

Example 2.3. Let us consider again the toy linear SVM C from Example 2.1. The abstraction $\mathcal{A}_C^{\mathrm{RAF}} : (\mathrm{RAF}_n)^n \to \mathrm{RAF}_n$ of C in the RAF abstract domain is $\mathcal{A}_C^{\mathrm{RAF}}(a) = -1 * 0.5(\mathbf{v}_1 \cdot^{\mathrm{RAF}} a) + 1 * 0.5(\mathbf{v}_2 \cdot^{\mathrm{RAF}} a)$. By performing the abstract computations of $\mathcal{A}_C^{\mathrm{RAF}}$ on the abstraction $a \in (\mathrm{RAF}_2)^2$ of its input space $X = [-1, 1]^2$ from Example 2.2, we obtain:

$$\mathcal{A}_C^{\mathrm{RAF}}(a) = -0.5(-0.5(0 + \epsilon_1) + 1(0 + \epsilon_2)) + 0.5(0.5(0 + \epsilon_1) - 1(0 + \epsilon_2))$$
$$= -0.5(0 + (-0.5)\epsilon_1 + \epsilon_2) + 0.5(0 + 0.5\epsilon_1 + (-1)\epsilon_2)$$
$$= 0 + 0.5\epsilon_1 + (-1)\epsilon_2 \qquad \qquad \square$$

3   Methodology

We delve into the four primary contributions, each elaborated within distinct subsections. Figure 1 delineates these contributions, categorizing them into improvements

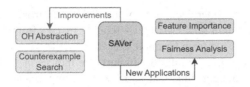

Fig. 1. Improvements to and new applications of SAVer

to SAVer and applications thereof. Section 3.1 introduces a method for assessing feature importance, drawing upon the RAF abstract transformers embedded in SAVer. Section 3.2 refines the aforementioned abstract transformers for both interval and RAF abstract domains by introducing One-Hot constraints. Section 3.3 exemplifies how SAVer can be used to verify individual fairness of SVM models leveraging the most common kernel functions. Lastly, Sect. 3.4 shows how to extend SAVer to achieve a counterexample search, thus providing additional insights on the analysis of fairness.

3.1 Abstract Feature Importance

Let us define our central notion of abstract feature importance (AFI).

Definition 3.1 (Abstract Feature Importance). Let $C : \mathbb{R}^n \to L$ be a SVM classifier and let $\mathcal{A}_C^{\mathrm{RAF}} : (\mathrm{RAF}_n)^n \to \mathrm{RAF}_n$ be the abstraction of C in the RAF abstract domain. Let $\mathcal{A}_C^{\mathrm{RAF}}(f_1, ..., f_n) \triangleq a_0 + \sum_{i=1}^{n} a_i \epsilon_i + a_r \epsilon_r$ be the abstract computation output for an abstract input $(f_1, ..., f_n)$, with $f_i \in \mathrm{RAF}_n$. The importance of every input feature $i \in [1, n]$ is defined as the absolute value $|a_i| \geq 0$. $\qquad\square$

This definition purposely ignores the accumulative error due to the approximations of all nonlinear operations performed by C, i.e., the term $a_r \epsilon_r$, influenced by all input features. When the input $(f_1, ..., f_n) \in (\mathrm{RAF}_n)^n$ abstracts the whole input space $X \subseteq \mathbb{R}^n$, AFI measures the *global* feature importance. Otherwise, AFI measures the *local* importance on the output label.

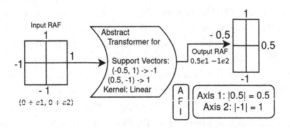

Fig. 2. Toy example of AFI for a linear SVM

Example 3.2. Let us consider again the toy linear SVM C from Example 2.1, and its abstract computation output $\mathcal{A}_C^{\mathrm{RAF}}(a) = 0 + 0.5\epsilon_1 + (-1)\epsilon_2$ from Example 2.3, for its abstract input $a \in (\mathrm{RAF}_2)^2$ from Example 2.2.

Based on Definition 3.1, we infer the importance indices $|a_1| = 0.5$ and $|a_2| = 1$ for, resp., $\mathbf{x}_1$ and $\mathbf{x}_2$, and conclude that $\mathbf{x}_2$ is twice as important as $\mathbf{x}_1$, as also shown by the picture in Fig. 2.

Note that, since C is a linear SVM, it can be rewritten in primal form as: $C'(\mathbf{x}) = -0.5(\mathbf{v_1} \cdot \mathbf{x}) + 0.5(\mathbf{v_2} \cdot \mathbf{x}) = -0.5(-0.5\mathbf{x}_1 + \mathbf{x}_2) + 0.5(0.5\mathbf{x}_1 - \mathbf{x}_2) = 0.5\mathbf{x}_1 - \mathbf{x}_2 = (0.5, -1) \cdot \mathbf{x}$, thus obtaining an explicit weight $\mathbf{w} = (0.5, -1)$ for the input features, whose absolute values 0.5 and 1 coincide with our importance indices. $\qquad\Box$

For linear SVMs, as in the above example, the abstraction in the RAF abstract domain is *sound* and *complete* yielding an output RAF that matches the primal form of the SVM, ensuring the full accuracy of AFI. Instead, nonlinear SVMs cannot be represented in primal form and thus the output RAF acts as a pseudo-primal form, soundly over-approximating the true output region of the SVM. In this case, the accuracy of AFI hinges upon the precision of the SVM abstraction. Our work builds upon the RAF abstract transformers for linear and nonlinear (polynomial and RBF) kernels initially introduced within SAVer [36] and refines them by incorporating further constraints to better handle categorical input features, as elaborated in the next Sect. 3.2. Our experimental evaluation in Sect. 4 shows that our refined abstractions offer higher accuracy, meanwhile remaining computationally efficient. The potential introduction of more precise SVM abstractions in the future will directly benefit the accuracy of AFI.

Example 3.3. Let us consider a toy SVM C similar to that of Example 2.1, sharing the same support vectors, labels and weights but using the polynomial kernel $k(\mathbf{z}, \mathbf{v}) \triangleq (\mathbf{z} \cdot \mathbf{v} + c)^d$, where $c = 1, d = 2$. Its abstract computation output turns out to be $\mathcal{A}_C^{\mathrm{RAF}}(a) = 0 + 1\epsilon_1 - 2\epsilon_2 + 1\epsilon_r$ for its abstract input $a \in (\mathrm{RAF}_2)^2$ from Example 2.2. We infer the importance indices $|a_1| = 1$ and $|a_2| = 2$ for, resp., $\mathbf{x}_1$ and $\mathbf{x}_2$, and, once again, infer that $\mathbf{x}_2$ is twice as important as $\mathbf{x}_1$. In this case, we also observe that the nonlinear noise accumulation term ϵ_r is greater than zero, meaning that some approximations happened through the computation, as expected for nonlinear kernels. $\Box$

Feature Grades. The AFI indices $a_i \geq 0$ (we assume that a_i implicitly denotes its absolute value) depend on the size of the abstract input, and this can make them harder to read and interpret, especially when the number of input features is high. To address this issue, we use a simple clustering strategy to assign an importance score to the input features. We consider the distribution of the AFI importance indices $(a_i)_{i=1}^n$ and compute its mean μ and standard deviation σ. Then, for each feature i, we consider its z-score $z_i \triangleq \frac{a_i - \mu}{\sigma}$, and the least integer greater than or equal to z_i as a corresponding rating, i.e., $score_i \triangleq \lceil \frac{a_i - \mu}{\sigma} \rceil$: this has the effect of standardizing the distribution into a normal distribution, slicing the distribution at every unit, and then labeling every slice with a progressive number. By doing so, features moderately influencing the result will have a score close to zero, relevant features will have higher scores, and those not influencing the outcome will have a negative score. Lastly, we shift and clip such distribution to achieve what we call *feature grades* ranging, e.g., in an interval $[a, b]$, as obtained by an easy transformation: $grade_i \triangleq \max(\min(b, score_i + shift), a)$.

For instance, let us consider a distribution of AFI indices for 10 features given by $(1, 6, 2, 5, 6, 1, 6, 7, 8, 9)$, where $\mu = 5.1$ and $\sigma = 2.85$: the corresponding scores are $(-1, 1, -1, 0, 1, -1, 1, 1, 2, 2)$, we shift them by $+6$ and then we clip these shifted values in $[3, 10]$, thus achieving the grades $(5, 7, 5, 6, 7, 5, 7, 7, 8, 8)$. Hence, e.g., x_1 and x_3 have similar impact on the classification, although they have different AFI indices.

Finally, let us remark that AFI does not require any knowledge on the ground truth values, nor the actual output of the SVM classifier, as AFI focuses on the computation process performed by the classifier, rather than how the result of such computation is used to assign a label to an input, thus making this approach applicable to scenarios where the correct prediction is not known in advance.

3.2 Abstracting One-Hot Encoding

Most ML algorithms need a way to represent categorical data in numerical form. Let $F = \{c_1, c_2, ..., c_k\}$ be the set of values of some categorical feature f. A naïve approach assigning a different number to each value in F introduces an unwanted ordering relation among features, that often induces bias or poor accuracy. A better and well-known method is *one-hot encoding*, that is, replacing f with k binary features $(x_1^f, x_2^f, ..., x_k^f) \in \{0, 1\}^k$ such that $\forall i \in [1, k].\, x_i^f = 1 \Leftrightarrow f = c_i$. This sequence $(x_i^f)_{i=1}^k$ of k bits is also referred to as a *tier* of f. Numerical abstractions, such as the hyperrectangle and RAF abstract domains, are likely to suffer from a significant loss of precision when dealing with these one-hot encoded features, as they are not able to keep track of the relation $\sum_{i=1}^k x_i^f = 1$ existing between the binary features resulting from one-hot encoding.

Example 3.4. Consider a categorical feature $f \in F = \{red, green, blue\}$, and let $(x_r, x_g, x_b) \in \{0, 1\}^3$ denote the corresponding one-hot encoded tiers. Consider the set of categories $\{red, green\}$, represented by the set of tiers $X = \{(1, 0, 0), (0, 1, 0)\}$. The most precise hyperrectangle abstraction of X is $h = (x_r \in [0, 1], x_g \in [0, 1], x_b \in [0, 0]) \in HR_3$. Observe that h also represents infinitely many vectors in $\mathbb{R}^3$ that do not belong to X and are illegal one-hot encodings, such as $(0.4, 0.8, 0)$, $(1, 1, 0)$ or $(0, 0, 0)$. □

To hinder this loss of precision, we define the *One-Hot abstraction*, a novel numerical abstraction tailored for one-hot encoded values.

Firstly, we recall the *constant propagation* abstract domain $CP \triangleq \mathbb{R} \cup \{\bot^{CP}, \top^{CP}\}$, ordinarily used for the constant folding optimization by modern compilers [1]. CP is a flat domain whose partial order $\sqsubseteq^{CP}$ is defined by $\bot^{CP} \sqsubseteq^{CP} z \sqsubseteq^{CP} \top^{CP}$, for all $z \in \mathbb{R}$. The concretization map $\gamma^{CP} : CP \to \wp(\mathbb{R})$ is as follows: $\gamma^{CP}(z) \triangleq \{z\}$, for all $z \in \mathbb{R}$, meaning that a given numerical feature can only assume a constant value z; $\gamma^{CP}(\top^{CP}) \triangleq \mathbb{R}$ representing no constancy information; $\gamma^{CP}(\bot^{CP}) \triangleq \varnothing$ encodes unfeasibility. CP also has an abstraction map $\alpha^{CP} : \wp(\mathbb{R}) \to CP$ that provides the best (i.e. least w.r.t. the order $\sqsubseteq^{CP}$) approximation in CP of a set of values, which is as follows: $\alpha^{CP}(\varnothing) \triangleq \bot^{CP}$, $\alpha^{CP}(\{z\}) \triangleq z$, and $\alpha^{CP}(X) \triangleq \top^{CP}$ otherwise.

The One-Hot abstract domain for a k-dimensional, i.e. with k original categories, one-hot encoded feature space is $OH_k \triangleq (CP \times CP)^k$. Thus, abstract values are k-tuples of pairs of values in CP, that keep track of the numerical information originated from a single one-hot k-encoded feature, both when this was originally false, i.e. set to 0, or true, i.e. set to 1. Given $a \in \text{OH}_k$ and one generic component $a_i \in \text{CP} \times \text{CP}$, with $i \in [1, k]$, let $a_{i,f/t} \in \text{CP}$ denote, resp., the first/second element of a_i, i.e., $a_i = (a_{i,f}, a_{i,t})$. The partial order $\sqsubseteq$ of OH_k is induced componentwise by $\sqsubseteq^{\text{CP}}$, namely, for all $a, b \in \text{OH}_k$, $a \sqsubseteq b \Leftrightarrow \forall i \in [1, k]. a_{i,f} \sqsubseteq^{\text{CP}} b_{i,f}$ & $a_{i,t} \sqsubseteq^{\text{CP}} b_{i,t}$. Then, for each component $i \in [1, k]$, the map $\hat{\gamma}_i : \text{OH}_k \to \wp(\mathbb{R}^k)$ is defined as:

$$\hat{\gamma}_i(a) \triangleq \{\mathbf{x} \in \mathbb{R}^k \mid \mathbf{x}_i \in \gamma^{\text{CP}}(a_{i,t}), \forall j \neq i. \mathbf{x}_j \in \gamma^{\text{CP}}(a_{j,f})\}.$$

Therefore, $a \in \text{OH}_k$ symbolically represents through $\hat{\gamma}_i$ the set of tiers whose i-th component was originally set to true. Note that if, for some $i \in [1, k]$, either $a_{i,f} = \bot^{\text{CP}}$ or $a_{i,t} = \bot^{\text{CP}}$, then $\hat{\gamma}_i(a) = \varnothing$. To retrieve all the concrete vectors represented by a, we collect all the vectors obtained by assuming that any component of the tier was originally set to true, namely, the concretization map $\gamma^{\text{OH}_k} : \text{OH}_k \to \wp(\mathbb{R}^k)$ is defined by $\gamma^{\text{OH}_k}(a) \triangleq \cup_{i=1}^k \hat{\gamma}_i(a)$.

Example 3.5. Let us continue Example 3.4 by considering the abstract element $a = ((0,1), (0,1), (0, \bot^{\text{CP}})) \in \text{OH}_3$. It turns out that a represents the set of tiers $X = \{(1,0,0), (0,1,0)\}$, which, in turn, is the one-hot encoding of $\{red, green\}$. In fact, its concretization is: $\gamma^{\text{OH}_3}(a) = \hat{\gamma}_1(a) \cup \hat{\gamma}_2(a) \cup \hat{\gamma}_3(a) = \{\mathbf{x} \in \mathbb{R}^3 \mid \mathbf{x}_1 \in \{1\}, \mathbf{x}_2 \in \{0\}, \mathbf{x}_3 \in \{0\}\} \cup \{\mathbf{x} \in \mathbb{R}^3 \mid \mathbf{x}_1 \in \{0\}, \mathbf{x}_2 \in \{1\}, \mathbf{x}_3 \in \{0\}\} \cup \{\mathbf{x} \in \mathbb{R}^3 \mid \mathbf{x}_1 \in \{0\}, \mathbf{x}_2 \in \{0\}, \mathbf{x}_3 \in \varnothing\} = \{(1,0,0), (0,1,0)\}$. Thus, a precisely represents X. □

Example 3.5 is not fortuitous. In fact, for any set X of one-hot encoded tiers there always exists an abstract value a in OH which precisely represents this set, i.e., such that $\gamma^{\text{OH}}(a) = X$.

Theorem 3.6. *If $X \in \wp(\mathbb{R}^k)$ is such that every vector of X is a one-hot encoded tier $(0, ..., 0, 1, 0, ...0) \in \{0, 1\}^k$, then the abstract value $a^X \in \text{OH}_k$ defined as $a_i^X \triangleq (\alpha^{\text{CP}}(\{\mathbf{x}_i \mid \mathbf{x} \in X, \mathbf{x}_i = 0\}), \alpha^{\text{CP}}(\{\mathbf{x}_i \mid \mathbf{x} \in X, \mathbf{x}_i = 1\}))$, for all $i \in [1, k]$, precisely represents X.*

It is worth remarking that this abstraction OH allows us to represent one-hot encoded information in a compact way. For example, given three categorical features with five possible values each, an abstract value in OH consists of $5 + 5 + 5 = 15$ pairs of type $(a_{i,f}, a_{i,t}) \in \text{CP}$, whereas the size of the concrete set of all the possible values for these three categorical features is $5^3 = 125$. In general, the size of an OH representation is in $O(NV)$ where N is the number of different categorical features and V is the maximum number of different categorical values for a given feature, whereas the size of the actual concrete values is in $O(V^N)$.

A value $a \in \text{OH}_k$ such that, for all $i \in [1, k]$ and $u \in \{f, t\}$, $a_{i,u} \in \text{CP} \setminus \{\top^{\text{CP}}\}$ holds, is called *top-less*. It is important to note that the abstract value $a^X \in \text{OH}_k$ defined in Theorem 3.6 is always top-less because the components of each pair a_i^X range in $\{0, 1, \bot^{\text{CP}}\}$.

Given a numerical function $f : \mathbb{R} \to \mathbb{R}$, its sound abstract counterpart $f^{\text{CP}} : \text{CP} \to \text{CP}$ on the CP abstraction is defined as follows:

$$f^{\text{CP}}(a) \triangleq \begin{cases} \bot^{\text{CP}} & \text{if } a = \bot^{\text{CP}} \\ f(z) & \text{if } a = z \text{ for some } z \in \mathbb{R} \\ \top^{\text{CP}} & \text{if } a = \top^{\text{CP}} \end{cases}$$

In turn, f^{CP} allows us to define a sound abstract counterpart of f on our OH abstract domain.

Theorem 3.7. *A sound approximation of f on OH_k is the function $f^{\mathrm{OH}} : \mathrm{OH}_k \to \mathrm{OH}_k$ defined, for all $i \in [1, k]$, as follows:*

$$(f^{\mathrm{OH}}(a))_i \triangleq \left(f^{\mathrm{CP}}(a_{i,f}), f^{\mathrm{CP}}(a_{i,t}) \right).$$

Example 3.8. Let us carry on Example 3.5. We assume a SVM with a polynomial kernel of degree two, whose abstract computation includes applying the nonlinear function $f(x) \triangleq x^2 - 3x + 1$ to each 0/1 component of the tiers in $\gamma^{\mathrm{OH}}(a) = \{(1, 0, 0), (0, 1, 0)\}$, so that $f(\gamma^{\mathrm{OH}}(a)) = \{(-1, 1, 1), (1, -1, 1)\}$. By Theorem 3.7, we have that $a' \triangleq f^{\mathrm{OH}}(a) = \left((f^{\mathrm{CP}}(0), f^{\mathrm{CP}}(1)), (f^{\mathrm{CP}}(0), f^{\mathrm{CP}}(1)), (f^{\mathrm{CP}}(0), f^{\mathrm{CP}}(\bot^{\mathrm{CP}})) \right) = ((1, -1), (1, -1), (1, \bot^{\mathrm{CP}}))$, whose concretization is:

$$
\begin{aligned}
\gamma^{\mathrm{OH}}(a') &= \hat{\gamma}_1(a') \cup \hat{\gamma}_2(a') \cup \hat{\gamma}_3(a') \\
&= \{\mathbf{x} \in \mathbb{R}^3 \mid \mathbf{x}_1 \in \{-1\}, \mathbf{x}_2 \in \{1\}, \mathbf{x}_3 \in \{1\}\} \\
&\quad \cup \{\mathbf{x} \in \mathbb{R}^3 \mid \mathbf{x}_1 \in \{1\}, \mathbf{x}_2 \in \{-1\}, \mathbf{x}_3 \in \{1\}\} \\
&\quad \cup \{\mathbf{x} \in \mathbb{R}^3 \mid \mathbf{x}_1 \in \{1\}, \mathbf{x}_2 \in \{1\}, \mathbf{x}_3 \in \varnothing\} \\
&= \{(-1, 1, 1), (1, -1, 1)\}.
\end{aligned}
$$

Notice that completeness holds because $f(\gamma^{\mathrm{OH}}(a)) = \gamma^{\mathrm{OH}}(f^{\mathrm{OH}}(a))$.

On the other hand, using the hyperrectangle abstraction, we noticed in Example 3.4 that $h = ([0, 1], [0, 1], [0, 0]) \in \mathrm{HR}_3$ is the initial hyperrectangle representation of the set of tiers $\{(1, 0, 0), (0, 1, 0)\}$. By applying the abstract transfer function f^{HR} to h, we compute:

$$
\begin{aligned}
h' = f^{\mathrm{HR}}(h) = &\left([0, 1]^{2^{\mathrm{HR}}} - 3 \cdot^{\mathrm{HR}} [0, 1] +^{\mathrm{HR}} [1, 1], \right. \\
&\quad [0, 1]^{2^{\mathrm{HR}}} - 3 \cdot^{\mathrm{HR}} [0, 1] +^{\mathrm{HR}} [1, 1], \\
&\quad \left. [0, 0]^{2^{\mathrm{HR}}} - 3 \cdot^{\mathrm{HR}} [0, 0] +^{\mathrm{HR}} [1, 1] \right) \\
= &\left([-2, 2], [-2, 2], [1, 1] \right).
\end{aligned}
$$

Thus, we have that $\gamma^{\mathrm{HR}}(h') = \{\mathbf{x} \in \mathbb{R}^3 \mid -2 \leq \mathbf{x}_1, \mathbf{x}_2 \leq 2, \mathbf{x}_3 = 1\}$. This latter concrete set of vectors is much larger than $\gamma^{\mathrm{OH}}(f^{\mathrm{OH}}(a))$: while soundness is nevertheless guaranteed in HR, we have lost a good deal of information due to the imprecision of the interval abstraction. In particular, observe that for $\mathbf{x}_1$ and $\mathbf{x}_2$, OH was able to derive precisely their values ranging in $\{-1, +1\}$, while the HR analysis computes much less precise lower/upper bounds for $\mathbf{x}_1$ and $\mathbf{x}_2$ such as -2 and 2, as a consequence of the abstract computation:

$$[0, 1]^{2^{\mathrm{HR}}} - 3 \cdot^{\mathrm{HR}} [0, 1] +^{\mathrm{HR}} [1, 1] = [0, 1] + [-3, 0] + [1, 1] = [-2, 2].$$

For instance, the lower bound -2 for $\mathbf{x}_1$ is obtained by adding the lower bounds 0 (for $\mathbf{x}_1^2$), -3 (for $-3\mathbf{x}_1$), 1 (for $+1$), namely, by requiring that $\mathbf{x}_1$ simultaneously assumes both values 0 and 1, which is an unfeasible spurious case.

The relational RAF abstraction induces more precise approximations w.r.t. HR, although this increase of precision is limited to linear dependencies only, including the intermediate values of the abstract computations. Thus, the OH abstraction shows a clear precision gain over both domains HR and RAF for abstract computations on one-hot encoded vectors.   □

In Example 3.8, we observed that f^{OH} is a complete approximation of f on a. This is a consequence of the following general result.

Corollary 3.9. *Let $a \in OH_k$ be top-less. Then:* (i) f^{OH} *is a complete abstraction of f on a;* (ii) *Given $f_1, f_2, ..., f_p : \mathbb{R} \to \mathbb{R}$, $f_1^{OH} \circ f_2^{OH} \circ ... \circ f_p^{OH}$ is a complete abstraction of $f_1 \circ f_2 \circ ... \circ f_p$ on a.*

Using OH in SAVer. We implemented in SAVer our OH abstraction for categorical features on top of the interval and RAF abstract domains for numerical features: these instances of SAVer are denoted in Sect. 4, resp., by Interval+OH and RAF+OH. In what follows, we focus on the RAF abstraction, since the interval domain conceptually follows the same pattern. Assume a vector of numerical features $\mathbf{x} \in \mathbb{R}^q$ and, for simplicity, a single categorical feature f having k categories[1]. We first perform the one-hot encoding of f, that is, f is transformed into a tier in $\{0,1\}^k$. We consider the so-called CAT perturbations of a categorical feature f (cf. Sect. 4), where f may take the value of any of its k categories: e.g., the CAT perturbation of *color* in Example 3.4 is $\{red, green, blue\}$. This means that in the CAT perturbation of the one-hot encoding of f, we allow every binary feature of the tier representing f to be either 0 or 1, so that the corresponding perturbed abstract value is always of the shape $oh = (0^{CP}, 1^{CP})^k \in OH_k$. For the numerical features $\mathbf{x}$, we consider a so-called NOISE perturbation (cf. Sect. 4), i.e., a maximum norm perturbation $P_\infty^\epsilon(\mathbf{x})$ for some magnitude $\epsilon > 0$. Hence, these NOISE and CAT perturbations define an initial RAF value $a \in RAF_q$ for $\mathbf{x}$ and an initial OH value $oh \in OH_k$ for f. The abstract kernel of the SVM is then computed on the RAF_q domain for $\mathbf{x}$ and on the OH_k domain for f, and we assume that the output of this abstract computation is $\langle a', oh' \rangle \in RAF_q \times OH_k$. For our purpose of certifying the robustness, the output OH value oh' is converted back to a hyperrectangle representing the lower and upper bounds of the k one-hot encoded components of f, namely, we compute k intervals $\alpha^{HR_k}(\gamma^{OH_k}(oh')) = \langle [\alpha_j, \beta_j] \rangle_{j=1}^k \in HR_k$ that provide lower and upper bounds of the numerical contributions to the SVM output of the k one-hot encoded components of f. On the other hand, the output $a' \in RAF_q$ is converted to the interval $[l, u] = \gamma^{RAF}(a') \in HR_1$ that gives a lower and upper bound to the sum of the contributions to the SVM output of the q numerical features $\mathbf{x}$. Hence, the sum of these $k+1$ intervals provides a single interval $[(l + (\sum_{j=1}^k \alpha_j), u + (\sum_{j=1}^k \beta_j)] \in HR_1$, which is a sound approximation of the sign of the SVM output.

Example 3.10. Consider a vector $\mathbf{x} \in \mathbb{R}^5$ where $\mathbf{x}_1, \mathbf{x}_2 \in \mathbb{R}$ are two numerical features and $\mathbf{x}_3, \mathbf{x}_4, \mathbf{x}_5 \in \{0,1\}$ are three binary values deriving from the one-hot encoding of a categorical feature. Assume an input perturbation of these features such that the output of the abstract computation for the numerical features is given

---

[1] For multiple categorical features, we keep track of the relation between all the categorical features and their corresponding tiers through a global lookup table.

by $a' = 1 + \epsilon_1 + 1.5\epsilon_2 + 0.5\epsilon_r \in \text{RAF}_2$, which is converted into the interval $\gamma^{\text{RAF}}(a') = [-2, 4] \in \text{HR}$, while on for the three one-hot encoded features x_3, x_4, x_5, the output on OH is $oh' = ((2, -1), (3, \perp^{\text{CP}}), (3, 2)) \in \text{OH}_3$. Thus, we have that $\gamma^{\text{OH}_3}(oh') = \{(-1, 3, 3), (2, 3, 2)\}$, which is then abstracted to the hyperrectangle $\alpha^{\text{HR}_3}(\gamma^{\text{OH}_3}(oh')) = ([-1, 2], [3, 3], [2, 3]) \in \text{HR}_3$. Hence, we derive a range interval of the SVM output which is $[-2, 4] +^{\text{HR}} [-1, 2] +^{\text{HR}} [3, 3] +^{\text{HR}} [2, 3] = [2, 12]$. Since $\gamma^{\text{HR}_1}([2, 12]) \subseteq \mathbb{R}_{>0}$, we have that $\text{sgn}([2, 12]) = +1$, meaning that the SVM classification is certified to be robust for the input perturbation. $\qquad\qquad\square$

3.3 Individual Fairness

Several formal models of fairness have been investigated in the literature. Dwork et al. [16] point out several weaknesses of the notions of group fairness and therefore study *individual fairness* defined as "the principle that two individuals who are similar w.r.t. a particular task should be classified similarly". This is formalized as a Lipschitz condition of the classifier, that is, by requiring that two individuals $x, y \in X$ whose distance is $\delta(x, y) \geq 0$, are mapped, resp., to distributions D_x and D_y whose distance is at most $\delta(x, y)$. Intuitively, the output distributions for x and y are indistinguishable up to their distance. Several distance metrics $\delta \colon X \times X \to \mathbb{R}_{\geq 0}$ can be used in this context, where Dwork et al. [16] study the total variation or relative ℓ_∞ distances.

Following [16], a classifier $C \colon X \to L$ is (*individually*) *fair* when C outputs the same label for all pairs of individuals $x, y \in X$ satisfying a similarity relation $S \subseteq X \times X$ between input samples. This relation S can be derived from a distance δ as follows: $(x, y) \in S \Leftrightarrow \delta(x, y) \leq \epsilon$, where $\epsilon \in \mathbb{R}$ is a similarity threshold.

Definition 3.11 (Individual Fairness). A classifier $C \colon X \to L$ is *fair* on an individual $x \in X$ w.r.t. a similarity relation $S \subseteq X \times X$, denoted by $\text{fair}(C, x, S)$, when $\forall z \in X. (x, z) \in S \Rightarrow C(z) = C(x)$. $\qquad\qquad\square$

To define a fairness metric for a classifier C, we compute how often C is fair on sets of similar individuals in a test set $T \subseteq X \times L$:

$$\text{fair}_{T,S}(C) \triangleq |\{(x, y) \in T \mid \text{fair}(C, x, S)\}|/|T|.$$

Hence, individual fairness for a similarity relation S boils down to robustness on the perturbation $P_S(x) \triangleq \{z \in X \mid (x, z) \in S\}$ induced by S, i.e., for all $x \in X$, $\text{fair}(C, x, S) \Leftrightarrow \text{robust}(C, P_S(x))$ holds.

3.4 Searching for Counterexamples

The abstract interpretation framework described in Sect. 2 is sound, thus a classifier C certified to be robust over a region $P(x)$ guarantees that all the inputs $x' \in P(x)$ actually receive the same label. The converse is generally not true for nonlinear kernels, due to lack of completeness: when the abstract verification is unable to certify robustness, it may be either due to a loss of precision or to a vector in $P(x)$ which truly receives a different label. We refer to the latter as a *counterexample* to the robustness of x. In case of an inconclusive analysis, we can mitigate the effect of incompleteness by searching for counterexamples: if at least one is found, then the classifier can be marked as being *not robust*. Catching a counterexample within a possibly infinite set of vectors, however, is a daunting task. Let $a \in \mathrm{RAF}_n$ be a sound abstraction for $P(\mathbf{x})$, C a classifier, and $\mathcal{A}_C^{\mathrm{RAF}}$ its abstraction on RAF. We define an informed heuristic search approach leveraging our AFI measure as follows.

Definition 3.12 (Counterexample Search).
(S_1) Let $a_{\mathrm{out}} \triangleq a_0 + \sum_{i=1}^{n} a_i \epsilon_i + a_r \epsilon_r$ be the output of the abstract computation $\mathcal{A}_C^{\mathrm{RAF}}(a)$ on RAF (cf. Sect. 3.1);
(S_2) If $C(\mathbf{x}) < 0$, we look for a potential counterexample $\mathbf{x}^*$ by maximizing a_{out}, i.e., by selecting the maximum possible value for every $\mathbf{x}_i$ when $a_i > 0$, and the minimum when $a_i < 0$ (the converse if $C(\mathbf{x}) > 0$);
(S_3) If $C(\mathbf{x}^*) \neq C(\mathbf{x})$, then $\mathbf{x}^*$ is a counterexample for $\mathbf{x}$, and the classifier C is proved not robust on $\mathbf{x}$;
(S_4) Otherwise, we select the most influential feature $\mathbf{x}_M$, and its mean value in $P(\mathbf{x})$, which is defined by $m \triangleq \left(\min\{\mathbf{y}_M \mid \mathbf{y} \in P(\mathbf{x})\} + \max\{\mathbf{y}_M \mid \mathbf{x} \in P(\mathbf{x})\}\right)/2$, and we partition $P(\mathbf{x})$ using the cutting hyperplane $\mathbf{x}_M \leq m$, thus obtaining left and right subsets $P_l(\mathbf{x}), P_r(\mathbf{x}) \subseteq P(\mathbf{x})$;
(S_5) We consider $a_l, a_r \in \mathrm{RAF}_n$ abstracting, resp., $P_l(\mathbf{x}), P_r(\mathbf{x})$, and we recursively repeat from step (S_1) until a counterexample is found, or a user-defined timeout is met. □

Maximization in step (S_2) requires additional care for one-hot encoded features, as exactly one of them must be set to 1: we set to 1 the most influential feature only. We also observe that computing a_l, a_r during step (S_5) does not introduce losses of precision, as RAFs represent hyperrectangles exactly, and partitioning a RAF by a cutting hyperplane of the form $\mathbf{x}_i \leq k$ yields two smaller hyperrectangles. Steps (S_1) and (S_2) correspond to looking for a counterexample in the vertices of the hyperrectangle represented by a, which have the greatest distance from the center and are therefore intuitively more likely to exhibit different labels. Since a hyperrectangle in $\mathbb{R}^n$ has 2^n vertices, it is not feasible to check them all, and we thus use our feature importance analysis to infer a gradient pointing towards the most promising one. If no counterexample is found, it may be due to the separation curve of C crossing $P(\mathbf{x})$ while leaving all the vertices on the same side. We therefore proceed to steps (S_4) and (S_5) that partition $P(\mathbf{x})$ into two smaller subsets $P_l(\mathbf{x})$ and $P_r(\mathbf{x})$ by cutting the former space in half along the axis of the most influential feature, and recursively repeating the process on the two components. Should a counterexample be found, C can be definitely marked as not robust. Otherwise, we set a timeout mechanism, such as a limit on the recursion depth, to avoid divergence.

Example 3.13. Let $\mathbf{s} = (0, -\sqrt{2})$, $\mathbf{t} = (-1, 1)$, $\mathbf{v} = (1, 1)$ be the support vectors of a SVM with polynomial kernel $k(\mathbf{x}, \mathbf{y}) = (\mathbf{x} \cdot \mathbf{y} + 1)^2$, $w_{\mathbf{s}} = -1$, $w_{\mathbf{t}} = w_{\mathbf{v}} = 1$ their weights, and $b = 0$ the bias. The SVM classifier is therefore defined as $C(\mathbf{x}) \triangleq -(\mathbf{s} \cdot \mathbf{x} + 1)^2 + (\mathbf{t} \cdot \mathbf{x} + 1)^2 + (\mathbf{v} \cdot \mathbf{x} + 1)^2$, and can be rewritten to its primal form $C(\mathbf{x}) = 2\mathbf{x}_1^2 + 2(2 + \sqrt{2})\mathbf{x}_2 + 1$, thus highlighting the separation curve as the parabola $\Gamma \triangleq \mathbf{x}_2 = -\frac{1}{2+\sqrt{2}}\mathbf{x}_1^2 - \frac{1}{2(2+\sqrt{2})}$. We now consider $\mathbf{x}' = (0.5, -0.5)$, and let $P(\mathbf{x}')$ be the hyperrectangle of radius 0.5 centered in $\mathbf{x}'$, that is, $P(\mathbf{x}') \triangleq \{\mathbf{x} \in \mathbb{R}^2 \mid 0 \leq$

$\mathbf{x}_1 \leq 1, -1 \leq \mathbf{x}_2 \leq 0$. We observe that $C(\mathbf{x}') \approx -1.91 < 0$, whereas every point $\mathbf{x} \in P(\mathbf{x}')$ having $\mathbf{x}_2 = 0$ evaluates to $2\mathbf{x}_1^2 + 1 > 0$, which is always positive, hence C is not robust over $P(\mathbf{x}')$. The parabola Γ crosses $P(\mathbf{x})$ leaving some vertices on different sides of the space.

We consider $a = (0.5 + 0.5\epsilon_1, -0.5 + 0.5\epsilon_2) \in \mathrm{RAF}_2^2$ that represents the perturbation $P(\mathbf{x}')$ exactly, and compute $a_{\mathrm{out}} = c + 0.5\epsilon_1 + (1 + \sqrt{2})\epsilon_2 + d\epsilon_r$, where the values of the center $c \in \mathbb{R}$ and the nonlinear accumulation term $d \in \mathbb{R}$ are omitted for simplicity, as they are not relevant for our purposes. Thus, the abstract feature importance vector is $(0.5, 1 + \sqrt{2})$ whose components are both positive. Since $C(\mathbf{x}') < 0$, we are looking for positive counterexamples and, following step (S_3) above, we select the candidate counterexample $\mathbf{x}^* \in P(\mathbf{x}')$ by considering the maximum values for $\mathbf{x}_1, \mathbf{x}_2$ for vectors ranging in $P(\mathbf{x}')$, that is, $\mathbf{x}^* = (1, 0)$. Then, it turns out that $C(\mathbf{x}^*) = 2(1)^2 + 2(2 + \sqrt{(2)})0 + 1 = 3 > 0$, so that $\mathbf{x}^*$ is indeed a counterexample that makes C not robust on $P(\mathbf{x}')$. □

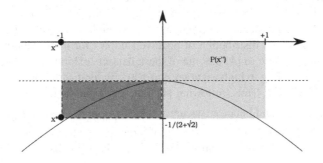

Fig. 3. Example of counterexample search

Example 3.14. Continuing with Example 3.13, we consider $\mathbf{x}'' = (-1, 0)$ and the region $P(\mathbf{x}'') = \{\mathbf{x} \in \mathbb{R}^2 \mid -1 \leq \mathbf{x}_1 \leq +1, -\frac{1}{2+\sqrt{2}} \leq \mathbf{x}_2 \leq 0\}$, as depicted in Fig. 3. Observe that every vertex of $P(\mathbf{x}'')$ is on the same side of Γ and thus receives the same class $+1$, while the subregion of $P(\mathbf{x}'')$ below the parabola lays on the other side, making C not robust. We iterate the search of counterexamples illustrated in Example 3.13, providing the AFI vector $\mathbf{i} = (0, \frac{3+2\sqrt{2}}{(2+\sqrt{2})^2}) = (0, 0.5)$, which can be viewed as a gradient vector guiding the counterexample search. In this case, we look for a counterexample $\mathbf{x}^* \in P(\mathbf{x}'')$ such that $C(\mathbf{x}^*) < 0$, hence moving in the opposite direction w.r.t. the gradient vector $\mathbf{i}$. By doing so, we consider the bottom-left corner $\mathbf{x}^* = (-1, -\frac{1}{2+\sqrt{2}})$ of $P(\mathbf{x}'')$ such that $C(\mathbf{x}^*) = 1$, thus meaning that we did not compute a counterexample. We therefore follow the steps (S_4) and (S_5) by partitioning $P(\mathbf{x}'')$ through the hyperplane $\mathbf{x}_2 \leq -\frac{1}{2(2+\sqrt{2})}$, as shown by the dotted line in Fig. 3, and then starting the recursive search. After the first recursive step, none of the two new (hyper)rectangles have counterexamples in their vertices, so another recursive step is applied to the lower rectangle which is partitioned by the hyperplane $\mathbf{x}_1 \leq 0$.

Table 1. Reference datasets.

Dataset	#Features	Training Set		Test Set	
		Size	Positive	Size	Positive
Adult	103	30162	24.9%	15060	24.6%
Compas	371	4222	53.3%	1056	55.6%
German	56	800	69.8%	200	71.0%

This generates two new (hyper)rectangles, and both have now counterexamples in their vertices, which can be found through steps (S_1) and (S_2), therefore proving, through a counterexample witness, that C is not robust. □

4   Experimental Evaluation

We consider the main reference datasets used in the fairness literature [30]:(i) **Adult** [15], which labels yearly incomes, above or below 50K US$, based on personal attributes; (ii) **Compas** [3], which labels recidivism risk based on personal attributes and criminal history; (iii) **German** [15], which labels good/bad credit scores. Table 1 displays size and distribution of positive samples for these datasets. The data is preprocessed according to [41]. Some of these datasets exhibit a highly unbalanced label distribution, leading to high accuracy and 100% individual fairness for a constant classifier like $C(\mathbf{x}) = 1$. Thus, following [41], we report in Table 2, for several SVM models, both accuracy and balanced accuracy, i.e., $\frac{1}{2}\left(\frac{truePos}{truePos+falseNeg} + \frac{trueNeg}{trueNeg+falsePos}\right)$.

Similarity Relations. Let $I \subseteq \mathbb{N}$ be a set of indices of features after one-hot encoding, and $\mathbf{x}, \mathbf{y} \in X$ be two individuals. Following [41], we consider three similarity relations.

NOISE: Given a subset of numerical features $I' \subseteq I$, let $S_{noise}(\mathbf{x}, \mathbf{y})$ iff $|\mathbf{x}_i - \mathbf{y}_i| \le \epsilon$ for all $i \in I'$, and $\mathbf{x}_i = \mathbf{y}_i$ for all $i \in I \smallsetminus I'$. This means that all the features of $\mathbf{x}$ in I' are subject to a maximum norm perturbation $P_\infty^\epsilon(\mathbf{x})$. For our experiments, we consider $\epsilon = 0.05$, namely, a $\pm 5\%$ perturbation for data normalised to $[0, 1]$.

CAT: Given a subset of sensitive categorical attributes $I' \subseteq I$, let $S_{cat}(\mathbf{x}, \mathbf{y})$ iff $\mathbf{x}_i = \mathbf{y}_i$ for all $i \in I \smallsetminus I'$. For Adult and German, we select the gender attribute, while for Compas, the race attribute.

NOISE-CAT: $S_{noise\text{-}cat}(\mathbf{x}, \mathbf{y}) \triangleq S_{noise}(\mathbf{x}, \mathbf{y}) \vee S_{cat}(\mathbf{x}, \mathbf{y})$.

Setup. We trained the SVMs used in our experiments with scikit-learn. Hyperparameters were chosen by hit-and-trial and observing trends. The final hyperparameters for each kernel and dataset were those that led to SVMs with high balanced accuracy.

Data-Availability Statement. Implementation, datasets, and scripts for reproducing the experimental results, are available on GitHub [33] and Zenodo with the identifier https://doi.org/10.5281/zenodo.10053395.

Individual Fairness. Table 2 shows accuracy, balanced accuracy, and verified individual fairness percentages $\text{fair}_{T,S}(\text{SVM})$ on the test sets T and similarity relations S for linear SVMs (denoted in the table as L(regularization parameter C)), nonlinear SVMs with RBF kernels (denoted as R(regularization parameter C, γ)) and polynomial kernels (denoted as P(regularization parameter C, degree, basis)) trained on

Table 2. Accuracy, balanced accuracy, and verified individual fairness percentages leveraging different abstractions.

Dataset	SVM Kernel	Acc.	Bal. Acc.	Interval			Interval+OH			RAF			RAF+OH		
				N	C	NC	N	C	NC	N	C	NC	N	C	NC
Adult	L(1)	84.6	75.6	**96.5**	**95.2**	**91.6**	**96.5**	**95.2**	**91.6**	**96.5**	**95.2**	**91.6**	**96.5**	**95.2**	**91.6**
	R(0.05,0.01)	83.8	72.0	2.4	2.8	0.0	2.4	2.8	0.0	94.8	42.4	37.2	**97.5**	**97.9**	**95.4**
	P(0.01,3,3)	83.9	76.7	0.0	0.0	0.0	0.0	0.0	0.0	**0.5**	**0.5**	**0.03**	**0.5**	**0.5**	**0.03**
Compas	L(1)	64.7	64.1	**95.5**	**99.5**	**94.9**	**95.5**	**99.5**	**94.9**	**95.5**	**99.5**	**94.9**	**95.5**	**99.5**	**94.9**
	R(0.05,0.01)	64.5	63.1	54.3	42.5	1.0	54.3	42.5	1.0	91.8	71.6	66.9	**94.4**	**97.5**	**89.3**
	P(0.01,3,3)	64.3	63.9	0.0	0.0	0.0	0.0	0.0	0.0	**0.5**	**0.6**	**0.1**	**0.5**	**0.6**	**0.1**
German	L(1)	79.0	70.8	**87.5**	**94.5**	**81.5**	**87.5**	**94.5**	**81.5**	**87.5**	**94.5**	**81.5**	**87.5**	**94.5**	**81.5**
	R(10,0.05)	79.5	74.1	0.0	0.0	0.0	0.0	0.0	0.0	**2.0**	0.0	0.0	**2.0**	**82.0**	0.0
	P(0.01,6,6)	75.5	71.8	0.0	0.0	0.0	0.0	0.0	0.0	**78.5**	**76.0**	**10.0**	**78.5**	**76.0**	**10.0**

each dataset. Verified individual fairness percentages are computed w.r.t. the NOISE (columns 'N'), CAT (columns 'C'), and NOISE-CAT (columns 'NC') similarity relations. This table compares results obtained using intervals and RAFs, with and without the OH abstraction. In every row corresponding to a SVM instance, we present the highest verified individual fairness percentages for each similarity relation in bold font and the lowest percentages in a faded shade. It turns out that the RAF abstraction typically outperforms intervals, and adding our OH abstraction always yields equal or higher (even much higher) verified individual fairness. For the same SVMs, in the following table:

Dataset	Linear		Polynomial		RBF	
	LB	UB	LB	UB	LB	UB
Adult	91.6	91.6	0.03	89.5	92.2	95.4
Compas	94.9	94.9	0.09	71.4	89.3	93.0
German	81.5	81.5	10.0	76.0	0.0	84.0

We provide lower and upper bounds on individual fairness w.r.t. the NOISE-CAT similarity relation using the RAF+OH abstraction. The lower bound (columns 'LB') is the verified individual fairness percentage as given in Table 2, while the upper bound (columns 'UB') is an estimate obtained through the counterexample search of Definition 3.12 without input partitioning, e.g., an upper bound of 76% for German/Polynomial means that we found concrete counterexamples to individual fairness for 48 over 200 test samples, thus entailing that at most 152 out of 200 test samples (i.e., 76%) are individually fair. The gap between these bounds is zero for linear SVMs and narrow for RBF kernels trained on the Adult and Compas datasets: in these cases, our RAF+OH abstraction turns out to be (very) precise and the counterexample search heuristics is strong. On the other hand, the gap is much wider in the remaining cases, notably for SVMs with polynomial kernels, mostly due to a lower precision of the abstraction. Using partitioning (i.e., step (S_4) of Definition 3.12) up to 3.125% of the original perturbation size, we get similar upper bounds, thus hinting the presence of a few additional

Table 3. Comparison of AFI and PFI on German.

		Grade for each feature										
Linear	Baseline (13.55s)	5	5	5	6	6	7	7	7	7	8	Distance
	AFI (0.01s)	5	5	5	6	6	7	8	7	7	8	1.0
	PFI (4.07s)	5	5	6	7	7	9	6	6	7	7	3.16
RBF	Baseline (17.98s)	5	5	5	6	6	7	7	7	8	8	Distance
	AFI (0.02s)	5	6	5	6	6	8	7	7	8	7	1.73
	PFI (6.23s)	6	7	5	6	7	8	7	6	7	5	4.24
Polynomial	Baseline (15.83s)	5	5	5	6	7	7	7	7	7	8	Distance
	AFI (0.01s)	7	6	7	7	5	7	6	6	5	8	4.47
	PFI (4.15s)	6	7	9	7	6	7	5	6	6	6	5.74

Table 4. Distances of AFI and PFI from several baselines for different SVMs.

	Baseline	$N=2k$ $\epsilon=0.2$	$N=10k$ $\epsilon=0.2$	$N=2k$ $\epsilon=0.4$	$N=10k$ $\epsilon=0.4$	$N=2k$ $\epsilon=0.6$	$N=5k$ $\epsilon=0.6$	$N=10k$ $\epsilon=0.6$	$N=2k$ $\epsilon=0.8$	$N=5k$ $\epsilon=0.8$	$N=10k$ $\epsilon=0.8$
Adult Linear	AFI (0.27s)	0.0	0.0	1.0	0.0	1.0	1.41	1.0	1.0	1.41	1.0
	PFI (10009s)	2.45	2.45	2.24	2.45	2.24	1.41	2.24	2.24	1.41	2.24
Adult RBF	AFI (0.48s)	1.0	1.41	1.41	1.41	1.73	1.73	1.41	1.41	1.41	1.41
	PFI (25221s)	1.73	2.45	2.45	2.0	2.65	2.65	2.45	2.45	2.45	2.45
Adult Polynomial	AFI (0.44s)	1.0	1.0	0.0	1.41	0.0	0.0	0.0	0.0	0.0	0.0
	PFI (9985s)	1.0	1.0	1.41	1.0	1.41	1.41	1.41	1.41	1.41	1.41
Compas Linear	AFI (0.22s)	1.41	1.41	1.73	1.73	1.41	1.73	1.41	1.41	1.41	1.73
	PFI (1953s)	1.73	1.73	2.0	2.0	2.24	2.0	2.24	2.24	2.24	2.83
Compas RBF	AFI (0.27s)	2.0	2.0	2.65	2.65	2.83	2.83	2.83	2.83	2.83	2.83
	PFI (6827s)	2.0	2.0	2.65	2.65	2.83	2.83	2.83	2.83	2.83	2.83
Compas Polynomial	AFI (0.22s)	4.24	4.24	4.12	4.12	4.24	4.24	4.24	4.24	4.24	4.24
	PFI (2069s)	2.45	2.45	3.0	3.0	3.74	3.74	3.74	3.74	3.74	3.74
German Linear	AFI (0.01s)	1.0	1.0	1.0	1.0	1.0	1.0	1.0	1.41	1.73	1.41
	PFI (4.07s)	3.16	3.46	3.16	3.16	3.16	3.16	3.16	3.6	3.74	3.0
German RBF	AFI (0.02s)	1.73	1.0	1.73	1.73	2.0	1.41	1.73	1.73	2.0	2.24
	PFI (6.23s)	4.0	3.46	4.24	4.24	4.36	3.61	4.24	4.24	4.36	4.47
German Polynomial	AFI (0.01s)	4.90	4.12	4.47	3.87	3.87	4.24	3.46	3.46	3.46	3.46
	PFI (4.15s)	5.74	5.10	5.74	4.69	4.69	5.0	4.58	4.58	4.58	4.58

counterexamples. Only partitioning up to 0.1% of the original input size, we could find substantially more counterexamples.

Global Feature Importance. We compare our abstract feature importance AFI, used as a *global* feature importance measure, with the popular global measure PFI, as implemented in the Python *sklearn.inspection* package with $n\_repeat = 10$. For the sake of comparison with an outside baseline, we uniformly sampled N points in the input space of the SVMs and determined how often a NOISE perturbation for a single numerical input feature changed the SVM classification: the more often the classification changed, the more important is the input feature. As a representative example, we show in Table 3, a comparison for the SVMs trained on the German dataset. In lines 'Baseline', 'AFI' and 'PFI', we show the feature grades, as defined in Sect. 3.1, of the 10 non-categorical

Table 5. Local Comparison of AFI and LIME.

Distance between LIME and ...	Adult			Compas			German		
	Lin	RBF	Poly	Lin	RBF	Poly	Lin	RBF	Poly
AFI ($\epsilon = 0.1$)	2.42	2.04	2.98	1.67	1.06	3.05	2.62	2.03	**5.31**
AFI ($\epsilon = 0.2$)	1.68	1.32	2.67	1.63	0.17	2.73	2.21	2.00	5.41
AFI ($\epsilon = 0.3$)	1.39	0.51	2.58	**1.57**	0.14	**2.62**	1.92	2.05	5.45
AFI (Global)	**1.37**	**0.01**	**1.01**	**1.57**	**0.13**	3.16	**1.90**	**1.89**	5.53

Table 6. Time Comparison (in sec) of AFI, PFI, LIME.

Dataset	Linear			Polynomial			RBF		
	AFI	PFI	LIME	AFI	PFI	LIME	AFI	PFI	LIME
Adult	**0.27**	$1 \cdot 10^4$	3.78	**0.45**	$1 \cdot 10^4$	6.21	**0.48**	$2 \cdot 10^4$	9.82
Compas	**0.22**	$2 \cdot 10^3$	2.72	**0.22**	$2 \cdot 10^3$	2.89	**0.27**	$6 \cdot 10^3$	8.97
German	**0.01**	4.07	0.198	**0.01**	4.15	0.355	**0.02**	6.23	0.223

(7 numerical plus 3 binary) input features of German based on the importance scores measured by, respectively, baseline, AFI and PFI. The baseline has been computed by considering $N = 10000$ samples and a NOISE perturbation with magnitude $\epsilon = 0.4$. We also indicate in parenthesis the time needed (in seconds) to compute these scores, where for our AFI measure, we used the RAF+OH abstraction. In column 'Distance' we show the Euclidean distance between the feature grades computed by AFI and PFI w.r.t. the baseline. We can observe that AFI better correlates with model variance to feature perturbations than PFI. In fact, the correlation is almost perfect for the linear SVM. For nonlinear SVMs, the abstraction RAF+OH loses more precision, so that the correlation decreases, nevertheless the distance to the baseline is still smaller than for PFI. Note that AFI is computed in a negligible fraction of time w.r.t. PFI.

Table 4 compares the Euclidean distance between the feature grades computed by AFI and PFI w.r.t. different choices of the number of samples N and magnitudes ϵ used for computing the baseline of SVMs trained on the Adult, Compas, and German datasets. For each AFI-baseline and PFI-baseline pair, the smaller distance is made bold and the larger has a faded shade. The data indicates AFI is closer to the baseline than PFI in most cases, except for the polynomial SVM trained on Compas: for this case, the likely reason is the low precision of our polynomial SVM abstraction, as also hinted by the low verified individual fairness scores in the entries for Compas/Polynomial/RAF+OH in Table 2.

Local Feature Importance. In Table 5, we present a comparative analysis between our measure AFI, used as a *local* feature importance measure, and the extensively used local feature importance measure LIME as implemented in the Python *lime.lime_tabular* package [39], for SVMs trained on the three different datasets. The local neighborhood used by LIME to determine the importance of features is obtained using a normal distribution, and the influence of a point in the neighborhood is dependent on its distance from the original point [38, Sect. 3.3]. Instead, AFI considers a local uniform distribution for the local neighborhood, i.e., a hyperrectangle abstraction where each point is equiprobable. Thus, it is not possible to find a common local neighborhood in which we can compare local AFI and LIME to a ground truth baseline. Instead, the comparison in Table 5 is based on computing the average Euclidean distance between the feature grades obtained with these two metrics for 200 inputs sampled from the test sets. For our local AFI measure, we took as input to the SVMs a local neighborhood determined by a NOISE perturbation with three different values of the magnitude ϵ. We also compared the distance of LIME with the global AFI computed over the entire input space of the SVMs. The results show that these distances are consistently minimal (except for the polynomial SVMs), implying a high degree of similarity between the two feature importance measures. Interestingly, this table shows that the local LIME feature grades are closer to the global AFI feature grades than to the local AFI feature grades. Furthermore, as the value of the magnitude ϵ used to compute local AFI increases, the distance between LIME and local AFI feature grades decreases.

Computation Time. Finally, Table 6 presents a comparison of the computation times (in seconds) of AFI, PFI, and LIME for our SVMs. As time taken by global and local AFI is similar, we only report global AFI time. The results shows that AFI always outperforms both PFI and LIME.

5   Conclusion

We put forward a novel feature importance measure based on an abstract interpretation of SVMs, which is tailored for achieving a precise symbolic representation of one-hot encoded features. We showed that our abstraction is effective for certifying robustness properties—notably, individual fairness—of SVMs, and that our abstract feature importance measure outperforms the state-of-the-art. As future work, we plan to extend our approach to certify alternative or stronger fairness notions. We also aim to design quantitative verification methods to provide probabilistic guarantees on the behavior of SVM models.

Acknowledgements. Francesco Ranzato and Marco Zanella were partially funded by the *Italian MIUR*, under the PRIN 2017 project no. 201784YSZ5. Francesco Ranzato was partially funded by: the *Italian MUR*, under the PRIN 2022 PNRR project no. P2022HXNSC; *Meta* (formerly *Facebook*) *Research*, under a "Probability and Programming Research Award" and under a *WhatsApp Research Award* on "Privacy-aware Program Analysis"; by an *Amazon Research Award* for "AWS Automated Reasoning". Caterina Urban was partially funded by the French PEPR Intelligence Artificielle SAIF project (ANR-23-PEIA-0006).

References

1. Aho, A.V., Lam, M.S., Sethi, R., Ullman, J.D.: Compilers: Principles, Techniques, and Tools, 2nd edn. Addison-Wesley Longman Publishing Co., Inc, USA (2006)
2. Albarghouthi, A.: Introduction to neural network verification. Found. Trends Program. Lang. 7(1–2), 1–157 (2021). https://doi.org/10.1561/2500000051

3. Angwin, J., Larson, J., Mattu, S., Kirchner, L.: Machine Bias. ProPublica 23 (2016), https://www.propublica.org/article/machine-bias-risk-assessments-in-criminal-sentencing

4. Apley, D.W., Zhu, J.: Visualizing the effects of predictor variables in black box supervised learning models. J. R. Stat. Soc. Ser. B Stat Methodol. **82**(4), 1059–1086 (2020). https://doi.org/10.1111/rssb.12377

5. Bhatt, U., et al.: Explainable machine learning in deployment. In: Proceedings of the 2020 Conference on Fairness, Accountability, and Transparency, FAT* 2020, pp. 648–657. ACM (2020). https://doi.org/10.1145/3351095.3375624

6. Breiman, L.: Random forests. Mach. Learn. **45**(1), 5–32 (2001). https://doi.org/10.1023/A:1010933404324

7. Carlini, N., Wagner, D.A.: Towards evaluating the robustness of neural networks. In: Proceedings of 38th IEEE Symposium on Security and Privacy (S & P 2017), pp. 39–57 (2017). https://doi.org/10.1109/SP.2017.49

8. Casalicchio, G., Molnar, C., Bischl, B.: Visualizing the feature importance for black box models. In: Machine Learning and Knowledge Discovery in Databases - Proceedings of the European Conference, ECML PKDD 2018. Lecture Notes in Computer Science, vol. 11051, pp. 655–670. Springer (2018). https://doi.org/10.1007/978-3-030-10925-7_40

9. Cervantes, J., Garcia-Lamont, F., Rodríguez-Mazahua, L., Lopez, A.: A comprehensive survey on support vector machine classification: applications, challenges and trends. Neurocomputing **408**, 189–215 (2020). https://doi.org/10.1016/j.neucom.2019.10.118

10. Chang, Y.W., Lin, C.J.: Feature ranking using linear SVM. In: Proceedings of the Workshop on the Causation and Prediction Challenge at WCCI 2008. Proceedings of Machine Learning Research, vol. 3, pp. 53–64. PMLR (2008), http://proceedings.mlr.press/v3/chang08a.html

11. Chouldechova, A.: Fair prediction with disparate impact: a study of bias in recidivism prediction instruments. Big Data **5**(2), 153–163 (2017). https://doi.org/10.1089/big.2016.0047

12. Cousot, P.: Principles of Abstract Interpretation. MIT Press (2021)

13. Cousot, P., Cousot, R.: Abstract interpretation: a unified lattice model for static analysis of programs by construction or approximation of fixpoints. In: Proceedings of the 4th ACM Symposium on Principles of Programming Languages (POPL 1977), pp. 238–252 (1977). https://doi.org/10.1145/512950.512973

14. Cristianini, N., Shawe-Taylor, J.: An Introduction to Support Vector Machines and Other Kernel-based Learning Methods. Cambridge University Press (2000). https://doi.org/10.1017/CBO9780511801389

15. Dua, D., Graff, C.: UCI Machine Learning repository (2017). https://archive.ics.uci.edu/ml

16. Dwork, C., Hardt, M., Pitassi, T., Reingold, O., Zemel, R.S.: Fairness through awareness. In: Innovations in Theoretical Computer Science 2012, pp. 214–226. ACM (2012). https://doi.org/10.1145/2090236.2090255

17. Fish, B., Kun, J., Lelkes, Á.D.: A confidence-based approach for balancing fairness and accuracy. In: Proceedings of the 2016 SIAM International Conference on Data Mining, pp. 144–152. SIAM (2016). https://doi.org/10.1137/1.9781611974348.17

18. Fisher, A., Rudin, C., Dominici, F.: All models are wrong, but many are useful: learning a variable's importance by studying an entire class of prediction models simultaneously. J. Mach. Learn. Res. **20**(177), 1–81 (2019). http://jmir.org/papers/v20/18-760.html

19. Friedman, J.H.: Greedy function approximation: a gradient boosting machine. Ann. Stat. **29**(5), 1189–1232 (2001). http://www.jstor.org/stable/2699986

20. Ghorbal, K., Goubault, E., Putot, S.: The zonotope abstract domain Taylor1+. In: Computer Aided Verification, 21st International Conference, CAV 2009. Proceedings. Lecture Notes in Computer Science, vol. 5643, pp. 627–633. Springer (2009). https://doi.org/10.1007/978-3-642-02658-4_47

21. Ghosh, B., Basu, D., Meel, K.S.: Algorithmic fairness verification with graphical models. In: Thirty-Sixth AAAI Conference on Artificial Intelligence, AAAI 2022, pp. 9539–9548 (2022). https://doi.org/10.1609/aaai.v36i9.21187

22. Goldstein, A., Kapelner, A., Bleich, J., Pitkin, E.: Peeking inside the black box: visualizing statistical learning with plots of individual conditional expectation. J. Comput. Graph. Stat. **24**(1), 44–65 (2015). https://doi.org/10.1080/10618600.2014.907095

23. Goodfellow, I., McDaniel, P., Papernot, N.: Making machine learning robust against adversarial inputs. Commun. ACM **61**(7), 56–66 (2018). https://doi.org/10.1145/3134599

24. Hechtlinger, Y.: Interpretation of prediction models using the input gradient. CoRR arXiv (2016). http://arxiv.org/abs/1611.07634

25. Hooker, G., Mentch, L., Zhou, S.: Unrestricted permutation forces extrapolation: variable importance requires at least one more model, or there is no free variable importance. Stat. Comput. **31**(6), 82 (2021). https://doi.org/10.1007/s11222-021-10057-z

26. Khandani, A.E., Kim, A.J., Lo, A.W.: Consumer credit-risk models via machine-learning algorithms. J. Bank. Finance **34**(11), 2767–2787 (2010). https://doi.org/10.1016/j.jbankfin.2010.06.001

27. Langenberg, P., Balda, E.R., Behboodi, A., Mathar, R.: On the robustness of support vector machines against adversarial examples. In: 13th International Conference on Signal Processing and Communication Systems, ICSPCS 2019, pp. 1–6. IEEE (2019). https://doi.org/10.1109/ICSPCS47537.2019.9008746

28. Liu, C., Arnon, T., Lazarus, C., Strong, C.A., Barrett, C.W., Kochenderfer, M.J.: Algorithms for verifying deep neural networks. Found. Trends Optim. **4**(3–4), 244–404 (2021). https://doi.org/10.1561/2400000035

29. Lundberg, S.M., Lee, S.: A unified approach to interpreting model predictions. In: Advances in Neural Information Processing Systems 30: Annual Conference on Neural Information Processing Systems 2017, pp. 4765–4774 (2017). https://proceedings.neurips.cc/paper/2017/hash/8a20a8621978632d76c43dfd28b67767-Abstract.html

30. Mehrabi, N., Morstatter, F., Saxena, N., Lerman, K., Galstyan, A.: A survey on bias and fairness in machine learning. ACM Comput. Surv. **54**(6), 1–35 (2021). https://doi.org/10.1145/3457607

31. Messine, F.: Extentions of affine arithmetic: application to unconstrained global optimization. J. Univ. Comput. Sci. **8**(11), 992–1015 (2002). https://doi.org/10.3217/jucs-008-11-0992

32. Mladenic, D., Brank, J., Grobelnik, M., Milic-Frayling, N.: Feature selection using linear classifier weights: interaction with classification models. In: SIGIR 2004: Proceedings of the 27th Annual International ACM SIGIR Conference on Research and Development in Information Retrieval, pp. 234–241. ACM (2004). https://doi.org/10.1145/1008992.1009034

33. Pal, A., Ranzato, F., Urban, C., Zanella, M.: Abstract Feature Importance for SVMs (2023). https://github.com/AFI-SVM

34. Park, S., Byun, J., Lee, J.: Privacy-preserving fair learning of support vector machine with homomorphic encryption. In: WWW 2022: The ACM Web Conference 2022, pp. 3572–3583. ACM (2022). https://doi.org/10.1145/3485447.3512252

35. Ranzato, F., Urban, C., Zanella, M.: Fairness-aware training of decision trees by abstract interpretation. In: Proceedings of the 30th ACM International Conference on Information & Knowledge Management, CIKM2021, pp. 1508–1517 (2021). https://doi.org/10.1145/3459637.3482342

36. Ranzato, F., Zanella, M.: Robustness verification of support vector machines. In: Proceedings of the 26th International Static Analysis Symposium (SAS 2019), pp. 271–295. LNCS vol. 11822 (2019). https://doi.org/10.1007/978-3-030-32304-2_14

37. Ranzato, F., Zanella, M.: Saver: SVM Abstract Verifier (2019). https://github.com/abstract-machine-learning/saver

38. Ribeiro, M.T., Singh, S., Guestrin, C.: "Why should I trust you?": explaining the predictions of any classifier. In: Proceedings of the 22nd ACM SIGKDD International Conference on Knowledge Discovery and Data Mining, 2016, pp. 1135–1144. ACM (2016). https://doi.org/10.1145/2939672.2939778

39. Ribeiro, M.T.C.: Local Interpretable Model-agnostic Explanations (LIME) (2016). https://lime-ml.readthedocs.io

40. Roh, Y., Lee, K., Whang, S., Suh, C.: Fr-train: a mutual information-based approach to fair and robust training. In: Proceedings of the 37th International Conference on Machine Learning (ICML 2020). Proceedings of Machine Learning Research, vol. 119, pp. 8147–8157. PMLR (2020). http://proceedings.mlr.press/v119/roh20a.html

41. Ruoss, A., Balunovic, M., Fischer, M., Vechev, M.T.: Learning certified individually fair representations. In: Advances in Neural Information Processing Systems 33: Annual Conference on Neural Information Processing Systems (NeurIPS 2020) (2020). https://proceedings.neurips.cc/paper/2020/hash/55d491cf951b1b920900684d71419282-Abstract.html

42. Shapley, L.S.: A value for n-person games. In: Kuhn, H.W., Tucker, A.W. (eds.) Contributions to the Theory of Games II, pp. 307–317. Princeton University Press, Princeton (1953)

43. Tjoa, E., Guan, C.: A survey on explainable artificial intelligence (XAI): toward medical XAI. IEEE Trans. Neural Networks Learn. Syst. **32**(11), 4793–4813 (2021). https://doi.org/10.1109/TNNLS.2020.3027314

44. Urban, C., Christakis, M., Wüstholz, V., Zhang, F.: Perfectly parallel fairness certification of neural networks. Proc. ACM Program. Lang. **4**(OOPSLA), 185:1-185:30 (2020). https://doi.org/10.1145/3428253

45. Urban, C., Miné, A.: A review of formal methods applied to machine learning. CoRR arXiv (2021). https://arxiv.org/abs/2104.02466

46. Verma, S., Rubin, J.: Fairness definitions explained. In: Proceedings of the International Workshop on Software Fairness, FairWare@ICSE 2018, pp. 1–7. ACM (2018). https://doi.org/10.1145/3194770.3194776

47. Xiao, H., Biggio, B., Nelson, B., Xiao, H., Eckert, C., Roli, F.: Support vector machines under adversarial label contamination. Neurocomputing **160**, 53–62 (2015). https://doi.org/10.1016/j.neucom.2014.08.081

48. Yurochkin, M., Bower, A., Sun, Y.: Training individually fair ML models with sensitive subspace robustness. In: Proceedings of the 8th International Conference on Learning Representations, ICLR 2020 (2020). https://openreview.net/forum?id=B1gdkxHFDH

Generation of Violation Witnesses by Under-Approximating Abstract Interpretation

Marco Milanese[✉] and Antoine Miné

Sorbonne Université, CNRS, LIP6, 75005 Paris,
France
{marco.milanese,antoine.mine}@lip6.fr

Abstract. This works studies abstract backward semantics to infer sufficient program preconditions, based on an idea first proposed in previous work [38]. This analysis exploits under-approximated domain operators, demonstrated in [38] for the polyhedra domain, to under-approximate Dijkstra's liberal precondition. The results of the analysis were implemented into a static analysis tool for a toy language. In this paper we address some limitations that hinder its applicability to C-like programs. In particular, we focus on two improvements: handling of user input and integer wrapping. For this, we extend the semantic and design sound and effective abstractions. Furthermore, to improve the precision, we explore an under-approximated version of the power-set construction. This in particular helps handling arbitrary union that is difficult to implement with under-approximated domains. The improved analysis is implemented and its performance is compared with other static analysis tools in SV-COMP23 using a selected subset of benchmarks.

Keywords: Abstract interpretation · Software verification · Program analysis · Bug catching · Under-approximation

1 Introduction

The focus of static analysis by abstract interpretation [16,17] has traditionally been on the assurance of program *correctness*. However, the dual problem of verifying the *presence* of bugs is equally intriguing, since in practice sound static analysis tools generate many false positives and checking them manually can be time-consuming. This calls for a novel kind of abstract semantics where domains and operators are under-approximated rather than over-approximated. A preliminary study in this area was done in our previous work [38], where we investigated an abstract backward semantic to infer *sufficient* program preconditions. This analysis builds on conventional abstract domains, but in this analysis they represent an *under*-approximation of the concrete invariant. To achieve this analysis, novel under-approximation domain operators are required, and in [38] we have shown how they can be designed for the polyhedra domain [21]. The results were implemented in a static analysis tool targeting a toy language.

R. Dimitrova et al. (Eds.): VMCAI 2024, LNCS 14499, pp. 50–73, 2024.
https://doi.org/10.1007/978-3-031-50524-9_3

```
unsigned int i = 10;

while (i >= 10) {
        i++;
}

i += input();
assert(i != 5);
```

(a) Program with integer overflow and input.

```
int i = input();
int j;
assume(i >= 0 && i <= 2);

if (i == 0) j = 0;
else if (i == 1) j = 1;
else if (i == 2) j = 0;

assert(j == 1);
```

(b) Simple disjunctive program.

Fig. 1. Programs that show the limits of the semantics of [38].

Motivation. Consider the program of Fig. 1a. The while loop starts with i = 10 and terminates when i overflows to 0. Therefore, if input() returns 5 the assertion will fail. Unfortunately the semantic proposed in [38] can not detect the overflow as it only supports mathematical numbers (i.e., numbers with infinite precision). Moreover in this semantic there is no built-in encoding for user input and thus the preconditions that it can find concern only program's arguments. Consequently, it can not find the sufficient precondition, input() = 5, for the assertion to fail.

Additionally, under-approximated operators were studied only for the polyhedra domain, but as for conventional abstract interpretation, other domains can be considered and the choice of the domain boils down to a precision-efficiency trade-off. As an example, the polyhedra domain fails to find the sufficient precondition, input() = 1, for the correctness of the program of Fig 1b as the invariant before the assertion, $(i \in \{0, 2\} \land j = 0) \lor (i = 1 \land j = 1)$, is not polyhedral.

Termination. The preconditions found by this analysis under-approximate Dijkstra's weakest liberal precondition, ensuring either divergence or termination within the post-condition. The former case can be problematic, such as when the analysis is used to find preconditions for bugs, as a non-empty precondition may be a symptom of an infinite loop rather than an actual bug. However, non-termination can be ruled out with various techniques, including termination checking with a ranking function and modern works on its synthesis have been quite successful [14,15]. For the sake of simplicity we do not implement those techniques and instead opt for a simple approach of checking termination by experimentally executing the program (with a time limit).

Related Works. Lately there has been an increase in interest on under-approximations following the seminal work of Peter O'Hearn on Reverse Hoare Logic/Incorrectness Logic [24,41]. Compared to our approach, this stream of works focuses on *forward*, not backward, analyses, thus they do not study preconditions for bugs but instead they find *post-conditions* for them. Moreover, as logic methods, they can *prove* that a post-condition is a valid under-approximation of the reachable states. Some hints on how post-conditions can be inferred were discussed in [41, Sect. 6], but they handle loops by unrolling, thus limiting the analysis to some loop bound. On the contrary, our approach can *infer* pre-conditions and loops are handled with widening operators, so that unrolling is not needed and unbounded loops can be handled. Incorrectness logic was made memory aware by Raad et al. [42] using ideas from separation logic [43]. In comparison, our work focuses exclusively on numeric programs and abstract domains for handling memory properties are left as a future work. Finally, reasoning with incorrectness logic can be made automatic with theorem provers as in [33,42], whereas in our work, reasoning occurs with abstract domains. This makes the analysis more scalable.

Counter-examples generation is also possible with several instances of model checking, e.g., symbolic execution [5,32] and CEGAR [13], where the state exploration is handled using SMT solvers (CEGAR is guided by counter-examples and utilizes other techniques besides SMT for the refinement phase, e.g., interpolants).

Traditional backward analyses based on abstract interpretation [12,18,19] focus on inferring *necessary* preconditions P, that is conditions such that no execution starting from $\neg P$ can succeed. For example Cousot et al. [20] propose a backward precondition analysis for code contracts. They differ from us in the handling of non-determinism as they keep states that succeed at least for one non-deterministic program path, whereas we keep states that succeed for all non-deterministic paths. Moreover, unlike us, they use symbolic reasoning, not numeric domains.

In this work we focus on sufficient preconditions P, that is conditions such that all executions from P must succeed: these require *under-approximated* operators. Designing under-approximation domains featuring optimal operators can be challenging [3] at least partially explaining why they are rarer than over-approximation ones. Several high-order constructions have been proposed in which conventional domains are used to construct under-approximation ones. Lev-Ami et al. [34] propose to use set-complements of abstract domains, but this yields shapes that are rarely interesting. Other methods based on existential quantification [44] and disjunctive completions [40] were proposed, but they incur in a too high complexity and are difficult to abstract away.

Under-approximations were used also in the work of Urban et al. [47], namely for the co-domain. However, the results are difficult to compare in theory due to different abstractions and different concrete semantics.

Contribution. In this paper we extend upon the backward sufficient preconditions analysis, addressing some of the limitations that hinder its applicability to C-like programs, namely: handling of user input and integer wrapping. To improve the precision of the polyhedra domain, we consider the well-known power-set construction and derive sound under-approximation operators for it.

We then proceed to implement the improved analysis in a static analysis tool and add support for extracting a violation witness in SV-COMP's format [8].

$$a ::= [x, y] \mid v \mid a_1 + a_2 \mid a_1 - a_2 \mid a_1 * a_2 \mid a_1/a_2$$
$$b ::= a_1 = a_2 \mid a_1 \neq a_2 \mid a_1 \leq a_2 \mid a_1 < a_2 \mid a_1 > a_2 \mid a_1 \geq a_2$$
$$\mid \mathbf{t} \mid \mathbf{f} \mid \neg b \mid b_1 \wedge b_2 \mid b_1 \vee b_2$$
$$s ::= \mathbf{skip}() \mid v := a \mid \mathbf{assume}(b)$$
$$\mid s_1; s_2 \mid \mathbf{if}\ b\ \mathbf{then}\ s_1\ \mathbf{else}\ s_2 \mid \mathbf{while}\ b\ \mathbf{do}\ s\ \mathbf{done}$$

Fig. 2. While language.

To the best of our knowledge, this is the first abstract interpretation based tool that can certify program incorrectness (at least among the ones participating in SV-COMP). We compare its performance with that of other static analysis tools participating in SV-COMP23 [7] on a selected subset of the competition's benchmarks.

2 Semantics

The semantic studied in [38] is limited to numeric variables and properties and is given for a toy language with mathematical numbers. In this paper, we address how it can be adapted to support fixed-precision integers and user input. Handling more advanced features of C-like languages, such as pointers, arrays, structures, dynamic memory allocation, remains a future work. Floating-point arithmetic is out of the scope of this work but it should not be a large problem as [38] shows that expression evaluation can be over-approximated and still leads to sound under-approximated statements, thus rounding errors can be abstracted as small non-deterministic error intervals, as in forward analysis, and easily supported in a polyhedral analysis.

This section is organized as follows: in 2.2 we recall the semantics of [38], then in 2.3 we extend it with user input and in 2.4 with finite-precision integers.

2.1 Notation

Given a set X, we denote with $\mathcal{P}(X)$ the set of all subsets of X. If f is a function, then $\mathrm{dom}(f)$ denotes its domain. If X is a poset and $f : X \to X$, we denote with $\mathrm{gfp}_R f$ the greatest fix-point of f less or equal than R.

2.2 Background on Sufficient Preconditions Semantic

We recall here the semantics from [38], on top of which we construct our new analysis. The analysis is given both in equational form (where the program is represented as a control flow graph) and in big-step form (where the program is represented with an inductive language), for our purposes we only consider the latter.

$$\tau[\![\mathbf{skip}()]\!]S \triangleq S$$

$$\overleftarrow{\tau}[\![v := a]\!]S \triangleq \{\rho \mid \forall x \in E[\![a]\!]\rho.\ \rho[v \mapsto x] \in S\}$$

$$\overleftarrow{\tau}[\![\mathbf{assume}(b)]\!]S \triangleq S \cup \{\rho \mid B[\![b]\!]\rho = \{\mathbf{f}\}\}$$

$$\overleftarrow{\tau}[\![s_1; s_2]\!]S \triangleq (\overleftarrow{\tau}[\![s_1]\!] \circ \overleftarrow{\tau}[\![s_2]\!])S$$

$$\overleftarrow{\tau}[\![\mathbf{if}\ b\ \mathbf{then}\ s_1\ \mathbf{else}\ s_2]\!]S \triangleq (\overleftarrow{\tau}[\![\mathbf{assume}(b)]\!] \circ \overleftarrow{\tau}[\![s_1]\!])S \cap (\overleftarrow{\tau}[\![\mathbf{assume}(\neg b)]\!] \circ \overleftarrow{\tau}[\![s_2]\!])S$$

$$\overleftarrow{\tau}[\![\mathbf{while}\ b\ \mathbf{do}\ s\ \mathbf{done}]\!]S \triangleq \mathrm{gfp}_{\overleftarrow{\tau}[\![\neg b]\!]}(\lambda X.\ X \cap (\overleftarrow{\tau}[\![\mathbf{assume}(b)]\!] \circ \overleftarrow{\tau}[\![s]\!])X)$$

Fig. 3. Backward semantic of statements.

Language and Forward Semantics. We assume a simple While programming language with $\mathbf{assume}(b)$, assignments and $\mathbf{skip}()$ atomic statements and sequencing, if-then-else and while loops inductive statements (see Fig. 2). The set of variables is denoted with $\mathcal{V}$ and it is assumed to be fixed. Variables are of mathematic integer type, hence program stores (or environments) are in $\mathcal{E} \triangleq \mathcal{V} \to \mathbb{Z}$. Arithmetic and boolean expressions are interpreted respectively by $E[\![a]\!] : \mathcal{P}(\mathcal{E}) \to \mathcal{P}(\mathbb{Z})$ and $B[\![b]\!] : \mathcal{P}(\mathcal{E}) \to \mathcal{P}(\{\mathbf{t}, \mathbf{f}\})$, whereas statements by $\tau[\![s]\!] : \mathcal{P}(\mathcal{E}) \to \mathcal{P}(\mathcal{E})$. For more details we refer the reader to [38,39].

Backward Semantic. Conventional backward analyses focus on inferring *necessary* preconditions for some post-condition. In particular, given a program $s \in$ While and a post-condition $S \in \mathcal{P}(\mathcal{E})$, they infer an over-approximation (an over-approximation of a necessary precondition is again a necessary precondition) of $P_n \triangleq \{\rho \mid \exists \rho' \in \tau[\![s]\!]\{\rho\}.\ \rho' \in S\}$. Notice that if $\tau[\![s]\!]\{\rho\} \in S$ then $\rho \in P_n$, i.e., if a store ρ transitions to S, then it is contained in P_n.

On the contrary, the backward analysis proposed in [38] infers *sufficient* preconditions. In particular, it infers an under-approximation (an under-approximation of a sufficient precondition is again a sufficient precondition) of $P_s \triangleq \{\rho \mid \forall \rho' \in \tau[\![s]\!]\{\rho\}.\ \rho' \in S\}$. Notice that necessary and sufficient preconditions can differ in the presence of non-determinism as demonstrated in the following example.

Example 1. Consider the program $s \equiv x := x + [-1, 1]$ with post-condition $S \triangleq [0, 5]^1$. The strongest necessary precondition is $P_n = [-1, 6]$ as for any $x \in P_n$ there *exists* a trace leading to the post-condition ($\forall x \in P_n.\ \tau[\![s]\!]\{x\} \cap S \neq \varnothing$). The weakest sufficient precondition is $P_s = [1, 4]$ as for any $x \in P_s$ *all* traces lead to the post-condition ($\forall x \in P_n.\ \tau[\![s]\!]\{x\} \subseteq S$). □

Let $f : \mathcal{P}(A) \to \mathcal{P}(B)$ be a function, we define the *backward* version of f, denoted $\overleftarrow{f}$, as

$$\overleftarrow{f}(B) \triangleq \{a \in A \mid f(\{a\}) \subseteq B\}.$$

In particular, letting $f \equiv \tau[\![s]\!]$ yields $\overleftarrow{\tau}[\![s]\!]$ that computes the *sufficient* precondition of s. Backward versions of functions enjoy several properties and in particular we can exploit them to compute $\overleftarrow{\tau}[\![s]\!]$ by induction on the syntax of

---

[1] With an abuse of notation we confuse the store $[x \mapsto z]$ with z.

s. We report the resulting backward semantic in Fig. 3 and refer the reader to Theorems 2 and 3 of [38] for further details (in particular, the soundness of this construction).

Abstract Semantic. As usual in abstract interpretation, we represent program properties with *abstract domains*. An abstract domain is a tuple $\langle D^\sharp, \gamma^\sharp, \sqsubseteq^\sharp,$ $\sqcup^\sharp, \sqcap^\sharp, \nabla^\sharp, \tau^\sharp[\![s]\!] \rangle$ where $\gamma^\sharp : D^\sharp \to \mathcal{P}(\mathcal{E})$ is a monotonic map, $\sqsubseteq^\sharp \colon D^\sharp \times D^\sharp$ is a partial order relation, $\sqcup^\sharp, \sqcap^\sharp$, are sound over-approximations of $\sqcup, \sqcap, \nabla^\sharp$ is a widening operator and $\tau^\sharp[\![s]\!]$ is a sound over-approximation of $\tau[\![s]\!]$ for atomic statements (compound statements are handled by induction on the syntax).

Whereas conventional reachability analysis is *sound* when it *over-approximates* concrete invariants, the sufficient precondition analysis is sound when it *under-approximates* them. For this reason, an abstraction of the concrete semantic of Fig. 3 can not be obtained by simply replacing the concrete operators with the abstract ones (as typically done in abstract interpretation), instead they must be replaced with a new special set of abstract operators that guarantee an under-approximation of the concrete computation. In particular, we need operators $\sqcup^\sharp, \sqcap^\sharp$, that under-approximate respectively $\sqcup, \sqcap$, a lower widening[2] $\underline{\nabla}^\sharp$ and $\overline{\tau}^\sharp[\![s]\!]$ that under-approximates $\overline{\tau}[\![s]\!]$ for atomic statements. As an example, in [38] it is shown how to design such operators for the polyhedra domain, with the exception of $\sqcup^\sharp$. However a simple, yet imprecise, definition for $\sqcup^\sharp$ is to just return one of its arguments. A more precise operator will be presented in Sect. 3, exploiting the powerset domain. Consequently, a sound abstraction of the backward semantic can be obtained by leveraging the under-approximated versions of domain operators.

Additionally, even though the concrete backward semantics depends solely on the post-condition, to design an abstract transfer function, it can be useful to have an over-approximation of the precondition (e.g., to linearize arithmetic expressions as in [37]). Fortunately this over-approximation can be easily computed (through a traditional forward reachability analysis) and stored for later usage in the backward pass. Hence, for the rest of this paper, we will assume the availability of an over-approximation of the result of each backward operator.

Correctness and Incorrectness. Program specifications can be modeled in the language with **assert**(b) statements. Their semantic changes depending on the goal of the analysis, whether it is for preconditions for program correctness or incorrectness. We can see this with an example.

Example 2. Consider the programs of Figs. 4a, 4b, 4c. Program 4a contains no assertion, thus it is trivially always correct (and never incorrect). In Program 4b, to compute a sufficient precondition for correctness we collect $x \geq 50$ from the assertion and then subtract 10, which yields $x \geq 40$ (likewise, for incorrectness

---

[2] A lower (or dual) widening $\underline{\nabla}^\sharp : D^\sharp \times D^\sharp \to D^\sharp$ is a binary operator such that: 1. for all $d_1^\sharp, d_2^\sharp \in D^\sharp$, $\gamma^\sharp(d_1^\sharp \underline{\nabla}^\sharp d_2^\sharp) \subseteq \gamma^\sharp(d_1^\sharp) \cap \gamma^\sharp(d_2^\sharp)$; 2. for any sequence $(x_i^\sharp)_{i \in \mathbb{N}}$, the sequence defined as $y_0^\sharp \triangleq x_0^\sharp$ and $y_{i+1}^\sharp \triangleq y_i^\sharp \underline{\nabla}^\sharp x_{i+1}^\sharp$ becomes stable in a finite number of iterations.

1: $x := x + 10$

(4a)

1: $x := x + 10$
2: **assert**$(x \geq 50)$

(4b)

1: $x := x + 10$
2: **assert**$(x \geq 50)$
3: $x := x - 10$
4: **assert**$(x \leq 50)$

(4c)

Fig. 4. Simple program instances with specifications encoded as assertions.

we obtain $x < 40$. In Program 4c the reasoning for correctness is the same as the previous program: we proceed backwards and for each assertion encountered we retain only the states satisfying its guard. This yields $x \in [40, 50]$. On the other hand, to find preconditions for program incorrectness we have two possibilities: the failure of the assertion at line 2 or 4. For the one at line 2 we proceed as in Program 4b (which yields $x < 40$). For the one at line 4 we collect $x > 50$ and proceed backwards. As soon as we encounter the assertion at line 2, we retain the states satisfying its guard (as for correctness), as in order to reach line 4, an execution must satisfy the assertion at line 2. Combining the preconditions yields $x \notin [40, 50]$. □

The previous example suggests that our semantics can infer preconditions for both correctness and incorrectness. In the former case the analysis has to start from ⊤ (or some other post-condition of interest) and compute $\overline{\tau}[\![\mathbf{assert}(b)]\!]S$ as $S \cap [b]$ where $[b]$ denotes the set of states satisfying b. In the latter it has to start from ⊥ and compute $\overline{\tau}[\![\mathbf{assert}(b)]\!]S$ as $(S \cap [b]) \cup [\neg b]$.

2.3   User Input

A crucial aspect of programming is I/O. The language we studied has a limited support for I/O in the form of input arguments and return values, but this is often not enough as real-world programs can perform I/O operations at arbitrary execution points. In particular, as we focus on finding preconditions, we are mostly interested in the effect of user *input*.

To address this issue, a new statement, $v := \mathbf{input}()$, is added to the language. When this statement is executed, the machine reads a value from an external source and stores it in v. As different input statements may read from different sources, we assume that finitely many sources are available, each identified with an index, and annotate each input statement with the index identifying the source, e.g., $v := \mathbf{input}_n()$ for an input from the nth source.

Remark 1 (Input versus non-determinism). It might appear that the following two programs

$x := \mathbf{input}_1()$
assert$(x > 100)$

$x := [-\infty, +\infty]$
assert$(x > 100)$

have the same semantics, but this is not the case: intuitively, $\mathbf{input}_1()$ differs from the non-deterministic interval $[-\infty, +\infty]$ in that the former depicts an user-controllable, input to the program, while the latter depicts an "internal" uncontrollable form of input. Consequently, the (correctness) precondition for the first program involves not only the value of x but also the value returned by $\mathbf{input}_1()$. Vice versa, the one of the second program only concerns the value of x. In the first case, the precondition is: $\mathbf{input}_1 > 100$. Indeed, if this condition holds the program satisfies the assertion. In the second one, the precondition is $\varnothing$. Indeed, there is no set of stores that for *any* non-deterministic execution ensures that the assertion is satisfied.   □

Concrete Semantic. In order to model user input, our representation of states as program stores is not sufficient anymore. Inspired by the input representation proposed in [25], we model external inputs as *streams*, i.e., pairs of an infinite sequence of integers and an index, where the index is used to store the position of the next number to be read from the sequence. The set of streams is denoted with $\mathcal{S} \triangleq \mathbb{Z}^\omega \times \mathbb{N}$ and $\mathrm{get}(s)$ indicates the current value of the stream s. We further assume that p different input sources are available, modeled as *multi-streams*, i.e., vectors of p streams, denoted with $\mathcal{M} \triangleq \mathcal{S}^p$. Therefore program states are modeled as pairs of a store (environment) and a multi-stream, $\mathcal{E}' \triangleq \mathcal{E} \times \mathcal{M}$. We further denote with $\mathrm{incr}_n(m)$ the multi-stream m in which the nth stream is equal to m_n but with its index incremented and the other streams left unchanged.

For non-input statements, the semantic $\tau[\![s]\!] : \mathcal{P}(\mathcal{E}') \to \mathcal{P}(\mathcal{E}')$ operates on the store as before, while leaving the streams untouched, e.g.,

$$\tau[\![v := a]\!]P \triangleq \{(\rho[v \mapsto x], m) \mid (\rho, m) \in P, x \in E[\![a]\!]\rho\}.$$

On the contrary, the semantic of $v := \mathbf{input}_n()$ stores in v the current value of the nth stream and increments its index:

$$\tau[\![v := \mathbf{input}_n()]\!]P \triangleq \{(\rho[v \mapsto \mathrm{get}(m_n)], \mathrm{incr}_n(m)) \mid (\rho, m) \in P\}.$$

Time-Invariant Stream Abstraction. In the concrete semantics, user inputs are modeled as reads from an infinite sequence, but since sequences are not directly representable by conventional numeric abstract domains, some further abstraction is necessary. Rather than providing directly an encoding of concrete states into numeric domains, we propose an intermediate abstraction allowing later an easier representation in numeric domains.

Abstraction. Input streams can be abstracted in several different ways, e.g., by retaining a finite prefix, or with an automaton, etc. In this work, we consider an abstraction that classifies streams as either time-invariant (e.g., $(111..., 0) \in \mathcal{S}$) or time-dependent (e.g., $(123..., 0) \in \mathcal{S}$). The set of time-invariant streams is denoted with $\mathcal{S}_i$ and the set of time-dependent streams with $\mathcal{S}_t$. In this abstraction, in the former case we track the value that is repeated in the stream, while in the latter all the information is discarded. In both cases the information regarding the current position on the stream (the index) is not preserved.

Example 3. Consider the following example.

```
i, j := 0
for x = 1 to 10 do
    i := i + input₁()
    j := j + input₂()
end for
assert(i = j)
```

There are several streams that can render the assertion true, for instance if **input**$_1$() returns 10 at the first iteration and 0 for other iterations and **input**$_2$() always returns 1. In this case the stream for **input**$_1$() is time-dependent and the one for **input**$_2$() is time-invariant.

Notice that the abstraction tracks sets of states, maintaining the value of both program variables and time-invariant streams, and thus it is able to express relationships between them. For instance, the set of preconditions where both streams are time-invariant, with the same value in $[0, 100]$. It can also infer relations between stream values and programs variables.   □

To formalize this abstraction we use a partial map from p *stream variables* to $\mathbb{Z}$. We denote the nth stream variable with v_n^s and with $\mathcal{V}_s$ the set of stream variables. If v_n^s is defined in the map, then the corresponding stream is time-invariant with value matching the variable's value, otherwise it is time-dependent. More formally:

Definition 1 (Time-invariant stream abstraction). *Let $\mathcal{E}'$ be a set of states. We define $\widehat{\mathcal{E}'} \triangleq (\mathcal{V} \cup \mathcal{V}_s) \rightharpoonup \mathbb{Z}$, the concretization $\widehat{\gamma} : \mathcal{P}(\widehat{\mathcal{E}'}) \to \mathcal{P}(\mathcal{E}')$ and abstraction $\widehat{\alpha} : \mathcal{P}(\mathcal{E}') \to \mathcal{P}(\widehat{\mathcal{E}'})$ functions as follows:*

- $\widehat{\gamma}(R) \triangleq \{(\rho, m) \mid \exists \widehat{\rho} \in R. \; \widehat{\rho}(v_k)|_{\mathcal{V}} = \rho(v_k)|_{\mathcal{V}}, \; \text{matchStream}(\widehat{\rho}, m)\}$
- $\widehat{\alpha}(R) \triangleq \{\widehat{\rho} \mid \exists (\rho, m) \in R. \; \widehat{\rho}(v_k)|_{\mathcal{V}} = \rho(v_k)|_{\mathcal{V}}, \; \text{matchStream}(\widehat{\rho}, m)\}$

where:

$$\text{matchStream}(\widehat{\rho}, m) \Leftrightarrow \forall n = 1, .., p. \; \widehat{\rho}(v_n^s) = \begin{cases} \text{get}(m_n) & \text{if } m_n \in S_i \\ undef & \text{if } m_n \in S_t \end{cases}$$ □

Theorem 1. *The following Galois Connection holds:* $(\mathcal{P}(\mathcal{E}'), \subseteq) \xleftrightarrow[\widehat{\alpha}]{\widehat{\gamma}} (\mathcal{P}(\widehat{\mathcal{E}'}), \subseteq).$

Semantic. The semantic of statements different from $v := \textbf{input}_n()$ coincides with the concrete one as those statements only operate on the store part of the state (not on the streams), and that part is not abstracted; for example

$$\widehat{\tau}[\![v := a]\!]P \triangleq \{\widehat{\rho}[v \mapsto x] \mid \widehat{\rho} \in P, x \in E[\![a]\!]\widehat{\rho}|_{\mathcal{V}}\}.$$

On the other hand, $\widehat{\tau}[\![v := \textbf{input}_n()]\!]$ affects the stream. In particular, if the stream is time-invariant (thus v_n^s is defined in $\widehat{\rho}$) then $\widehat{\tau}[\![v := \textbf{input}_n()]\!]$ copies

v_n^s (which is equal to the stream's value) to v. Otherwise, if the stream is time-dependent, v gets $[-\infty, +\infty]$ as no information is retained in the abstraction and $[-\infty, +\infty]$ is always a sound choice. More formally:

$$\widehat{\tau}[\![v := \mathbf{input}_n()]\!]P \triangleq \{\widehat{\rho}[v \mapsto x] \mid \widehat{\rho} \in P, x \in \mathbb{Z}. (v_n^s \in \mathrm{dom}(\widehat{\rho}) \Rightarrow x = \widehat{\rho}(v_n^s))\}.$$

As in the forward semantic, the backward semantic of non-input statements can be easily derived from the concrete semantic; for example

$$\widehat{\overleftarrow{\tau}}[\![v := a]\!]S = \{\widehat{\rho} \mid \forall x \in E[\![a]\!]\widehat{\rho}. \widehat{\rho}[v \mapsto x] \in S\}.$$

The backward semantic of input statements ensures that if the stream is time-dependent (v_n^s undefined) then for *all* substitutions of v in $\widehat{\rho}$ the resulting store is in the post-condition. Otherwise if the stream is time-independent, then $\widehat{\rho}[v \mapsto \widehat{\rho}(v_n^s)]$ must be in the post-condition. More formally we have:

$$\widehat{\overleftarrow{\tau}}[\![v := \mathbf{input}_n()]\!]S = \{\widehat{\rho} \mid \forall x \in \mathbb{Z}. (v_n^s \notin \mathrm{dom}(\widehat{\rho}) \wedge \widehat{\rho}[v \mapsto x] \in S) \vee$$
$$(v_n^s \in \mathrm{dom}(\widehat{\rho}) \wedge (x = \widehat{\rho}(v_n^s) \Leftrightarrow \widehat{\rho}[v \mapsto x] \in S))\}$$

Theorem 2. *The semantic of statements, both forward and backward, is sound:*

$$\tau[\![s]\!]\widehat{\gamma}(R) \subseteq \widehat{\gamma}(\widehat{\tau}[\![s]\!]R) \qquad\qquad \widehat{\gamma}(\widehat{\overleftarrow{\tau}}[\![s]\!]S) \subseteq \overleftarrow{\tau}[\![s]\!]\widehat{\gamma}(S).$$

Abstract Semantic. In the previous section, we demonstrated an abstraction of input streams into environments where some variables can be defined (time-invariant streams) or not (time-dependent streams). The usual numeric domains can not be used directly as they assume that *all* variables are defined. This issue was already studied in the context of abstracting *heterogeneous* environments (i.e., environments where some variables are optional). One simple approach is to partition the environments according to the defined variables, but this scales poorly as there can be an exponential number of partitions in the worst case.

We adopt instead the method proposed in [31], though in a simplified version. This approach lifts a numeric domain $D^\sharp$, to a domain $\widehat{D}^\sharp$ consisting of pairs $\langle d^\sharp, l \rangle$, where $\mathcal{V} \subseteq l \subseteq \mathcal{V} \cup \mathcal{V}_s$ and $d^\sharp \in D^\sharp$. The element $d^\sharp$ is defined on $\mathcal{V} \cup \mathcal{V}_s$ and the concretization of $\langle d^\sharp, l \rangle$ yields states with domain subsuming l and satisfying the constraints of $d^\sharp$. In the original approach [31], the elements of $\widehat{D}^\sharp$ contained an additional set $u \supseteq l$ representing an upper bound for the domain of the states (here $u = \mathcal{V} \cup \mathcal{V}_s$), but in our case this additional flexibility is not needed since stream variables can not be added or removed explicitly (e.g., with ad-hoc statements) but they can only be *added* as a side-effect of input statements, hence only the lower bound can vary.

More formally the concretization is defined as $\widehat{\gamma}^\sharp(\langle d^\sharp, l \rangle) \triangleq \{\widehat{\rho} \mid \exists \widehat{\rho}' \in \gamma^\sharp(d^\sharp), \mathcal{V} \subseteq l \subseteq \mathrm{dom}(\widehat{\rho}) \subseteq \mathcal{V} \cup \mathcal{V}_s, \widehat{\rho} = \widehat{\rho}'|_{\mathrm{dom}(\widehat{\rho})}\}$. Details on the construction of over-approximation domain operators can be found in [31]. Here instead we focus on under-approximation operators. If $\sqcap^\sharp, \sqcup^\sharp, \underline{\nabla}^\sharp$ are under-approximation operators for the base domain $D^\sharp$, then they can be lifted to $\widehat{D}^\sharp$:

- *Join:* $\langle d_1^\sharp, l_1 \rangle \; \widehat{\sqcup}^\sharp \; \langle d_2^\sharp, l_2 \rangle \triangleq \langle d_1^\sharp \sqcup^\sharp d_2^\sharp, l_1 \cup l_2 \rangle$;
- *Meet:* $\langle d_1^\sharp, l_1 \rangle \; \widehat{\sqcap}^\sharp \; \langle d_2^\sharp, l_2 \rangle \triangleq \langle d_1^\sharp \sqcap^\sharp d_2^\sharp, l_1 \cup l_2 \rangle$;
- *Widening:* $\langle d_1^\sharp, l_1 \rangle \; \widehat{\nabla}^\sharp \; \langle d_2^\sharp, l_2 \rangle \triangleq \begin{cases} \langle d_1^\sharp \; \nabla^\sharp \; d_2^\sharp, l_1 \rangle & \text{if } l_2 \subseteq l_1 \\ \langle d_1^\sharp \sqcap^\sharp d_2^\sharp, l_2 \rangle & \text{if } l_1 \subset l_2 \end{cases}$.

Proposition 1. $\widehat{\sqcup}^\sharp$, $\widehat{\sqcap}^\sharp$ *are sound under-approximations of* $\cup$, $\cap$ *and* $\widehat{\nabla}^\sharp$ *is a lower widening.*

Semantic. The semantic of input statements can be handled as follows:

$$\widehat{\tau}^\sharp [\![v := \mathbf{input}_n()]\!] \langle D^\sharp, l \rangle \triangleq \begin{cases} \langle \tau^\sharp [\![v := [-\infty, \infty]]\!] D^\sharp, l \rangle & \text{if } v_n^s \notin l \\ \langle \tau^\sharp [\![v := v_n^s]\!] D^\sharp, l \rangle & \text{if } v_n^s \in l \end{cases}$$

Indeed, if $v_n^s \notin l$ then the concretization contains *both* time-invariant and time-dependent streams: for the latter no information is stored, thus v gets $\top$. If instead $v_n^s \in l$ then the concretization contains *only* time-invariant streams and thus the assignment copies the value from the stream variable.

The backward semantic is computed as:

$$\widehat{\overleftarrow{\tau}}^\sharp [\![v := \mathbf{input}_n()]\!] \langle D^\sharp, l \rangle \triangleq \begin{cases} \langle D^\sharp, l \rangle & \text{if } v_n^s \notin l \wedge \widehat{\tau}^\sharp [\![v := [-\infty, \infty]]\!] D^\sharp = D^\sharp \\ \langle \overleftarrow{\tau}^\sharp [\![v := v_n^s]\!] D^\sharp, l \cup \{ v_n^s \} \rangle & \text{otherwise} \end{cases}$$

Indeed, we can distinguish three cases:

1. If $v_n^s \in l$ then we only have time-invariant streams in the post-condition. In this case the forward transfer function performs the assignment $v := v_n^s$, thus the backward precondition can simply invert this assignment;
2. If $v_n^s \notin l$ and $\widehat{\tau}^\sharp [\![v := [-\infty, \infty]]\!] D^\sharp = D^\sharp$ then we have *both* time-invariant and time-dependent streams. In addition, as the backward projection leaves $D^\sharp$ unmodified, for *all* states $\widehat{\rho} \in \gamma^\sharp(D^\sharp)$ and $x \in \mathbb{Z}$, $\widehat{\rho}[v \mapsto x] \in \gamma^\sharp(D^\sharp)$. Therefore the backward precondition is simply $\langle D^\sharp, l \rangle$. Notice that the condition on the backward projection is crucial to ensure the soundness of time-dependent streams: the projection over-approximates any assignment, thus [38, Theorem 2.6] ensures that $\widehat{\rho}[v \mapsto x] \in \gamma^\sharp(D^\sharp)$.
3. If $v_n^s \notin l$ and $\widehat{\tau}^\sharp [\![v := [-\infty, \infty]]\!] D^\sharp \neq D^\sharp$ then, as before, we have *both* time-invariant and time-dependent streams, but, unlike the previous case, there exist $\widehat{\rho} \in \gamma^\sharp(D^\sharp)$ and $x \in \mathbb{Z}$ such that $\widehat{\rho}[v \mapsto x] \notin \gamma^\sharp(D^\sharp)$. Consequently, time-dependent streams can not be included in the precondition as they would be unsound. For this reason the precondition adds v_n^s to l (thus under-approximating the precondition) and transforms $D^\sharp$ as in the first case.

Theorem 3. *The abstract semantic is sound, i.e., for any* $s \in$ While *and* $\widehat{d}^\sharp \in \widehat{D}^\sharp$ *the following holds:*

$$\widehat{\tau}[\![s]\!]\widehat{\gamma}^\sharp(\widehat{d}^\sharp) \subseteq \widehat{\gamma}^\sharp(\widehat{\tau}^\sharp [\![s]\!]\widehat{d}^\sharp) \qquad \widehat{\overleftarrow{\tau}}[\![s]\!]\widehat{\gamma}^\sharp(\widehat{d}^\sharp) \supseteq \widehat{\gamma}^\sharp(\widehat{\overleftarrow{\tau}}^\sharp [\![s]\!]\widehat{d}^\sharp)$$

2.4   Integer Wrapping

In this section, we generalize our framework to support fixed precision integers (i.e., with wrap-around), typically found in C-like languages. This is important as some analyzers detect integer overflows but do not handle wrap-around: they either stop the analysis for the traces that overflow (which is not sound for programs that do wrap-around on purpose) or put the variable to the full range of their type (which is sound but imprecise). For the sake of brevity, we limit our presentation to unsigned 8-bit integers, but it is easy to generalize this framework to other types.

Arithmetic and boolean semantics are replaced with versions that operate with 8-bit unsigned integers, e.g., $E[\![x + 10]\!]\{x \mapsto 250\} = \{[x \mapsto 4]\}$. To do so, it suffices to replace the usual arithmetic operators with versions that take care of integer wrapping. Unfortunately this requires new wrap aware operators to be designed. To avoid this difficulty, we prefer a modular approach in which firstly the result is computed with infinite precision operators (i.e., the usual unwrapped ones), and then it is wrapped with a wrapping operator.

Definition 2 (Wrapping operator). *Define* wrap : $\mathbb{Z} \to [0, 255]$ *as* wrap$(z) \triangleq z$ mod 256, *where* mod *computes the Euclidean remainder.*   □

Consequently, the abstract semantic must take into account wrapping of integers. Several approaches have been proposed in the literature to handle this problem [27, 29, 45, 46].

Case Study: Polyhedra Domain. As an example, we show how to instantiate the abstract semantic to the case of the polyhedra domain. Our work is based on the work of Simon et al. [46]: they demonstrate how to design sound abstract operators for the polyhedra domain that take into account integer wrapping. For this purpose, they presented an algorithm for computing a wrap$^\sharp$ operator. It takes in input a polyhedron P, a variable v to wrap, and as result it produces a new polyhedron P' in which v lies in $[0, 255]$. Figure 5 shows an example of the computation of wrap$^\sharp$.

We extend their work (which only tackles over-approximation forward operators) to handle under-approximation backward operators. Intuitively, the backward version of wrap$^\sharp$ for a polyhedron P (along a variables v) should compute a polyhedral representation of the points that, after wrapping, end up in P. This boils down to replicating P infinitely many times, where each copy is translated by a integer multiple of 256 along v, i.e., the sequence $\{P + 256k\mathbf{e}_v\}_{k \in \mathbb{Z}}$. Figure 6 shows an example of this computation.

Although all the polyhedra of the sequence above are valid unwrappings, not necessarily all of them represent reachable states. In particular, if *pre* is an over-approximation of the input of wrap$^\sharp$, then the valid polyhedra are only the ones intersecting *pre*. This information can be used to guide the unwrapping of a polyhedron. We present in Algorithm 1 a procedure for computing $\overleftarrow{\text{wrap}}^\sharp$. The auxiliary function *quadrantIndices*(*pre*, *v*) computes the indices of

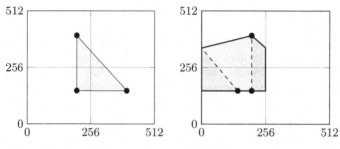

(a) Polyhedron before wrapping.   (b) Polyhedron after wrapping.

Fig. 5. Wrapping of a polyhedron along the horizontal axis. The polyhedron on the left is split in two parts: one that does not need wrapping ($x \in [0, 255]$) and another that does ($x \in [256, 511]$). To compute the result (green polyhedron), the first part is joined with the translation of the second one by -256 (along x). (Color figure online)

the quadrants spanned by v in *pre* (see [46, Algorithm 1]). Notice that the polyhedra of the sequence are merged together with an under-approximating join, but this can incur a loss of precision if the polyhedra are separated (as in Fig. 6) since in this case the set union is not convex, and thus to under-approximate it with a polyhedral shape, only one polyhedron can be retained (hence in Fig. 6, the result must be either one of the polyhedra in green or the one in blue). A robust solution to this kind of imprecisions will be addressed in the next section. Additionally, the meet with *pre* in the algorithm excludes the polyhedra that are surely not reachable, thus increasing the odds of retaining an appropriate polyhedron.

3   Powerset Domain

As noted in [38], designing an under-approximating join for polyhedra can be challenging. This problem was sidestepped by designing such an operator only in some specific cases, namely on joins occurring in the analysis of backward filters of if-then-else statements and while loops. This simplifies the design as only under-approximations of the join of an arbitrary polyhedron with a half-space are handled. This is carried out using special heuristics, tailored to handle many practical cases. Unfortunately, they are not robust and may cause losses of precision in other cases. This is especially true for our semantic, as, unlike the one in [38], we use $\sqcup^\sharp$ to handle arbitrary joins (e.g., in $\overleftarrow{\text{wrap}}$ and later in this section for widenings).

Furthermore, even if a perfect join heuristic could be designed, the polyhedra domain would still not be precise if the concrete union is non-convex. This can often occur in real world programs (e.g., in the unwrapping of the polyhedron of Fig. 6).

Algorithm 1. Calculate $\overleftrightarrow{wrap}^{\sharp}$ (P, v, t, pre)

Require: Parameter $m > 0$: maximum number of copies.
$q_l, q_u \leftarrow quadrantIndices(pre, v, t)$
if $q_l = -\infty \wedge q_u = +\infty$ **then**
    return P
else if $q_l = -\infty$ **then**
    $q_l \leftarrow q_u - m$
else if $q_u = +\infty$ **then**
    $q_u \leftarrow q_l + m$
else
    {Retain at most m copies}
    $q_u \leftarrow q_l + \min(q_u - q_l, m)$
end if
$Q \leftarrow \bot^{\sharp}$
for $k \leftarrow q_l$ **to** $q_u - 1$ **do**
    $Q \leftarrow Q \sqcup^{\sharp} ((\tau^{\sharp}[\![v := v - 256k]\!]P) \sqcap^{\sharp} pre)$
end for
return Q

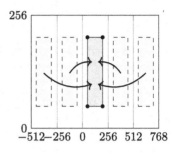

Fig. 6. Unwrapping of x. The unwrapping of the blue polyhedron produces infinitely many (here only four are shown) copies of it, separated by the integer's size. The wrapping of each polyhedron in green (depicted with the arrows) coincides with P. (Color figure online)

3.1 Under-Approximated Powerset

A robust approach to addressing this issue is to leverage the powerset [26] construction: in this construction, a base domain $D^{\sharp}$ is lifted to a finite set of abstract elements $\mathcal{P}_{finite}(D^{\sharp})$. The concretization of a set yields the union of the concretizations of all its elements. Notably, the join operator becomes exact, and thus it is a sound under-approximation of $\sqcup$. Consequently the under-approximation join $\sqcup_p^{\sharp}$ can coincide with the over-approximating one $\sqcup_p^{\sharp}$.

Definition 3 (Powerset domain). *Let $D^{\sharp}$ be an abstract domain. We let $P(D^{\sharp}) \triangleq \mathcal{P}_{finite}(D^{\sharp})$ be its powerset lifting. $P(D^{\sharp})$ is partially ordered by $S_1^{\sharp} \sqsubseteq_p^{\sharp} S_2^{\sharp} \Leftrightarrow \forall d_1^{\sharp} \in S_1^{\sharp}. \exists d_2^{\sharp} \in S_2^{\sharp}. d_1^{\sharp} \sqsubseteq^{\sharp} d_2^{\sharp}$. Moreover, join and meet are respectively defined as $S_1^{\sharp} \sqcup_p^{\sharp} S_2^{\sharp} \triangleq S_1^{\sharp} \cup S_2^{\sharp}$ and $S_1^{\sharp} \sqcap_p^{\sharp} S_2^{\sharp} \triangleq \{d_1^{\sharp} \sqcap^{\sharp} d_2^{\sharp} \mid d_1^{\sharp} \in S_1^{\sharp}, d_2^{\sharp} \in S_2^{\sharp}\}$.* □

Additionally, if the base meet is exact (which is the case in many numeric domains, including polyhedra), also $\sqcap_p^{\sharp}$ is, thus it is a sound under-approximating operator as well.

Widening. A trivial widening can be obtained by joining all elements of the powerset (for each argument) and then applying the base widening (hence the result is a singleton). Likewise to get a trivial lower widening, it is possible to apply the base lower widening on just one element of each argument and discard

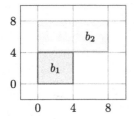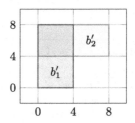

Fig. 7. Powerset refinement. The blue polyhedron (left figure) can be extended over the green one (right figure) without changing the overall concretization of the powerset. (Color figure online)

all the others. Unfortunately, these operators are quite imprecise, as shown in the following example.

Example 4. Consider the following example suggested by Gopan and Reps [28].

$i, j := 0$
for $i := 1$ to 100 **do**
    if $i \leq 50$ **then** $j \leftarrow j + 1$ **else** $j \leftarrow j - 1$
end for

The while loop presents two phases: one in which j is incremented (then branch), and one in which it is decremented (else branch). This induces a non-convex loop invariant, that requires at least two polyhedra to be represented precisely. It is clear that the trivial widenings can not find such a result as they yield a singleton. □

A simple, yet useful, improvement consists in retaining *stable* elements, i.e., elements that are shared in both arguments, and widen only the *remaining* (unstable) elements. More formally:

Definition 4 (Improved Powerset Widening). Let $S_1^\sharp = \langle d_{1,1}^\sharp, .., d_{1,n}^\sharp \rangle$ and $S_2^\sharp = \langle d_{2,1}^\sharp, .., d_{2,m}^\sharp \rangle$. *Define:*

$$S_1^\sharp \, \nabla_p^\sharp \, S_2^\sharp \triangleq \begin{cases} \langle d_1^\sharp \, \nabla^\sharp \, (\sqcup^\sharp S_{u_2}^\sharp) \rangle \cup (S_s^\sharp \setminus \{d_1^\sharp\}) & \text{if } S_{u_1}^\sharp = \varnothing \wedge S_{u_2}^\sharp \neq \varnothing \wedge d_1^\sharp \in S_1^\sharp \\ \langle (\sqcup^\sharp S_{u_1}^\sharp) \, \nabla^\sharp \, (\sqcup^\sharp S_{u_2}^\sharp) \rangle \cup S_s^\sharp & \text{otherwise} \end{cases}$$

$$S_1^\sharp \, \underline{\nabla}_p^\sharp \, S_2^\sharp \triangleq \begin{cases} S_s^\sharp & \text{if } S_{u_1}^\sharp = \varnothing \vee S_{u_2}^\sharp = \varnothing \\ \langle d_{u_1}^\sharp \, \underline{\nabla}^\sharp \, d_{u_2}^\sharp \rangle \cup S_s^\sharp & \text{if } d_{u_1}^\sharp \in S_{u_1}^\sharp \wedge d_{u_2}^\sharp \in S_{u_2}^\sharp \end{cases}$$

where $S_s^\sharp \triangleq S_1^\sharp \cap S_2^\sharp$ *is the set of stable elements,* $S_{u_1}^\sharp \triangleq S_1^\sharp \setminus S_s^\sharp$ *the set of unstable elements of* $S_1^\sharp$ *and* $S_{u_2}^\sharp \triangleq S_2^\sharp \setminus S_s^\sharp$ *the set of unstable elements of* $S_2^\sharp$. □

Proposition 2. $\nabla_p^\sharp \, (\underline{\nabla}_p^\sharp)$ *is an upper (lower) widening operator for* $P(D^\sharp)$.

Algorithm 2. Integer polyhedral refinement: adjacent constraints

---

Require: $d_1^\sharp, d_2^\sharp$ polyhedra in constraint representation.
Ensure: $d^\sharp$ refines $d_1^\sharp$ with $d_2^\sharp$.

$\quad d^\sharp \leftarrow d_1^\sharp$
$\quad$**for** c **matching** $a \cdot v \geq b$ **in** $d_1^\sharp$ **do**
$\quad\quad d_{1,nc}^\sharp \leftarrow d_1^\sharp \setminus \{c\}$
$\quad\quad v_1, r_1 \leftarrow sat(d_1^\sharp, c)$ $\{sat(d^\sharp, c)$ returns the vertices and rays of $d^\sharp$ saturating $c\}$
$\quad\quad c' \leftarrow a \cdot v \geq b - 1$
$\quad\quad v_2, r_2 \leftarrow sat(d_2^\sharp, c')$
$\quad\quad d_m^\sharp \leftarrow gen(v_1 \cup v_2, r_1 \cap r_2)$ $\{gen(v, r)$ returns the polyhedron generated by vertices
$\quad\quad v$ and rays $r\}$
$\quad\quad$**if** $c' \in d_m^\sharp$ **then**
$\quad\quad\quad d_{m,nc'}^\sharp \leftarrow d_m^\sharp \setminus \{c'\}$
$\quad\quad\quad d_i^\sharp \leftarrow d_{1,nc}^\sharp \cap d_{m,nc'}^\sharp \cap d_2^\sharp$
$\quad\quad$**else**
$\quad\quad\quad d_i^\sharp \leftarrow d_{1,nc}^\sharp \cap d_m^\sharp \cap d_2^\sharp$
$\quad\quad$**end if**
$\quad\quad d_h^\sharp \leftarrow d_1^\sharp \sqcup^\sharp d_i^\sharp$
$\quad\quad \{rays(\cdot)$ computes the set of rays of a polyhedron$\}$
$\quad\quad$**if** $rays(d_1^\sharp) \subseteq rays(d_h^\sharp)$ **then**
$\quad\quad\quad d^\sharp \leftarrow d_h^\sharp$
$\quad\quad\quad$**break**
$\quad\quad$**end if**
$\quad$**end for**

Refinement. Consider the program s computing $y := y + [0, 8]$ and the powerset $S \triangleq \{b_1, b_2\}$ of Fig. 7. The backward (concrete) semantic of S, computed point-wise, is $\{\varnothing\}$ as both $\overleftarrow{\top}[\![s]\!]b_1 = \varnothing$ and $\overleftarrow{\top}[\![s]\!]b_2 = \varnothing$. But the set of states represented by S, that is $\cup_{b \in S} b$, does admit a non-empty precondition since $\overleftarrow{\top}[\![s]\!] \cup_{b \in S} b = [x \mapsto [0, 4], y \mapsto 0] \neq \varnothing$. This is possible as the backward semantic (unlike the forward one) is not a $\cup$−morphism, but instead only the inclusion holds, i.e., $\bigcup_i \overleftarrow{\top}[\![s]\!]S_i \subseteq \overleftarrow{\top}[\![s]\!]\bigcup_i S_i$ for any family of states $\{S_i\}_{i \in \mathbb{N}}$ (whereas the equality holds for $\cup$−morphisms).

However, the powerset $S' \triangleq \{b_1', b_2'\}$ admits a non-empty backward semantic as $\{\overleftarrow{\top}[\![s]\!]b_1', \overleftarrow{\top}[\![s]\!]b_2'\} = \{[x \mapsto [0, 4], y \mapsto 0], \varnothing\}$, even if S' represents the same states as S (the only difference is the internal composition of the powerset). For this reason the elements of the powerset domain should be kept as large as possible, so that $\overleftarrow{\top}^\sharp[\![s]\!]$ is maximized (notice that in the forward analysis setting, we strive for the opposite goal, namely keeping the elements as small as possible). In particular, we can allow some sharing of states among elements of the set, provided that this does not affect the overall concretization of the powerset.

For this purpose we use a *refinement* operator: an under-approximation join $\sqcup^\sharp$ is a refinement operator if $d^\sharp \triangleq d_1^\sharp \sqcup^\sharp d_2^\sharp \sqsupseteq^\sharp d_1^\sharp$, meaning that $d^\sharp$ *refines* $d_1^\sharp$ with states from $d_2^\sharp$. Then, we can refine a powerset by replacing each element with its refinement with all the other elements. Additionally, refinements can

---

Algorithm 3. Integer polyhedral refinement: adjacent singleton variables

---

Require: $d_1^\sharp, d_2^\sharp$ polyhedra in constraint representation.
Ensure: $d^\sharp$ refines $d_1^\sharp$ with $d_2^\sharp$.
    $d^\sharp \leftarrow d_1^\sharp$
    for c **matching** $v = n$ **in** $d_1^\sharp$ **do**
        if $v = n + 1 \in d_2^\sharp \vee v = n - 1 \in d_2^\sharp$ **then**
            $\{rays(\cdot)$ computes the set of rays of a polyhedron$\}$
            if $rays(d_1^\sharp) = rays(d_2^\sharp)$ **then**
                $d^\sharp \leftarrow d_1^\sharp \sqcup^\sharp d_2^\sharp$
                break
            end if
        end if
    end for

---

help mitigate the computational cost of the powerset as, after refinement, some elements may become redundant and thus can be removed.

3.2 Case Study: Polyhedra Refinement

To design a refinement operator, it is possible to leverage a procedure for checking if the over-approximation join is exact. Indeed, an exact join is also a valid refinement. For the polyhedra domain, this problem has been studied by Bemporad et al. [6] and Bagnara et al. [4], but they focus on polyhedra representing real-valued environments.

On the other hand, if variables are of integer type (as in our semantic), the join can be exact even if the union of the polyhedra is not convex. For example the union of the polyhedra (in constraint representation) $d_1^\sharp \triangleq \{0 \leq x \leq 1\}$ and $d_2^\sharp \triangleq \{2 \leq x \leq 3\}$ is not a convex set, but still the join is exact: $\gamma(d_1^\sharp) \cup \gamma(d_2^\sharp) = \{0, 1, 2, 3\} = \gamma(d_1^\sharp \sqcup^\sharp d_2^\sharp)$.[3] Since this kind of polyhedra appears frequently in practice (e.g., in loops incrementing variables by one unit), we propose two refinement algorithms tailored for these cases.

Consider the bi-dimensional polyhedra $d_1^\sharp \triangleq \{x \geq 0, y \geq 0, x + y \leq 4\}$ and $d_2^\sharp \triangleq \{x \leq 3, y \leq 3, x + y \geq 5\}$. It is easy to check that there exists a part of $d_2^\sharp$ that can be exactly joined with $d_1^\sharp$ (the triangle with vertices $(2, 3)$, $(3, 2)$, $(\frac{8}{3}, \frac{8}{3})$), and thus can refine $d_1^\sharp$. This is the case as the strip $4 < x + y < 5$ separating it from $d_1^\sharp$ does not contain any integer. Algorithm 2 tackles this case by scanning for constraints of this kind and if they are found, computes parts of the second argument that can refine the first.

Algorithm 3 scans for a variable in $d_1^\sharp$ and $d_2^\sharp$ that is fixed in the two polyhedra to constants differing by one unit. If such a variable is found, then the join between the two polyhedra is exact. As an example, let $d_1^\sharp \triangleq \{x = 0, 0 \leq y \leq 2\}$ and $d_2^\sharp \triangleq \{x = 1, 4 \leq y \leq 6\}$. Since the strip $0 < x < 1$ does not contain any integer, the join is exact.

---

[3] With an abuse of notation, we confuse $\{x\} \to \mathbb{Z}$ with $\mathbb{Z}$.

4   Implementation and Experiments

In addition to the theoretical foundations, the contribution [38] included a PoC static analyzer, Banal [1]. This analyzer targets a toy language with a semantic not compatible with the one of C (e.g., it assumes mathematical integers instead of machine integers). We extended it with the features presented in this work: an implementation of a subset of the C semantic and a frontend for a significant subset of the language, user input (Sect. 2.3), machine integers (Sect. 2.4), a powerset domain (Sect. 3) and improved operators. As a consequence, our new prototype is able to analyze benchmarks from the SV-COMP competition. As our work focuses on incorrectness, we report only the results of the analysis of incorrect programs.

Witnesses Generation. To analyze SV-COMP benchmarks, Banal translates each call site to `__VERIFIER_nondet_int()` with an input statement (each with a distinct stream). Consequently, it computes preconditions in the form of an abstract element relating the input variables. Then, as all states in the abstract element are valid sufficient preconditions for the violation of some assertion, we extract one concrete vector of values. Notice that for the purpose of SV-COMP, the quality of a violation witness [8] is measured by how much it restricts the state-space exploration. The more restricted it is, the less states the validator has to explore in order to check the witness. By picking a concrete vector (which represents only one execution path) we obtain the most precise kind of witness.

Furthermore, as previously discussed, the preconditions generated by our analysis may simply lead to an infinite loop, rather than a true bug. To rule out this possibility Banal replaces each input call site with its concrete value, compiles the benchmark and runs it with a time limit (2 s). If an assertion fails, then the counter-example is confirmed.

Finally, a witness is generated in SV-COMP's `graphml` format: we make a control flow automaton resembling the control flow graph of the benchmark and specify for each input site the corresponding concrete value. Moreover, to certify the correctness of our result using independent techniques, we validate the witness with the CPA-W2T [9] validator (which is specifically tailored for checking concrete witnesses) and declare the benchmark to be successfully analyzed only if successfully validated.

Experimental Evaluation. To asses the performance of our analysis, we run our tool and three leading tools from SV-COMP23: CPAChecker [10,22], UAutomizer [30] and Veriabs [2,23] on a selected subset of the competition's benchmarks. In particular, we built our set of benchmarks from the `ReachSafety-Loops` set of the competition, as it comprises several simple numerical programs, from which we removed the `nla-digbench` and `nla-digbench-scaling` folders as they contain programs with polynomial invariants that require special analysis techniques that are out of scope of this work. Our set of benchmarks contains 63 C files (35172 LOC) corresponding to 61% (in terms of LOC) of the `ReachSafety-Loops` set.

Table 1. Outcome of the analysis of `ReachSafety-Loops` excluding `nla-digbench` and `nla-digbench-scaling` folders.

Analyzer	Count				
	Success	Unknown	Timeout/OOM	Unsupported	Other
Banal	16	16	6	25	0
CPAChecker	44	0	18	0	1
UAutomizer	37	0	25	0	1
Veriabs	43	3	17	0	0

Table 2. Analysis time for increasing number of loop iterations, for some selected benchmarks. ✗ denotes a timeout.

Benchmark	Analyzer	Time [s]					
		10^1	10^2	10^3	10^4	10^5	10^6
Mono3_1	Banal	0.94	1	0.84	0.96	0.84	0.92
	CPAChecker	10	22	83	✗	✗	✗
	UAutomizer	50	✗	✗	✗	✗	✗
	Veriabs	42	47	50	92	190	✗
Mono4_1	Banal	0.79	0.81	0.8	0.82	0.85	0.85
	CPAChecker	9.7	19	63	✗	✗	✗
	UAutomizer	36	97	✗	✗	✗	✗
	Veriabs	43	43	46	64	✗	✗
Mono5_1	Banal	1.2	1.3	1.2	1.2	1.2	1.3
	CPAChecker	10	20	69	✗	✗	✗
	UAutomizer	42	✗	✗	✗	✗	✗
	Veriabs	42	43	45	78	✗	✗
Mono6_1	Banal	1.4	1.2	1.2	1.3	1.2	1.2
	CPAChecker	9.9	21	76	✗	✗	✗
	UAutomizer	75	✗	✗	✗	✗	✗
	Veriabs	41	42	46	70	✗	✗
const_1-2	Banal	0.38	0.41	0.4	0.42	0.39	0.41
	CPAChecker	9.5	19	55	✗	✗	✗
	UAutomizer	36	✗	✗	✗	✗	✗
	Veriabs	30	32	32	32	32	32

Benchmark	Analyzer	Time [s]					
		10^1	10^2	10^3	10^4	10^5	10^6
count_up_down-2	Banal	0.49	0.5	0.51	0.51	0.5	0.49
	CPAChecker	9.6	19	62	✗	✗	✗
	UAutomizer	37	✗	✗	✗	✗	✗
	Veriabs	31	32	30	28	29	28
multivar_1-2	Banal	0.61	0.65	0.71	0.67	0.63	0.61
	CPAChecker	7.7	8	7.6	7.7	7.8	7.8
	UAutomizer	28	27	27	27	27	27
	Veriabs	27	28	27	29	28	28
simple_2-2	Banal	0.31	0.33	0.32	0.31	0.34	0.34
	CPAChecker	✗	✗	✗	✗	✗	✗
	UAutomizer	✗	✗	✗	✗	✗	✗
	Veriabs	26	26	27	26	27	27
simple_nested	Banal	11	11	11	11	11	11
	CPAChecker	27	✗	✗	✗	✗	✗
	UAutomizer	83	✗	✗	✗	✗	✗
	Veriabs	39	100	✗	✗	✗	✗
assert_loop	Banal	0.46	0.41	0.38	0.41	0.46	0.41
	CPAChecker	9.8	21	73	✗	✗	✗
	UAutomizer	34	110	✗	✗	✗	✗
	Veriabs	36	35	37	88	✗	✗

All tests were conducted on an Intel Core i7-8550U CPU with 3GiB memory limit and 300 s time limit using the BenchExec [11] platform. We report the results in Table 1, where *unknown* indicates an inconclusive result and *unsupported* indicates a failure due to missing support for some C features (e.g., arrays, pointers).

Despite some encouraging results, Banal performs worse than the other tools due to several imprecisions (e.g., widening failure, non-linear arithmetic) in the analysis and missing support for several C features. However, since it is based on abstract interpretation, Banal is faster than the other tools. In particular, on the successfully analyzed tasks Banal is 22x faster than CPAChecker, 50x than UAutomizer and 50x than Veriabs. This performance gap becomes even wider if we consider programs where bugs are reached after many loop iterators (so called deep bugs) as Banal uses widening operators whereas other tools often necessitate loop unrolling and thus are limited to bugs reachable in few loop iterations (so called shallow bugs). To assess this, we selected some benchmarks from the previous set and re-run the analysis fixing the number of loop iterations with different values. The results are reported in Table 2. CPAChecker and UAutomizer hit timeouts when loops require > 1000 unrollings, while Banal's execution time is not affected. In all cases (including shallow bugs, e.g., < 10 iterations) we observe that Banal is much faster than the other tools. Interestingly, also Veriabs can scale thanks to *loop summarization* [48] techniques allowing it to replace loops with expressions summarizing their effect. However these techniques only work with special loop structures. To exhibit this, we added two synthetic benchmarks `simple_nested` and `assert_loop` (see Fig 8a, 8b) for which the summarization fails (thus forcing Veriabs to unroll the loop) but Banal succeeds.

```
for (int a = 0; a < 1000; ++a) {
        for (int b = 0; b < 1000; ++b) {
                assert(a != 1000-1 || b !=
↪   1000-1);
        }
}
```

```
int i = 0;

while (i < 1000) {
        i++;
        assert(i < 1000);
}
```

(a) `simple_nested`   (b) `assert_loop`

Fig. 8. Simple programs with deep bugs.

5   Conclusion

In this article, we built on top of the preliminary work of [38], studying how to improve it to construct a more effective analysis. It supports more varied and realistic semantics (such as wrap-around) as well as classic abstract domain constructions (such as powerset domains, improved widenings, etc.), to the point where it can provide encouraging results on realistic analysis problems. Our implementation targeted C programs, but the semantics are agnostic with respect to the language and can used to analyze any language with machine integers data types.

Future Work. Although this work displays promising results, much work is still needed to analyze real-world programs. Firstly, we believe that more precise abstractions are needed to analyze numeric properties (e.g., domains for constants, congruences, bit-wise operations). The semantic should be extended to handle more features of the C language (e.g., memory allocation, arrays, structs). We do support non-linear integer arithmetic, thus adding support for floating point arithmetic should not be a too large effort. Moreover, we proposed an abstraction modeling streams as returning always the same value: this may suffice in loop-free programs (as each stream is read only once), but can be imprecise in other cases. Whereas the polyhedra domain (and even its power-set) is precise, it comes with a significant computational cost. We believe that more lightweight domains (like intervals [16] or octagons [36]) and packing techniques will play a crucial role in making this analysis more scalable to real world programs.

Acknowledgments. This work was supported by the SECURVAL project. The SECUREVAL project was funded by the "France 2030" government investment plan managed by the French National Research Agency, under the reference ANR-22-PECY-0005.

Data Availability Statement. All the software used for the experimental part of this work was released in an artifact [35]. It includes not only the source code of the Banal static analyzer, but also the benchmarks and scripts used to produce the Tables 1, 2.

References

1. The banal static analyzer prototype. http://www.di.ens.fr/~mine/banal. Accessed 11 Sep 2023
2. Afzal, M., et al.: Veriabs: verification by abstraction and test generation. In: 2019 34th IEEE/ACM International Conference on Automated Software Engineering (ASE), pp. 1138–1141. IEEE (2019)
3. Ascari, F., Bruni, R., Gori, R.: Limits and difficulties in the design of under-approximation abstract domains. In: FoSSaCS 2022. LNCS, vol. 13242, pp. 21–39. Springer, Cham (2022). https://doi.org/10.1007/978-3-030-99253-8_2
4. Bagnara, R., Hill, P.M., Zaffanella, E.: Exact join detection for convex polyhedra and other numerical abstractions. Comput. Geom. **43**(5), 453–473 (2010)
5. Baldoni, R., Coppa, E., D'elia, D.C., Demetrescu, C., Finocchi, I.: A survey of symbolic execution techniques. ACM Comput. Surv. (CSUR) **51**(3), 1–39 (2018)
6. Bemporad, A., Fukuda, K., Torrisi, F.D.: Convexity recognition of the union of polyhedra. Comput. Geom. **18**(3), 141–154 (2001)
7. Beyer, D.: Competition on software verification and witness validation: Sv-comp 2023. In: International Conference on Tools and Algorithms for the Construction and Analysis of Systems, pp. 495–522. Springer (2023). https://doi.org/10.1007/978-3-031-30820-8_29
8. Beyer, D., Dangl, M., Dietsch, D., Heizmann, M., Stahlbauer, A.: Witness validation and stepwise testification across software verifiers. In: Proceedings of the 2015 10th Joint Meeting on Foundations of Software Engineering, pp. 721–733 (2015)
9. Beyer, D., Dangl, M., Lemberger, T., Tautschnig, M.: Tests from witnesses. In: Dubois, C., Wolff, B. (eds.) TAP 2018. LNCS, vol. 10889, pp. 3–23. Springer, Cham (2018). https://doi.org/10.1007/978-3-319-92994-1_1
10. Beyer, D., Keremoglu, M.E.: CPACHECKER: a tool for configurable software verification. In: Gopalakrishnan, G., Qadeer, S. (eds.) CAV 2011. LNCS, vol. 6806, pp. 184–190. Springer, Heidelberg (2011). https://doi.org/10.1007/978-3-642-22110-1_16
11. Beyer, D., Löwe, S., Wendler, P.: Reliable benchmarking: requirements and solutions. Int. J. Softw. Tools Technol. Transfer **21**, 1–29 (2019)
12. Bourdoncle, F.: Abstract debugging of higher-order imperative languages. In: Proceedings of the ACM SIGPLAN 1993 Conference on Programming Language Design and Implementation, pp. 46–55 (1993)

13. Clarke, E., Grumberg, O., Jha, S., Lu, Y., Veith, H.: Counterexample-guided abstraction refinement for symbolic model checking. J. ACM (JACM) **50**(5), 752–794 (2003)

14. Colón, M.A., Sipma, H.B.: Synthesis of linear ranking functions. In: Margaria, T., Yi, W. (eds.) TACAS 2001. LNCS, vol. 2031, pp. 67–81. Springer, Heidelberg (2001). https://doi.org/10.1007/3-540-45319-9_6

15. Cook, B., Podelski, A., Rybalchenko, A.: TERMINATOR: beyond safety. In: Ball, T., Jones, R.B. (eds.) CAV 2006. LNCS, vol. 4144, pp. 415–418. Springer, Heidelberg (2006). https://doi.org/10.1007/11817963_37

16. Cousot, P., Cousot, R.: Abstract interpretation: a unified lattice model for static analysis of programs by construction or approximation of fixpoints. In: Proceedings of the 4th ACM SIGACT-SIGPLAN symposium on Principles of programming languages, pp. 238–252 (1977)

17. Cousot, P., Cousot, R.: Systematic design of program analysis frameworks. In: Proceedings of the 6th ACM SIGACT-SIGPLAN Symposium on Principles Of Programming Languages, pp. 269–282 (1979)

18. Cousot, P., Cousot, R.: Abstract interpretation and application to logic programs. J. Logic Program. **13**(2–3), 103–179 (1992)

19. Cousot, P., Cousot, R.: Refining model checking by abstract interpretation. Autom. Softw. Eng. **6**(1), 69–95 (1999)

20. Cousot, P., Cousot, R., Logozzo, F.: Precondition inference from intermittent assertions and application to contracts on collections. In: Jhala, R., Schmidt, D. (eds.) VMCAI 2011. LNCS, vol. 6538, pp. 150–168. Springer, Heidelberg (2011). https://doi.org/10.1007/978-3-642-18275-4_12

21. Cousot, P., Halbwachs, N.: Automatic discovery of linear restraints among variables of a program. In: Proceedings of the 5th ACM SIGACT-SIGPLAN symposium on Principles of programming languages, pp. 84–96 (1978)

22. Dangl, M., Löwe, S., Wendler, P.: CPACHECKER with support for recursive programs and floating-point arithmetic. In: Baier, C., Tinelli, C. (eds.) TACAS 2015. LNCS, vol. 9035, pp. 423–425. Springer, Heidelberg (2015). https://doi.org/10.1007/978-3-662-46681-0_34

23. Darke, P., Agrawal, S., Venkatesh, R.: VeriAbs: a tool for scalable verification by abstraction (competition contribution). In: TACAS 2021. LNCS, vol. 12652, pp. 458–462. Springer, Cham (2021). https://doi.org/10.1007/978-3-030-72013-1_32

24. de Vries, E., Koutavas, V.: Reverse hoare logic. In: Barthe, G., Pardo, A., Schneider, G. (eds.) SEFM 2011. LNCS, vol. 7041, pp. 155–171. Springer, Heidelberg (2011). https://doi.org/10.1007/978-3-642-24690-6_12

25. Delmas, D., Miné, A.: Analysis of software patches using numerical abstract interpretation. In: Chang, B.-Y.E. (ed.) SAS 2019. LNCS, vol. 11822, pp. 225–246. Springer, Cham (2019). https://doi.org/10.1007/978-3-030-32304-2_12

26. Filé, G., Ranzato, F.: The powerset operator on abstract interpretations. Theoret. Comput. Sci. **222**(1–2), 77–111 (1999)

27. Gange, G., Navas, J.A., Schachte, P., Søndergaard, H., Stuckey, P.J.: Interval analysis and machine arithmetic: Why signedness ignorance is bliss. ACM Trans. Programming Lang. Syst. (TOPLAS) **37**(1), 1–35 (2015)

28. Gopan, D., Reps, T.: Lookahead widening. In: Ball, T., Jones, R.B. (eds.) CAV 2006. LNCS, vol. 4144, pp. 452–466. Springer, Heidelberg (2006). https://doi.org/10.1007/11817963_41

29. Gotlieb, A., Leconte, M., Marre, B.: Constraint solving on modular integers. In: ModRef Worksop, Associated to CP 2010 (2010)

30. Heizmann, M., Hoenicke, J., Podelski, A.: Software model checking for people who love automata. In: Sharygina, N., Veith, H. (eds.) CAV 2013. LNCS, vol. 8044, pp. 36–52. Springer, Heidelberg (2013). https://doi.org/10.1007/978-3-642-39799-8_2

31. Journault, M., Miné, A., Ouadjaout, A.: An abstract domain for trees with numeric relations. In: Caires, L. (ed.) ESOP 2019. LNCS, vol. 11423, pp. 724–751. Springer, Cham (2019). https://doi.org/10.1007/978-3-030-17184-1_26

32. King, J.C.: Symbolic execution and program testing. Commun. ACM **19**(7), 385–394 (1976)

33. Le, Q.L., Raad, A., Villard, J., Berdine, J., Dreyer, D., O'Hearn, P.W.: Finding real bugs in big programs with incorrectness logic. Proc. ACM Programming Lang. **6**(OOPSLA1), 1–27 (2022)

34. Lev-Ami, T., Sagiv, M., Reps, T., Gulwani, S.: Backward analysis for inferring quantified preconditions. Tr-2007-12-01, Tel Aviv University (2007)

35. Marco, M., Miné, A.: Artifact of paper: Generation of Violation Witnesses by Under-Approximating Abstract Interpretation (Oct 2023). https://doi.org/10.5281/zenodo.8399723

36. Miné, A.: The octagon abstract domain. Higher-Order Symbolic Comput. **19**, 31–100 (2006)

37. Miné, A.: Symbolic methods to enhance the precision of numerical abstract domains. In: Emerson, E.A., Namjoshi, K.S. (eds.) VMCAI 2006. LNCS, vol. 3855, pp. 348–363. Springer, Heidelberg (2005). https://doi.org/10.1007/11609773_23

38. Miné, A.: Backward under-approximations in numeric abstract domains to automatically infer sufficient program conditions. Sci. Comput. Program. **93**, 154–182 (2014)

39. Miné, A., et al.: Tutorial on static inference of numeric invariants by abstract interpretation. Foundation Trends® Programming Lang. **4**(3–4), 120–372 (2017)

40. Moy, Y.: Sufficient preconditions for modular assertion checking. In: Logozzo, F., Peled, D.A., Zuck, L.D. (eds.) VMCAI 2008. LNCS, vol. 4905, pp. 188–202. Springer, Heidelberg (2008). https://doi.org/10.1007/978-3-540-78163-9_18

41. O'Hearn, P.W.: Incorrectness logic. Proc. ACM Programming Lang. **4**(POPL), 1–32 (2019)

42. Raad, A., Berdine, J., Dang, H.-H., Dreyer, D., O'Hearn, P., Villard, J.: Local reasoning about the presence of bugs: incorrectness separation logic. In: Lahiri, S.K., Wang, C. (eds.) CAV 2020. LNCS, vol. 12225, pp. 225–252. Springer, Cham (2020). https://doi.org/10.1007/978-3-030-53291-8_14

43. Reynolds, J.C.: Separation logic: a logic for shared mutable data structures. In: Proceedings 17th Annual IEEE Symposium on Logic in Computer Science, pp. 55–74. IEEE (2002)

44. Schmidt, D.A.: A calculus of logical relations for over-and underapproximating static analyses. Sci. Comput. Program. **64**(1), 29–53 (2007)

45. Sen, R., Srikant, Y.: Executable analysis using abstract interpretation with circular linear progressions. In: 2007 5th IEEE/ACM International Conference on Formal Methods and Models for Codesign (MEMOCODE 2007), pp. 39–48. IEEE (2007)

46. Simon, A., King, A.: Taming the wrapping of integer arithmetic. In: Nielson, H.R., Filé, G. (eds.) SAS 2007. LNCS, vol. 4634, pp. 121–136. Springer, Heidelberg (2007). https://doi.org/10.1007/978-3-540-74061-2_8

47. Urban, C., Ueltschi, S., Müller, P.: Abstract interpretation of CTL properties. In: Podelski, A. (ed.) SAS 2018. LNCS, vol. 11002, pp. 402–422. Springer, Cham (2018). https://doi.org/10.1007/978-3-319-99725-4_24

48. Xie, X., Chen, B., Liu, Y., Le, W., Li, X.: Proteus: computing disjunctive loop summary via path dependency analysis. In: Proceedings of the 2016 24th ACM SIGSOFT International Symposium on Foundations of Software Engineering, pp. 61–72 (2016)

Correctness Witness Validation
by Abstract Interpretation

Simmo Saan[1]([✉]) [ID], Michael Schwarz[2] [ID],
Julian Erhard[2] [ID], Helmut Seidl[2] [ID],
Sarah Tilscher[2] [ID], and Vesal Vojdani[1] [ID]

[1] University of Tartu, Tartu, Estonia
{simmo.saan,vesal.vojdani}@ut.ee
[2] Technische Universität München, Garching, Germany
{m.schwarz,julian.erhard,helmut.seidl,sarah.tilscher}@tum.de

Abstract. Witnesses record automated program analysis results and make them exchangeable. To validate correctness witnesses through abstract interpretation, we introduce a novel abstract operation unassume. This operator incorporates witness invariants into the abstract program state. Given suitable invariants, the unassume operation can accelerate fixpoint convergence and yield more precise results. We demonstrate the feasibility of this approach by augmenting an abstract interpreter with unassume operators and evaluating the impact of incorporating witnesses on performance and precision. Using manually crafted witnesses, we can confirm verification results for multi-threaded programs with a reduction in effort ranging from 7% to 47% in CPU time. More intriguingly, we discover that using witnesses from model checkers can guide our analyzer to verify program properties that it could not verify on its own.

Keywords: Correctness Witness · Witness Validation · Software Verification · Program Analysis · Abstract Interpretation

1 Introduction

Automated software verifiers can be faulty and may produce incorrect results. To increase trust in their verdicts, verifiers may produce *witnesses* that expose their reasoning. Such proof objects allow independent validators to confirm analysis results. The use of witnesses as a standardized way to communicate between different automated software verifiers was pioneered by Beyer et al. [17]. They introduced an analyzer-agnostic automaton-based format for explaining property violations. The witness automaton guides the validator towards a feasible counterexample. This witness format was later extended to explain program correctness using invariants [15]. Witnesses form a cornerstone of the annual software verification competition SV-COMP [14] and have played a key role in the emergence of Cooperative Verification [19,38], where independent verifiers collaborate by exchanging witnesses [24].

© The Author(s), under exclusive license to Springer Nature Switzerland AG 2024
R. Dimitrova et al. (Eds.): VMCAI 2024, LNCS 14499, pp. 74–97, 2024.
https://doi.org/10.1007/978-3-031-50524-9_4

This paper aims to show how *correctness witnesses* can be validated using abstract interpretation. Existing validators are based on model checking [16], test execution [18], interpretation [7,59], or SMT-based verification [43]; whereas, validators for correctness witnesses at SV-COMP 2023 were all based on model checking [14,15]. In that year's competition report, the long-time organizer Dirk Beyer highlighted the scarcity of validators—which leaves many verification outcomes without independent confirmation—as a "remarkable gap in software-verification research" [14]. Abstract interpretation, originally proposed by Cousot and Cousot [29], has proven successful in the efficient verification of large real-world software [9,31] and multi-threaded programs [33,46,48,53,54]. To complement existing validators, we propose enhancing the framework of abstract interpretation to incorporate invariants from witnesses.

For communication across technological boundaries, correctness witnesses must be restricted to invariants that do not expose internal abstractions of tools. For example, as each tool may abstract dynamically allocated memory differently, invariants about the content of such memory may only be expressed indirectly, e.g., via C invariants such as $*p \geq 0$. The challenge is how to incorporate such tool-independent invariants into an abstract interpreter. A key technical contribution of this paper are techniques to incorporate witness invariants, given as expressions, into abstract domains without relying on those invariants to actually hold. This differs from existing work on witnesses for abstract interpretation (detailed in Sect. 7), which does not allow for or aim at the exchange of witnesses across tool boundaries.

Our solution is to introduce a new abstract operation *unassume*. This operator can be used to selectively inject imprecision (hence the name) to speed up fixpoint computation. Suitably increasing abstract values during fixpoint computation can also improve the precision of an existing analysis, most notably due to the non-monotonicity of widening operators. The following example illustrates both the speedup and increase in precision.

Example 1. Consider the example (shown right) from Miné [47]. An abstract interpreter using interval abstraction first reaches the loop head on line 2 with the interval $[40, 40]$ for x. After one iteration, the loop head is reached with $[39, 39]$, so the abstract value at that point is $[39, 40]$. To accelerate

```
1  int x = 40;
2  while (x != 0) {
3      x--;
4  }
```

fixpoint iteration for termination, standard interval widening is applied, which abandons the unstable lower bound, resulting in $[-\infty, 40]$. Another iteration with this interval reaches the loop head with $[-\infty, 39]$, which is subsumed by the previous abstract value. Subsequent standard interval narrowing cannot improve the inferred invariant $[-\infty, 40]$ at the loop head; therefore, the analysis fails to establish a lower bound for x.

Now, suppose that a witness provides the invariant $0 \leq x \leq 40$ for the loop head at line 2. When guiding the fixpoint iteration with this witness, the loop head is again first reached with the singleton interval $[40, 40]$. Using the provided invariant, the unassume operator relaxes the lower bound of the interval:

$$[\![\mathrm{unassume}(0 \le x \le 40)]\!]^{\sharp}\{x \mapsto [40, 40]\} = \{x \mapsto [0, 40]\}.$$

After one iteration, the loop head is now reached with $[0, 39]$, which makes the abstract value at that point $[0, 40]$. Thus, a fixpoint is reached without the need for widening or narrowing, and the stronger invariant $[0, 40]$ is confirmed. This demonstrates how the same analysis, when guided, can validate an invariant that it could not infer on its own. A well-chosen witness invariant can prevent precision loss during widening that cannot be recovered by narrowing, serving as a proxy for providing known widening thresholds. Additionally, the witness-guided analysis required fewer steps (transfer function evaluations and fixpoint iterations). Using the same invariant as a widening threshold does not yield such speedup.

After introducing relevant terms (Sect. 2) and discussing the shortcomings of intuitive approaches to validate witnesses with abstract interpretation (Sect. 3), the paper presents the following main contributions:

- a specification of the unassume operator, and a general realization for relational abstract domains using dual-narrowing (Sect. 4);
- an efficient algorithm for unassuming in non-relational abstract domains (Sect. 5), with generalization to pointer variables [51, Appendix A];
- an implementation of an abstract-interpretation–based witness validator, which is evaluated using hand-crafted invariants for multi-threaded programs and invariants produced by state-of-the-art model checkers for intricate literature examples (Sect. 6).

Our evaluation results provide practical evidence of the unassumed witness invariants making the analysis faster and more precise.

2   Preliminaries

In the following, we formally introduce the notion of location-based correctness witnesses, subsequently referred to simply as *witnesses*, and recall the basics of abstract interpretation.

2.1   Witnesses

Following the refined definitions of Beyer and Strejček [23], a correctness witness should contain hints for the proof of program correctness. Witness automata [15] are a powerful way to provide such hints, but Strejček [57] has observed that their control-flow semantics are ambiguous, impairing interoperability. In practice, however, invariants per program location are often sufficient [15,22], which has led the SV-COMP community to adopt them [58]. Therefore, we consider here correctness witnesses consisting of location-based invariants.

Let $\mathcal{N}$ denote the set of program locations. For clarity of exposition, we consider a fixed set $\mathcal{V}$ of program variables. Invariants from some language $\mathcal{E}$, which we do not fix, are used to specify properties of the program executions reaching a particular program location. We assume that there is a trivial invariant $e_{\text{true}} \in \mathcal{E}$ that always holds. Since the goal is to exchange invariants between tools, the choice of an invariant language involves a trade-off:

1. Beyer et al. [15] use boolean-valued side-effect–free C expressions for their invariants. The chief advantage is its conceptual simplicity: the semantics of such assertions is well-known, and analyzers already come with the necessary facilities to manipulate these expressions, as they appear in the analyzed program. In C expressions, pointers allow exchanging information also about the heap between verifiers. Nevertheless, the expressivity of such invariants is limited, especially regarding more complex data structures.
2. ACSL [10] has more expressive power than plain C expressions by offering quantification, memory predicates, etc. On the downside, considerably fewer analyzers support it, limiting exchange possibilities.
3. Custom invariant languages can be arbitrarily expressive, at the cost of restricting communication to few similar tools whose re-verification to boost.

With these notions, we introduce the definitions of a witness, its validation, and witness-guided verification.

Definition 1. *A* witness *for a safety property Φ of a program P is a tuple (W, P, Φ), where W is a total mapping $W : \mathcal{N} \to \mathcal{E}$ from the program locations of P to invariants from $\mathcal{E}$.*

The textual format in which witnesses are exchanged is not required to provide invariants for all program locations—we implicitly assume that if the witness contains no information for some program location, then this location is mapped to e_{true}. If the invariant language contains a contradictory expression $e_{\text{false}} \in \mathcal{E}$ that never holds, then it can be used to convey unreachability of a program location. This notion of a witness is generic and can be instantiated to different programming and invariant languages. For our examples, we use an invariant language of arithmetic and boolean expressions, enriched with basic pointers, address-taking (&) and dereferencing ($\star$). The pointer constructs pose practical challenges, as will become apparent in subsequent sections.

Definition 2. *A* witness (W, P, Φ) *is* valid *if*

1. *P satisfies the property Φ;*
2. *whenever the execution of P reaches the location $n \in \mathcal{N}$, the invariant $W\,n$ holds.*

A *witness validator* attempts to prove that a witness is valid; specifically, it tries to recreate the proof that the program satisfies the property Φ, and checks that the witness makes only true claims about the program. However, the validation track at SV-COMP 2023 scored participants according to a limited form of validation which only confirms the first condition [14].

A *witness-guided verifier* uses the witness as guidance towards the verification of Φ. A sound verifier can perform this task without assuming the witness invariants to be true; therefore, it qualifies as a sound validator of the first condition. It may additionally verify the invariants in W to perform full witness validation.

2.2   Abstract Interpretation

We rely on the framework of abstract interpretation as introduced by Cousot and Cousot [29,30], and briefly recall relevant notions here. Let S denote the set of all concrete program states. Its subsets are abstracted by an *abstract domain* $\mathbb{D}$ satisfying the following properties [47]:

- a *partial order* $\sqsubseteq$, modeling the relative precision of abstract states;
- a monotonic *concretization* function $\gamma : \mathbb{D} \to 2^S$, mapping an abstract element to the set of concrete states it represents;
- a least element $\bot$, representing unreachability, i.e. $\gamma \bot = \varnothing$;
- a greatest element $\top$, representing triviality, i.e. $\gamma \top = S$;
- sound abstractions *join* ($\sqcup$) and *meet* ($\sqcap$) of $\cup$ and $\cap$ on S, respectively, i.e. $\gamma x \cup \gamma y \subseteq \gamma (x \sqcup y)$ and $\gamma x \cap \gamma y \subseteq \gamma (x \sqcap y)$ for all $x, y \in \mathbb{D}$;
- a *widening* (∇) operator, computing upper bounds that ensure termination in abstract domains with infinite ascending chains, i.e. $x \sqsubseteq x \nabla y$ and $y \sqsubseteq x \nabla y$ for all $x, y \in \mathbb{D}$, and for every sequence $(y_i)_{i \in \mathbb{N}}$ from $\mathbb{D}$, the sequence $(x_i)_{i \in \mathbb{N}}$ defined by $x_0 = y_0$, $x_{i+1} = x_i \nabla y_{i+1}$ is ultimately stable;
- a *narrowing* (Δ) operator, recovering some precision given up by widening, i.e. $x \sqcap y \sqsubseteq x \Delta y \sqsubseteq x$ for all $x, y \in \mathbb{D}$, and for every sequence $(y_i)_{i \in \mathbb{N}}$ from $\mathbb{D}$, the sequence $(x_i)_{i \in \mathbb{N}}$ defined by $x_0 = y_0$, $x_{i+1} = x_i \Delta y_{i+1}$ is ultimately stable.

An abstract interpreter uses an abstract domain $\mathbb{D}$ and sound abstractions $[\![s]\!]^\sharp : \mathbb{D} \to \mathbb{D}$ of primitive statements s to model the abstract semantics of a program. Fixpoint iteration (potentially with widening and narrowing) is used to compute for each program location an abstract state, which represents a superset of all reaching concrete program states. The resulting abstract states may be used to check whether the program satisfies a given safety property.

For *validating* a witness by abstract interpretation, we assume that the analyzer provides us with a mapping $\sigma : \mathcal{N} \to \mathbb{D}$ from locations to abstract values. A witness (W, P, Φ) is validated by the abstract interpreter, if

1. σ is sufficient to verify that Φ holds for program P;
2. for each $n \in \mathcal{N}$, the invariant $W\,n$ is true in every state of $\gamma(\sigma n)$.

In practice, the second condition may not be easy to check since computing γ is not always feasible. Thus, abstract expression evaluation is used instead to perform the validity check, although this is possibly less precise, as the following example shows.

Example 2. Assume the non-relational abstract domain $\mathbb{D} = \mathcal{V} \to \mathbb{V}$ of environments, where $\mathbb{V}$ is the abstract domain of individual values. Using intervals for the latter, let $d = \{\mathsf{x} \mapsto [1, 2]\}$ be the computed abstract state at some program location where $\gamma d = \{\{\mathsf{x} \mapsto 1\}, \{\mathsf{x} \mapsto 2\}\}$ is *exact*. Consider the validation of the following two logically equivalent invariants at this location:

1. The invariant $1 \leq \mathsf{x} \wedge \mathsf{x} \leq 2$ holds for each concrete state in γd. It also evaluates abstractly to true on d using standard syntax-driven evaluation (see Sect. 5.1), because both conjuncts are true for the interval $[1, 2]$.

2. The invariant $x = 1 \lor x = 2$ also holds for each concrete state in $\gamma\, d$. However, when evaluated abstractly and syntax-driven on d, it evaluates to an unknown boolean, because both disjuncts evaluate to an unknown boolean for the interval $[1, 2]$.

Hence, the abstract interpreter is not complete, i.e., it may fail to validate witnesses which are indeed valid, due to imprecision arising from abstraction or fixpoint acceleration. Nevertheless, validating witnesses using abstract expression evaluation is sound, i.e., all witnesses claimed to be validated by the abstract interpreter, are indeed valid.

A witness which maps all locations to the invariant e_{true} is called *trivial*. The validation of such a witness trivially passes the second validity condition, and checking of the first condition falls entirely to the analyzer itself. To be useful, a witness has to be non-trivial and aid the analyzer in proving that the program P satisfies the property Φ, either by improving the precision or the performance of the verification process. For *witness-guided verification*, it suffices if the analyzer can show that Φ holds—even if the invariants of the witness cannot be validated. Given a witness (W, P, Φ), the challenge of witness-guided abstract interpretation is to simultaneously achieve the following:

1. to use the invariants in W to reach a fixpoint $\sigma_W : \mathcal{N} \to \mathbb{D}$ in fewer iterations;
2. to avoid overshooting the required property, i.e., σ_W suffices to prove Φ;
3. to *not trust* the witness, i.e., σ_W should remain sound even in presence of *wrong* invariants.

Subsequently, we first consider some intuitive approaches to motivate our unassume operator for soundly speeding up abstract interpretation with the help of untrusted witnesses.

3   Initialization-Based Approaches

Given a witness (W, P, Φ), one natural idea is to extract from the mapping $W : \mathcal{N} \to \mathcal{E}$ a mapping $w : \mathcal{N} \to \mathbb{D}$ of initial abstract values as non-$\bot$ start points for constructing inductive invariants for the program. We discuss two flavors for realizing this idea, along with their shortcomings.

Total Initial Values. In the first approach, the initial value $w\, n$ for program location n is chosen such that $\gamma\,(w\, n)$ includes every concrete state where $W\, n$ holds. For example, by choosing $w\, \ell_2 = \{x \mapsto [0, 40]\}$ in Example 1. Such a value, however, is only suitable if *all* relevant information for program location n is formalized in the invariant $W\, n$ and expressible by the abstract domain. These requirements limit the applicability, scalability and practicality of this approach.

For example, consider the invariant $*p \geq 0$ involving a pointer p dereference for some program location n. It provides no information about which variables p may point to, thus nothing can be concluded about any integer variables it intends to describe. Therefore, $w\, n = \top$ which leads to a complete loss of precision at location n during the analysis.

This approach also makes silent assumptions about the way in which the analyzer computes values, namely how such initial abstract values are incorporated into analysis, if at all. For example, TD fixpoint solvers [55] only use initial values at *dynamically* identified widening points for starting fixpoint iteration. Additionally, context-sensitive interprocedural analysis is known to give rise to *infinite* constraint systems [5], requiring dedicated changes to the analyzer to ensure that all accessed constraint variables associated with a given location are appropriately initialized.

Partial Initial Values. In order to remedy the problem that all relevant information must be provided in the invariants for program locations, one may instead rely on *partial* initialization. For that to work, we require here a non-relational abstract domain $\mathbb{D} = \mathcal{V} \rightarrow \mathbb{V}$, which does not reduce to $\bot$ if the value of any variable is $\bot$. Assuming that the invariant $W\,n$ only speaks of variables from $V \subseteq \mathcal{V}$, the partial initial value is the same as the total initial value $w\,n$ except all unmentioned variables $x \in \mathcal{V} \setminus V$ are assigned $\bot$.

Example 3. Consider for a particular program location n, two integer variables i and j and a pointer variable p. Let the abstract domain $\mathbb{V}$ of values consist of intervals for abstracting integers and points-to sets for abstracting pointers. Consider two invariants:

1. The witness invariant $\text{i} \geq 0 \wedge \text{j} \geq 0$ can be represented by the partial state $\{\text{p} \mapsto \bot, \text{i} \mapsto [0, \infty], \text{j} \mapsto [0, \infty]\}$.
2. The witness invariant $*\text{p} \geq 0$, on the other hand, results in the partial state $\{\text{p} \mapsto \top, \text{i} \mapsto \bot, \text{j} \mapsto \bot\}$.

Now assume that during analysis of the program, the complete abstract state $\{\text{p} \mapsto \{\&\text{i}, \&\text{j}\}, \text{i} \mapsto [0, 0], \text{j} \mapsto [0, 0]\}$, where p may point to either i or j, reaches the program location n. In order to exploit the witness, this value is *joined* with the partial state constructed from the witness in the corresponding transfer function. For the two invariants above, we respectively obtain:

1. $\{\text{p} \mapsto \{\&\text{i}, \&\text{j}\}, \text{i} \mapsto [0, \infty], \text{j} \mapsto [0, \infty]\}$,
2. $\{\text{p} \mapsto \top, \text{i} \mapsto [0, 0], \text{j} \mapsto [0, 0]\}$.

The first may be useful to guide the analysis since the information for i and j is maximally relaxed such that the witness invariant can still be validated, while the information for the pointer variable p is retained. On the other hand, the second state loses all information about p, which is problematic if memory is accessed through p later in the program. At the same time, here, the values for the variables i and j remain overly precise. Instead, one would have liked to obtain the former abstract state also when using the invariant $*\text{p} \geq 0$.

By joining initial values within transfer functions, this approach is more general: it works for all program locations regardless of the analysis engine and can be seamlessly applied to infinite constraint systems. Nevertheless, the incorporation of witnesses via partial initial values is only applicable to non-relational domains and cannot depend on analysis state. Therefore, in the next section, we propose a more general solution that overcomes these issues.

4   Unassuming

We introduce new statements unassume(e) to the programming language for all invariants $e \in \mathcal{E}$. Given a witness (W, P, Φ), we insert at every location n, the statement unassume($W\,n$) if it is different from e_{true}. In case the invariant is not a legal program expression, we may instead insert the statement at the location into the internal representation used by the analyzer (e.g., the control-flow graph). In the concrete semantics, unassume statements have no effect, i.e., their arguments are not evaluated and thus does not cause runtime errors or undefined behavior.

During the abstract interpretation of the program, the abstract state transformer for the statement unassume(e) for location n is meant to inject the desired imprecision into the abstract state for n. Intuitively, the abstract semantics of unassume is dual to the *assume* operation, i.e., it relaxes a state instead of refining it. Thus, e.g.,

$$\{x \mapsto [0, \infty]\} \xrightleftharpoons[\text{unassume}(x \geq 0)]{\text{assume}(x=0)} \{x \mapsto [0, 0]\}.$$

Note that unassume is not the inverse of assume because the used expressions are different. By integrating unassume operations as statements, they can be treated path- and context-sensitively – just like all other statements – if the abstract interpreter supports such sensitivity, yielding a general approach.

4.1   Specification

Subsequently, we provide abstract operators $[\![\text{unassume}_V(e)]\!]^\sharp : \mathbb{D} \to \mathbb{D}$ which we use to abstractly interpret the corresponding unassume statement. The abstract operators are parameterized by the set of variables $V \subseteq \mathcal{V}$ whose values are relaxed up to the constraining invariant e. The abstract unassume operator $[\![\text{unassume}_V(e)]\!]^\sharp$ is *sound* if it abstracts the concrete no-op operator, i.e.,

$$\gamma\,d \subseteq \gamma\,([\![\text{unassume}_V(e)]\!]^\sharp\,d)$$

for all $d \in \mathbb{D}$. In particular, this is the case if the operator is *extensive*, i.e.,

$$d \sqsubseteq [\![\text{unassume}_V(e)]\!]^\sharp\,d.$$

Given that the abstract interpreter is sound w.r.t. the original program, and sound unassume operations are inserted, we conclude that the resulting abstract interpreter is sound w.r.t. the modified program. Since the newly inserted statements have no effects in the concrete, the resulting abstract interpreter remains sound also w.r.t. the original program. This implies the soundness of our validation approach.

Theorem 1 (Sound witness validation). *Assume a witness (W, P, Φ) is used to insert unassume statements and $\sigma_W : \mathcal{N} \to \mathbb{D}$ is the result of analyzing the instrumented program. If the sound analyzer confirms Φ and all invariants of $W : \mathcal{N} \to \mathcal{E}$ abstractly evaluate to true in σ_W, then the witness must be valid.*

Example 4. The desired behavior of unassume operators is illustrated by the following examples.

1. Unmentioned parts of the abstract state should be retained:

$$[\![\mathrm{unassume}_{\{x\}}(x \geq 0)]\!]^{\sharp} \{x \mapsto [0,0], y \mapsto [0,0]\} = \{x \mapsto [0,\infty], y \mapsto [0,0]\}.$$

2. Information on variables used in the invariant but not contained in V should also be retained:

$$[\![\mathrm{unassume}_{\{i\}}(i \leq n)]\!]^{\sharp} \{i \mapsto [0,0], n \mapsto [10,10]\} =$$
$$= \{i \mapsto [-\infty,10], n \mapsto [10,10]\}.$$

3. Relational invariants between relaxed and not relaxed variables should be preserved whenever possible without restricting the unassumed invariant; e.g., relaxing the state $0 = x \leq y$ with $0 \leq x$ should result in $0 \leq x \leq y$:

$$[\![\mathrm{unassume}_{\{x\}}(x \geq 0)]\!]^{\sharp} \{x \leq 0, -x \leq 0, -y \leq 0, -x - y \leq 0, x - y \leq 0\} =$$
$$= \{-x \leq 0, -y \leq 0, -x - y \leq 0, x - y \leq 0\}$$

   when using the octagon domain [45].[1] More specifically, this is the most precise result which, when restricted to V, still includes (here equals) the abstract state $\{-x \leq 0\}$, defined only by the unassumed invariant on V.
4. Information provided by the input abstract state should be leveraged to propagate imprecision to further variables not mentioned in the invariant (cf. Example 3):

$$[\![\mathrm{unassume}_{\{i,j\}}(*p \geq 0)]\!]^{\sharp} \{p \mapsto \{\&i, \&j\}, i \mapsto [0,0], j \mapsto [0,0]\} =$$
$$= \{p \mapsto \{\&i, \&j\}, i \mapsto [0,\infty], j \mapsto [0,\infty]\}.$$

We remark that Items 2 and 4 illustrate cases where V differs from the set of variables syntactically occurring in e. Since the behavior with relational invariants depends on the expressivity and representation of the *abstract* domain, we do not characterize the operator on sets of concrete states.

4.2   Naïve Definition

We present the first unassume operator in terms of the abstract operators for non-deterministic assignments and guards. In this section, we assume the invariant language $\mathcal{E}$ is a subset of the side-effect–free expressions used for conditional branching in the programming language.

For an expression e, let assume(e) denote the concrete operation which only continues execution if the condition e is true, and aborts otherwise. Let $[\![\mathrm{assume}(e)]\!]^{\sharp} : \mathbb{D} \to \mathbb{D}$ be a sound abstraction.

For a set of variables $V \subseteq \mathcal{V}$, let havoc(V) denote the concrete operation which non-deterministically assigns arbitrary values to all $x \in V$, and $[\![\mathrm{havoc}(V)]\!]^{\sharp} : \mathbb{D} \to \mathbb{D}$ be a sound abstraction.

---

[1] Redundant constraints are grayed out. They can be derived from non-redundant (non-grayed out) constraints using the octagon closure algorithm.

Definition 3 (Naïve unassume). *Let $V \subseteq \mathcal{V}$, $e \in \mathcal{E}$ and $d \in \mathbb{D}$. Then the* naïve unassume *is defined as*

$$[\![\text{unassume}_V(e)]\!]_1^\sharp \, d = d \sqcup ([\![\text{assume}(e)]\!]^\sharp \circ [\![\text{havoc}(V)]\!]^\sharp) \, d.$$

Intuitively, the argument state is relaxed by joining with an additional value. This value is obtained by first forgetting all information about the variables from V and then assuming the information provided by e. Due to the join, this unassume operator is *sound by construction*. The naïve unassume operator is already sufficient to gain the improvements illustrated by Example 1 when choosing $V = \{x\}$.[2] This operator also succeeds for Items 1 and 2 in Example 4, but fails when there are relations between elements of V and $\mathcal{V} \setminus V$, e.g., for Item 3 it returns $\{-x \leq 0, -y \leq 0, -x - y \leq 0\}$. In this case the octagon constraint $x - y \leq 0$ is lost by havocing and cannot be recovered by assuming.

4.3   Dual-Narrowing

We will address the above challenge by relying on additional insights from abstract interpretation. Let us recall the term dual-narrowing, which is the lattice analogue of Craig interpolation [28]. A dual-narrowing operator $\tilde{\Delta} : \mathbb{D} \to \mathbb{D} \to \mathbb{D}$ returns for every $d_1, d_2 \in \mathbb{D}$ with $d_1 \sqsubseteq d_2$, a value between both of them, i.e., $d_1 \sqsubseteq d_1 \, \tilde{\Delta} \, d_2 \sqsubseteq d_2$.

Using such an operator, we can define an abstract unassume that, given d, may return an abstract value in the range from d to $[\![\text{unassume}_V(e)]\!]_1^\sharp \, d$:

Definition 4 (Dual-narrowing unassume). *Let $\tilde{\Delta} : \mathbb{D} \to \mathbb{D} \to \mathbb{D}$ be a dual-narrowing. Let $V \subseteq \mathcal{V}$, $e \in \mathcal{E}$ and $d \in \mathbb{D}$. Then the* dual-narrowing unassume *is defined as a wrapper around the naïve unassume:*

$$[\![\text{unassume}_V(e)]\!]_2^\sharp \, d = d \, \tilde{\Delta} \, [\![\text{unassume}_V(e)]\!]_1^\sharp \, d.$$

Example 5. A dual-narrowing for relational domains can be defined using *heterogeneous* environments and *strengthening* [42]. Let $\text{dom}(d) \subseteq \mathcal{V}$ denote the environment of the abstract value $d \in \mathbb{D}$. Let $d|_V$ denote the restriction of the abstract value $d \in \mathbb{D}$ to the program variables $V \subseteq \mathcal{V}$.

An environment-aware order $\underline{\sqsubseteq}$ is defined for $d_1, d_2 \in \mathbb{D}$ by

$$d_1 \, \underline{\sqsubseteq} \, d_2 \iff \text{dom}(d_1) \subseteq \text{dom}(d_2) \wedge d_1 \sqsubseteq d_2|_{\text{dom}(d_1)}.$$

Let $\underline{\sqcup} : \mathbb{D} \to \mathbb{D} \to \mathbb{D}$ be an upper bound operator w.r.t. $\underline{\sqsubseteq}$, such that the resulting environment is minimal, i.e., $\text{dom}(d_1 \, \underline{\sqcup} \, d_2) = \text{dom}(d_1) \cup \text{dom}(d_2)$. Specifically, Journault et al. [42] define $\underline{\sqcup}$ as follows. The result of joining d_1 and d_2 in their common environment $\text{dom}(d_1) \cap \text{dom}(d_2)$ is extended to $\text{dom}(d_1) \cup \text{dom}(d_2)$ by adding unconstrained dimensions. A *strengthening* operator refines this result by iteratively adding back constraints from both arguments which would not cause

---

[2] Complete computations for this and the following examples can be found in the extended version [51].

the upper-boundedness w.r.t. $\sqsubseteq$ to be violated. Note that this definition is not semantic, i.e., the result depends on the constraints representing the arguments and their processing order.

By defining $d_1 \mathbin{\tilde{\Delta}} d_2 = d_1 \mathbin{\underline{\vee}} d_2 |_V$, which is *parametrized* by V, dual-narrowing unassume yields the desired result for Item 3 from Example 4 [51, Example 5]:

$$\{x \le 0, -x \le 0, -y \le 0, -x - y \le 0, x - y \le 0\} \mathbin{\tilde{\Delta}} \{-x \le 0, -y \le 0, -x - y \le 0\} =$$
$$= \{x \le 0, -x \le 0, -y \le 0, -x - y \le 0, x - y \le 0\} \mathbin{\underline{\vee}} \{-x \le 0\} =$$
$$= \{-x \le 0, -y \le 0, -x - y \le 0, x - y \le 0\}.$$

Although the restriction to V first destroys relations between V and $\mathcal{V} \setminus V$, the subsequent strengthening join can restore original relations which are compatible with e on V.

5   Unassuming Indirectly

We now turn to the unassuming of more complex invariants, which include indirection via pointers and dependent subexpressions. Naïve unassume is unable to achieve the desired precision for Item 4 from Example 4 for which it returns $\{p \mapsto \{\&i, \&j\}, i \mapsto \top, j \mapsto \top\}$. This is due to both integer variables being havoced and the assume operator not being able to soundly refine via ambiguous points-to sets [51, Appendix A]. Technically, there exists a dual-narrowing that yields the desired result, but it would be ad-hoc.

To address the disjunctive nature of the ambiguous points-to set, we propose an improved unassume operator. Suppose we are provided a family of mappings $\mathrm{explode}_V(e) : \mathbb{D} \to 2^{\mathbb{D}}$ which explode any given abstract state d into a non-empty finite subset $\mathrm{explode}_V(e)\, d \subseteq \mathbb{D}$ of abstract states where for each resulting element d' we have $d' \sqsubseteq d$. The *explode* operator can be used to make disjunctive information in abstract states explicit, e.g., resolve non-singleton points-to sets for pointer variables not contained in V.

Definition 5 (Exploding unassume). *Let* $V \subseteq \mathcal{V}$, $e \in \mathcal{E}$ *and* $d \in \mathbb{D}$. *Let* $\mathrm{explode}_V(e)$ *be an explode operator. Then the* exploding unassume *is defined as*

$$[\![\mathrm{unassume}_V(e)]\!]_3^{\sharp}\, d = \bigsqcap_{d' \in \mathrm{explode}_V(e)\, d} d \sqcup ([\![\mathrm{assume}(e)]\!]^{\sharp} \circ [\![\mathrm{havoc}(V)]\!]^{\sharp})\, d'.$$

This improved unassume operator is extensive and therefore sound for any choice of $\mathrm{explode}_V(e)$. One might want to establish that $\bigsqcup \mathrm{explode}_V(e)\, d = d$ holds, but this is not necessary for soundness. Whereas $\mathrm{explode}_V(e)\, d = \{\bot\}$ would make the unassume a no-op.

Example 6. Consider an explode operator, which splits ambiguous points-to sets:

$$\mathrm{explode}_{\{i,j\}}(\texttt{*p} \ge 0)\, \{p \mapsto \{\&i, \&j\}, i \mapsto [0,0], j \mapsto [0,0]\} =$$
$$= \{\{p \mapsto \{\&i\}, i \mapsto [0,0], j \mapsto [0,0]\}, \{p \mapsto \{\&j\}, i \mapsto [0,0], j \mapsto [0,0]\}\}.$$

Using this explode operator, Item 4 from Example 4 is handled as desired [51, Example 6].

Example 7. However, consider the following, where different subexpressions depend on each other (here through p):

$$[\![\text{unassume}_{\{p,i,j\}}((p = \&i \lor p = \&j) \land {*}p \geq 0)]\!]^{\sharp} \{p \mapsto \{\&i\}, i \mapsto [0,0], j \mapsto [0,0]\}.$$

In contrast to Example 6, there is no ambiguous points-to set in the abstract state supplied as the argument. All possible explosions lead to the same issue as when using the naïve unassume on this example. After havocing, the environment contains $p \mapsto \top$, thus, in the assume a top pointer needs to be dereferenced and its targets refined. The semantics of this is unclear and imprecise at best, when one has to consider assignments to *all* possible (unrelated) memory locations.

5.1 Propagating Unassume

The *HC4-revise* algorithm by Benhamou et al. [11] can be used to implement the assume operation for complex expressions on non-relational domains in a syntax-directed manner [47,61]. It is also known as *backwards evaluation* [27]. We describe the algorithm and then apply it to construct an unassume operator.

We loosely follow the presentation by Cousot [27]. Let the languages of expressions e and logical conditions c be defined by the grammars in Fig. 1. For each $n \in \mathbb{N}$, let $\mathcal{O}_n$ be the set of n-ary operators. For simplicity of presentation, assume that the condition is in negation normal form (NNF), i.e., negations in conditions have been "pushed down" into binary comparisons according to boolean logic. The logical conditions form an invariant language (see Sect. 2). The following algorithms generalize from just variables to lvalues, allowing for languages with pointers like our example invariant language from before. This generalization is formalized in the extended version [51, Appendix A].

Evaluation. Let $\mathbb{V}$ be the abstract domain for individual values and $\mathbb{D} = \mathcal{V} \to \mathbb{V}$ the abstract domain for non-relational environments. Let $\mathbb{B}$ be the flat boolean domain, where $\bot \sqsubseteq \{\text{true}^{\sharp}, \text{false}^{\sharp}\} \sqsubseteq \top$. The standard abstract forward evaluation of expressions $\mathbb{E}[\![e]\!]^{\sharp}$ and conditions $\mathbb{C}[\![c]\!]^{\sharp}$ in the non-relational environment $d \in \mathbb{D}$ is shown in Fig. 2. For a constant k, let $k^{\sharp}$ be its corresponding abstraction, and $\Box^{\sharp}, \bowtie^{\sharp}, \wedge^{\sharp}, \vee^{\sharp}$ be abstract versions of the corresponding operators.

$$
\begin{array}{llll}
e ::= k & \text{(constant)} & c ::= e \bowtie e & \text{(binary comparison,} \\
\mid x & \text{(variable, } x \in \mathcal{V}) & & \bowtie \in \{=, \neq, <, \leq, >, \geq\}) \\
\mid \Box\,(e)_{i=1}^{n} & \text{(}n\text{-ary operator,} & \mid c \land c & \text{(conjunction)} \\
& n \in \mathbb{N}, \Box \in \mathcal{O}_n) & \mid c \lor c & \text{(disjunction)}
\end{array}
$$

Fig. 1. Syntax of expressions and conditions.

$$\mathbb{E}[\![e]\!]^\sharp : \mathbb{D} \to \mathbb{V} \qquad\qquad\qquad \mathbb{C}[\![c]\!]^\sharp : \mathbb{D} \to \mathbb{B}$$

$$\mathbb{E}[\![k]\!]^\sharp d = k^\sharp \qquad\qquad\qquad\qquad \mathbb{C}[\![e_1 \bowtie e_2]\!]^\sharp d = \mathbb{E}[\![e_1]\!]^\sharp d \bowtie^\sharp \mathbb{E}[\![e_2]\!]^\sharp d$$

$$\mathbb{E}[\![x]\!]^\sharp d = d\,x \qquad\qquad\qquad\qquad \mathbb{C}[\![c_1 \wedge c_2]\!]^\sharp d = \mathbb{C}[\![c_1]\!]^\sharp d \wedge^\sharp \mathbb{C}[\![c_2]\!]^\sharp d$$

$$\mathbb{E}[\![\square\,(e_i)_{i=1}^n]\!]^\sharp d = \square^\sharp\,(\mathbb{E}[\![e_i]\!]^\sharp d)_{i=1}^n \qquad \mathbb{C}[\![c_1 \vee c_2]\!]^\sharp d = \mathbb{C}[\![c_1]\!]^\sharp d \vee^\sharp \mathbb{C}[\![c_2]\!]^\sharp d$$

Fig. 2. Forward evaluation of expressions and conditions.

Assume. The HC4-revise algorithm for the assume operation has two phases:

1. Bottom-up forward propagation on the expression tree abstractly evaluates the expression, as usual.
2. Top-down backward propagation refines each abstract value with the expected result of the sub-expression. This relies on backward abstract operators, which refine each argument based on the other arguments and the expected result, while variables are refined at the leaves.

The algorithm $[\![\mathrm{assume}(e)]\!]^\sharp$ with its abstract backward evaluation of expressions $\overleftarrow{\mathbb{E}}\,[\![e]\!]^\sharp$ and conditions $\overleftarrow{\mathbb{C}}\,[\![c]\!]^\sharp$ is shown in Fig. 3. Instead of evaluating to an abstract value, they refine values of variables in the abstract environment. For each $n \in \mathbb{N}$, $\square \in \mathcal{O}_n$, let $\overleftarrow{\square}^\sharp : \mathbb{V} \to \mathbb{V}^n \to \mathbb{V}^n$ be the abstract backward version of the n-ary operator $\square$. It returns abstract values for its arguments under the assumption that the operator evaluates to the given abstract value v' and the other arguments have the given abstract values. For example, if $n = 2$, then

$$\overleftarrow{\square}^\sharp v'\,(v_1, v_2) = (v_1', v_2') \implies \{x_1 \in \gamma\,\mathbb{V} \mid \exists x_2 \in \gamma\,v_2 : \square\,(x_1, x_2) \in \gamma\,v'\} \subseteq \gamma\,v_1' \wedge$$
$$\wedge\,\{x_2 \in \gamma\,\mathbb{V} \mid \exists x_1 \in \gamma\,v_1 : \square\,(x_1, x_2) \in \gamma\,v'\} \subseteq \gamma\,v_2'.$$

Unlike Cousot [27] and Miné [47], we require that the backward operators *do not* intersect an argument's backward-computed value with its current value. Instead, we make this explicit in the algorithm like Benhamou et al. [11]. Similarly, let $\overleftarrow{\bowtie}^\sharp : \mathbb{V} \to \mathbb{V} \to \mathbb{V} \times \mathbb{V}$ be the abstract backward version of the comparison $\bowtie$. Since conditions are in NNF, the expected result is always $\mathsf{true}^\sharp$ and no v' argument is needed for it. The evaluations $\mathbb{E}[\![e]\!]^\sharp d$ should all be cached and reused from a single forward evaluation as the argument environment is passed around without changes [11,47].

Unassume. This algorithm can be adapted into a *propagating unassume* operator $[\![\mathrm{unassume}(e)]\!]^\sharp$ as shown in Fig. 4. Changes are required to achieve the following properties:

Variable set variance. In Example 7 the first conjunct should relax $\{\mathsf{p}\}$, while the second should relax $\{\mathsf{i}, \mathsf{j}\}$ via the relaxed pointer. In order to allow different sub-expressions to relax different variable sets, the abstract environments returned by $\tilde{\mathbb{E}}[\![e]\!]^\sharp$ and $\tilde{\mathbb{C}}[\![c]\!]^\sharp$ are partial: they only contain variables which have been relaxed at leaves in the corresponding sub-expression.

$$[\![\text{assume}(e)]\!]^{\sharp}\, d = \overleftarrow{\mathbb{C}}\,[\![e]\!]^{\sharp}\, d$$

where

$\overleftarrow{\mathbb{E}}\,[\![e]\!]^{\sharp} : \mathbb{V} \to \mathbb{D} \to \mathbb{D}$

$\overleftarrow{\mathbb{E}}\,[\![k]\!]^{\sharp}\, v'\, d = \textbf{if } k^{\sharp} \sqsubseteq v' \textbf{ then } d \textbf{ else } \bot$

$\overleftarrow{\mathbb{E}}\,[\![x]\!]^{\sharp}\, v'\, d = d[x \mapsto d\,x \sqcap v']$

$\overleftarrow{\mathbb{E}}\,[\![\square\,(e_i)_{i=1}^{n}]\!]^{\sharp}\, v'\, d =$

$\quad \textbf{let } (v_i)_{i=1}^{n} = (\mathbb{E}[\![e_i]\!]^{\sharp}\, d)_{i=1}^{n} \textbf{ in}$

$\quad \textbf{let } (v_i')_{i=1}^{n} = \overleftarrow{\square}^{\sharp}\, v'\, (v_i)_{i=1}^{n} \textbf{ in}$

$$\prod_{i=1}^{n} \overleftarrow{\mathbb{E}}\,[\![e_i]\!]^{\sharp}\, (v_i \sqcap v_i')\, d$$

$\overleftarrow{\mathbb{C}}\,[\![c]\!]^{\sharp} : \mathbb{D} \to \mathbb{D}$

$\overleftarrow{\mathbb{C}}\,[\![e_1 \bowtie e_2]\!]^{\sharp}\, d =$

$\quad \textbf{let } (v_1, v_2) = (\mathbb{E}[\![e_1]\!]^{\sharp}\, d, \mathbb{E}[\![e_2]\!]^{\sharp}\, d) \textbf{ in}$

$\quad \textbf{let } (v_1', v_2') = v_1 \overleftarrow{\bowtie}^{\sharp} v_2 \textbf{ in}$

$\quad \overleftarrow{\mathbb{E}}\,[\![e_1]\!]^{\sharp}\, (v_1 \sqcap v_1')\, d \sqcap \overleftarrow{\mathbb{E}}\,[\![e_2]\!]^{\sharp}\, (v_2 \sqcap v_2')\, d$

$\overleftarrow{\mathbb{C}}\,[\![c_1 \wedge c_2]\!]^{\sharp}\, d = \overleftarrow{\mathbb{C}}\,[\![c_1]\!]^{\sharp}\, d \sqcap \overleftarrow{\mathbb{C}}\,[\![c_2]\!]^{\sharp}\, d$

$\overleftarrow{\mathbb{C}}\,[\![c_1 \vee c_2]\!]^{\sharp}\, d = \overleftarrow{\mathbb{C}}\,[\![c_1]\!]^{\sharp}\, d \sqcup \overleftarrow{\mathbb{C}}\,[\![c_2]\!]^{\sharp}\, d$

Fig. 3. Assume via backward evaluation of expressions and conditions by the propagation algorithm.

Thus heterogeneous lattice join ⊍ from Example 5 is used. However, here in the non-relational case its definition is simpler: values are joined pointwise while using $\bot$ for missing variables. The heterogeneous lattice meet ⩔ is defined analogously, using $\top$ for missing variables. Note that ⩔ is *not* the meet w.r.t. $\sqsubseteq$, because ⩔ must preserve all relaxed variables from both operands, not just the common ones.

Soundness. The result is joined with the pre-unassume environment to ensure soundness.

Relaxation. Backward propagation only propagates backward values v_i' and does not refine them using abstract values v_i computed by forward propagation. Otherwise, sub-expressions cannot be relaxed at all from their current values. Note that the forward values are still necessary for evaluating the backward operators. This modification on its own yields the HC4-revise* algorithm described by Benhamou et al. [11].

Unlike our previous unassume operators, this algorithm implicitly chooses V to be those variables which are relaxed in the process. Note that it is different from the set of variables syntactically occurring in e in a more complex invariant language, such as our example language with pointers, which is described in the extended version [51, Appendix A]. This is illustrated by Example 8 below.

Local Iteration. Repeated application of the propagation algorithm for assuming can improve precision in the presence of dependent subexpressions, i.e., when the same variable occurs multiple times in the condition [27,47]. Analogously, repeated application of the propagation algorithm for unassuming can perform more relaxation in the presence of dependent subexpressions. Both repetitions can be iterated to a local fixpoint.

$$[\![\text{unassume}(e)]\!]^\sharp d = d \uplus (\tilde{\mathbb{C}}[\![e]\!]^\sharp d)$$

where

$$\tilde{\mathbb{E}}[\![e]\!]^\sharp : \mathbb{V} \to \mathbb{D} \to \mathbb{D}$$

$$\tilde{\mathbb{E}}[\![k]\!]^\sharp v' d = \varnothing$$

$$\tilde{\mathbb{E}}[\![x]\!]^\sharp v' d = \{x \mapsto v'\}$$

$$\tilde{\mathbb{E}}[\![\square\,(e_i)_{i=1}^n]\!]^\sharp v' d =$$

$$\textbf{let } (v_i)_{i=1}^n = (\mathbb{E}[\![e_i]\!]^\sharp d)_{i=1}^n \textbf{ in}$$

$$\textbf{let } (v_i')_{i=1}^n = \overleftarrow{\square}^\sharp v' (v_i)_{i=1}^n \textbf{ in}$$

$$\boxed{\bullet}_{i=1}^n \tilde{\mathbb{E}}[\![e_i]\!]^\sharp v_i' d$$

$$\tilde{\mathbb{C}}[\![c]\!]^\sharp : \mathbb{D} \to \mathbb{D}$$

$$\tilde{\mathbb{C}}[\![e_1 \bowtie e_2]\!]^\sharp d =$$

$$\textbf{let } (v_1, v_2) = (\mathbb{E}[\![e_1]\!]^\sharp d, \mathbb{E}[\![e_2]\!]^\sharp d) \textbf{ in}$$

$$\textbf{let } (v_1', v_2') = v_1 \overleftarrow{\bowtie}^\sharp v_2 \textbf{ in}$$

$$\tilde{\mathbb{E}}[\![e_1]\!]^\sharp \boxed{v_1'}\, d \,\boxed{\bullet}\, \tilde{\mathbb{E}}[\![e_2]\!]^\sharp \boxed{v_2'}\, d$$

$$\tilde{\mathbb{C}}[\![c_1 \wedge c_2]\!]^\sharp d = \tilde{\mathbb{C}}[\![c_1]\!]^\sharp d \,\boxed{\bullet}\, \tilde{\mathbb{C}}[\![c_2]\!]^\sharp d$$

$$\tilde{\mathbb{C}}[\![c_1 \vee c_2]\!]^\sharp d = \tilde{\mathbb{C}}[\![c_1]\!]^\sharp d \uplus \tilde{\mathbb{C}}[\![c_2]\!]^\sharp d$$

Fig. 4. Unassume via backward evaluation of expressions and conditions by the propagation algorithm (changes from the assume algorithm are highlighted).

Example 8. Consider using the above algorithm for the case from Example 7:

$$[\![\text{unassume}((p = \&i \vee p = \&j) \wedge *p \geq 0)]\!]^\sharp \{p \mapsto \{\&i\}, i \mapsto [0,0], j \mapsto [0,0]\}.$$

As formalized in the extended version [51, Appendix A], the first backward propagation returns for $p = \&i \vee p = \&j$ the partial map $\{p \mapsto \{\&i, \&j\}\}$ and uses $*p \geq 0$ to relax $*p$. To do so, backward operator $\overleftarrow{\geq}^\sharp$ uses the expected true result and its forward-evaluated right argument $[0,0]$ to propagate the expected value $[0, \infty]$ into its left argument. Using forward evaluated $p \mapsto \{\&i\}$, backward propagation of the lvalue $*p$, acting as a leaf, returns the partial map $\{i \mapsto [0, \infty]\}$. The new constructed environment is $\{p \mapsto \{\&i, \&j\}, i \mapsto [0, \infty], j \mapsto [0,0]\}$.

The second backward propagation does all of the above and also includes $j \mapsto [0, \infty]$, due to the new points-to set. The final fixpoint environment is $\{p \mapsto \{\&i, \&j\}, i \mapsto [0, \infty], j \mapsto [0, \infty]\}$.

In Example 8 the two iterations induced $V = \{p, i, j\}$, which used with naïve unassume yields the issue described in Example 7. Therefore propagating unassume is not equivalent to simply using its induced variable set with a naïve unassume. Propagating unassume fuses the multiple steps involved together into one algorithm which avoids intermediate imprecision and undefined behavior. We do not give an exact characterization of the result computed by the modified algorithm as it has remained an open problem for HC4-revise itself [11, 36].

In case the value lattice has infinite chains, the local iteration of propagating assume must use narrowing to ensure termination [27, 47]. Similarly, the local iteration of propagating unassume must instead use widening.

6   Evaluation

We implement the unassume operator in a state-of-the-art abstract interpreter. Since the analyzer is sound, this yields a sound witness validator. However, a sound validator can trivially be obtained by replacing the unassume operator with the identity operator and ignoring the witnesses entirely. Therefore, our experimental evaluation aims to demonstrate that our witness-guided verifier effectively uses witnesses. More specifically, we seek to confirm that "the effort and feasibility of validation depends on witness content" [15]. To assess the analyzer's dependency on witness content, we pose the following questions:

Precision. Can the witness-guided verifier leverage witnesses to validate verification results that it could not confirm without a witness?

Performance. Do witnesses influence the verification effort in the application domain of the analyzer?

It is worth noting that the performance improvement from technology-agnostic correctness witnesses is expected to be modest. In fact, Beyer et al. [15] observed no consistent trend in performance gains.

Experimental Setup and Data. Our benchmarks are executed on a laptop running Ubuntu 22.04.3 on an AMD RYZEN 7 PRO 4750U processor. For reliable measurements, all the experiments are carried out using the BENCHEXEC framework [21], where each tool execution is limited to 900 s of CPU time on one core and 4 GB of RAM. The benchmarks, tools and scripts used, as well as the raw results of the evaluation, are openly archived on Zenodo [52].

Implementation. GOBLINT is an abstract interpretation framework for C programs [60]. We have extended the framework with unassume operators and YAML witness support. The correctness witnesses proposed by Beyer et al. [15] and subsequently used in SV-COMP [14] provide invariants using an automaton in the GraphML format. The witnesses we consider (defined in Sect. 2.1) are much simpler and, thus, we use the newly-proposed YAML format [56,58], which directly matches our notion. To this end, our implementation includes parsing of YAML witnesses and matching provided invariants to program locations such that the unassume operator can be applied. Our implementation contains two unassume operators:

1. Propagating unassume (Sect. 5.1) for non-relational domains. The existing propagating assume in GOBLINT could be generalized and directly reused, yielding an unassume operator capable of handling, e.g., C lvalues, not just variables, with no extra effort [51, Appendix A].
2. Strengthening-based dual-narrowing unassume (Sect. 4.3) for relational domains. Although APRON [41], which GOBLINT uses for its relational domains, does not provide dual-narrowing, the generic approach described in Example 5 works for, e.g., octagons and convex polyhedra. Since the relational analysis is just numeric, V is collected syntactically.

To prevent unintended precision loss when widening from initially reached abstract values to the unassumed ones, we must take care to delay the application of widening. We tag abstract values with the identifiers of incorporated witness invariants (UUIDs from the YAML witness) and delay the widening if this set increases [44]. Such widening tokens ensure that each witness invariant can be incorporated without immediate overshooting.

6.1 Precision Evaluation

We collected and provide a set of 11 example programs (excluding duplicates) from literature [3,26,37,47] where more advanced abstract interpretation techniques are developed to infer certain invariants, where standard accelerated solving strategies fail. We configure GOBLINT the same as in SV-COMP [49,50], except autotuning is disabled and relational analysis using polyhedra is unconditionally enabled. We manually created YAML witnesses containing suitable loop invariants for these programs. We also used two state-of-the-art verifiers from SV-COMP 2023 to generate real witnesses: CPACHECKER [20,32] and UAUTOMIZER [39,40]. Both verifiers are able to verify these programs and produce GraphML witnesses. Following Beyer et al. [22], we use CPACHECKER in its `witness2invariant` configuration to convert them into YAML witnesses that GOBLINT can consume.

Table 1. Evaluation results on literature examples (excluding duplicates). The GOBLINT column indicates whether it can verify the program without any witness. Remaining columns indicate results with corresponding witnesses: witness validated (✓), program verified with witness-guidance but witness not validated (✓✗) or program not verified with witness-guidance (✗).

Author(s)	Example	GOBLINT w/o witness	GOBLINT w/ witness from Manual	CPACHECKER	UAUTOMIZER
Miné [47]	4.6	✗	✓	✓✗	✓
	4.7	✗	✓	✓✗	✓
	4.8	✓	✓	✓	✓
	4.10	✓	✓	✓✗	✓
Halbwachs and Henry [37]	1.b	✓	✓	✓✗	✓✗
	2.b	✗	✓	✓✗	✓
	3	✗	✓	✓✗	✓✗
Boutonnet and Halbwachs [26]	1 (polyhedra)	✗	✓	✓✗	✓✗
	3	✗	✓	✗	✓
	"additional"	✗	✓	✗	✓
Amato and Scozzari [3]	hybrid	✗	✓	✓✗	✓
Total	11	✓: 3	✓: 11	✓✗: 8 , ✓: 1	✓✗: 3 , ✓: 8

The results are summarized in Table 1. GOBLINT manages to verify the desired property for 3 of these programs without any witness, but can validate all handwritten witnesses, despite not implementing any of the advanced techniques needed for their inference. With CPACHECKER witnesses our validator can verify 9 out of 11 programs and validate 1 out of 11 witnesses. With UAU-TOMIZER witnesses our validator can verify all 11 programs and validate 8 out of 11 witnesses. Furthermore, our abstract interpreter can validate the witnesses from model checkers *orders of magnitude* faster than it took to generate them.

The evaluation, however, shows many instances where the program was only verified thanks to witness-guidance, but not all witness invariants could be validated, especially for CPACHECKER. This is precisely due to the phenomenon described in Example 2: in these small programs bounded model checking is successful and yields disjunctive invariants over all the finitely-many cases. Surprisingly, an invariant is useful for witness-guided verification, even when it cannot be proven to hold abstractly.

Example 9. The invariant from Example 2 relaxes an abstract state:

$$[\![\text{unassume}_{\{x\}}(x = 1 \vee x = 2)]\!]^{\sharp} \{x \mapsto [1,1]\} = \{x \mapsto [1,2]\}.$$

6.2 Performance Evaluation

To explore whether a suitable witness can reduce verification effort, we consider larger programs, as runtimes for the literature examples are negligible. Since GOBLINT specializes in the analysis of multi-threaded programs, we examine a set of multi-threaded POSIX programs previously used to evaluate GOB-LINT [53,54]. We manually construct witnesses that contain core invariants for these programs, based on how widenings were applied during fixpoint solving. We configure GOBLINT as described earlier, but with relational analysis disabled. In addition to CPU time, we measure analysis effort without a witness and with witness-guidance via transfer function evaluation counts. This metric of evaluations is proportional to CPU time, but excludes irrelevant pre- and post-processing, and is independent of hardware.

The results, aggregated in Table 2, show a noticeable performance improvement in the abstract interpreter when guided by a witness. However, the fixpoint-solving process still requires numerous widening iterations. This is due to various abstractions used by GOBLINT that cannot be expressed as C expressions, including but not limited to array index ranges in abstract addresses and various concurrency aspects. Nevertheless, the average 1.23× CPU time speedup is relatively close to the average 1.63× improvement achieved by Albert et al. [2] when using analyzer-specific certificates (see Sect. 7).

Admittedly, we have used a limited set of benchmarks and hand-crafted witnesses because our automatically generated witnesses produce excessive information. Large witnesses that express full proofs with numerous invariants can be problematic for a validator [1,15], which must manipulate, use, and/or verify them. In our case, the validator performs an unassume operation for each invariant each time the corresponding transfer function is evaluated. The speedup

Table 2. Evaluation results on GOBLINT benchmarks. The LLoC column counts logical
lines of code, i.e., only lines with executable code, excluding declarations.

Program	LLoC	w/o witness Evals	CPU time (s)	w/ witness Evals	CPU time (s)	Reduction Evals	CPU time
pfscan	559	4194	.86	2919	.73	30.4%	15.4%
aget	587	7932	2.23	4683	1.68	41.0%	24.7%
knot	981	29588	4.92	21432	4.54	27.6%	7.7%
smtprc	3037	48559	15.00	24091	7.95	50.4%	47.0%
Average						37.3%	23.7%

gained from using the witness must outweigh the overhead to truly benefit from
witnesses. This amounts to witnesses containing partial proofs, like loop invariants [15]. Moreover, our approach does not take advantage of exact invariants,
such as equalities outside of disjunctions, since these do not relax a reached
state already representing such exact values. Even if such invariants are useful
for some validators, they do not benefit our witness-guided verifier. Therefore,
the challenge remains for us, in collaboration with other tool developers, to
develop methods for generating suitable witnesses.

7 Related Work

Fixpoint iterations involving widening and narrowing are well-studied [3,4,28,29,
55], but focus mostly on improving precision and ensuring termination. Halbwachs and Henry [37] extend fixpoint iteration with partial restarting, which
derives from the narrowing result a new initial value for the following widening
iteration, hoping it improves the result. Their restarted value is analogous to
our partial initialization and could be used as such. They focus on finding such
values automatically, while we focus on using them to avoid all the computation leading up to it. Boutonnet and Halbwachs [26] improve the technique for
finding good restarting candidates. Cousot [28] extends fixpoint iteration with
dual-narrowing, hoping it improves the result further. Both approaches focus
purely on improving precision with more iterations, while we aim to skip that
iteration and arrive at the same result quicker, knowing the invariant. Hence,
the techniques can be combined: use theirs to find a precise invariant and use
ours to directly reuse it.

Arceri et al. [6] swap the abstract domain for a more precise one when switching from widening to narrowing. This can be considered a precision improvement
technique, which makes the narrowing phase more expensive. However, it can
also be viewed as an optimization, which makes the widening phase cheaper.
Either way, it can be combined with our approach: immediately using the more
precise domain with the final invariant, ideally skipping iteration in both ways.

Widening operators themselves are also well-studied [8,25,35,44,47]. Widening *up to* or *with threshold* use candidate invariants as intermediates to avoid

irrecoverably losing precision. Such automated techniques can identify which candidates are true invariants. Using these as input to our approach is an effective way of supplying known good thresholds at specific program points, removing the need for retrying all the candidates on re-verification. This is one instance of dealing with the inherent non-monotonicity of widening operators [28]. Furthermore, widening thresholds require domain-specific implementation, whereas our approach is more generic.

Our naïve unassume with its havoc and assume bears some similarity to the generation of *verification conditions* from user-supplied loop invariants [34]. However, there one additionally refines the state with the loop condition and then checks that the loop invariant is preserved. We avoid the former to remain sound by construction, but effectively do the latter when validating witness invariants.

Albert et al. [2] introduce *Abstraction-Carrying Code* (ACC) as an abstract-interpretation–based instance of Proof-Carrying Code (PCC). For validation they use a simplified analyzer which only performs a single pass of abstract interpretation and no fixpoint iteration. Thus, it requires certificates to supply invariants for all loops. This is more restrictive than our validation approach, which runs in a single pass if all necessary invariants are provided, but also allows some fixpoint iteration if this is not the case. Nevertheless, we could handicap our analyzer with this stronger restriction.

Albert et al. [1] develop a notion of reduced certificates which can be smaller and are used by the validator to reconstruct full certificates. Besson et al. [12] propose a fixpoint compression algorithm to further compact the certificates. In follow-up work, Besson et al. [13] develop a theory for studying the issue of certificate size. Rather than using the strongest information from the least fixpoint of an analysis, they seek the weakest information still sufficient for implying correctness. This omits irrelevant information, leading to smaller witnesses. Such techniques could also be used when generating witnesses for our validator.

8 Conclusion

We have demonstrated how to turn abstract-interpretation–based tools into witness-guided verifiers and witness validators, by equipping them with unassume operations. These can be constructed from abstract transformers for assumes, non-deterministic assignments, joins and dual-narrowings, which allow retaining more precision for relational abstract interpretation. A powerful syntax-directed unassume operation for non-relational domains can be derived from a classical algorithm with minimal changes. Our implementation and evaluation demonstrate that unassuming invariants from witnesses can both speed up the analysis and make it more precise. The experiments further show that the abstract interpreter can benefit from witnesses produced by model checkers, and thus indicate that the approach is suited even for cross-technology collaboration.

Acknowledgements. This work was supported by Deutsche Forschungsgemeinschaft (DFG) - 378803395/2428 CᴏɴVᴇY and Shota Rustaveli National Science Foundation of Georgia under the project FR-21-7973.

Data Availability Statement. The benchmarks, tools and scripts used, as well as the raw results of the evaluation, are openly archived on Zenodo [52].

References

1. Albert, E., Arenas, P., Puebla, G., Hermenegildo, M.: Reduced certificates for abstraction-carrying code. In: Etalle, S., Truszczyński, M. (eds.) ICLP 2006. LNCS, vol. 4079, pp. 163–178. Springer, Heidelberg (2006). https://doi.org/10.1007/11799573_14
2. Albert, E., Puebla, G., Hermenegildo, M.: Abstraction-carrying code. In: Baader, F., Voronkov, A. (eds.) LPAR 2005. LNCS (LNAI), vol. 3452, pp. 380–397. Springer, Heidelberg (2005). https://doi.org/10.1007/978-3-540-32275-7_25
3. Amato, G., Scozzari, F.: Localizing widening and narrowing. In: Logozzo, F., Fähndrich, M. (eds.) SAS 2013. LNCS, vol. 7935, pp. 25–42. Springer, Heidelberg (2013). https://doi.org/10.1007/978-3-642-38856-9_4
4. Amato, G., Scozzari, F., Seidl, H., Apinis, K., Vojdani, V.: Efficiently intertwining widening and narrowing. Sci. Comput. Program. **120**, 1–24 (2016). https://doi.org/10.1016/j.scico.2015.12.005
5. Apinis, K., Seidl, H., Vojdani, V.: Side-effecting constraint systems: a swiss army knife for program analysis. In: Jhala, R., Igarashi, A. (eds.) APLAS 2012. LNCS, vol. 7705, pp. 157–172. Springer, Heidelberg (2012). https://doi.org/10.1007/978-3-642-35182-2_12
6. Arceri, V., Mastroeni, I., Zaffanella, E.: Decoupling the ascending and descending phases in abstract interpretation. In: Sergey, I. (ed.) APLAS 2022. LNCS, vol. 13658, pp. 25–44. Springer, Switzerland (2022). https://doi.org/10.1007/978-3-031-21037-2_2
7. Ayaziová, P., Chalupa, M., Strejček, J.: SYMBIOTIC-WITCH: a KLEE-based violation witness checker. In: TACAS 2022. LNCS, vol. 13244, pp. 468–473. Springer, Cham (2022). https://doi.org/10.1007/978-3-030-99527-0_33
8. Bagnara, R., Hill, P.M., Ricci, E., Zaffanella, E.: Precise widening operators for convex polyhedra. Sci. Comput. Program. **58**(1–2), 28–56 (2005). https://doi.org/10.1016/j.scico.2005.02.003
9. Baudin, P., et al.: The dogged pursuit of bug-free C programs: the Frama-C software analysis platform. Commun. ACM **64**(8), 56–68 (2021). https://doi.org/10.1145/3470569
10. Baudin, P., Cuoq, P., Filliâtre, J.C., Marché, C., Monate, B., Moy, Y., Prevosto, V.: ANSI/ISO C specification language version 1.19 (2023). http://frama-c.com/download/acsl.pdf
11. Benhamou, F., Goualard, F., Granvilliers, L., Puget, J.F.: Revising hull and box consistency. In: Logic Programming, pp. 230–244. The MIT Press (1999). https://doi.org/10.7551/mitpress/4304.003.0024
12. Besson, F., Jensen, T., Pichardie, D.: Proof-carrying code from certified abstract interpretation and fixpoint compression. Theor. Comput. Sci. **364**(3), 273–291 (2006). https://doi.org/10.1016/j.tcs.2006.08.012
13. Besson, F., Jensen, T., Turpin, T.: Small witnesses for abstract interpretation-based proofs. In: De Nicola, R. (ed.) ESOP 2007. LNCS, vol. 4421, pp. 268–283. Springer, Heidelberg (2007). https://doi.org/10.1007/978-3-540-71316-6_19
14. Beyer, D.: Competition on software verification and witness validation: SV-COMP 2023. In: Sankaranarayanan, S., Sharygina, N. (eds.) TACAS 2023. LNCS, vol.

13994, pp. 495–522. Springer, Switzerland (2023). https://doi.org/10.1007/978-3-031-30820-8_29

15. Beyer, D., Dangl, M., Dietsch, D., Heizmann, M.: Correctness witnesses: exchanging verification results between verifiers. In: Proceedings of the 2016 24th ACM SIGSOFT International Symposium on Foundations of Software Engineering, pp. 326–337. ACM (2016). https://doi.org/10.1145/2950290.2950351

16. Beyer, D., Dangl, M., Dietsch, D., Heizmann, M., Lemberger, T., Tautschnig, M.: Verification witnesses. ACM Trans. Softw. Eng. Methodol. 31(4), 1–69 (2022). https://doi.org/10.1145/3477579

17. Beyer, D., Dangl, M., Dietsch, D., Heizmann, M., Stahlbauer, A.: Witness validation and stepwise testification across software verifiers. In: Proceedings of the 2015 10th Joint Meeting on Foundations of Software Engineering, pp. 721–733, ACM (2015). https://doi.org/10.1145/2786805.2786867

18. Beyer, D., Dangl, M., Lemberger, T., Tautschnig, M.: Tests from witnesses. In: Dubois, C., Wolff, B. (eds.) TAP 2018. LNCS, vol. 10889, pp. 3–23. Springer, Cham (2018). https://doi.org/10.1007/978-3-319-92994-1_1

19. Beyer, D., Kanav, S.: CoVeriTeam: on-demand composition of cooperative verification systems. In: TACAS 2022. LNCS, vol. 13243, pp. 561–579. Springer, Cham (2022). https://doi.org/10.1007/978-3-030-99524-9_31

20. Beyer, D., Keremoglu, M.E.: CPACHECKER: a tool for configurable software verification. In: Gopalakrishnan, G., Qadeer, S. (eds.) CAV 2011. LNCS, vol. 6806, pp. 184–190. Springer, Heidelberg (2011). https://doi.org/10.1007/978-3-642-22110-1_16

21. Beyer, D., Löwe, S., Wendler, P.: Reliable benchmarking: requirements and solutions. Int. J. Softw. Tools Technol. Transf. 21(1), 1–29 (2017). https://doi.org/10.1007/s10009-017-0469-y

22. Beyer, D., Spiessl, M., Umbricht, S.: Cooperation between automatic and interactive software verifiers. In: Software Engineering and Formal Methods, pp. 111–128. Springer, Cham (2022), https://doi.org/10.1007/978-3-031-17108-6_7

23. Beyer, D., Strejček, J.: Case study on verification-witness validators: where we are and where we go. In: Singh, G., Urban, C. (eds.) SAS 2022. LNCS, vol. 13790, pp. 160–174. Springer, Cham (2022). https://doi.org/10.1007/978-3-031-22308-2_8

24. Beyer, D., Wehrheim, H.: Verification artifacts in cooperative verification: survey and unifying component framework. In: Margaria, T., Steffen, B. (eds.) ISoLA 2020. LNCS, vol. 12476, pp. 143–167. Springer, Cham (2020). https://doi.org/10.1007/978-3-030-61362-4_8

25. Blanchet, B., et al.: A static analyzer for large safety-critical software. ACM SIGPLAN Not. 38(5), 196–207 (2003). https://doi.org/10.1145/780822.781153

26. Boutonnet, R., Halbwachs, N.: Improving the results of program analysis by abstract interpretation beyond the decreasing sequence. Formal Methods Syst. Des. 53(3), 384–406 (2017). https://doi.org/10.1007/s10703-017-0310-y

27. Cousot, P.: The calculational design of a generic abstract interpreter. In: Calculational System Design, NATO ASI Series F. IOS Press, Amsterdam (1999). https://www.di.ens.fr/cousot/COUSOTpapers/publications.www/Cousot-Marktoberdorf98.pdf.gz

28. Cousot, P.: Abstracting induction by extrapolation and interpolation. In: D'Souza, D., Lal, A., Larsen, K.G. (eds.) VMCAI 2015. LNCS, vol. 8931, pp. 19–42. Springer, Heidelberg (2015). https://doi.org/10.1007/978-3-662-46081-8_2

29. Cousot, P., Cousot, R.: Abstract interpretation: a unified lattice model for static analysis of programs by construction or approximation of fixpoints. In: Proceedings

of the 4th ACM SIGACT-SIGPLAN symposium on Principles of programming languages, pp. 238–252. ACM Press (1977). https://doi.org/10.1145/512950.512973

30. Cousot, P., Cousot, R.: Abstract interpretation frameworks. J. Log. Comput. **2**(4), 511–547 (1992). https://doi.org/10.1093/logcom/2.4.511

31. Cousot, P., Cousot, R., Feret, J., Mauborgne, L., Miné, A., Rival, X.: Why does Astrée scale up? Formal Methods Syst. Des. **35**(3), 229–264 (2009). https://doi.org/10.1007/s10703-009-0089-6

32. Dangl, M., Löwe, S., Wendler, P.: CPACHECKER with support for recursive programs and floating-point arithmetic. In: Baier, C., Tinelli, C. (eds.) TACAS 2015. LNCS, vol. 9035, pp. 423–425. Springer, Heidelberg (2015). https://doi.org/10.1007/978-3-662-46681-0_34

33. Farzan, A., Kincaid, Z.: Verification of parameterized concurrent programs by modular reasoning about data and control. In: Proceedings of the 39th Annual ACM SIGPLAN-SIGACT Symposium on Principles of Programming Languages, pp. 297–308. ACM (2012). https://doi.org/10.1145/2103656.2103693

34. Flanagan, C., Saxe, J.B.: Avoiding exponential explosion: generating compact verification conditions. In: Proceedings of the 28th ACM SIGPLAN-SIGACT Symposium on Principles of Programming Languages, pp. 193–205. ACM (2001). https://doi.org/10.1145/360204.360220

35. Gopan, D., Reps, T.: Lookahead widening. In: Ball, T., Jones, R.B. (eds.) CAV 2006. LNCS, vol. 4144, pp. 452–466. Springer, Heidelberg (2006). https://doi.org/10.1007/11817963_41

36. Goualard, F., Granvilliers, L.: Controlled propagation in continuous numerical constraint networks. In: Proceedings of the 2005 ACM Symposium on Applied Computing. ACM (2005). https://doi.org/10.1145/1066677.1066765

37. Halbwachs, N., Henry, J.: When the decreasing sequence fails. In: Miné, A., Schmidt, D. (eds.) SAS 2012. LNCS, vol. 7460, pp. 198–213. Springer, Heidelberg (2012). https://doi.org/10.1007/978-3-642-33125-1_15

38. Haltermann, J., Wehrheim, H.: Information exchange between over- and underapproximating software analyses. In: Software Engineering and Formal Methods, pp. 37–54. Springer, Heidelberg (2022). https://doi.org/10.1007/978-3-031-17108-6_3

39. Heizmann, M., et al.: Ultimate Automizer and the CommuHash normal form. In: Sankaranarayanan, S., Sharygina, N. (eds.) TACAS 2023. LNCS, vol. 13994, pp. 577–581. Springer, Cham (2023). https://doi.org/10.1007/978-3-031-30820-8_39

40. Heizmann, M., Hoenicke, J., Podelski, A.: Software model checking for people who love automata. In: Sharygina, N., Veith, H. (eds.) CAV 2013. LNCS, vol. 8044, pp. 36–52. Springer, Heidelberg (2013). https://doi.org/10.1007/978-3-642-39799-8_2

41. Jeannet, B., Miné, A.: APRON: a library of numerical abstract domains for static analysis. In: Bouajjani, A., Maler, O. (eds.) CAV 2009. LNCS, vol. 5643, pp. 661–667. Springer, Heidelberg (2009). https://doi.org/10.1007/978-3-642-02658-4_52

42. Journault, M., Miné, A., Ouadjaout, A.: An abstract domain for trees with numeric relations. In: Caires, L. (ed.) ESOP 2019. LNCS, vol. 11423, pp. 724–751. Springer, Cham (2019). https://doi.org/10.1007/978-3-030-17184-1_26

43. Ponce-de-León, H., Haas, T., Meyer, R.: DARTAGNAN: SMT-based violation witness validation (competition contribution). In: TACAS 2022. LNCS, vol. 13244, pp. 418–423. Springer, Cham (2022). https://doi.org/10.1007/978-3-030-99527-0_24

44. Mihaila, B., Sepp, A., Simon, A.: Widening as abstract domain. In: Brat, G., Rungta, N., Venet, A. (eds.) NFM 2013. LNCS, vol. 7871, pp. 170–184. Springer, Heidelberg (2013). https://doi.org/10.1007/978-3-642-38088-4_12

45. Miné, A.: The octagon abstract domain. Higher-Order Symb. Comput. **19**(1), 31–100 (2006). https://doi.org/10.1007/s10990-006-8609-1

46. Miné, A.: Static analysis of run-time errors in embedded real-time parallel C programs. Logical Methods Comput. Sci. **8**(1), 1–63 (2012). https://doi.org/10.2168/lmcs-8(1:26)2012

47. Miné, A.: Tutorial on static inference of numeric invariants by abstract interpretation. Found. Trends® Program. Lang. **4**(3–4), 120–372 (2017). https://doi.org/10.1561/2500000034. https://hal.sorbonne-universite.fr/hal-01657536/document

48. Monat, R., Miné, A.: Precise thread-modular abstract interpretation of concurrent programs using relational interference abstractions. In: Bouajjani, A., Monniaux, D. (eds.) VMCAI 2017. LNCS, vol. 10145, pp. 386–404. Springer, Cham (2017). https://doi.org/10.1007/978-3-319-52234-0_21

49. Saan, S., et al.: GOBLINT: thread-modular abstract interpretation using side-effecting constraints. In: TACAS 2021. LNCS, vol. 12652, pp. 438–442. Springer, Cham (2021). https://doi.org/10.1007/978-3-030-72013-1_28

50. Saan, S., et al.: GOBLINT: autotuning thread-modular abstract interpretation. In: Sankaranarayanan, S., Sharygina, N. (eds.) TACAS 2023. LNCS, vol. 13994, pp. 547–552. Springer, Cham (2023). https://doi.org/10.1007/978-3-031-30820-8_34

51. Saan, S., Schwarz, M., Erhard, J., Seidl, H., Tilscher, S., Vojdani, V.: Correctness witness validation by abstract interpretation (2023). https://doi.org/10.48550/arXiv.2310.16572

52. Saan, S., Schwarz, M., Erhard, J., Seidl, H., Tilscher, S., Vojdani, V.: Correctness witness validation by abstract interpretation (2023). https://doi.org/10.5281/zenodo.8253000, artifact

53. Schwarz, M., Saan, S., Seidl, H., Apinis, K., Erhard, J., Vojdani, V.: Improving thread-modular abstract interpretation. In: Drăgoi, C., Mukherjee, S., Namjoshi, K. (eds.) SAS 2021. LNCS, vol. 12913, pp. 359–383. Springer, Cham (2021). https://doi.org/10.1007/978-3-030-88806-0_18

54. Schwarz, M., Saan, S., Seidl, H., Erhard, J., Vojdani, V.: Clustered relational thread-modular abstract interpretation with local traces. In: ESOP 2023. LNCS, vol. 13990, pp. 28–58. Springer, Cham (2023). https://doi.org/10.1007/978-3-031-30044-8_2

55. Seidl, H., Vogler, R.: Three improvements to the top-down solver. Math. Struct. Comput. Sci. **31**(9), 1090–1134 (2021). https://doi.org/10.1017/s0960129521000499

56. SoSy-Lab: YAML-based exchange format for correctness witnesses (2021). https://gitlab.com/sosy-lab/benchmarking/sv-witnesses/-/blob/main/README-YAML.md

57. Strejček, J.: Issues related to the fact that the semantics of witnesses are defined over CFAs and the translation from C programs to CFAs is undefined (2022). https://gitlab.com/sosy-lab/benchmarking/sv-witnesses/-/blob/main/GraphML_witness_format_issues.pdf

58. SV-COMP community: Community meeting (2023)

59. Švejda, J., Berger, P., Katoen, J.-P.: Interpretation-based violation witness validation for C: NITWIT. In: TACAS 2020. LNCS, vol. 12078, pp. 40–57. Springer, Cham (2020). https://doi.org/10.1007/978-3-030-45190-5_3

60. Vojdani, V., Apinis, K., Rõtov, V., Seidl, H., Vene, V., Vogler, R.: Static race detection for device drivers: the Goblint approach. In: Proceedings of the 31st IEEE/ACM International Conference on Automated Software Engineering. ACM (2016). https://doi.org/10.1145/2970276.2970337

61. Ziat, G.: A combination of abstract interpretation and constraint programming. Theses, Sorbonne Université (2019). https://theses.hal.science/tel-03987752

Infinite-State Systems

Project and Conquer: Fast Quantifier Elimination for Checking Petri Net Reachability

Nicolas Amat$^{(\boxtimes)}$, Silvano Dal Zilio ,
and Didier Le Botlan

LAAS-CNRS, Université de Toulouse,
CNRS, INSA, Toulouse, France
nicolas.amat@laas.fr

Abstract. We propose a method for checking generalized reachability properties in Petri nets that takes advantage of structural reductions and that can be used, transparently, as a pre-processing step of existing model-checkers. Our approach is based on a new procedure that can project a property, about an initial Petri net, into an equivalent formula that only refers to the reduced version of this net. Our projection is defined as a variable elimination procedure for linear integer arithmetic tailored to the specific kind of constraints we handle. It has linear complexity, is guaranteed to return a sound property, and makes use of a simple condition to detect when the result is exact. Experimental results show that our approach works well in practice and that it can be useful even when there is only a limited amount of reductions.

Keywords: Petri nets · Quantifier elimination · Reachability problems

1 Introduction

We describe a method to accelerate the verification of reachability properties in Petri nets by taking advantage of structural reductions [13]. We focus on the verification of generalized properties, that can be expressed using a Boolean combination of linear constraints between places, such as $(2\,p_0 + p_1 = 5) \wedge (p_1 \geqslant p_2)$ for example. This class of formulas corresponds to the reachability queries used in the Model Checking Contest (MCC) [9], a competition of Petri net verification tools that we use as a benchmark.

In essence, net reductions are a class of transformations that can simplify an initial net, (N_1, m_1), into another, residual net (N_2, m_2), while preserving a given class of properties. This technique has become a conventional optimization integrated into several model-checking tools [15,17,44]. A contribution of our paper is a procedure to transform a property F_1, about the net N_1, into a property F_2 about the reduced net N_2, while preserving the verdict. We have implemented this procedure into a new tool, called Octant [1], that can act as a pre-processor allowing any model-checker to transparently benefit from our

© The Author(s), under exclusive license to Springer Nature Switzerland AG 2024
R. Dimitrova et al. (Eds.): VMCAI 2024, LNCS 14499, pp. 101–123, 2024.
https://doi.org/10.1007/978-3-031-50524-9_5

optimization. Something that was not possible in our previous works. In practice, it means that we can use our approach as a front-end to accelerate any model-checking tool that supports generalized reachability properties, without modifying them.

Our approach relies on a notion, called *polyhedral reduction* [2–4,16], that describes a linear dependence relation, E, between the reachable markings of a net and those of its reduced version. This equivalence, denoted $(N_1, m_1) \equiv_E (N_2, m_2)$, preserves enough information in E so that we can rebuild the state space of N_1 knowing only the one of N_2. An interesting application of this relation is the following *reachability conservation theorem* [2]: assume we have $(N_1, m_1) \equiv_E (N_2, m_2)$, then property F is reachable in N_1 if and only if $E \wedge F$ is reachable in N_2. This property is interesting since it means that we can apply more aggressive reduction techniques than, say, *slicing* [31,37,41], *cone of influence* [18], or other methods [24,30] that seek to remove or gather together places that are not relevant to the property we want to check. We do not share this restriction in our approach, since we reduce nets beforehand and can therefore reduce places that occur in the initial property. We could argue that approaches similar to slicing only simplify a model with respect to a formula, whereas, with our method, we simplify the model as much as possible and then simplify formulas as needed. This is more efficient when we need to check several properties on the same model and, in any case, nothing prevents us from applying slicing techniques on the result of our projection.

However, there is a complication, arising from the fact that the formula $E \wedge F$ may include variables (places) that no longer occur in the reduced net N_2, and therefore act as existentially quantified variables. This can complicate some symbolic verification techniques, such as k-induction [42], and impede the use of explicit, enumerative approaches. Indeed, in the latter case, it means that we need to solve an integer linear problem for each new state, instead of just evaluating a closed formula. To overcome this problem, we propose a new method for projecting the formula $E \wedge F$ into an equivalent one, F', that only refers to the places of N_2. We define our projection as a procedure for quantifier elimination in Presburger Arithmetic (PA) that is tailored to the specific kind of constraints we handle in E. Whereas quantifier elimination has an exponential complexity in general for existential formulas, our construction has linear complexity and can only decrease the size of a formula. It also always terminates and returns a result that is guaranteed to be sound; meaning it under-approximates the set of reachable models and, therefore, a witness of F' in N_2 necessarily corresponds to a witness of F in N_1. Additionally, our approach includes a simple condition on F that is enough to detect when our result is exact, meaning that if F' is unreachable in N_2, then F is unreachable in N_1. We show in Sect. 6 that our projection is complete for 80% of the formulas used in the MCC.

Outline and Contributions. We start by giving some technical background about Petri nets and the notion of polyhedral abstraction in Sect. 2, then describe how to use this equivalence to accelerate the verification of reachability properties (Theorem 1 in Sect. 3). We also use this section to motivate our need to find methods to eliminate (or project) variables in a linear integer system. We define our fast projection method in Sect. 4, which is based on a dedicated graph structure, called Token Flow Graph (TFG), capturing the particular form of constraints occurring with polyhedral reductions. We prove the correctness of this method in Sect. 5. Our method has been implemented, and we report on the results of our experiments in Sect. 6. We give quantitative evidence about several natural questions raised by our approach. We start by proving the effectiveness of our optimization on both k-induction and random walk. Then, we show that our method can be transparently added as a preprocessing step to off-the-shelf verification tools. This is achieved by testing our approach with the three best-performing tools that participated in the reachability category of the MCC—ITS-Tools [43] (or ITS for short); LoLA [46]; and TAPAAL [21]—which are already optimized for the type of models and formulas used in our benchmark. Our results show that reductions are effective on a large set of queries and that their benefits do not overlap with other existing optimizations, an observation that was already made in [3,17]. We also prove that our procedure often computes an exact projection and compares favorably well with the variable elimination methods implemented in isl [45] and Redlog [22]. This supports our claim that we are able to solve non-trivial quantifier elimination problems.

2   Petri Nets and Polyhedral Abstraction

Most of our results involve non-negative integer solutions to constraints expressed in Presburger Arithmetic, the first-order theory of the integers with addition [25]. We focus on the quantifier-free fragment of PA, meaning Boolean combinations (using $\wedge$, $\vee$ and $\neg$) of atomic propositions of the form $\alpha \sim \beta$, where $\sim$ is one of $=, \leqslant$ or $\geqslant$, and α, β are linear expressions with coefficients in $\mathbb{Z}$. Without loss of generality, we can consider only formulas in disjunctive normal form (DNF), with *linear predicates* of the form $(\sum k_i\, x_i) + b \geqslant 0$. We deliberately do not add a divisibility operator $k \mid \alpha$, which requires that k evenly divides α, since it can already be expressed with linear predicates, though at the cost of an extra existentially quantified variable. This fragment corresponds to the set of reachability formulas supported by many model-checkers for Petri nets, such as [10,14,21,43,46].

We use $\mathbb{N}^V$ to denote the space of mappings over $V = \{x_1, \ldots, x_n\}$, meaning total mappings from V to $\mathbb{N}$. We say that a mapping m in $\mathbb{N}^V$ is a *model* of a quantifier-free formula F if the variables of F, denoted $\mathsf{fv}(F)$, are included in V and the closed formula $F\{m\}$ (the substitution of m in F) is true. We denote this relation $m \models F$.

$$F\{m\} \triangleq F\{x_1 \leftarrow m(x_1)\} \ldots \{x_n \leftarrow m(x_n)\} \tag{1}$$

We say that a Presburger formula is *consistent* when it has at least one model. We can use this notion to extend the definition of models in the case where F is over-specified; i.e. it has a larger support than m. If $m' \in \mathbb{N}^U$ with $U \subseteq \mathsf{fv}(F)$, we write $m' \models F$ when $F\{m'\}$ is consistent.

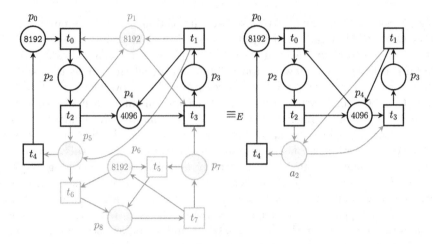

Fig. 1. An example of Petri net, M_1 (left), and one of its polyhedral abstractions, M_2 (right), with $E \triangleq (p_1 = p_4 + 4096) \wedge (p_6 = p_0 + p_2 + p_3 + p_5 + p_7) \wedge (a_1 = p_7 + p_8) \wedge (a_2 = a_1 + p_5)$. Colors are used to emphasize places that are either removed or added.

Petri Nets and Reachability Formulas. A *Petri net* N is a tuple $(P, T, \text{Pre}, \text{Post})$ where $P = \{p_1, \ldots, p_n\}$ is an ordered set of places, $T = \{t_1, \ldots, t_k\}$ is a finite set of transitions (disjoint from P), and $\text{Pre} : T \to \mathbb{N}^P$ and $\text{Post} : T \to \mathbb{N}^P$ are the pre- and post-condition functions (also called the flow functions of N). A state m of a net, also called a *marking*, is a mapping of $\mathbb{N}^P$. A marked net (N, m_0) is a pair composed of a net and its initial marking m_0.

We extend the comparison $(=, \geqslant)$ and arithmetic operations $(-, +)$ to their point-wise equivalent. With our notations, a transition $t \in T$ is said *enabled* at marking m when $m \geqslant \text{Pre}(t)$. A marking m' is reachable from a marking m by firing transition t, denoted $m \xrightarrow{t} m'$, if: (1) transition t is enabled at m; and (2) $m' = m - \text{Pre}(t) + \text{Post}(t)$. When the identity of the transition is unimportant, we simply write this relation $m \to m'$. More generally, a marking m' is reachable from m in N, denoted $m \to^* m'$ if there is a (possibly empty) sequence of transitions such that $m \to \ldots \to m'$. We denote $R(N, m_0)$ the set of markings reachable from m_0 in N.

We are interested in the verification of properties over the reachable markings of a marked net (N, m_0), with a set of places P. Given a formula F with variables in P, we say that F is reachable if there exists at least one reachable marking, $m \in R(N, m_0)$, such that $m \models F$. We call such marking a *witness* of F. Likewise, F is said to be an *invariant* when all the reachable markings of (N, m_0) are models of F. This corresponds to the two classes of queries found in our benchmark: EF F, which is true only if F is reachable; and AG F, which is true when F is an invariant, with the classic relationship that AG $F \equiv \neg (\text{EF} \neg F)$. Examples of properties we can express in this way include: checking if some transition can possibly be enabled, checking if there is a deadlock, checking whether some linear invariant between places is always true, etc.

We use a standard graphical notation for nets where places are depicted with circles and transitions with squares. We give an example in Fig. 1, where net M_1 depicts the SmallOperatingSystem model, borrowed from the MCC benchmark [33]. This net abstracts the lifecycle of a task in a simplified operating system handling several memory segments (place p_0), disk controller units (p_4), and cores (p_6). The initial marking of the net gives the number of resources available (e.g., there are 8 192 available memory segments in our example).

We chose this model since it is one of the few examples in our benchmark that fits on one page. This is not to say that the example is simple. Net M_1 has about 10^{17} reachable states, which means that it is out of reach of enumerative methods, and only one symbolic tool in the MCC is able to generate its whole state space[1] [32]. For comparison, the reduced net M_2 has about 10^{10} states.

Polyhedral Abstraction. We recently defined an equivalence relation that describes linear dependencies between the markings of two different nets, N_1 and N_2 [3]. In the following, we reserve F for formulas about a single net and use E to refer to relations. Assume m is a mapping of $\mathbb{N}^V$. We can associate m to the linear predicate $\underline{m}$, which is a formula with a unique model m.

$$\underline{m} \triangleq \bigwedge\{x = m(x) \mid x \in V\} \tag{2}$$

By extension, we say that m is a (partial) solution of E if the system $E \wedge \underline{m}$ is consistent. In some sense, we use $\underline{m}$ as a substitution, since the formulas $E\{m\}$ and $E \wedge \underline{m}$ have the same models. Given two mappings $m_1 \in \mathbb{N}^{V_1}$ and $m_2 \in \mathbb{N}^{V_2}$, we say that m_1 and m_2 are *compatible* when they have equal values on their shared domain: $m_1(x) = m_2(x)$ for all x in $V_1 \cap V_2$. This is a necessary and sufficient condition for the system $\underline{m_1} \wedge \underline{m_2}$ to be consistent. Finally, we say that m_1 and m_2 are related up-to E, denoted $m_1 \equiv_E m_2$, when $E \wedge \underline{m_1} \wedge \underline{m_2}$ is consistent.

$$m_1 \equiv_E m_2 \quad \Leftrightarrow \quad \exists m \in \mathbb{N}^V . \, m \models E \wedge \underline{m_1} \wedge \underline{m_2} \tag{3}$$

This relation defines an equivalence between markings of two different nets ($\equiv_E \,\subseteq\, \mathbb{N}^{P_1} \times \mathbb{N}^{P_2}$) and, by extension, can be used to define an equivalence between nets themselves, that we call *polyhedral equivalence*.

Definition 1 (E-equivalence). *We say that (N_1, m_1) is E-equivalent to (N_2, m_2), denoted $(N_1, m_1) \equiv_E (N_2, m_2)$, if and only if:*

(A1) $E \wedge \underline{m}$ *is consistent for all markings m in $R(N_1, m_1)$ or $R(N_2, m_2)$;*
(A2) *initial markings are compatible: $m_1 \equiv_E m_2$;*
(A3) *assume m_1', m_2' are markings of N_1, N_2, such that $m_1' \equiv_E m_2'$, then m_1' is reachable iff m_2' is reachable: $m_1' \in R(N_1, m_1) \iff m_2' \in R(N_2, m_2)$.*

By definition, given the equivalence $(N_1, m_1) \equiv_E (N_2, m_2)$, every marking m_2' reachable in N_2 can be associated to a subset of markings in N_1, defined from

---

[1] It is **Tedd**, part of the Tina toolbox, which also uses polyhedral reductions.

the solutions to $E \wedge m_2'$ (by condition (A1) and (A3)). In practice, this gives a partition of the reachable markings of (N_1, m_1) into "convex sets"—hence the name polyhedral abstraction—each associated with a reachable marking in N_2. This approach is particularly useful when the state space of N_2 is very small compared to the one of N_1.

We can prove that the two marked nets in our running example satisfy $M_1 \equiv_E M_2$, for the relation E defined in Fig. 1. Net M_2 is obtained automatically from M_1 by applying a set of reduction rules, iteratively, and in a compositional way. This process relies on the reduction system defined in [3, 16]. As a result, we manage to remove five places: p_1, p_5, p_6, p_7, p_8, and only add a new one, a_2. The "reduction system" (E) also contains an extra variable, a_1, that does not occur in any of the nets. It corresponds to a place that was introduced and then removed in different reduction steps.

Polyhedral abstractions are not necessarily derived from reductions, but reductions provide a way to automatically find interesting instances of abstractions. Also, the equation systems obtained using structural reductions exhibit a specific structure, that we exploit in Sect. 4.

3   Combining Polyhedral Abstraction with Reachability

We can define a counterpart to our notion of polyhedral abstraction which relates to reachability formulas. We show that this equivalence can be used to speed up the verification of properties by checking formulas on a reduced net instead of the initial one (see Theorem 1 and its corollary, below). In the following, we assume that we have two marked nets such that $(N_1, m_1) \equiv_E (N_2, m_2)$. Our goal is to define a relation $F_1 \equiv_E F_2$, between reachability formulas, such that F_1 and F_2 have the same truth values on equivalent models, with respect to E.

Definition 2 (Equivalence between formulas). *Assume F_1, F_2 are reachable formulas with respective sets of variables, V_1 and V_2, in the support of E. We say that formula F_2 implies F_1 up-to E, denoted $F_2 \sqsubseteq_E F_1$, if for every marking $m_2' \in \mathbb{N}^{V_2}$ such that $m_2' \models E \wedge F_2$ there exists at least one marking $m_1' \in \mathbb{N}^{V_1}$ such that $m_1' \equiv_E m_2'$ and $m_1' \models E \wedge F_1$.*

$$F_2 \sqsubseteq_E F_1 \quad iff \quad \forall m_2'. (m_2' \models E \wedge F_2) \Rightarrow \exists m_1'. (m_1' \equiv_E m_2' \wedge m_1' \models E \wedge F_1) \quad (4)$$

We say that F_1 and F_2 are equivalent, denoted $F_1 \equiv_E F_2$, when both $F_1 \sqsubseteq_E F_2$ and $F_2 \sqsubseteq_E F_1$.

This notion is interesting when F_1, F_2 are reachability formulas on the nets N_1, respectively N_2. Indeed, we prove that when $F_2 \sqsubseteq_E F_1$, it is enough to find a witness of F_2 in N_2 to prove that F_1 is reachable in N_1.

Theorem 1 (Finding Witnesses). *Assume $(N_1, m_1) \equiv_E (N_2, m_2)$ and $F_2 \sqsubseteq_E F_1$, and take a marking m_2' reachable in (N_2, m_2) such that $m_2' \models F_2$. Then there exists $m_1' \in R(N_1, m_1)$ such that $m_1' \equiv_E m_2'$ and $m_1' \models F_1$.*

Proof. Assume we have m_2' reachable in N_2 such that $m_2' \models F_2$. By property (A1) of E-equivalence (Definition 1), formula $E \wedge m_2'$ is consistent, which gives $m_2' \models E \wedge F_2$. By definition of the E-implication $\overline{F_2} \sqsubseteq_E F_1$, we get a marking m_1' such that $m_1' \models F_1$ and $m_1' \equiv_E m_2'$. We conclude that m_1' is reachable in N_1 thanks to property (A3). □

Hence, when $F_2 \sqsubseteq_E F_1$ holds, F_2 reachable in N_2 implies that F_1 is reachable in N_1. We can derive stronger results when F_1 and F_2 are equivalent.

Corollary 1. *Assume* $(N_1, m_1) \equiv_E (N_2, m_2)$ *and* $F_1 \equiv_E F_2$, *with* $\mathsf{fv}(F_i) \subseteq P_i$ *for all* $i \in 1..2$, *then:* (CEX) *property* F_1 *is reachable in* N_1 *if and only if* F_2 *is reachable in* N_2 *; and* (INV) F_1 *is an invariant on* N_1 *if and only if* F_2 *is an invariant on* N_2.

Theorem 1 means that we can check the reachability (or invariance) of a formula on the net N_1 by checking instead the reachability of another formula (F_2) on N_2. But it does not indicate how to compute a good candidate for F_2. By Definition 2, a natural choice is to select $F_2 \triangleq E \wedge F_1$. We can actually do a bit better. It is enough to choose a formula F_2 that has the same (integer points) solution as $E \wedge F_1$ over the places of N_2. More formally, let $A \triangleq \mathsf{fv}(E) \setminus P_2$ be the set of "additional variables" from E; variables occurring in E which are not places of the reduced net N_2. Then if F_2 has the same integer solutions over $\mathbb{N}^{P_2}$ than the Presburger formula $\exists A. (E \wedge F_1)$, we have $F_1 \equiv_E F_2$. We say in this case that F_2 is the projection of $E \wedge F_1$ on the set P_2, by eliminating the variables in A.

In the next section, we show how to compute a candidate projection formula without resorting to a classical, complete variable elimination procedure on $E \wedge F_1$. This eliminates a potential source of complexity blow-up.

We can use Fourier-Motzkin elimination (FM) as a point of reference. Given a system of linear inequalities S, with variables in V, we denote $\mathrm{FM}_A(S)$ the system obtained by FM elimination of variables in A from S. (We do not describe the construction of $\mathrm{FM}_A(S)$ here, since there exists many good references [28, 39] on the subject.) Borrowing an intuition popularized by Pugh in its Omega test [40], we can define two distinct notions of "shadows" cast by the projection of S. On the one hand, we have the *real shadow*, relative to A, which are the integer points (in $\mathbb{N}^{V \setminus A}$) solutions of $\mathrm{FM}_A(S)$. On the other hand, the *integer shadow* of S is the set of markings m' with an integer point antecedent in S. We need the latter to check a query on N_1. A main source of complexity is that the (real) shadow is only exact on rational points and may contain strictly more models than the integer shadow. Moreover, while the real shadow of a convex region will always be convex, it may not be the case with the integer shadow. Like with the real shadow, the set of equations computed with our fast projection will always be convex. Unlike FM, our procedure will compute an under-approximation of the integer shadow, not an over-approximation. Also, we never rearrange or create more inequalities than the one contained in S; but instead rely on variable substitution.

We illustrate the concepts introduced in this section on our running example, with the reduction system from Fig. 1. With our notations, we try to eliminate variables in $A \triangleq \{a_1, p_1, p_5, p_6, p_7, p_8\}$ and keep only those in $P_2 \triangleq \{a_2, p_0, p_2, p_3, p_4\}$.

Take the formula $G_1 \triangleq (p_5 + p_6 \leqslant p_8)$. Using substitutions from constraints in E, namely the fact that $(a_2 = p_7 + p_8 + p_5)$ and $(p_6 = p_0 + p_2 + p_3 + p_5 + p_7)$, we can remove occurrences of a_1, p_1, p_6, p_8 from $E \wedge G_1$, leaving the resulting equation $(3\, p_5 + 2\, p_7 + p_0 + p_2 + p_3 \leqslant a_2) \wedge (a_2 = p_5 + p_7 + p_8)$, that still refers to p_5 and p_7. We observe that non-trivial coefficients (like $3\, p_5$) can naturally occur during this process, even though all the coefficients are either 1 or -1 in the initial constraints. We can remove the remaining variables to obtain an exact projection of G_1 using our fast projection method, described below. The result is the formula $G_2 \triangleq (p_0 + p_2 + p_3 \leqslant a_2)$.

Another example is $H_1 \triangleq (p_6 = p_8)$. We can prove that the integer shadow of $E \wedge H_1$, after projecting the variables in A, are the solutions to the PA formula $(a_2 - p_0 - p_2 - p_3 \equiv 0 \bmod 2) \wedge (a_2 \geqslant p_0 + p_2 + p_3)$. This set is not convex, since $(a_2 = 0 \wedge p_0 = p_1 = p_2 = p_3 = 0)$ and $(a_2 = 2 \wedge p_0 = p_1 = p_2 = p_3 = 0)$ are in the integer shadow, but not $(a_2 = 1 \wedge p_0 = p_1 = p_2 = p_3 = 0)$ for instance. Our fast projection method will compute the formula $H_2 \triangleq (a_2 + p_0 + p_2 + p_3 = 0)$ and flag it as an under-approximation.

4   Projecting Formulas Using Token Flow Graphs

We describe a formula projection procedure that is tailored to the specific kind of constraints occurring in polyhedral reductions. The example in Fig. 1 is representative of the "shape" of reduction systems: it mostly contains equalities of the form $x = \sum x_i$, over a sparse set of variables, but may also include some inequalities; and it can have a very large number of literals (often proportional to the size of the initial net). Another interesting feature is the absence of cyclic dependencies, which underlines a hierarchical relationship between variables.

We can find a more precise and formal description of these constraints in [4], with the definition of a *Token Flow Graph* (TFG). Basically, a TFG for a reduction system E is a directed acyclic graph (DAG) with one vertex for each variable occurring in E. We consider two kinds of arcs, redundancy ($\rightarrow\!\bullet$) and agglomeration ($\circ\!\rightarrow$), that correspond to two main classes of reduction rules.

Arcs for *redundancy equations*, $q \rightarrow\!\bullet\, p$, correspond to equations of the form $p = q + r + \dots$, expressing that the marking of place p can be reconstructed from the marking of $q, r, \dots$ In this case, we say that place p is *removed* by arc $q \rightarrow\!\bullet\, p$, because the marking of q may influence the marking of p, but not necessarily the other way round.

Arcs for *agglomeration equations*, $a \circ\!\rightarrow p$, represent equations of the form $a = p + q + \dots$, generated when we agglomerate several places into a new one. In this case, we expect that if we can reach a marking with k tokens in a, then we can certainly reach a marking with k_1 tokens in p, and k_2 tokens in q, $\dots$ such that $k = k_1 + k_2 + \dots$. Hence, the possible markings of p and q can be reconstructed from the markings of a. In this case, it is $p, q, \dots$ which are removed. We also say that node a is *inserted*; it does not exist in N_1 but may

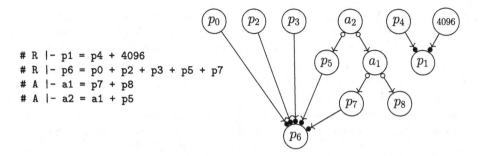

```
# R |- p1 = p4 + 4096
# R |- p6 = p0 + p2 + p3 + p5 + p7
# A |- a1 = p7 + p8
# A |- a2 = a1 + p5
```

Fig. 2. Equations from our example in Fig. 1 and the associated TFG.

appear as a new place in N_2 unless it is removed by a subsequent reduction. We can have more than two places in an agglomeration, see the rules in [16].

A TFG can also include nodes for *constants*, used to express invariant statements on the markings of the form $p + q = k$. To this end, we assume that we have a family of disjoint sets $K(n)$ for each n in $\mathbb{N}$, such that the "valuation" of a node $v \in K(n)$ is always n. We use K to denote the set of all constants.

Definition 3 (Token Flow Graph). *A TFG with set of places P is a directed graph $(P, S, R^\bullet, A^\circ)$ such that:*

- *$V = P \cup S$ is a set of vertices (or nodes) with $S \subset K$ a finite set of constants,*
- *$R^\bullet \in V \times V$ is a set of redundancy arcs, $v \rightarrowtail v'$,*
- *$A^\circ \in V \times V$ is a set of agglomeration arcs, $v \circ\!\!\rightarrow v'$, disjoint from R.*

The main source of complexity in this approach arises from the need to manage interdependencies between A° and $R^\bullet$ arcs, that is situations where redundancies and agglomerations are combined. This is not something that can be easily achieved by looking only at the equations in E and thus motivates the need for a specific data structure.

We define several notations that will be useful in the following. We use the notation $v \rightarrow v'$ when we have $(v \rightarrowtail v')$ or $(v \circ\!\!\rightarrow v')$. We say that a node v is a *root* if it is not the target of an arc. A sequence of nodes $(v_1, \ldots, v_n)$ in V^n is a *path* if for all $1 \leqslant i < n$ we have $v_i \rightarrow v_{i+1}$. We use the notation $v \rightarrow^\star v'$ when there is a path from v to v' in the graph, or when $v = v'$. We write $v \circ\!\!\rightarrow X$ when X is the largest subset $\{v_1, \ldots, v_k\}$ of V such that $X \neq \emptyset$ and $v \circ\!\!\rightarrow v_i$ for all $i \in 1..k$. And similarly with reductions, $X \rightarrowtail v$. Finally, the notation $\downarrow v$ denotes the set of successors of v, that is: $\downarrow v \triangleq \{v' \in V \setminus \{v\} \mid v \rightarrow^\star v'\}$. We extend it to a set of variables X with $\downarrow X = \bigcup_{x \in X} \downarrow x$.

We display an example of TFG in Fig. 2 (right), which corresponds to the reduction equations generated on our running example, where annotations R and A indicate if an equation is a redundancy or an agglomeration. TFGs were initially defined in [4,5] to efficiently compute the set of concurrent places in a net, that is all pairs of places that can be marked simultaneously in some reachable marking. We reuse this concept here to project reachability formulas.

High Literal Factor. The projection procedure, described next, applies to *cubes* only, meaning a conjunction of literals $\bigwedge_{i\in 1..n} \alpha_i$. Given a formula F_1, assumed in DNF, we can apply the projection procedure to each of its cubes, separately. Then the projection of F_1 is the disjunction of the projected cubes. We assume from now on that F_1 is a cube formula.

We assume that every literal is in *normal form*, $\alpha_i \triangleq (\sum_{p_j\in B} k_j^i\, p_j) + b_i \geqslant 0$, where the k_j^i's and b_i are in $\mathbb{Z}$. In the following, we denote $\alpha_i(q)$ the coefficient associated with variable q in α_i. We also use $\max_X \alpha_i$ and $\min_X \alpha_i$ for the maximal (resp. minimal) coefficient associated with variables in $X \subseteq B$.

$$\alpha_i = \sum_{p\in B} \alpha_i(p)\, p + b_i \qquad \text{and} \qquad \max_X \alpha_i = \max\, \{\alpha_i(p) \mid p \in X\}$$

We define the *Highest Literal Factor* (HLF) of a set of variables X with respect to a set of normalized literals $(\alpha_i)_{i\in I}$. In the simplest case, the HLF of X with respect to a single literal, α, is the subset of variables in X with the highest coefficients in α. Then, the HLF of X with respect to a set of literals is the (possibly empty) intersection of the HLFs of X with respect to each literal. When non-empty, it means that at least one variable in X always has the highest coefficient, and we say then that the whole set X is *polarized* with respect to the literals (α_i).

$$\mathrm{HLF}_X(\alpha_i) = \{p \in X \mid \alpha_i(p) = \max_X \alpha_i\}$$

$$\mathrm{HLF}_X(\alpha_i)_{i\in I} = \bigcap_{i\in I} \mathrm{HLF}_X(\alpha_i)$$

Definition 4 (Polarized Set of Constraints). *A set of variables* $X \subseteq \mathrm{fv}(C)$ *is said polarized with respect to a set of normalized literals* C *when* $\mathrm{HLF}_X(C) \neq \emptyset$.

We prove in the next section that our procedure is exact when the variables we eliminate are polarized. While this condition seems very restrictive, we observe that it is often true with the queries used in our experiments.

Example. Let us illustrate our approach with two examples. Assume we want to eliminate an agglomeration $a \multimap \{q,r\}$, meaning that we have the condition $a = q + r$ and that both q and r must disappear. We consider two examples of systems, each with only two literals, and with free variables $\{p,q\}$.

$$
\begin{array}{ll}
3p + 2q - 1r \geqslant 0 & \qquad 3p + 2q - r \geqslant 0 \\
2p + 1q + 1r - 5 \geqslant 0 & \qquad -p + q + 2r - 5 \geqslant 0 \\
\qquad\quad \downarrow & \qquad\qquad\quad \downarrow \\
3p + 2a \geqslant 0 & \qquad 3p - a \geqslant 0 \\
2p + 1a - 5 \geqslant 0 & \qquad -p + a - 5 \geqslant 0
\end{array}
$$

In the left example, the set $\{q,r\}$ is polarized with respect to the initial system (top), with the highest literal factor being q. So we replace q with a in both literals and eliminate r. Uninvolved variables (the singleton $\{p\}$ in this case) are left unchanged. We can prove that both systems, before and after substitution, are equivalent. For instance, every solution in the resulting system can be associated with a solution of the initial one by taking $q = a$ and $r = 0$.

The initial system on the right (top) is non-polarized: the HLF relative to $\{q, r\}$ is $\{q\}$ for the first literal ($+2\,q$ versus $-r$) and $\{r\}$ in the second ($+q$ versus $2\,r$). So we substitute a to the variable with the lowest literal factor, in each literal, and remove the other variable (r in the first literal and q in the second). This is sound because we take into account the worst case in each case. But this is not complete, because we may be too pessimistic. For instance, the resulting system has no solution for $p = 2$; because it entails $a \leqslant 6$ and $a \geqslant 7$. But $p = 2, q = 3, r = 2$ is a model of the initial system.

Next, we give a formal description of our projection procedure as a sequence of formula rewriting steps and prove that the result is exact (we have $F_2 \equiv_E F_1$) when all the reduction steps corresponding to an agglomeration are on polarized variables.

5   Formal Procedure and Proof of Correctness

In all the results of this section, we assume that N_1 and N_2 are two nets, with respective sets of places P_1, P_2 and initial markings m_1, m_2, such that $(N_1, m_1) \equiv_E (N_2, m_2)$. Given a formula F_1 with support on N_1, we describe a procedure to project formula $E \wedge F_1$ into a new formula, F_2, with support on N_2. Our projection will always lead to a sound formula, meaning $F_2 \sqsubseteq_E F_1$. It is also able in many cases (see some statistics in Sect. 6) to result in an exact formula, such that $F_2 \equiv_E F_1$.

Constraints on TFGs. To ensure that a TFG preserves the semantics of the system E we must introduce a set of constraints on it. A *well-formed TFG G built from E* is a graph with one node for every variable and constant occurring in E, such that we can find one set of arcs, either $X \twoheadrightarrow v$ or $v \circ\!\!\rightarrow X$, for every equation of the form $v = \sum_{v_i \in X} v_i$ in E. We deal with inequalities by adding slack variables. We also impose additional constraints which reflect that the same place cannot be removed more than once. Note that the places of N_2 are exactly the root of G (if we forget about constants).

Definition 5 (Well-formed TFG). *A TFG $G = (P, S, R^\bullet, A^\circ)$ for the equivalence statement $(N_1, m_1) \equiv_E (N_2, m_2)$ is well-formed when the following constraints are met, with P_1, P_2 the set of places in N_1, N_2:*

(T1) nodes in S are roots: if $v \in S$ then v is a root of G;
(T2) nodes can be removed only once: it is not possible to have $v \circ\!\!\rightarrow w$ and $v' \twoheadrightarrow w$ with $v \neq v'$, or to have both $v \twoheadrightarrow w$ and $v \circ\!\!\rightarrow w$;
(T3) G contains all and only the equations in E: we have $v \circ\!\!\rightarrow X$ or $X \twoheadrightarrow v$ if and only if the equation $v = \sum_{v_i \in X} v_i$ is in E;
(T4) G is acyclic and roots in $P \setminus S$ are exactly the set P_2.

Given a relation $(N_1, m_1) \equiv_E (N_2, m_2)$, the well-formedness conditions are enough to ensure the unicity of a TFG (up-to the choice of constant nodes) when we set each equation to be either in A or in R. In this case, we denote the graph $\mathrm{T}(E)$. In practice, we use the tool Reduce [34] to generate the reduction system E.

Formula Rewriting. We assume given a relation $(N_1, m_1) \equiv_E (N_2, m_2)$, and its associated well-formed TFG written $\mathrm{T}(E)$. We consider that F_1 is a cube of n literals, $F_1 \triangleq \bigwedge_{i\in 1..n} \alpha_i^0$. Our algorithm rewrites each α_i^0 by applying iteratively an elimination step, described next, according to the constraints expressed in $\mathrm{T}(E)$. The final result is a conjunction $F_2 = \bigwedge_{i\in 1..n} \beta_i$, where each literal β_i has support in N_2. Rewriting can only replace a variable with a group of other variables that are its predecessors in the TFG, which ensures termination in polynomial time (in the size of E). Although the result has the same number of literals, it usually contains many redundancies and trivial constant comparisons, so that, after simplification, F_2 can actually be much smaller than F_1.

A reduction step (to be applied repeatedly) takes as input the current set of literals, $C = (\alpha_i)_{i\in 1..n}$, and modifies it. To ease the presentation, we also keep track of a set of variables, B such that $\bigcup_{i\in 1..n} \mathsf{fv}(\alpha_i) \subseteq B$.

An elimination step is a reduction written $(B, C) \mapsto (B', C')$ where $C = (\alpha_i)_{i\in 1..n}$ and $B' \subsetneq B$, defined as one of the three cases below (one for redundancy, and two for agglomerations, depending on whether the removed variables are polarized or not). We assume that literals are in normal form and that X is a set of variables $\{x_1, \ldots, x_k\}$. Note the precondition $\downarrow X \cap B = \emptyset$ on all rules, which forces them to be applied bottom-up on the TFG (remember it is a DAG). We gave a short example of how to apply rules (AGP) and (AGD) at the end of the previous section.

(RED) If $X \twoheadrightarrow p$ and $\downarrow p \cap B = \emptyset$ then $(B, C) \mapsto (B', C')$ holds, where $B' = B \setminus \{p\}$ and C' is the set of literals α_i' obtained by normalizing the linear constraint $\alpha_i\{p \leftarrow x_1 + \cdots + x_k\}$. That is, we substitute p with $\sum_{x_i\in X} x_i$ in C, which is the meaning of the redundancy equation (constraint (T3) in Definition 5).

(AGP) If $a \circ\!\!\rightarrow X$ with $\downarrow X \cap B = \emptyset$, $a \in B$, and X polarized with respect to C. Then $(B, C) \mapsto (B', C')$ holds, where $B' = B \setminus X$, and, by taking $x_j \in \mathrm{HLF}_X(C)$, we define C' as the set of literals α_i' obtained by normalizing the linear constraint $\alpha_i\{x_l \leftarrow 0\}_{l\neq j}\{x_j \leftarrow a\}$. That is, we eliminate the variables x_l, different from x_j, from C and replace x_j with a; where x_j is a variable of X that always have the highest coefficient in each literal (among the ones of X).

(AGD) If $a \circ\!\!\rightarrow X$ with $\downarrow X \cap B = \emptyset$, $a \in B$, and X is not polarized with respect to C. Then $(B, C) \mapsto (B', C')$ holds, where $B' = B \setminus X$ and C' is the set of literals α_i' obtained by normalizing the linear constraint $\alpha_i\{x_l \leftarrow 0\}_{l\neq j}\{x_j \leftarrow a\}$ such that $\alpha_i(x_j) = \min_X \alpha_i$. Meaning we eliminate the variables x_l different from x_j from α_i and replace x_j with a, where x_j is a variable with the smallest coefficient in α_i (among the ones of X). Note that the chosen variable x_j is not necessarily the same in every literal of C.

Our goal is to preserve the semantics of formulas at each reduction step, in the sense of the relations $\sqsubseteq_E$ and $\equiv_E$. In the following, we use C to represent both a set of literals $(\alpha_i)_{i \in I}$ and the cube formula $\bigwedge_{i \in I} \alpha_i$. We can prove that the elimination steps corresponding to the redundancy (RED) and polarized agglomeration (AGP) cases preserve the semantics of the formula C. On the other hand, a non-polarized agglomeration step (AGD) may lose some markings.

Proof of the Algorithm. We prove the main result of the paper, namely that fast quantifier elimination preserves the integer solutions of a system when we only have polarized agglomerations. To this end, we need to prove two theorems. First, Theorem 2, which entails the soundness of one step of elimination. It also entails completeness for rules (RED) and (AGP). Second, we prove a progress property (Theorem 3 below), which guarantees that we can apply elimination steps until we reach a set of literals C' with support on the reduced net N_2.

Theorem 2 (Projection Equivalence). *If $(B, C) \mapsto (B', C')$ is a (RED) or (AGP) reduction then $C' \equiv_E C$; otherwise $C' \sqsubseteq_E C$.*

We prove Theorem 2 in two steps. We start by proving that elimination steps are sound, meaning that the integer solutions of C' are also solutions of C (up-to E). Then we prove that elimination is also complete for rules (RED) and (AGP). In the following, we use C to represent both a set of literals $(\alpha_i)_{i \in I}$ and the cube formula $\bigwedge_{i \in I} \alpha_i$.

Lemma 1 (Soundness). *If $(B, C) \mapsto (B', C')$ then $C' \sqsubseteq_E C$.*

Proof. Take a valuation m' of $\mathbb{N}^{B'}$ such that $m' \models E \wedge C'$. We want to show that there exists a marking m of $\mathbb{N}^B$ such that $m \equiv_E m'$ satisfying $E \wedge C$.

We have three possible cases, corresponding to rule (RED), (AGP) or (AGD). In each case, we provide a marking m built from m'. Since $m \equiv_E m'$ is enough to prove $m \models E$, we only need to check two properties: first that $m \equiv_E m'$ (∗), then that $m \models \alpha$ for every literal α in C (∗∗).

(RED) In this case we have $X \twoheadrightarrow p$ and $B' = B \setminus \{p\}$, with $X = \{x_1, \dots, x_k\}$. We can extend m' into the unique valuation m of $\mathbb{N}^B$ such that $m(p) = m'(x_1) + \cdots + m'(x_k)$ and $m(v) = m'(v)$ for all other nodes v in $B \setminus \{p\}$. Since $p = x_1 + \cdots + x_k$ is an equation of E (condition (T3)) we obtain that $m' \equiv_E m$ and therefore also $m \models E$ (∗).
We now prove that $m \models C$. The literals in C' are of the form $\alpha\sigma$ with σ the substitution $\{p \leftarrow x_1 + \cdots + x_k\}$ and α in C. Remember that, with our notations (e.g. Equation (1) in page 3), we have $m \models \alpha$ if and only if $\alpha\{m\}$ SAT (is satisfiable). By hypothesis, $m' \models \alpha\sigma$. Hence, $\alpha\sigma\{m'\}$ SAT, which is equivalent to $\alpha\{m\}$ SAT, and therefore $m \models \alpha$ (∗∗), as required.

(AGP) In this case we have $a \circ\!\!\rightarrow X$ with $X = \{x_1, \ldots, x_k\}$, polarized relative to C, and $B' = B \setminus X$. We consider x_j in X the variable in $\mathrm{HLF}_X(C)$ that was chosen in the reduction; meaning that C' is a conjunction of literals of the form $\alpha\{x_l \leftarrow 0\}_{l \neq j}\{x_j \leftarrow a\}$, with α a literal of C. Given m' a model of C', we define m the unique marking on $\mathbb{N}^B$ such that $m(x_j) = m(a)$, $m(x_l) = 0$ for all $l \neq j$, and $m(v) = m'(v)$ for all other variables v in $B \setminus X$.
From Lemma 2 of [4] (the "token propagation" property of TFGs), we know that any distribution of $m(a)$ tokens, in place a, over the $(x_i)_{i\in1..k}$, is also a model of E. Which means that $m \models E$ (*). Note that the token propagation Lemma does not imply that the value of $m(v)$, for the nodes "below X" (v in $\downarrow X$), is unchanged. This is not problematic, since the side condition $\downarrow X \cap B = \emptyset$ ensures that these nodes are not in B, and therefore cannot influence the value of $\alpha\{m\}$.
Consider a literal α in C. Since $m' \models C'$, we have that $\alpha\{x_l \leftarrow 0\}_{l\neq j}\{x_j \leftarrow a\}\{m'\}$ SAT, which is exactly $\alpha\{m\}$, since $\downarrow X \cap B = \emptyset$, as needed (**).

(AGD) In this case we have $a \circ\!\!\rightarrow X$ with $X = \{x_1, \ldots, x_k\}$, non-polarized relative to C, and $B' = B \setminus X$. We know that $m' \models E$, therefore there is a marking m of $\mathbb{N}^B$ that extends m' such that $m \equiv_E m'$ (*).
Consider a literal α in C. By definition of (AGD), we have an associated literal $\alpha' \triangleq \alpha\{x_l \leftarrow 0\}_{l\neq j}\{x_j \leftarrow a\}$ in C' such that $\alpha(x_j) = \min_X \alpha_i$. Since the coefficient of x_j is minimal, we have that $\sum_{i\in1..k} \alpha(x_i)\, m(x_i) \geqslant \alpha(x_j) \sum_{i\in1..k} m(x_i) = \alpha(x_j)\, m'(a)$, and therefore $\sum_{v\in B} \alpha(v)\, m(v) \geqslant \sum_{v\in B'} \alpha'(v)\, m'(v)$. The result follows from the fact that $\alpha\{m\}$ SAT (**).

$\square$

Now we prove that our quantifier elimination step, for the (RED) and (AGP) cases, leads to a complete projection, that is any solution of the initial formula corresponds to a projected solution in the projected formula.

Lemma 2 (Completeness). *If $(B, C) \mapsto (B', C')$ is a (RED) or (AGP) reduction then $C \sqsubseteq_E C'$.*

Proof. Take a marking m of $\mathbb{N}^B$ such that $m \models E \wedge C$. We want to show that there exists a valuation m' of $\mathbb{N}^{B'}$ such that $m \equiv_E m'$ (*) and $m' \models C'$ (**). This is enough to prove $m' \models E \wedge C'$. We have two different cases, corresponding to the rules (RED) and (AGP).

(RED) In this case we have $X \twoheadrightarrow\!\!\bullet\, p$ with $X = \{x_1, \ldots, x_k\}$ and $B' = B \setminus \{p\}$. We define m' as the (unique) projection of m on B'. Since $m \models E$ we have that $m' \equiv_E m$ (*).
Also, literals in C' are of the form $\alpha' \triangleq \alpha\{p \leftarrow x_1 + \cdots + x_k\}$ where α is a literal of C. Since $m(p) = \sum_{i\in1..k} m(x_i)$ and m is a model of α, it is also the case that m' is a model of α' (**).

(AGP) In this case we have $a \circ\!\!\rightarrow X$ with $X = \{x_1, \ldots, x_k\}$ and $B' = B \setminus X$. We define m' as the (unique) projection of m on B', by taking $m'(a) =$

$\sum_{i\in 1..k} m(x_i)$. Since $m \models E$ we have that $m' \equiv_E m$ (∗).

We consider x_j in X the variable in $\mathrm{HLF}_X(C)$ that was chosen in the reduction; meaning that C' is a conjunction of literals of the form $\alpha\{x_l \leftarrow 0\}_{l\neq j}\{x_j \leftarrow a\}$, with α a literal of C. Since $\sum_{i\in 1..k} \alpha(x_i) m(x_i) \leqslant \alpha(x_j) \sum_{i\in 1..k} m(x_i) = \alpha(x_j) m'(a)$, we have m' is a model of α' (∗∗).   □

The final step of our proof relies on a *progress property*, meaning there is always a reduction step to apply except when all the literals have their support on the reduced net, N_2. This property relies on relation $\mapsto^*$, which is the transitive closure of $\mapsto$. Together with Theorem 2, the progress theorem ensures the existence of a sequence $(P, C) \mapsto^* (P_2, C')$, such that $C \equiv_E C'$ (or $C' \sqsubseteq_E C$ if we have at least one non-polarized agglomeration). In this context, P is the set of all variables occurring in the TFG of E, and therefore it contains $P_1 \cup P_2$.

Theorem 3 (Progress). *Assume* $(P, F_1) \mapsto^* (B, C)$ *then either* $B \subseteq P_2$, *the set of places of* N_2, *or there is an elimination step* $(B, C) \mapsto (B', C')$ *such that* $\mathsf{fv}(C') \subseteq B'$ *and the places removed from* B *have no successors in* B': *for all places* p *in* $B \setminus B'$, *we have* $\downarrow p \cap B = \emptyset$.

Proof. Assume we have $(P, F_1) \mapsto^* (B, C)$ and $B \nsubseteq P_2$.

By condition (T4) in Definition 5, we know that P_2 are roots in the TFG $T(E)$. We consider the set of nodes in $B \setminus P_2$, which corresponds to nodes in B with at least one parent. Also, by condition (T4), we know that $T(E)$ is acyclic, then there are nodes in $B \setminus P_2$ that have no successors in B. We call this set L. Hence, $L \triangleq \{v \mid v \in B \setminus P_2 \land \downarrow v \cap B = \emptyset\}$.

Take a node p in L. We have two possible cases. If there is a set X such that $X \twoheadrightarrow p$, we can apply the (RED) elimination rule. Otherwise, there exists a node a and a set $X \subseteq L$ (by condition (T2)) such that $a \multimap X$ with $p \in X$. In this case, apply rule (AGP) or (AGD), depending on whether the agglomeration is polarized or not.   □

Remark. We have designed the rule (AGD) to obtain at least $F_2 \sqsubseteq_E F_1$ when the procedure is not complete (instead of $F_2 \equiv_E F_1$), which is useful for finding witnesses (see Theorem 1). Alternatively, we could propose a variant rule, say (AGD'), which chooses the variable x_j having the highest coefficient in α_i, that is $\alpha_i(x_j) = \max_X \alpha_i$. This variant guarantees a dual result, that is $F_1 \sqsubseteq_E F_2$. In this case, if F_2 is not reachable then F_1 is not reachable, which is useful to prove invariants.

6 Experimental Results

Data-Availability Statement. We have implemented our fast quantifier elimination procedure in a new, open-source tool called Octant [1], available on GitHub. All the tools, scripts and benchmarks used in our experiments are part of our artifact [6].

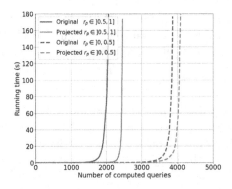

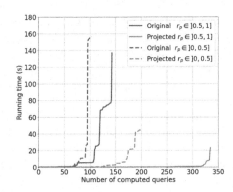

Fig. 3. Random walk w/wo reductions.   **Fig. 4.** k-induction w/wo reductions.

We use an extensive, and independently managed, set of models and formulas collected from the 2022 edition of the Model Checking Contest (MCC) [32]. The benchmark is built from a collection of 128 models. Most models are parametrized and can have several instances. This amounts to about 1 400 different instances of Petri nets whose size varies widely, from 9 to 50 000 places, and from 7 to 200 000 transitions. This collection provides a large number of examples with various structural and behavioral characteristics, covering a large variety of use cases. Each year, the MCC organizers randomly generate 16 reachability formulas for each instance. The pair of a Petri instance and a formula is a *query*; which means we have more than 22 000 queries overall.

We do not compute reductions ourselves but rely on the tool Reduce, part of the latest public release of Tina [34]. We define the reduction ratio (r_p) of an instance as the ratio $(p_{init} - p_{red})/p_{init}$ between the number of places before (p_{init}) and after (p_{red}) reduction. We only consider instances with a ratio, r_p, between 1% and 100% (meaning the net is *fully reduced*), which still leaves about 17 000 queries, so about 77% of the whole benchmark. More information about the distribution of reductions can be found in [3,16], where we show that almost half the instances (48%) can be reduced by a factor of 25% or more.

The size of the reduction system, E, is proportional to the number of places that can be removed. To give a rough idea, the mean number of variables in E is 1 375, with a median value of 114 and a maximum of about 62 000. The number of literals is also rather substantial: a mean of 869 literals (62% of agglomerations and 38% of redundancies), with a median of 27 and a maximum of about 38 000.

We report on the results obtained on two main categories of experiments: first with model-checking, to evaluate if our approach is effective in practice, using real tools; then to assess the precision and performance of our fast quantifier elimination procedure.

Model-Checking. We start by showing the effectiveness of our approach on both k-induction and random walk. This is achieved by comparing the computation time, with and without reductions, on a model-checker that provides a "reference" implementation of these techniques. (Without any other optimizations that could interfere with our experiments.) It is interesting to test the results of our optimization separately on these two techniques. Indeed, each technique is adapted to a different category of queries: properties that can be decided by finding a witness, meaning true EF formulas or false AG ones, can often be checked more efficiently using a random exploration of the state space. On the other hand, symbolic verification methods are required when we want to check invariants.

We display our results using the two "cactus plots" in Figs. 3 and 4. We distinguish between two categories of instances, depending on their reduction ratio. We use dashed lines for models with a low or moderate reduction ratio (value of r_p less than 50%) and solid lines for models that can be reduced by more than half. The first category amounts to roughly 10 700 queries (62% of our benchmark), while the second category contains about 6 000 queries. The most interesting conclusion we can draw from these results is the fact that our approach is beneficial even when there is only a limited amount of reductions.

Our experiments were performed with a maximal timeout of 180 s and integrated the projection time into the total execution time. The "vertical asymptote" that we observe in these plots is a good indication that we cannot expect further gains, without projection, for timeout values above 60 s. Hence, our choice of timeout value does not have an overriding effect. We observe moderate performance gains with random exploration (with ×1.06 more computed queries on low-reduction instances and ×1.19 otherwise) and good results with k-induction (respectively ×1.94 and ×2.33).

We obtain better results if we focus on queries that take more than 1 s on the original formula, which indicates that reductions are most effective on "difficult problems" (there is not much to gain on instances that are already easy to solve). With random walk, for instance, the gain becomes ×1.48 for low-reduction instances and ×1.93 otherwise. The same observation is true with k-induction, with performance gains of ×3.78 and ×3.51 respectively.

Model-Checking Under Real Conditions. We also tested our approach by transparently adding polyhedral reductions as a front-end to three different model-checkers: TAPAAL [21], ITS [43], and LoLA [46], that implement portfolios of verification techniques. All three tools regularly compete in the MCC (on the same set of queries that we use for our benchmark), TAPAAL and ITS share the top two places in the reachability category of the 2022 and 2023 editions.

We ran each tool on our set of complete projections, which amounts to almost 100 000 runs (one run for each tool, once on both the original and the projected query). We obtained a 100% reliability result, meaning that all tools gave compatible results on all the queries; and therefore also compatible results on the original and the projected formulas.

A large part of the queries can be computed by all the tools in less than 100 ms and can be considered as easy. These queries are useful for testing relia-

Table 1. Impact of projections on the challenging queries.

Tool	# Queries		Speed-up		# Exclusive Queries
	Original	Projected	Mean	Median	
ITS	302	352	1.42	1.00	78
LoLA	143	205	14.97	1.44	76
TAPAAL	134	216	1.87	1.17	99

bility but can skew the interpretation of results when comparing performances. This is why we decided to focus our results on a set of 809 *challenging queries*, that we define as queries for which either TAPAAL or ITS, or both, are not able to compute a result before projection. The 809 challenging queries (4% of queries) are well distributed, since they cover 209 different instances (14% of all instances), themselves covering 43 different models (33% of the models).

We display the results obtained on the challenging queries, for a timeout of 180 s, in Table 1. We provide the number of computed queries before and after projection, together with the mean and median speed-up (the ratio between the computation time with and without projection). The "Exclusive" column reports, for each tool, the number of queries that can only be computed using the projected formula. Note that we may sometimes time out with the projected query, but obtain a result without. This can be explained by cases where the size of the formula blows up during the transformation into DNF.

We observe substantial performance gains with our approach and can solve about half of the challenging queries. For instance, we are able to compute ×1.6 more challenging queries with TAPAAL using projections than without. (We display more precise results on TAPAAL, the winner of the MCC 2022 edition, in Fig. 5.) We were also able to compute 62 queries, using projections, that no tool was able to solve during the last MCC (where each tool has 3 600 s to answer 16 queries on a given model instance). All these results show that polyhedral reductions are effective on a large set of queries and that their benefits do not significantly overlap with other existing optimizations, an observation that was already made, independently, in [3] and [17].

The approach implemented in Octant was partially included in the version of our model-checker, called SMPT [8], that participated in the MCC 2023 edition. We mainly left aside the handling of under-approximated queries, when the formula projection is not complete. While SMPT placed third in the reachability category, the proportion of queries it was able to solve raised by 5.5% between 2022 (without the use of Octant) and 2023, to reach a ratio of 93.6% of all queries solved with our tool. A 5% gain is a substantial result, taking into account that the best-performing tools are within 1% of each other; the ratios for ITS and TAPAAL in 2023 are respectively 94.6% and 94.3%.

Performance of Fast Elimination. Our last set of experiments is concerned with the accuracy and performance of our quantifier elimination procedure. We decided to compare our approach with Redlog [22] and isl [45] (we give more details on these two tools in Sect. 7).

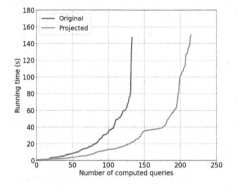

Fig. 5. TAPAAL w/wo polyhedral reductions.

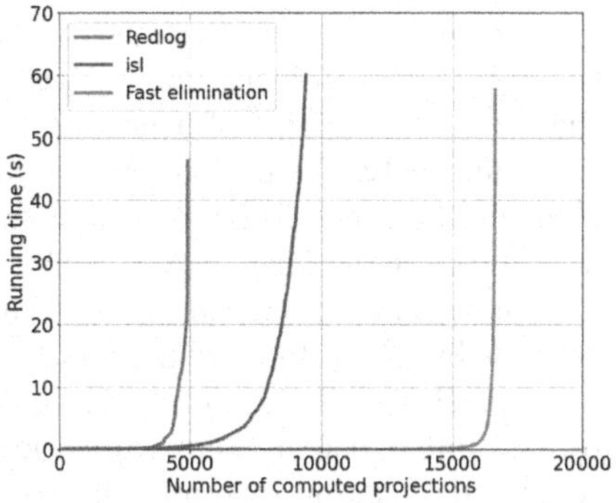

Fig. 6. Redlog vs isl vs Fast elimination.

We display our results in the cactus plot of Fig. 6, where we compare the number of projections we can compute given a fixed timeout. We observe a significant performance gap. For instance, with a timeout of 60 s, we are able to compute 16 653 projections, out of 16 976 queries (98%), compared to 9 407 (55%) with isl or 4 915 (29%) with Redlog. So an increase of ×1.77 over the better of the two tools. This provides more empirical evidence that the class of linear systems we manage is not trivial, or at least does not correspond to an easy case for the classical procedure implemented in Redlog and isl. We also have good results concerning the precision of our approach, since we observe that about 80% of the projections are complete. Furthermore, projections are inexpensive. For instance, the computation time is less than 1 s for 96% of the formulas. We also obtained a median reduction ratio (computed as for the number of places) of 0.2 for the number of cubes and their respective number of literals.

7   Discussion and Related Works

We proposed a quantifier elimination procedure that can take benefit of polyhedral reductions and can be used, transparently, as a pre-processing step of existing model-checkers. The main characteristic of our approach is to rely on a graph structure, called Token Flow Graph, that encodes the specific shape of our reduction equations.

The idea of using linear equations to keep track of the effects of structural reductions originates from [15,16], as a method for counting the number of reachable markings. We extended this approach to Bounded Model Checking (BMC) in [3] where we defined our *polyhedral abstraction* equivalence, $\equiv_E$, and in [7] where we proposed a procedure to automatically prove when such abstractions are correct. The idea to extend this relation to formulas is new (see Definition 2). In this paper, we broaden our approach to a larger set of verification methods; most particularly k-induction, which is useful to prove invariants, and simulation (or *random walk*), which is useful for finding counter-examples. We introduced the notion of a Token Flow Graph (TFG) in [4], as a new method to compute the *concurrent places* of a net; meaning all pairs of places that can be marked simultaneously. We find a new use for TFGs here, as the backbone of our variable elimination algorithm, and show that we can efficiently eliminate variables in systems of the form $E \wedge F$, for an arbitrary F.

We formulated our method as a variable elimination procedure for a restricted class of linear systems. There exist some well-known classes of linear systems where variable elimination has a low complexity. A celebrated example is given by the link between unimodular matrices and integral polyhedra [26], which is related to many examples found in abstract domains used in program verification, such as *systems of differences* [11] or octagon [29,38]. To the best of our knowledge, none of the known classes correspond to what we define using TFGs. There is also a rich literature about quantifier elimination in Presburger arithmetic, such as Cooper's algorithm [19,25] or the Omega test [40] for instance, and how to implement it efficiently [27,35,39]. These algorithms have been implemented in several tools, using many different approaches: automata-based, e.g. TaPAS [36]; inside computer algebra systems, like with Redlog [22]; or in program analysis tools, like isl [45], part of the Barvinok toolbox. Another solution would have been to retrieve "projected formulas" directly from SMT solvers for linear arithmetic, which often use quantifier elimination internally. Unfortunately, this feature is not available, even though some partial solutions have been proposed recently [12]. All the exact methods that we tested have proved impractical in our case. This was to be expected. Quantifier elimination can be very complex, with an exponential time complexity in the worst case (for existential formulas as we target); it can generate very large formulas; and it is highly sensitive to the number of variables, when our problem often involves several hundreds and sometimes thousands of variables. Also, quantifier elimination often requires the use of a divisibility operator (also called *stride format* in [40]), which is not part of the logic fragment that we target.

Another set of related works is concerned with *polyhedral techniques* [23], used in program analysis. For instance, our approach obviously shares similarities with works that try to derive linear equalities between variables of a program [20], and polyhedral abstractions are very close in spirit to the notion of linear dependence between vectors of integers (markings in our case) computed in compiler optimizations. Another indication of this close relationship is the fact that isl, the numerical library that we use to compare our performances, was developed to support polyhedral compilation. We need to investigate this relationship further and see if our approach could find an application with program verification. From a more theoretical viewpoint, we are also looking into ways to improve the precision of our projection in the cases where we find non-polarized sets of constraints.

References

1. Amat, N.: Octant (version 1.0): projection of Petri net reachability properties (2023). https://github.com/nicolasAmat/Octant
2. Amat, N., Berthomieu, B., Dal Zilio, S.: On the combination of polyhedral abstraction and SMT-based model checking for Petri Nets. In: Buchs, D., Carmona, J. (eds.) PETRI NETS 2021. LNCS, vol. 12734, pp. 164–185. Springer, Cham (2021). https://doi.org/10.1007/978-3-030-76983-3_9
3. Amat, N., Berthomieu, B., Dal Zilio, S.: A polyhedral abstraction for Petri Nets and its application to SMT-based model checking. Fundamenta Informaticae **187**(2–4) (2022). https://doi.org/10.3233/FI-222134
4. Amat, N., Dal Zilio, S., Le Botlan, D.: Accelerating the computation of dead and concurrent places using reductions. In: Laarman, A., Sokolova, A. (eds.) SPIN 2021. LNCS, vol. 12864, pp. 45–62. Springer, Cham (2021). https://doi.org/10.1007/978-3-030-84629-9_3
5. Amat, N., Dal Zilio, S., Le Botlan, D.: Leveraging polyhedral reductions for solving Petri net reachability problems. Int. J. Softw. Tools Technol. Transf. **25** (2022). https://doi.org/10.1007/s10009-022-00694-8
6. Amat, N., Dal Zilio, S., Le Botlan, D.: Artifact for VMCAI 2024 Paper "Project and Conquer: Fast Quantifier Elimination for Checking Petri Net Reachability" (2023). https://doi.org/10.5281/zenodo.7935153
7. Amat, N., Dal Zilio, S., Le Botlan, D.: Automated polyhedral abstraction proving. In: Gomes, L., Lorenz, R. (eds.) PETRI NETS 2023. LNCS, vol. 13929, pp. 324–345. Springer, Cham (2023). https://doi.org/10.1007/978-3-031-33620-1_18
8. Amat, N., Zilio, S.D.: SMPT: a testbed for reachability methods in generalized Petri Nets. In: Gomes, L., Lorenz, R. (eds.) Formal Methods (FM). LNCS, vol. 13929, pp. 324–345. Springer, Cham (2023). https://doi.org/10.1007/978-3-031-27481-7_25
9. Kordon, F., et al.: MCC'2017 – the seventh model checking contest. In: Koutny, M., Kristensen, L.M., Penczek, W. (eds.) Transactions on Petri Nets and Other Models of Concurrency XIII. LNCS, vol. 11090, pp. 181–209. Springer, Heidelberg (2018). https://doi.org/10.1007/978-3-662-58381-4_9
10. Amparore, E.G., Balbo, G., Beccuti, M., Donatelli, S., Franceschinis, G.: 30 years of GreatSPN. In: Fiondella, L., Puliafito, A. (eds.) Principles of Performance and Reliability Modeling and Evaluation. SSRE, pp. 227–254. Springer, Cham (2016). https://doi.org/10.1007/978-3-319-30599-8_9
11. Aspvall, B., Shiloach, Y.: A polynomial time algorithm for solving systems of linear inequalities with two variables per inequality. SIAM J. Comput. **9**(4) (1980). https://doi.org/10.1137/0209063
12. Barth, M., Dietsch, D., Heizmann, M., Podelski, A.: Ultimate Eliminator at SMT-COMP (2022)
13. Berthelot, G.: Transformations and decompositions of nets. In: Brauer, W., Reisig, W., Rozenberg, G. (eds.) ACPN 1986. LNCS, vol. 254, pp. 359–376. Springer, Heidelberg (1987). https://doi.org/10.1007/978-3-540-47919-2_13
14. Berthomieu, B., Ribet, P.O., Vernadat, F.: The tool TINA - construction of abstract state spaces for Petri nets and time Petri nets. Int. J. Prod. Res. **42**(14) (2004). https://doi.org/10.1080/00207540412331312688

15. Berthomieu, B., Le Botlan, D., Dal Zilio, S.: Petri Net reductions for counting markings. In: Gallardo, M.M., Merino, P. (eds.) SPIN 2018. LNCS, vol. 10869, pp. 65–84. Springer, Cham (2018). https://doi.org/10.1007/978-3-319-94111-0_4

16. Berthomieu, B., Le Botlan, D., Dal Zilio, S.: Counting Petri net markings from reduction equations. Int. J. Softw. Tools Technol. Transf. **22** (2019). https://doi.org/10.1007/s10009-019-00519-1

17. Bønneland, F.M., Dyhr, J., Jensen, P.G., Johannsen, M., Srba, J.: Stubborn versus structural reductions for Petri nets. J. Log. Algebr. Methods Program. **102** (2019). https://doi.org/10.1016/j.jlamp.2018.09.002

18. Clarke, E.M., Jr., Grumberg, O., Kroening, D., Peled, D., Veith, H.: Model Checking. MIT Press, Cambridge (2018)

19. Cooper, D.C.: Theorem proving in arithmetic without multiplication. Mach. Intell. **7**(91–99) (1972)

20. Cousot, P., Halbwachs, N.: Automatic discovery of linear restraints among variables of a program. In: Proceedings of the 5th ACM SIGACT-SIGPLAN Symposium on Principles of Programming Languages (1978). https://doi.org/10.1145/512760.512770

21. David, A., Jacobsen, L., Jacobsen, M., Jørgensen, K.Y., Møller, M.H., Srba, J.: TAPAAL 2.0: integrated development environment for timed-arc Petri Nets. In: Flanagan, C., König, B. (eds.) TACAS 2012. LNCS, vol. 7214, pp. 492–497. Springer, Heidelberg (2012). https://doi.org/10.1007/978-3-642-28756-5_36

22. Dolzmann, A., Sturm, T.: REDLOG: computer algebra meets computer logic. ACM SIGSAM Bull. **31**(2) (1997). https://doi.org/10.1145/261320.261324

23. Feautrier, P., Lengauer, C.: Polyhedron model. Encycl. Parallel Comput. **1** (2011). https://doi.org/10.1007/978-0-387-09766-4_502

24. Ganty, P., Raskin, J.F., Van Begin, L.: From many places to few: automatic abstraction refinement for Petri nets. Fundamenta Informaticae **88**(3) (2008)

25. Haase, C.: A survival guide to Presburger arithmetic. ACM SIGLOG News **5**(3) (2018). https://doi.org/10.1145/3242953.3242964

26. Hoffman, A.J., Kruskal, J.B.: Integral boundary points of convex polyhedra. In: Jünger, M., et al. (eds.) 50 Years of Integer Programming 1958-2008, pp. 49–76. Springer, Heidelberg (2010). https://doi.org/10.1007/978-3-540-68279-0_3

27. Huynh, T., Lassez, C., Lassez, J.L.: Practical issues on the projection of polyhedral sets. Ann. Math. Artif. Intell. **6**(4) (1992). https://doi.org/10.1007/BF01535523

28. Imbert, J.L.: Fourier's elimination: which to choose? In: PPCP, vol. 1 (1993)

29. Jeannet, B., Miné, A.: APRON: a library of numerical abstract domains for static analysis. In: Bouajjani, A., Maler, O. (eds.) CAV 2009. LNCS, vol. 5643, pp. 661–667. Springer, Heidelberg (2009). https://doi.org/10.1007/978-3-642-02658-4_52

30. Kang, J., Bai, Y., Jiao, L.: Abstraction-based incremental inductive coverability for Petri Nets. In: Buchs, D., Carmona, J. (eds.) PETRI NETS 2021. LNCS, vol. 12734, pp. 379–398. Springer, Cham (2021). https://doi.org/10.1007/978-3-030-76983-3_19

31. Khan, Y.I., Konios, A., Guelfi, N.: A survey of Petri nets slicing. ACM Comput. Surv. (CSUR) **51**(5) (2018). https://doi.org/10.1145/3241736

32. Kordon, F., et al.: Complete Results for the 2022 Edition of the Model Checking Contest (2022). http://mcc.lip6.fr/2022/results.php

33. Kordon, F.: Model SmallOperatingSystem, Model Checking Contest benchmark (2015). https://mcc.lip6.fr/2023/pdf/SmallOperatingSystem-form.pdf

34. LAAS-CNRS: Tina Toolbox (2023). http://projects.laas.fr/tina

35. Lasaruk, A., Sturm, T.: Weak quantifier elimination for the full linear theory of the integers: a uniform generalization of Presburger arithmetic. Appl. Algebra Eng. Commun. Comput. **18** (2007). https://doi.org/10.1007/s00200-007-0053-x

36. Leroux, J., Point, G.: TaPAS: the talence Presburger arithmetic suite. In: Kowalewski, S., Philippou, A. (eds.) TACAS 2009. LNCS, vol. 5505, pp. 182–185. Springer, Heidelberg (2009). https://doi.org/10.1007/978-3-642-00768-2_18

37. Llorens, M., Oliver, J., Silva, J., Tamarit, S.: An integrated environment for Petri net slicing. In: van der Aalst, W., Best, E. (eds.) PETRI NETS 2017. LNCS, vol. 10258, pp. 112–124. Springer, Cham (2017). https://doi.org/10.1007/978-3-319-57861-3_8

38. Miné, A.: The octagon abstract domain. Higher-Order Symb. Comput. **19**(1) (2006). https://doi.org/10.1007/s10990-006-8609-1

39. Monniaux, D.: Quantifier elimination by lazy model enumeration. In: Touili, T., Cook, B., Jackson, P. (eds.) CAV 2010. LNCS, vol. 6174, pp. 585–599. Springer, Heidelberg (2010). https://doi.org/10.1007/978-3-642-14295-6_51

40. Pugh, W.: The omega test: a fast and practical integer programming algorithm for dependence analysis. In: Proceedings of the ACM/IEEE Conference on Supercomputing. ACM (1991). https://doi.org/10.1145/125826.125848

41. Rakow, A.: Safety slicing Petri Nets. In: Haddad, S., Pomello, L. (eds.) PETRI NETS 2012. LNCS, vol. 7347, pp. 268–287. Springer, Heidelberg (2012). https://doi.org/10.1007/978-3-642-31131-4_15

42. Sheeran, M., Singh, S., Stålmarck, G.: Checking safety properties using induction and a SAT-solver. In: Hunt, W.A., Johnson, S.D. (eds.) FMCAD 2000. LNCS, vol. 1954, pp. 127–144. Springer, Heidelberg (2000). https://doi.org/10.1007/3-540-40922-X_8

43. Thierry-Mieg, Y.: Symbolic model-checking using ITS-tools. In: Baier, C., Tinelli, C. (eds.) TACAS 2015. LNCS, vol. 9035, pp. 231–237. Springer, Heidelberg (2015). https://doi.org/10.1007/978-3-662-46681-0_20

44. Thierry-Mieg, Y.: Symbolic and structural model-checking. Fundamenta Informaticae **183**(3–4) (2021). https://doi.org/10.3233/FI-2021-2090

45. Verdoolaege, S.: *isl*: an integer set library for the polyhedral model. In: Fukuda, K., Hoeven, J., Joswig, M., Takayama, N. (eds.) ICMS 2010. LNCS, vol. 6327, pp. 299–302. Springer, Heidelberg (2010). https://doi.org/10.1007/978-3-642-15582-6_49

46. Wolf, K.: Petri Net model checking with LoLA 2. In: Khomenko, V., Roux, O.H. (eds.) PETRI NETS 2018. LNCS, vol. 10877, pp. 351–362. Springer, Cham (2018). https://doi.org/10.1007/978-3-319-91268-4_18

Parameterized Verification of Disjunctive Timed Networks

Étienne André[1] , Paul Eichler[2] , Swen Jacobs[2] ,
and Shyam Lal Karra[2(✉)]

[1] Université Sorbonne Paris Nord, LIPN, CNRS UMR 7030,
93430 Villetaneuse, France
`Etienne.Andre@lipn.univ-paris13.fr`
[2] CISPA Helmholtz Center for Information Security,
Saarbrücken, Germany
`{paul.eichler,jacobs,shyam.karra}@cispa.de`
https://lipn.univ-paris13.fr/ andre/, https://swenjacobs.github.io/

Abstract. We introduce new techniques for the parameterized verification of disjunctive timed networks (DTNs), i.e., networks of timed automata (TAs) that communicate via *location guards* that enable a transition only if there is another process in a given location. We address the minimum-time reachability problem (MINREACH) in DTNs, and show how to efficiently solve it based on a novel zone-graph algorithm. We further show that solving MINREACH allows us to construct a "summary" TA capturing exactly the possible behaviors of a single TA within a DTN of arbitrary size. The combination of these two results enables the parameterized verification of DTNs, while avoiding the construction of an exponential-size cutoff system required by existing results. Additionally, we develop sufficient conditions for solving MINREACH and parameterized verification problems even in certain cases where locations that appear in location guards can have clock invariants, a case that has usually been excluded in the literature. Our techniques are also implemented, and experiments show their practicality.

Keywords: Networks of Timed Automata · Parameterized Verification · Cutoffs · Minimum-time Reachability

1 Introduction

Many computer systems today are distributed and rely on some form of synchronization between largely independent processes to reach a common goal. Formally reasoning about the correctness of such systems is difficult, since correctness guarantees are expected to hold regardless of the number of processes —a problem that is also known as *parameterized verification* [1,5], since the

---

This work is partially supported by ANR-NRF ProMiS (ANR-19-CE25-0015/2019 ANR NRF 0092) and by ANR BisoUS (ANR-22-CE48-0012).

R. Dimitrova et al. (Eds.): VMCAI 2024, LNCS 14499, pp. 124–146, 2024.
https://doi.org/10.1007/978-3-031-50524-9_6

number of processes is considered as a parameter of the system. Parameterized verification is undecidable even in very restricted settings, e.g., for safety properties of systems composed of finite-state processes with rather limited communication, such as token-passing or transition guards [23,39]. However, many classes of systems and properties have been identified that have decidable parameterized verification problems [9,17,27], usually with finite-state processes.

Systems and properties that involve timing constraints, such as clock synchronization algorithms or planning of time-critical tasks, cannot be adequately modeled with finite-state processes. A natural model for such processes are timed automata (TAs) [7], and the parameterized verification of systems of TAs, also called *timed networks*, has already received some attention in the literature. For models with powerful communication primitives, such as k-process synchronization or broadcast, safety properties are decidable if every process has a single clock, but they are undecidable if processes have multiple clocks, and liveness properties are undecidable regardless of the number of clocks [2–4].

In *disjunctive timed networks* (DTNs), communication is restricted to disjunctive guards that enable a transition only if at least one process is in a given location. This a commonly met communication primitive in the literature, having been studied for finite-state systems under the notion of guarded protocols [13,23,24,33], and more recently in immediate observation Petri nets [28,37] and immediate observation population protocols [26]. For DTNs, decidability is obtained also for multiple clocks per process and for liveness properties [38].

However, existing results on timed networks have no or very limited support for *location invariants* (which can force an action to happen before a time limit runs out), and the decidability result for DTNs relies on the construction of a product system that can be hard to check in practice. Moreover, to the best of our knowledge, no techniques exist for the computation of real-time properties such as minimum-time reachability in timed networks (for single TAs, see [8,20,40]).

In this paper, we show that minimum-time reachability can be effectively computed in DTNs, which also leads to a more efficient parameterized verification technique for safety and liveness properties, and we provide conditions under which location invariants can be supported. DTNs are an interesting computational model, since (even without clock invariants) they can express classes of models where a resource, encoded as a location, is produced or unlocked once for all; this is the case of the whole class of planning problems, and some problems in chemical reaction networks. Moreover, disjunctive networks have been used to verify security problems in grid computing [36], and real-time properties are also of interest in such applications [35]. Another natural application area are clock synchronization protocols, which we consider in the following.

Motivating Example: Gossiping Clock Synchronization. Consider the example in Fig. 1, depicting a simple clock synchronization protocol. The semantics of a single process is the same as for standard TAs. Invariants are depicted using dotted yellow boxes; $x \leftarrow 0$ denotes reset of clock x. As a synchronization mechanism, some transitions are *guarded* by a location h_i (such location guards are highlighted in light blue), i.e., they may only be taken by one process if another process is in h_i. Arrows marked with h_0, h_1 stand for two transitions, each guarded by one of the h_i.

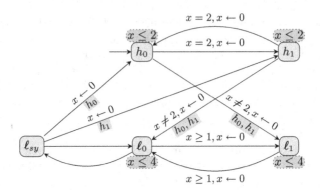

Fig. 1. TA with disjunctive guards for gossiping clock synchronization

Processes are synchronized when they are in a location h_i for $i \in \{0, 1\}$, and they should move to $h_{(i+1 \bmod 2)}$ after two time units. However, they can non-deterministically lose synchronization and move to a location ℓ_i, where they move according to different timing constraints. Via ℓ_{sy}, they can re-synchronize with processes that are still in a location h_i.

A version of this example was considered by Spalazzi and Spegni [38]. Notably, their version came *without* clock invariants on locations h_i, i.e., processes can stay in these locations forever, which is clearly not the intended behavior. Besides not supporting location invariants, their parameterized verification requires the construction of a product system, which is fine for this small example but quickly becomes impractical for bigger examples.

Contributions. In this paper, we provide novel decidability results for DTNs:

1. We state the *minimum-time reachability problem* (MINREACH) for DTNs, i.e., computing the minimal time needed to reach any location of a TA, in networks with an arbitrary number of processes. We develop a technique to efficiently solve the problem for TAs without location invariants, based on a specialized zone-graph computation.
2. We show that solving MINREACH allows us to construct a *summary automaton* that captures the local semantics of a process, i.e., has the same set of possible executions as any single process in the DTN. This allows us to decide parameterized verification problems without resorting to the global semantics in a product system that grows exponentially with the cutoff.
3. We show that under certain conditions, parameterized verification is still decidable even in the presence of location invariants for 1-clock TAs. We develop new domain-specific proof techniques that allow us to compute summary automata and cutoffs for these cases.

4. We experimentally evaluate our techniques on variants of the clock synchronization case study and a hand-crafted example, demonstrating its practicality and its benefits compared to the cutoff-based approach.

Related Work. Infinite-state systems have been of high interest to the verification community. Their infiniteness can stem from two causes. First, the state space of a single process can be infinite, as in TAs [7]. In order to obtain decidability results for single TAs, abstraction techniques have been proposed (e.g., [14,19,32]). Second, infinite-state systems effectively arise if we consider parameterized systems that consist of an arbitrary number of components. In general, reachability is undecidable in parameterized systems composed of an arbitrary number of finite-state systems [12], and even for uniform finite-state components with very limited means of communication [39]. However, the parameterized verification problem remains decidable for many interesting classes of systems with finite-state processes [9,17,18,21,23,25,29].

According to this classification, the parameterized verification of networks of TAs deals with infiniteness in two dimensions [3,4], and similarly the verification of timed Petri nets [34]. Here, the notion of urgency (which is closely related to location invariants) makes the reachability problem undecidable, even for the case where each TA is restricted to a single clock. Our setting also shares similarities with timed *ad hoc* networks [2], where processes are TAs that communicate by broadcast in a given topology. However, even simple problems like parameterized reachability quickly become undecidable [2], e.g., when processes contain two clocks if the topology is a clique (i.e., all nodes can communicate with each other), or even when the nodes are equipped with a single clock if the topology is a star of diameter five. In our setting, the topology is fixed to a clique, and communication is not based on actions, but on location guards. In addition, [2,4] can only decide reachability properties and do not seem to be usable to derive minimum-time reachability results.

As the closest work to ours, Spalazzi and Spegni [38] consider DTNs, and use the cutoff method to obtain decidability results. They show that, while cutoffs do not always exist, they do exist for a subclass where the underlying TAs do not have invariants on locations that appear in guards. In our work, we show how to decide parameterized verification problems by constructing a summary automaton instead of an exponentially larger cutoff system, and we present new decidability results that address the limitation on location invariants.

A similar idea to our summary automaton for parameterized verification of systems communicating via pairwise rendez-vous has also appeared in [9], but in their case this is a Büchi automaton, whereas we need to use a TA.

Parameterized verification *with timing parameters* was considered in [10], with mostly undecidability results; in addition, the few decidability results only concern *emptiness* of parameter valuations sets, and not synthesis, and therefore these parametric results cannot be used to encode a minimum-time reachability using a parameter minimization.

Outline. Section 2 recalls TAs and DTNs, and states the parameterized verification problem. Section 3 introduces a technique for efficient parameterized verification of DTNs where locations that appear in location guards cannot have invariants, while Sect. 4 proposes another technique that supports such invariants under certain conditions. Section 5 shows the practical evaluation of our algorithms on several examples, and Sect. 6 concludes the paper.

2   Preliminaries

We define here TAs with location guards (also: guarded TAs) and networks of TAs, followed by parameterized verification problems and cutoffs (Sect. 2.1), and finally the standard symbolic semantics of TAs (Sect. 2.2).

Let C be a set of clock variables. A *clock valuation* is a mapping $v : C \to \mathbb{R}_{\geq 0}$. We denote by $\mathbf{0}$ the clock valuation that assigns 0 to every clock, and by $v + \delta$ for $\delta \in \mathbb{R}_{\geq 0}$ the valuation such that $(v + \delta)(c) = v(c) + \delta$ for all $c \in C$. We call *clock constraints* $\mathcal{CC}(C)$ the terms of the following grammar:

$$\varphi ::= \top \mid \varphi \wedge \varphi \mid c \sim c' + d \mid c \sim d \text{ with } d \in \mathbb{N}, c, c' \in C, \sim \in \{<, \leq, =, \geq, >\}.$$

A clock valuation v is said to *satisfy* a clock constraint φ, written as $v \models \varphi$, if φ evaluates to true after replacing every $c \in C$ with its value $v(c)$.

Guarded Timed Automaton (gTA). A *gTA* A is a tuple $(Q, \hat{q}, C, \mathcal{T}, Inv)$:

- Q is a finite set of locations with *initial location* $\hat{q}$,
- C is a finite set of clock variables,
- $\mathcal{T} \subseteq Q \times \mathcal{CC}(C) \times 2^C \times (Q \cup \{\top\}) \times Q$ is a transition relation, and
- $Inv : Q \to \mathcal{CC}(C)$ assigns to every location q an *invariant* $Inv(q)$.

Intuitively, a transition $\tau = (q, g, r, \gamma, q') \in \mathcal{T}$ takes the automaton from location q to q', it can only be taken if *clock guard* g and *location guard* γ are both satisfied, and it resets all clocks in r. We also write $lguard(\tau)$ for γ. Note that satisfaction of location guards is only meaningful in a *network* of TAs, formally defined subsequently. Intuitively, a location guard is satisfied if it is $\top$ or if another automaton in the network currently occupies location γ. We say that γ is *trivial* if $\gamma = \top$, and we write $Guards(A) \subseteq Q$ for the set of non-trivial location guards that appear in A.[1] We call $q \in Guards(A)$ also a *guard location*. We say that a location q *does not have an invariant* if $Inv(q) = \top$.

Example 1. Figure 2 shows an example gTA with 5 locations and one clock x. Location $\hat{q}$ is the initial location, with a transition to q_0 that has a clock constraint $x = 2$, and a transition to q_2 that resets x (which in figures is denoted as $x \leftarrow 0$ for readability) and a location guard q_1. Location q_0 has an invariant $x \leq 4$. The transition from q_0 to $\hat{q}$ has location guard "q_0", requiring another process to be in location q_0 to take this transition.

Fig. 2. A gTA example

Given a gTA A, we denote by $UG(A)$ the unguarded version of A i.e., ignoring all location guards (or equivalently, replacing each location guard with $\top$). Note that $UG(A)$ is a TA according to standard definitions in the literature [7].

A *configuration* of a gTA A is a pair (q, v), where $q \in Q$ and $v : \mathsf{C} \to \mathbb{R}_{\geq 0}$ is a *clock valuation*. When considering a gTA A in isolation, its semantics is the usual semantics of the TA $UG(A)$, i.e., ignoring location guards. That is, a *delay transition* is of the form $(q, v) \to^\delta (q, v + \delta)$ for some $\delta \in \mathbb{R}_{\geq 0}$ such that $\forall \delta' \in [0, \delta] : v + \delta' \models Inv(q)$. A *discrete transition* is of the form $(q, v) \to^\tau (q', v')$, where $\tau = (q, g, r, \gamma, q') \in \mathcal{T}$, $v \models g$, $v'[c] = 0$ if $c \in r$ and $v[c]$ otherwise, and $v' \models Inv(q')$. We say that the transition is *based on* τ. We write $(q, v) \to (q', v')$ if either $(q, v) \to^\delta (q', v')$ or $(q, v) \to^\tau (q', v')$. For convenience, we assume that for every location q there is a discrete *stuttering transition* $\epsilon = (q, \top, \emptyset, \top, q)$ that can always be taken and does not change the configuration.

We write $(q, v) \xrightarrow{\delta, \tau} (q', v')$ if there is a delay transition $(q, v) \to^\delta (q, v + \delta)$ followed by a discrete transition $(q, v + \delta) \to^\tau (q', v')$. Then, a *timed path* of A is a (finite or infinite) sequence[2] $\rho = (q_0, v_0) \xrightarrow{\delta_0, \tau_0} (q_0, v_0) \xrightarrow{\delta_1, \tau_1} \dots$.

For a finite timed path $\rho = (q_0, v_0) \xrightarrow{\delta_0, \tau_0} \dots \xrightarrow{\delta_{l-1}, \tau_{l-1}} (q_l, v_l)$, let $\mathsf{lastloc}(\rho) = q_l$ be the final location of ρ, $\delta(\rho) = \sum_{0 \leq i < l} \delta_i$ the total time delay of ρ. We write $(q, v) \to^* (q', v')$ if there is a (finite) timed path $\rho = (q_0, v_0) \xrightarrow{\delta_0, \tau_0} \dots \xrightarrow{\delta_{l-1}, \tau_{l-1}} (q_l, v_l)$. A timed path is a *computation* if $q_0 = \hat{q}$ and $v_0 = \mathbf{0}$. The *language* of A, denoted $\mathcal{L}(A)$, is the set of all of its computations.

Example 2. Consider $UG(A)$ for the gTA A in Fig. 2. From its initial configuration $(\hat{q}, \mathbf{0})$, there can be arbitrary delay transitions (since $\hat{q}$ does not have an invariant), and after a delay of $\delta \geq 0$ we can take a transition to q_2 which resets the clock x, i.e., we arrive at configuration $(q_2, \mathbf{0})$. The location guard on this transition is ignored, since we consider $UG(A)$. In contrast, the transition from $\hat{q}$ to q_0 has a clock constraint that needs to be observed, i.e., we can only take the transition after we reach a configuration $(\hat{q}, v)$ with $v(x) = 2$. Then, the gTA can only stay in q_0 as long as the invariant $x \leq 4$ is satisfied.

Network of TAs. For a given gTA A, we denote by A^n the parallel composition $A \parallel \dots \parallel A$ of n copies of A, also called a *network of TAs* (NTA for short). A copy of A in the NTA will also be called a *process*.

---

[1] Note that since $Guards(A)$ is effectively a set of locations, we also use it as such.

[2] Wlog, we assume that the first transition of a timed path is a delay transition and that delay transitions and discrete transitions alternate.

A *configuration* $\mathfrak{c}$ of an NTA A^n is a tuple $((q_1, v_1), \ldots, (q_n, v_n))$, where every (q_i, v_i) is a configuration of A. The semantics of A^n can be defined as a *timed transition system* $(\mathfrak{C}, \hat{\mathfrak{c}}, T)$, where $\mathfrak{C}$ denotes the set of all configurations of A^n, $\hat{\mathfrak{c}}$ is the unique initial configuration $(\hat{q}, \mathbf{0})^n$, and the transition relation T is the union of the following delay and discrete transitions:

delay transition $((q_1, v_1), \ldots, (q_n, v_n)) \xrightarrow{\delta} ((q_1, v_1 + \delta), \ldots, (q_n, v_n + \delta))$
if $\forall i \in [1, n] : \forall \delta' \leq \delta : v_i + \delta' \models \mathit{Inv}(q_i)$, i.e., we can delay $\delta \in \mathbb{R}_{\geq 0}$ units of time if all invariants $\mathit{Inv}(q_i)$ will be satisfied until the end of the delay;

discrete transition $((q_1, v_1), \ldots, (q_n, v_n)) \xrightarrow{(i, \tau)} ((q_1', v_1'), \ldots, (q_n', v_n'))$ if

1. $(q_i, v_i) \xrightarrow{\tau} (q_i', v_i')$ is a discrete transition of A with $\tau = (q_i, g, r, \gamma, q_i')$,
2. $\gamma = \top$ or $q_j = \gamma$ for some $j \in [1, n] \setminus \{i\}$, and
3. $q_j' = q_j$ and $v_j' = v_j$ for all $j \in [1, n] \setminus \{i\}$.

That is, location guards γ are interpreted as disjunctive guards: unless $\gamma = \top$, at least one of the other processes needs to occupy the location γ.

We write $\mathfrak{c} \xrightarrow{\delta, (i, \tau)} \mathfrak{c}''$ for a delay transition $\mathfrak{c} \xrightarrow{\delta} \mathfrak{c}'$ followed by a discrete transition $\mathfrak{c}' \xrightarrow{(i, \tau)} \mathfrak{c}''$. Then, a *timed path* of A^n is a (finite or infinite) sequence $\pi = \mathfrak{c}_0 \xrightarrow{\delta_0, (i_0, \tau_0)} \mathfrak{c}_1 \xrightarrow{\delta_1, (i_1, \tau_1)} \cdots$. For a finite timed path $\pi = \mathfrak{c}_0 \xrightarrow{\delta_0, (i_0, \tau_0)} \mathfrak{c}_1 \xrightarrow{\delta_1, (i_1, \tau_1)} \cdots \xrightarrow{\delta_{l-1}, (i_{l-1}, \tau_{l-1})} \mathfrak{c}_l$, let $\mathsf{lastconf}(\pi) = \mathfrak{c}_l$ be the final configuration of π and $\delta(\pi) = \sum_{0 \leq i < l} \delta_i$ the total time delay of π. A timed path π of A^n is a *computation* if $\mathfrak{c}_0 = \hat{\mathfrak{c}}$. The *language* of A^n, denoted $\mathcal{L}(A^n)$, is the set of all of its computations.

We will also use *projections* of these global objects onto subsets of the processes. That is, if $\mathfrak{c} = ((q_1, v_1), \ldots, (q_n, v_n))$ and $\mathcal{I} = \{i_1, \ldots, i_k\} \subseteq [1, n]$, then $\mathfrak{c}|_{\mathcal{I}}$ is the tuple $((q_{i_1}, v_{i_1}), \ldots, (q_{i_k}, v_{i_k}))$, and we extend this notation in the natural way to computations $\pi|_{\mathcal{I}}$ and the language $\mathcal{L}_{|\mathcal{I}}(A^n)$.[3] Similarly, for timed paths π_1 of A^{n_1} and π_2 of A^{n_2} we denote by $\pi_1 \parallel \pi_2$ their (non-necessarily unique) *composition* into a timed path of $A^{n_1 + n_2}$. We write $q \in \mathfrak{c}$ if $\mathfrak{c} = ((q_1, v_1), \ldots, (q_n, v_n))$ and $q = q_i$ for some $i \in [1, n]$, and similarly $(q, v) \in \mathfrak{c}$.

We say that a location q is *reachable* in A^n if there exists a reachable configuration $\mathfrak{c}$ s.t. $q \in \mathfrak{c}$. We denote by $\mathit{ReachL}(A)$ the set of reachable locations in $UG(A)$, and by $\mathit{ReachL}(A^n)$ the set of reachable locations in A^n.

If π is a timed path and $d \in \mathbb{R}_{\geq 0}$, then by $\pi^{\leq d}$ we denote the maximal prefix of π with $\delta(\pi^{\leq d}) \leq d$, and similarly for timed paths $\rho^{\leq d}$ of a single gTA.

Example 3. Consider a network with 2 processes that execute the gTA in Fig. 2. When both processes are in the initial configuration $(\hat{q}, \mathbf{0})$, the transition to q_2 is disabled because of the location guard q_1. However, after a delay of $\delta = 2$, one of them can move to q_0, and after another delay of $\delta = 2$ from q_0 to q_1. As q_1 is now occupied, the process that stayed in $\hat{q}$ can now take the transition to q_2.

---

[3] In particular, in the projection $\pi|_{\mathcal{I}}$ of a timed path π, any discrete transition of a process $j \notin \mathcal{I}$ is replaced by a stuttering transition ϵ of one of the processes $i \in \mathcal{I}$, and $\mathcal{L}_{|\mathcal{I}}(A^n)$ contains the projection for every possible choice of i.

Disjunctive Timed Network. A given gTA A induces a *disjunctive timed network* (DTN)[4] A^∞, defined as the following family of NTAs: $A^\infty = \{A^n : n \in \mathbb{N}_{>0}\}$. We define $\mathcal{L}(A^\infty) = \bigcup_{n \in \mathbb{N}} \mathcal{L}(A^n)$ and, for $\mathcal{I} = [1, i]$ with $i \in \mathbb{N}$, let $\mathcal{L}_{|\mathcal{I}}(A^\infty) = \bigcup_{n \geq i} \mathcal{L}_{|\mathcal{I}}(A^n)$.

We say that a gTA A *has persistent guard locations* if every location $q \in Guards(A)$ has $Inv(q) = \top$. We denote by DTN$^-$ the class of disjunctive timed networks where A has persistent guard locations. That is, in a DTN$^-$ a location can only appear in a location guard if it does not have an invariant.

2.1   Parameterized Verification Problems

In this work, we are mostly interested in determining $\mathcal{L}_{|\mathcal{I}}(A^\infty)$ for some fixed finite $\mathcal{I}$, since this allows us to solve many parameterized verification problems. This includes any local safety or liveness properties of a single process (with $\mathcal{I} = [1, 1]$), as well as mutual exclusion properties (with $\mathcal{I} = [1, 2]$) and variants of such properties for larger $\mathcal{I}$. Note that, even though we are interested in the language of a small number of processes, these processes still interact with an arbitrary number of identical processes in the network.[5]

Cutoffs. We call a *family of gTAs* a collection $\mathcal{A}$ of gTAs, expressed in some formalism. E.g., let $\mathcal{A}_\top$ be the collection of all gTAs with persistent guard locations and, for any k, let $\mathcal{A}_k$ be the collection of all gTAs with $|Guards(A)| \leq k$.

Then, a *cutoff* for a family of TAs $\mathcal{A}$ and a number of processes $m \in \mathbb{N}$ is a number $c \in \mathbb{N}$ such that for every $A \in \mathcal{A}$:

$$\mathcal{L}_{|[1,m]}(A^\infty) = \mathcal{L}_{|[1,m]}(A^c)$$

Note that our definition of cutoffs requires language equality (for a fixed number m of processes) between A^c and A^∞, which implies that c is also a cutoff under other definitions from the literature which only require equivalence of A^c and A^∞ wrt. certain logic fragments that define subsets of the language.

In particular, we can immediately state the following generalization of an existing result, which gives a cutoff for gTAs with k locations used in guards, and without invariants on these locations. Let $\mathcal{A}_{\top,k} = \mathcal{A}_\top \cap \mathcal{A}_k$.

Theorem 1 (follows from [38]). *For any $m \in \mathbb{N}$, $m + k$ is a cutoff for $\mathcal{A}_{\top,k}$.*

2.2   Symbolic Semantics of Timed Automata

We build on the standard zone-based symbolic semantics of TAs [16]. In the following, let $A = (Q, \hat{q}, \mathsf{C}, \mathcal{T}, Inv)$ be a gTA, interpreted as a standard TA.

A *zone* is a clock constraint φ (as defined in Sect. 2), representing all clock valuations that satisfy φ. In the following we will use the constraint notation

---

[4] We reuse terminology and abbreviations from [38].

[5] One notable exception, i.e., a parameterized verification problem that cannot be answered based on $\mathcal{L}_{|\mathcal{I}}(A^\infty)$, is the (global) deadlock detection problem, not considered here, or similar problems of simultaneous behavior of all processes.

and the set (of clock valuations) notation interchangeably. We denote by $\mathcal{Z}$ the set of all zones.

A *symbolic configuration* of A is a pair (q, z), where $q \in Q$ and z is a zone over C. For a given symbolic configuration (q, z), let $z^\uparrow = \{u + \delta \mid u \in z, \delta \in \mathbb{R}_{\geq 0}\}$, and $(q, z)^\uparrow = (q, z^\uparrow \cap Inv(q))$. For a zone z and a clock constraint g, let $z \wedge g = \{v \mid v \in z \text{ and } v \models g\}$.

Zone Graph. The *zone graph* $ZG(A) = (S, S_0, Act, \Rightarrow)$ of A is a transition system where *i)* S is the (infinite) set of nodes of the form (q, z) where $q \in Q$ and z is a zone; *ii)* The initial node is $S_0 = (\hat{q}, \mathbf{0})^\uparrow$; *iii)* For any two nodes (q, z) and (q', z'), there is a *symbolic transition*, $(q, z) \overset{\tau}{\Rightarrow} (q', z')$ if there exists a transition $\tau = (q, g, r, \gamma, q') \in T$ such that $(q', z') = \left(q', \{v' \in \mathbb{R}_{\geq 0}^X \mid \exists v \in z : (q, v) \to^\tau (q', v')\}\right)^\uparrow$ and $z' \neq \emptyset$. *iv)* $\Rightarrow$ is the union of all $\overset{\tau}{\Rightarrow}$, and $\Rightarrow^*$ is the transitive closure of $\Rightarrow$. A *symbolic path* in a zone graph is a (finite or infinite) sequence of symbolic transitions $\bar{\rho} = (q_0, z_0) \overset{\tau_0}{\Rightarrow} (q_1, z_1) \overset{\tau_1}{\Rightarrow} \cdots$.

3   Minimum-Time Reachability Algorithm for DTN$^-$

In this section, we show how we can solve the *minimum-time reachability problem* (MINREACH), i.e., determine reachability of every $q \in Q$ in a DTN$^-$ and compute minimal reachability times $\delta_{min}^\infty(q)$ for every reachable location q. Solving MINREACH is essential since, in a DTN with $A \in \mathcal{A}_T$, the minimal reachability times completely determine when each disjunctive guard can be satisfied. We will show that this also allows us to determine $\mathcal{L}|_{[1,1]}(A^\infty)$ by "filtering" $\mathcal{L}(UG(A))$ with respect to the $\delta_{min}^\infty(q)$-values.

Formally, we define

$$\delta_{min}(q) = \min\left(\{d \in \mathbb{R}_{\geq 0} \mid \exists \rho \in \mathcal{L}(UG(A)) \text{ s.t. } \rho^{\leq d} \text{ is finite and } q = \mathsf{lastloc}(\rho^{\leq d})\}\right),$$

i.e., the minimal global time to reach q in $UG(A)$, and for $i \in \mathbb{N} \cup \{\infty\}$

$$\delta_{min}^i(q) = \min\left(\{d \in \mathbb{R}_{\geq 0} \mid \exists \pi \in \mathcal{L}(A^i) \text{ s.t. } \pi^{\leq d} \text{ is finite and } q \in \mathsf{lastconf}(\pi^{\leq d})\}\right),$$

i.e., the minimal global time such that one process reaches q in a DTN of size i (if $i \in \mathbb{N}$), or a network of any size (if $i = \infty$).

Then, MINREACH is the problem of determining $\delta_{min}^\infty(q)$ for every $q \in Q$.

Example 4. For the gTA in Fig. 2 we have $\delta_{min}(q_3) = 2$, as the transition from $\hat{q}$ to q_2 is immediately enabled, and we then need to wait 2 time units for the transition from q_2 to q_3 to be enabled. But we have $\delta_{min}^2(q_3) = \delta_{min}^\infty(q_3) = 6$, as one process needs to move to q_1 before another can take the transition to q_2. Also note that in a network with one process, q_3 cannot be reached i.e., $\delta_{min}^1(q_3) = \infty$.

Remark 1. Note that the *minimum* may not always be defined, as the smallest reachable time may be an *infimum* (e.g., of the form $\delta_{min}^i(q) > d$ for some $d \in \mathbb{N}$). To ease the exposé, we assume that the minimum is always defined, but all our constructions can be extended to work for both cases of minimum and infimum.

A "naive" way to solve MINREACH would be based on Theorem 1, our generalization of existing cutoff results for DTNs.[6] I.e., we can consider the cutoff system $A^{1+|Guards(A)|}$ and determine $\delta_{min}^{1+|Guards(A)|}(q)$ with standard techniques for TAs [22], but this approach does not scale well in the size of $Guards(A)$, as each additional clock (in the product system) may result in an exponentially larger zone graph [30,31]. Instead, we will show how to solve MINREACH more efficiently, based on a specialized zone-graph construction that works on a single copy of the given gTA, and determines minimal reachability times sequentially.

To this end, note that all reachable locations can be totally ordered based on the order in which they can be reached. Formally, define $\leq_{reach} \subseteq ReachL(A^\infty) \times ReachL(A^\infty)$ as follows: we have $q \leq_{reach} q'$ if either $i)$ $\delta_{min}^\infty(q) < \delta_{min}^\infty(q')$, or $ii)$ $\delta_{min}^\infty(q) = \delta_{min}^\infty(q')$ and there exists a computation $\pi \in \mathcal{L}(A^\infty)$ that reaches q before it reaches q' (i.e., q is the last location of a prefix of π that does not contain q'). Now consider $RGuards(A) = Guards(A) \cap ReachL(A^\infty)$, the reachable guard locations of A. Then the following lemma states that, if $\{q_1, \ldots, q_n\}$ are the reachable guard locations of a DTN$^-$ ordered by $\leq_{reach}$, then for every $i \in \{1, \ldots, n\}$ there is a computation of A^i such that each process j with $j \leq i$ reaches q_j at minimal time, and can stay there forever.

Lemma 1. *Let A be a DTN$^-$. Let $RGuards(A) = \{q_1, \cdots, q_n\}$ such that $q_i \leq_{reach} q_{i+1}$ for $i = 1, \ldots, n-1$. Then for every $i \in \{1, \ldots, n\}$, there exists a computation $\pi \in \mathcal{L}(A^i)$ such that, for each $j \in \{1, \ldots, i\}$, the projection $\rho_j = \pi|_{[j,j]}$ reaches q_j at time $\delta_{min}^\infty(q_j)$, and stays there forever.*

Proof. Fix $i \in \{1, \ldots, n\}$. First consider $j = 1$. By definition of $\leq_{reach}$, q_1 is a guard location that is reachable at the earliest time, and any guard location q' with $\delta_{min}^\infty(q') = \delta_{min}^\infty(q_1)$ can be reached on a computation that reaches q_1 first. Thus, there exists a computation ρ of $UG(A)$ reaching q_1 at time $\delta_{min}^\infty(q_1)$ without passing any non-trivial location guard. Note that ρ is also a computation π of A^1 with the desired properties.

Now, for any $1 < j \leq i$, we can assume that there is a computation π of A^{j-1} where for each of the guard locations $\{q_1, \ldots, q_{j-1}\}$ there is a process that reaches it in minimal time and stays there forever (which is possible because A is a DTN$^-$). By definition of $\delta_{min}^\infty(q_j)$ and $\leq_{reach}$, there exists a computation π_j of A^m, for some m, that reaches q_j in minimal time, and the transitions along π_j only depend on guard locations in $\{q_1, \ldots, q_{j-1}\}$. If ρ_j is the local computation of the process that reaches q_j in π_j, then $\pi' = \pi \parallel \rho_j$ is a computation of A^j with the desired properties. $\square$

The following lemma formalizes the connection between computations of A and A^∞, stating that a computation of A is also the projection of a computation of A^∞ if no transition with a location guard q is taken before q can be reached.

---

[6] Minimum-time reachability cannot be expressed in their specification language, but it can be shown that their proof constructions preserve minimum-time reachability.

Lemma 2. *Let $A \in \mathcal{A}_\top$. A computation $\rho \in UG(A)$ is in $\mathcal{L}|_{[1,1]}(A^\infty)$ iff for every prefix ρ' of ρ that ends with a discrete transition $(q,v) \to^\tau (q',v')$ with $lguard(\tau) \neq \top$, we have $\delta^\infty_{min}(lguard(\tau)) \leq \delta(\rho')$.*

Proof. For the "if" direction, assume that every discrete transition τ in ρ happens at a time that is greater than $\delta^\infty_{min}(lguard(\tau))$. By Lemma 1 there exists a computation π of $A^{|Guards(A)|}$ in which every $q \in RGuards(A)$ is reached at time $\delta^\infty_{min}(q)$, and the process that reaches q stays there forever. Thus, we have $(\rho \parallel \pi) \in \mathcal{L}(A^\infty)$, and therefore $\rho \in \mathcal{L}|_{[1,1]}(A^\infty)$.

The "only if" direction is simple: if ρ contains a discrete transition $(q,v) \to^\tau (q',v')$ that happens at a time strictly smaller than $\delta^\infty_{min}(lguard(\tau))$, then by definition no other local run can be in location $lguard(\tau)$ at this time (in any computation of A^∞), and therefore ρ cannot be in $\mathcal{L}|_{[1,1]}(A^\infty)$.   □

Example 5. Consider again the gTA in Fig. 2 (ignoring the location invariant of q_0, as otherwise it is not a DTN$^-$). Considering the possible behaviors of a process in a DTN$^-$, we can assume that transition $\hat{q} \to q_2$ is enabled whenever the global time is at least $\delta^\infty_{min}(q_1)$. This is because we can assume that another process moves to that location on the minimum-time path (and stays there).

We will show subsequently that Lemma 2 allows us to solve MINREACH by

1. considering a variant A' of A with a global clock variable never reset, and
2. applying a modified zone-graph algorithm on a *single instance* of A'.

Working on a single instance of A' will lead to an exponential reduction of time and memory when compared to a naive method based on cutoff results.

3.1 Symbolic Semantics for DTN$^-$

As Lemma 2 shows, to decide whether a path with location guards can be executed in a DTN$^-$, it is important to keep track of global time. The natural way to do this is by introducing an auxiliary clock t that is never reset [6,15]. We capture this idea by proposing the following definition of a DTN-Zone Graph.

DTN-Zone Graph. Given a gTA $A = (Q, \hat{q}, \mathcal{C}, \mathcal{T}, Inv)$, let $A' = (Q, \hat{q}, \mathcal{C} \cup \{t\}, \mathcal{T}, Inv)$. Then the *DTN-zone graph* $ZG^\infty(A')$ is a transition system where

- the set S of nodes and the initial node $(\hat{q}, z_0)$ are as in $ZG(A')$.
- For any two nodes (q, z) and (q', z'), there is
  - a *guarded transition* $(q, z) \xrightarrow{\tau, \gamma} (q', z')$ if there is a symbolic transition $(q, z) \xRightarrow{\tau} (q', z')$ in $ZG(A')$ and $\gamma = \top$;
  - a *guarded transition* $(q, z) \xrightarrow{\tau, \gamma} (q', z')$ if there exists a $\tau \in \mathcal{T}$ such that $(q', z') = (q', \{v' \in \mathbb{R}^X_{\geq 0} \mid \exists v \in z: v(t) \geq \delta^\infty_{min}(\gamma) \wedge (q,v) \to^\tau (q',v')\})^\top$, $z' \neq \emptyset$ and $\gamma \neq \top$. I.e., in this case we effectively add $t \geq \delta^\infty_{min}(\gamma)$ to the clock constraint of τ.
- $\xRightarrow{\gamma}$ is the union of all $\xRightarrow{\tau, \gamma}$, and $\xRightarrow{\gamma}^*$ is the transitive closure of $\xRightarrow{\gamma}$

Let v_{min} be a function that takes as input a zone z and a clock c, and returns the lower bound of c in z. Then we want (a finite version of) $ZG^\infty(A')$ to satisfy the following properties:

- *soundness with respect to* MINREACH, i.e., if $(\hat{q}, z_0) \overset{\gamma}{\Rightarrow}{}^* (q', z')$ is such that $v_{min}(z', t) \leq v_{min}(z'', t)$ for all nodes (q', z'') reachable in $ZG^\infty(A')$, then there exist $n \in \mathbb{N}$ and a timed path $\pi = \hat{\mathfrak{c}} \xrightarrow{\delta_0, (i_0, \tau_0)} \mathfrak{c}_1 \xrightarrow{\delta_1, (i_1, \tau_1)} \cdots \xrightarrow{\delta_{l-1}, (i_{l-1}, \tau_{l-1})} \mathfrak{c}_l$ of A^n with $(q', v') \in \mathfrak{c}_l$ such that $v' \in z'$ and $v'(t) = v_{min}(z', t)$.
- *completeness with respect to* MINREACH, i.e., if q' is reachable in A^∞ then there exists $(\hat{q}, z_0) \overset{\gamma}{\Rightarrow}{}^* (q', z')$ with $v_{min}(z', t) = \delta^\infty_{min}(q')$.

Note that for a gTA A, and A' defined as above, the zone graph $ZG(UG(A'))$ is sound w.r.t. MINREACH when considering executions of a single copy of $UG(A)$. Moreover, it is known that completeness in this setting can be preserved under a time-bounded exploration with a bound B such that $B > \delta_{min}(q)$ for every $q \in Q$. For $ZG^\infty(A')$ and executions of A^n however, having B only slightly larger than $\delta_{min}(q)$ for every $q \in Q$ may not be sufficient to preserve completeness, as the following example shows:

Example 6. We have already seen that for the gTA in Fig. 2 we have $\delta^\infty_{min}(q_3) = 6$, even though $\delta_{min}(q) \leq 4$ for all q. Thus, if we choose $B = 4$, a time-bounded exploration will not find a path to q_3 in $ZG^\infty(A')$.

Bounding Minimal Reachability Times. In the following, we compute an upper-bound on the minimum-time reachability in a DTN$^-$, which will allow us to perform a time-bounded exploration, thus rendering the zone graph finite.

Let $\delta_{max} = \max\{\delta_{min}(q) \mid q \in ReachL(A)\}$. Our upper bound is defined as $UB(A) = \delta_{max} \cdot (|Guards(A)| + 1)$, i.e., the maximum over the locations of the minimum to reach that location, times the number of location guards plus one.

Lemma 3. *For any given gTA $A = (Q, \hat{q}, \mathcal{C}, \mathcal{T}, Inv)$, we have*

1. *$ReachL(A) \supseteq ReachL(A^\infty)$, and*
2. *for all $q \in ReachL(A^\infty)$, $\delta^\infty_{min}(q) \leq UB(A)$.*

Proof. Point 1 directly follows from the fact that $\mathcal{L}|_{[1,1]}(A^\infty) \subseteq \mathcal{L}(A)$. For point 2, let $ReachL(A^\infty) = \{q_1, \ldots, q_k\}$ and assume wlog. $q_1 \leq_{\mathsf{reach}} q_2 \leq_{\mathsf{reach}} \cdots \leq_{\mathsf{reach}} q_k$. Then the minimal timed path to reach q_1 must be such that it only uses discrete transitions with trivial location guards. Therefore, we have $\delta^\infty_{min}(q_1) = \delta_{min}(q_1) \leq \delta_{max}$. Inductively, we get for every q_i with $i > 1$ that it can only use transitions that rely on $\{q_1, \ldots, q_{i-1}\} \cap Guards(A)$, and therefore $\delta^\infty_{min}(q_i) \leq (i - j) \cdot \delta_{max}$, where $j = |\{q_1, \ldots, q_{i-1}\} \setminus Guards(A)|$. $\square$

Thus, it is sufficient to perform a time-bounded analysis with $UB(A)$ as a time horizon. Formally, let $ZG^\infty_{UB(A)}(A')$ be the finite zone graph obtained from $ZG^\infty(A')$ by intersecting every zone with $t \leq UB(A)$.

Lemma 4. *For $A = (Q, \hat{q}, \mathcal{C}, \mathcal{T}, Inv) \in \mathcal{A}_\top$, let $A' = (Q, \hat{q}, \mathcal{C} \cup \{t\}, \mathcal{T}, Inv)$. Then $ZG^\infty_{UB(A)}(A')$ is sound and complete with respect to* MINREACH.

Proof. Soundness w.r.t. MINREACH: let $\overline{p} = (\hat{q}, z_0) \xrightarrow{\tau_0, \gamma} (q_1, z_1) \xrightarrow{\tau_1, \gamma} \cdots \xrightarrow{\tau_{l-1}, \gamma}$ (q_l, z_l) in $ZG^\infty_{UB(A)}(A')$ with $v_{min}(z_l, t) \leq v_{min}(z, t)$ for all (q_l, z) reachable in $ZG^\infty_{UB(A)}(A')$.

We prove existence of a computation $\pi = \hat{\mathfrak{c}} \xrightarrow{\delta_0, (i_0, \tau_0)} \mathfrak{c}_1 \xrightarrow{\delta_1, (i_1, \tau_1)}$ $\cdots \xrightarrow{\delta_{l-1}, (i_{l-1}, \tau_{l-1})} \mathfrak{c}_l$ of A^n (for some n) with $v \in z_l$ and $v(t) = v_{min}(z_l, t)$ for some $(q, v) \in \mathfrak{c}_l$ based on induction over the number k of different non-trivial location guards along π.

Induction base $(k = 0)$ In this case $\overline{p}$ also exists in $ZG(A')$, and a corresponding computation of A' exists by MINREACH-soundness of $ZG(A')$. After projecting away t, this is a computation of A^1 with the desired properties.

Induction step $(k \rightarrow k+1)$ Assume that $\overline{p}$ has $k+1$ different non-trivial location guards, and let $(\hat{q}, z_0) \xrightarrow{\tau_0, \gamma} (q_1, z_1) \xrightarrow{\tau_1, \gamma_1} \cdots \xrightarrow{\tau_{i-1}, \gamma_{i-1}} (q_i, z_i)$ be the maximal prefix that has only k non-trivial location guards, i.e., it is followed by a transition $(q_i, z_i) \xrightarrow{\tau_i, \gamma_i} (q_{i+1}, z_{i+1})$ where $\gamma_i \neq \top$ and γ_i has not appeared as a location guard on the prefix. Then by definition of $ZG^\infty_{UB(A)}(A')$, this guarded transition can only be taken if there is a valuation $v \in z_i$ with $v(t) \geq \delta^\infty_{min}(\gamma_i)$. By induction hypothesis, there is a computation $\pi_k = \hat{\mathfrak{c}} \xrightarrow{\delta_0, (i_0, \tau_0)} \mathfrak{c}_1 \xrightarrow{\delta_1, (i_1, \tau_1)}$ $\cdots \xrightarrow{\delta_{i-1}, (i_{i-1}, \tau_{i-1})} \mathfrak{c}_i$ of some A^{n_1} with $v \in z_i$ and $v(t) = v_{min}(z_i, t)$ for some $(q_i, v) \in \mathfrak{c}_i$. Moreover, by (a variant of the proof of) Lemma 1, since not all guard locations that are reachable in this time may appear on the path, there exists a computation π_{min} of A^{k+1} that reaches each of the $k + 1$ location guards at minimal time, and each process stays in its guard location forever. Then the desired timed path is $\pi_k \parallel \pi_{min}$.

Completeness w.r.t. MINREACH: follows by construction of $ZG^\infty_{UB(A)}(A')$ and from Lemma 3. □

3.2   An Algorithm for Solving MINREACH

To solve MINREACH in practice, it is usually not necessary to construct $ZG^\infty_{UB(A)}(A')$ completely. For $ReachL(A^\infty) = \{q_1, \ldots, q_k\}$, assume wlog. that $q_1 \preceq_{reach} q_2 \preceq_{reach} \cdots \preceq_{reach} q_k$. Then the timed path to q_i with minimal global time for every q_i can only have location guards that are in $\{q_j \mid j < i\}$. If we explore zones in order of increasing $v_{min}(z, t)$, we will find $\delta^\infty_{min}(q_1)$ without using any transitions with non-trivial location guards. Then, we enable transitions guarded by q_1, and will find $\delta^\infty_{min}(q_2)$ using only the trivial guard and q_1, etc.

Algorithm 1: Algorithm to solve MINREACH

---

1 **foreach** q **do** $MinReach(q) = \infty, visited(q) = \bot$;

2 $W = \{(q_0, z_0)\}$, $D = \emptyset$ // Waiting nodes and Disabled transitions

3 **while** $W \neq \emptyset$ **and** $\exists q.visited(q) = \bot$ **do**

4   Remove (q, z) from W with the least $v_{min}(z, t)$ value

5   **if** $z \neq \emptyset$ **and** $v_{min}(z, t) \leq UB(A)$ **then**

6   Set $visited(q) = \top$

         // compute successors along enabled transitions:

7          $S = \{(q', z') \mid \exists \tau.lguard(\tau) = \gamma \wedge (q, z) \xrightarrow{\tau} (q', z') \wedge (visited(\gamma) \vee \gamma = \top)\}$

         // new disabled transitions:

8          $D' = \{(\tau, (q, z)) \mid (q, z) \xrightarrow{\tau, \gamma} (q', z') \wedge \neg(visited(\gamma))\}$

9          $D = D \cup D'$, $W = W \cup S$

10   **if** $v_{min}(z, \delta) < MinReach(q)$ **then**

11              $MinReach(q) = v_{min}(z, t)$ // happens only once for each q

             // transitions enabled by q:

12              $E = \{(\tau, (q', z')) \in D \mid lguard(\tau) = q\}$

             // compute successors along newly enabled transitions:

13              $S = \{(q'', z'') \mid (\tau, (q', z')) \in E$

                         $\wedge (q', (z' \wedge t \geq v_{min}(z, t))) \xrightarrow{\tau} (q'', z'')\}$

14              $D = D \setminus E$, $W = W \cup S$

15 **return** $MinReach$

An incremental algorithm that constructs the zone graph[7] has to keep track of guarded transitions that cannot be taken at the current global time, but that may be possible to take at some later time.

Algorithm 1 takes A as an input, constructs the relevant parts of $ZG^\infty_{UB(A)}(A')$, and returns a mapping of locations q to their $\delta^\infty_{min}(q)$. As soon as we have discovered timed paths to all $q \in Q$, the algorithm terminates (line 3). Otherwise, as long as we have graph nodes (q, z) to explore, we pick the minimal node (w.r.t. $z(\delta)$ in line 4) and check that its zone z is non-empty. If this is the case, we mark q as visited and add any successor nodes. Furthermore, we remember any transitions that are currently not enabled, and store them with the current node in D.

Finally, if the current node (q, z) is such that $v_{min}(z, t) < MinReach(q)$ (line 10), then we have discovered the minimal reachability time of q. In this case we store it in $MinReach$(line 11), and we compute the successor along τ for every tuple $(\tau, (q', z')) \in D$ with $lguard(\tau) = q$, representing a transition that has just become enabled, after intersecting z' with $t \geq v_{min}(z, t)$, as the transition is only enabled now (line 13).

Correctness of the algorithm follows directly from Lemma 4, and termination follows from finiteness of $ZG^\infty_{UB(A)}(A')$.

---

[7] As usual, for efficient zone graph construction we encode zones using DBMs, see [16].

3.3   Verification of DTNs

For a given gTA $A = (Q, \hat{q}, \mathcal{C}, \mathcal{T}, \mathit{Inv})$, the *summary automaton of A* is the gTA $\hat{A} = (Q, \hat{q}, \mathcal{C} \cup \{t\}, \hat{\mathcal{T}}, \mathit{Inv})$ with $\hat{\tau} = (q, \hat{g}, r, \top, q') \in \hat{\mathcal{T}}$ if $\tau = (q, g, r, \gamma, q') \in \mathcal{T}$ and either $\gamma = \top \wedge \hat{g} = g$ or $\gamma \in \mathit{ReachL}(A^{\infty}) \wedge \hat{g} = (g \wedge t \geq \delta^{\infty}_{min}(\gamma))$.

Theorem 2 (Summary Automaton). *Let $A = (Q, \hat{q}, \mathcal{C}, \mathcal{T}, \mathit{Inv}) \in \mathcal{A}_{\top}$. Then*
$$\mathcal{L}|_{[1,i]}(A^{\infty}) = \mathcal{L}\big((UG(\hat{A}))^i\big) \text{ for all } i \in \mathbb{N}.$$

Proof. For $i = 1$, the statement directly follows from Lemma 2 and the definition of $\hat{A}$. For $i > 1$, it follows from the fact that $\mathcal{L}|_{[j,j]}(A^{\infty}) = \mathcal{L}\big(UG(\hat{A})\big)$ for each $j \in [1, i]$.   □

Theorem 2 tells us that the summary automaton $\hat{A}$ can be used to answer any verification question that is based on $\mathcal{L}|_{[1,1]}(A^{\infty})$, i.e., the local runs of a single gTA in a DTN^- A^{∞}. This includes standard problems like reachability of locations, but also (global) timing properties, as well as liveness properties. Moreover, the same holds for behaviors of multiple processes in a DTN^-, such as mutual exclusion properties, by simply composing copies of $\hat{A}$. In particular, any model checking problem in a DTN^- that can be reduced to checking a cutoff system by Theorem 1 can be reduced to a problem on the summary automaton, which is exponentially smaller than the cutoff system.

4   Conditions for Decidability with Location Invariants

In Sect. 1, we argued that our motivating example would benefit from invariants that limit how long a process can stay in any of the locations. Neither the results presented so far nor existing cutoff results support such a model, since it would have invariants on locations that appear in location guards. To see this, note that the correctness of Theorem 1, like the MINREACH-soundness argument of the DTN-zone graph, relies on the fact that in a DTN^-, if we know that a location q can be reached (within some time or on some timed path) and we argue about the existence of a local behavior of a process, we can always assume that there are other processes in the system that reach q *and stay there forever*. This proof construction is called *flooding* of location q, but it is in general not possible if q has a non-trivial invariant.

In this section we generalize this flooding construction and provide sufficient conditions under which we can obtain a reduction to a summary automaton or a cutoff system, even in the presence of invariants. For the rest of the section, we assume that A has a single clock x.

General Flooding Computations. We say that $\pi \in \mathcal{L}(A^{\infty})$ is a *flooding computation for q* if $q \in \mathsf{lastconf}(\pi^{\leq d})$ for every $d \geq \delta^{\infty}_{min}(q)$, i.e., q is reached in minimal time and will be occupied by at least one process at every later point in time. Then, we obtain the following as a consequence of Lemma 4:[8]

---

[8] To see this, consider the π_{min} from the soundness part of the proof, but instead of letting all processes stay in the guard locations, use our new flooding computation to determine their behaviour afterwards.

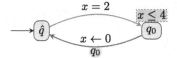

(a) A loop that can flood location q_0

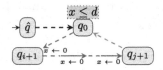

(b) A general lasso for flooding q_0, showing ψ_1 in red, ψ_2 in green, and ψ_3 in brown

Fig. 3. Loops and color codings for flooding computations

Corollary 1. *Let $A = (Q, \hat{q}, C, T, Inv)$ be a gTA, not necessarily from A_T, and let $A' = (Q, \hat{q}, C \cup \{t\}, T, Inv)$. If there exists a flooding computation for every $q \in RGuards(A)$, then $ZG_{UB(A)}^{\infty}(A')$ is correct.*

Note that a flooding computation trivially exists if $q \in ReachL(A^\infty)$ and $Inv(q) = \top$. Thus, the next question is how to determine whether flooding computations exist for locations with non-trivial invariants.

Identifying Flooding Computations Based on Lassos. We aim to find lasso-shaped local computations of a single process that visit a location q infinitely often and can be composed into a flooding computation for q.

Since any flooding computation π for q must use one of the minimal paths to q that we can find in $ZG_{UB(A)}^{\infty}(A')$, and any local computation in π must also be a computation of the summary automaton $\hat{A}$ of A, our analysis will be based on the summary automaton $\hat{A}$ instead of A.[9] Since furthermore every possible prefix of ρ, including its final configuration (q, v), is already computed in $ZG_{UB(A)}^{\infty}(A')$, what remains to be found is the loop of the lasso, where one or more processes start from a configuration (q, v) and always keep q occupied. To this end, the basic idea is that each process spends as much time as possible in q. To achieve this with a small number c of processes, we also want to minimize times where more than one process is in q. We illustrate the idea on an example, and formalize it afterwards.

Example 7. Figure 3a shows a subgraph of the gTA in Fig. 2. To find a flooding computation for q_0, we let two processes $p1$ and $p2$ start in location $\hat{q}$ and have both of them move to q_0 at global time 2. Then, $p1$ immediately moves to $\hat{q}$, which is possible as the location guard is enabled by $p2$. After 2 time units, $p2$ has to leave q_0 due to the invariant, so we let $p1$ return to q_0, which allows $p2$ to take the transition to $\hat{q}$, where stays for 2 time units. In this way, both processes can keep taking turns traversing the loop, keeping q_0 occupied forever.

---

[9] Note that for this argument, we *assume* that there are flooding computations for all $q \in RGuards(A)$, and therefore $ZG^\infty(A')$ and $\hat{A}$ are correct. If we identify flooding lassos under this assumption then we have also shown that the assumption was justified.

Syntactic Paths and Timed Loops. A (finite) *path* in $\hat{A}$ is a sequence of transitions $\psi = q_0 \xrightarrow{\tau_0} q_1 \xrightarrow{\tau_1} \ldots \xrightarrow{\tau_{l-1}} q_l$, with $\tau_i = (q_i, g_i, r_i, \gamma_i, q_i') \in \mathcal{T}$ and $q_i' = q_{i+1}$ for $i \in [0, l[$. We also denote q_0 as firstloc(ψ), and call ψ *initial* if firstloc(ψ) = $\hat{q}$. A path ψ is a *loop* if $q_0 = q_l$. We call a path *resetting* if it has at least one clock reset.

We call a path $\psi = q_0 \xrightarrow{\tau_0} \ldots \xrightarrow{\tau_{l-1}} q_l$ *executable from* (q_0, v_0) if there is a timed path $\rho = (q_0, v_0) \xrightarrow{\delta_0, \tau_0} \ldots \xrightarrow{\delta_{l-1}, \tau_{l-1}} (q_l, v_l)$ of $\hat{A}$. We say that ρ is *based on* ψ. For a path $\psi = q_0 \xrightarrow{\tau_0} \ldots \xrightarrow{\tau_{l-1}} q_l$ in $\hat{A}$ that is executable from (q_0, v_0), we denote by $\rho_{\mathsf{asap}}(\psi, v_0)$ the unique timed path $(q_0, v_0) \xrightarrow{\delta_0, \tau_0} \ldots \xrightarrow{\delta_{l-1}, \tau_{l-1}} (q_l, v_l)$ such that in every step, δ_i is minimal. A timed path $\rho = (q_0, u_0) \xrightarrow{\delta_0, \tau_0} \ldots \xrightarrow{\delta_{l-1}, \tau_{l-1}} (q_l, u_l)$ is called a *timed loop* if $q_0 = q_l$ and $u_0(x) = u_l(x)$.

Sufficient Conditions for Existence of a Flooding Computation. Let ψ be a resetting loop, where τ_i is the first resetting transition and τ_j is the last resetting transition on ψ. As depicted in Fig. 3b, we split ψ into 1) $\psi_1 = q_0 \xrightarrow{\tau_0} \ldots \xrightarrow{\tau_i} q_{i+1}$; 2) $\psi_2 = q_{i+1} \xrightarrow{\tau_{i+1}} \ldots \xrightarrow{\tau_j} q_{j+1}$; and 3) $\psi_3 = q_{j+1} \xrightarrow{\tau_{j+1}} \ldots \xrightarrow{\tau_{l-1}} q_l$. Note that ψ_2 is empty if $i = j$, and ψ_3 is empty if $j = l - 1$. If ψ is executable from (q_0, v_0), then $\rho_{\mathsf{asap}}(\psi, v_0) = (q_0, v_0) \xrightarrow{\delta_0, \tau_0} \ldots \xrightarrow{\delta_{l-1}, \tau_{l-1}} (q_l, v_l)$ exists and we can compute the time needed to execute its parts as

- $d_1 = \delta\big((q_0, v_0) \xrightarrow{\delta_0, \tau_0} \ldots \xrightarrow{\delta_i, \tau_i} (q_{i+1}, v_{i+1})\big),$
- $d_2 = \delta\big((q_{i+1}, v_{i+1}) \xrightarrow{\delta_{i+1}, \tau_{i+1}} \ldots \xrightarrow{\delta_j, \tau_j} (q_{j+1}, v_{j+1})\big),$
- $d_3 = \delta\big((q_{j+1}, v_{j+1}) \xrightarrow{\delta_{l-1}, \tau_{l-1}} \ldots \xrightarrow{\delta_{l-1}, \tau_{l-1}} (q_0, v_l)\big).$

Note that the first process p_1 that traverses the loop along $\rho_{\mathsf{asap}}(\psi, v_0)$ will return to q_0 after time $d_1 + d_2 + d_3$. Thus, if a process p_2 starts in the same configuration (q_0, v_0), and stays in q_0 while p_1 traverses the loop, then p_1 will return to q_0 before p_2 has to leave due to $Inv(q_0)$ if $U_x^{Inv}(q_0) \geq d_1 + d_2 + d_3 + v(x)$, where $U_x^{Inv}(q_0)$ denotes the upper bound imposed on x by $Inv(q_0)$. Moreover, p_2 can still execute ψ from a configuration (q_0, v') with $v' = v + \delta$ if and only if $v(x) + \delta \leq T$, where $T = \min\{U_x^{Inv}(q_k) \mid 0 \leq k \leq i\}$. Generally, traversing ψ is possible from any configuration (q_0, v') with $v(x) \leq v'(x) \leq T$ and $v(t) \leq v'(t)$. In particular, we have $\delta(\rho_{\mathsf{asap}}(\psi, v')) \leq d_1 + d_2 + d_3$ and in the reached configuration (q_0, v'') we have $v''(x) \leq d_3$. Thus, if p_2 has to leave before p_1 returns, we can add more processes that start traversing the loop in regular intervals, such that p_{i+1} always arrives in q_0 before p_i has to leave.

Example 8. Consider again the flooding computation constructed in Example 7, depicted in Fig. 4. In this case, $\psi_1 = q_0 \to \hat{q}$ (depicted in red), ψ_2 is empty, and $\psi_3 = \hat{q} \to q_0$ (in brown). Note that we have a repetition of the global configuration after a bounded time, and therefore can keep q_0 occupied forever by repeating the timed loop.

Based on these observations, we can show that the conditions mentioned above are sufficient to guarantee that a flooding computation for q_0 exists, provided that we also have a flooding computation for all other guard locations.

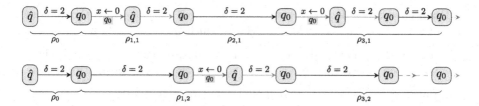

Fig. 4. Flooding computation for location q_0 from Fig. 3a

Lemma 5. *Let $\rho_0 = (\hat{q}, \mathbf{0}) \to \ldots \to (q_0, v_0)$ be a computation of $\hat{A}$ that reaches q_0 after minimal time, and $\psi = q_0 \xrightarrow{\tau_0} \ldots \xrightarrow{\tau_{l-1}} q_0$ a resetting loop in $\hat{A}$ executable from (q_0, v_0). Let d_1, d_2, d_3, T as defined above. If $T \geq d_1 + d_2 + d_3 + v(x)$, $T > d_3$, and there exists a flooding computation for all $q \in RGuards(A) \setminus \{q_0\}$, then there exists a flooding computation for q_0.*

Proof. To find a flooding computation for q_0 in $\mathcal{L}(A^\infty)$, we construct a number c of local timed paths of $\hat{A}$, and show that there exists a computation π of A^n with $n \geq c$ such that $\pi|_{[1,c]}$ is the composition of these local timed paths, and therefore π is a flooding computation of q_0. Note that in both cases below, all location guards except q_0 are implicitly taken care of, since ρ_0 and our computation of the d_i are based on $\hat{A}$ (and therefore take the minimal time to reach guard locations into account), and by assumption all other locations have a flooding path.

Case 1: $d_2 = d_3 = 0$, i.e., if $x = T$ then the loop can be traversed from q_0 in 0 time. In this case the flooding computation lets two processes move to q_0 in minimal time, lets both of them stay in q_0 until $x = T$, then lets process 1 traverse the loop in 0 time, and finally does the same for process 2. This can be repeated until infinity, always keeping q_0 occupied. Note that a flooding computation with a single process may not be possible, as q_0 could appear as a location guard.

Case 2: The loop cannot be traversed in 0 time. In this case, let $c = \lceil \frac{T+d_2}{T-d_3} \rceil$, and for each $p \in [1, c]$, define $\rho(p) = \rho_{1,p} \cdot \rho_{2,p} \cdot \rho_{3,p}$ as follows:

- $\rho_{1,p} = (q_0, v_0) \xrightarrow{\delta_p} (q_0, v_0 + \delta_p) \cdot \rho_{\mathsf{asap}}(\psi, v_0 + \delta_p)$ with $\delta_p = (p-1)(T - d_3)$

- $\rho_{2,p} = (q_0, v_1) \xrightarrow{\delta'_p} (q_0, v_1 + \delta'_p)$ with v_1 the resulting clock valuation of $\rho_{\mathsf{asap}}(\psi, v_0 + \delta_p)$ and δ'_p such that $(v_0 + \delta'_p)(x) = T$

- $\rho_{3,p} = \rho_{\mathsf{asap}}(\psi, v_2) \cdot (q_0, v'_2) \xrightarrow{\delta''_p} (q_0, v'_2 + \delta''_p)$ with $v_2 = v_1 + \delta'_p$, v'_2 the resulting clock valuation of $\rho_{\mathsf{asap}}(\psi, v_2)$, and δ''_p such that $(v'_2 + \delta''_p)(x) = T$

By construction, $\rho(p) \in \mathcal{L}(\hat{A})$ for every p. Moreover, note that, up to the valuation of t, $\rho_{3,p}$ can be appended an arbitrary number of times to $\rho(p)$ (since for any v with $v(x) = T$ and $v(t) \geq v_0(t) + d_1 + d_2 + d_3$, the time $\delta(\rho_{\mathsf{asap}}(q_0, v))$ and the resulting clock valuation for x will always be the same). Thus, let $\rho^\omega(p)$ be the projection of $\rho_{1,p} \cdot \rho_{2,p} \cdot (\rho_{3,p})^\omega$ to A, i.e., without valuations of t.

Now, note that at any time $d \geq 0$, q_0 is occupied in at least one of the $\rho^\omega(p)$: for $d \leq d_1 + d_2 + d_3$, process c has not left q_0 yet, and at any time $d > d_1 + d_2 + d_3$, process $\lceil \frac{d - (d_1 + d_2 + d_3)}{T - d_3} \rceil \bmod c$ occupies q_0.

Both cases: The desired π is the composition of $\rho_0 \cdot \rho^\omega(p)$ with the flooding computations for all $q \in ReachL(A^\infty)\backslash\{q_0\}$, which we assumed to exist.   □

Note that this property depends on our assumption that A has a single clock. It remains open whether a more complex construction works for multiple clocks.

Lemma 6. *If for every $q \in RGuards(A)$ we either have $Inv(q) = \top$ or there exists a flooding computation, then there exists a computation of A^∞ that floods all of these locations at the same time.*

New Cutoff Results. In addition to witnessing the correctness of the summary automaton for A, the proofs of Lemmas 5 and 6 also allow us to compute a cutoff for the given gTA. For a location q, let $w(q) = 1$ if $Inv(q) = \top$ and $w(q) = max\{2, \lceil\frac{T+d_2}{T-d_3}\rceil\}$, where T, d_2, d_3 are as in Lemma 5, if there exists a flooding computation based on the lemma. Intuitively, $w(q)$ is the *width* of the flooding computation, i.e., how many processes need to be dedicated to q.

Corollary 2. *For any gTA A that satisfies the conditions of Lemma 6 and any $m \in \mathbb{N}$, $m + \sum_{q \in RGuards(A)} w(q)$ is a cutoff.*

Sufficient and Necessary Conditions for Decidability. Note that above, we give sufficient but not necessary conditions to establish that a guard location q can always remain occupied after $\delta^\infty_{min}(q)$. Further note that there are DTNs where a guard location q *cannot* always remain occupied after it is reached, and in such cases the language $\mathcal{L}|_{[1,i]}(A^\infty)$ can be determined iff one can determine all (global-time) intervals in which q can be occupied in the DTN, for all such q. While it is known that in this case cutoffs for parameterized verification do not exist [38], an approach based on a summary automaton would work whenever these intervals can be determined. Whether parameterized verification is decidable in the presence of location invariants in general remains an open question.

5   Evaluation

We compare the performance of parameterized verification of DTNs based on our new techniques to the existing cutoff-based techniques (according to Theorem 1, using UPPAAL). To this end, we implemented Algorithm 1 and the detection of flooding lassos from Sect. 4, and constructed three parametric examples: *i*) parametric versions of the TA in Fig. 1 with locations $h_0, \cdots, h_{k-1}, \ell_0, \cdots, \ell_{k-1}, h_{sy}, \ell_{sy}$ for some k, without location invariants on the h_i, denoted $GCS^\top(k)$ in the following, *ii*) versions of $GCS^\top(k)$ with invariants on the h_i, denoted $GCS(k)$, *iii*) hand-crafted examples $Star(k)$, where some q_{final} can only be reached after all $q \in Guards(A)$ with $|Guards(A)| = k$ have been reached for $k = 4$ and 5).

All experiments have been conducted on a machine with Intel® Core™ i7-8565U CPU @ 1.80 GHz and 16 GiB RAM, and our implementation as well as the benchmarks can be found at https://doi.org/10.5281/zenodo.8337446.[10]

---

[10] For the latest version of the tool refer to the GitHub repository.

Table 1. Performance comparisons, all times in seconds, timeout (TO) is 3000 s.

(a) Comparison of Algorithm 1 and UPPAAL (on the cutoff system A^c) for detection of minimal reachability times on $Star(k)$.

Benchmark		Time	
Name	Cutoff c	Alg. 1	UPPAAL
$Star(4)$	5	4.8	36.0
$Star(5)$	6	9.1	TO
$Star(6)$	7	15.7	TO

(b) Comparison of verification times for different versions of Fig. 1 and properties ϕ_i, based on the summary automaton ($\hat{A}$) or the cutoff system (A^c).

Benchmark					Time	
Name	$|Q|$	Cutoff c	Property	Result	$\hat{A}$	A^c
$GCS^\top(3)$	8	4	ϕ_1	False	< 0.1	< 0.1
			ϕ_3	True	< 0.1	26.0
$GCS(3)$	8	7	ϕ_1	False	< 0.1	< 0.1
			ϕ_3	True	< 0.1	TO

On $Star(k)$, Algorithm 1 significantly outperforms the cutoff-based approach for solving MINREACH with UPPAAL, as can be seen in Table 1a. On $GCS^\top(3)$, solving MINREACH and constructing the summary automaton takes 0.23 s, and 1.13 s for $GCS(3)$. Solving MINREACH using cutoffs is even faster, which is not surprising since in this example location guards do not influence the shortest paths. However, we can also use the summary automaton to check more complex temporal properties of the DTN, and two representative examples are shown in Table 1b: ϕ_1 states that a process that is in a location h_i will eventually be in a location ℓ_j, and ϕ_3 states that a process in a location h_i will eventually be in a location h_j. For ϕ_1, both approaches are very fast, while for ϕ_3 our new approach significantly outperforms the cutoff-based approach. Note that even if we add the time for construction of the summary automaton to the verification time, we can solve most queries significantly faster than the cutoff-based approach. Additional experimental results can be found in [11].

6   Conclusion

In this work, we proposed a novel technique for parameterized verification of disjunctive timed networks (DTNs), i.e., an unbounded number of timed automata, synchronizing on disjunctive location guards. Our technique to solve MINREACH in a network of arbitrary size relies on an extension of the zone graph of a *single* TA of the network—leading to an exponential reduction of the model to be analyzed, when compared to classical cutoff techniques.

If guard locations can always remain occupied after first reaching them, solving MINREACH allows us to construct a *summary automaton* that can be used for parameterized verification of more complex properties, which is again exponentially more efficient than existing cutoff techniques. This is the case for the full class of DTNs without invariants on guard locations, and we give a sufficient condition for correctness of the approach on the larger class with invariants, but with a single clock per automaton. Moreover, our *ad-hoc* prototype implementation already outperforms cutoff-based verification in UPPAAL on tasks that significantly rely on location guards.

Data Availability Statement. The program (including the source code) and bench-mark files as evaluated in Sect. 5 are available at https://doi.org/10.5281/zenodo. 8337446. Additionally, we plan to track all future development of our tool in the GitHub repository at https://github.com/cispa/Verification-Disjunctive-Time-Networks.

References

1. Abdulla, P.A., Delzanno, G.: Parameterized verification. Int. J. Softw. Tools Technol. Transfer **18**(5), 469–473 (2016). https://doi.org/10.1007/s10009-016-0424-3
2. Abdulla, P.A., Delzanno, G., Rezine, O., Sangnier, A., Traverso, R.: Parameterized verification of time-sensitive models of ad hoc network protocols. Theoret. Comput. Sci. **612**, 1–22 (2016). https://doi.org/10.1016/j.tcs.2015.07.048
3. Abdulla, P.A., Deneux, J., Mahata, P.: Multi-clock timed networks. In: LiCS, pp. 345–354. IEEE Computer Society (2004). https://doi.org/10.1109/LICS.2004. 1319629
4. Abdulla, P.A., Jonsson, B.: Model checking of systems with many identical timed processes. Theoret. Comput. Sci. **290**(1), 241–264 (2003). https://doi.org/10.1016/ S0304-3975(01)00330-9
5. Abdulla, P.A., Sistla, A.P., Talupur, M.: Model checking parameterized systems. In: Handbook of Model Checking, pp. 685–725. Springer, Cham (2018). https:// doi.org/10.1007/978-3-319-10575-8_21
6. Al-Bataineh, O.I., Reynolds, M., French, T.: Finding minimum and maximum termination time of timed automata models with cyclic behaviour. Theoret. Comput. Sci. **665**, 87–104 (2017). https://doi.org/10.1016/j.tcs.2016.12.020
7. Alur, R., Dill, D.L.: A theory of timed automata. Theoret. Comput. Sci. **126**(2), 183–235 (1994). https://doi.org/10.1016/0304-3975(94)90010-8
8. Alur, R., La Torre, S., Pappas, G.J.: Optimal paths in weighted timed automata. Theoret. Comput. Sci. **318**(3), 297–322 (2004). https://doi.org/10.1016/j.tcs.2003. 10.038
9. Aminof, B., Kotek, T., Rubin, S., Spegni, F., Veith, H.: Parameterized model checking of rendezvous systems. Distrib. Comput. **31**(3), 187–222 (2018). https:// doi.org/10.1007/s00446-017-0302-6
10. André, É., Delahaye, B., Fournier, P., Lime, D.: Parametric timed broadcast protocols. In: Enea, C., Piskac, R. (eds.) VMCAI 2019. LNCS, vol. 11388, pp. 491–512. Springer, Cham (2019). https://doi.org/10.1007/978-3-030-11245-5_23
11. André, É., Eichler, P., Jacobs, S., Karra, S.L.: Parameterized verification of disjunctive timed networks (2023). https://doi.org/10.48550/arXiv.2305.07295
12. Apt, K.R., Kozen, D.: Limits for automatic verification of finite-state concurrent systems. Inf. Process. Lett. **22**(6), 307–309 (1986). https://doi.org/10.1016/0020-0190(86)90071-2
13. Außerlechner, S., Jacobs, S., Khalimov, A.: Tight cutoffs for guarded protocols with fairness. In: Jobstmann, B., Leino, K.R.M. (eds.) VMCAI 2016. LNCS, vol. 9583, pp. 476–494. Springer, Heidelberg (2016). https://doi.org/10.1007/978-3-662-49122-5_23
14. Behrmann, G., Bouyer, P., Larsen, K.G., Pelánek, R.: Lower and upper bounds in zone-based abstractions of timed automata. Int. J. Softw. Tools Technol. Transfer **8**(3), 204–215 (2006). https://doi.org/10.1007/s10009-005-0190-0
15. Behrmann, G., Fehnker, A., Hune, T., Larsen, K., Pettersson, P., Romijn, J.: Efficient guiding towards cost-optimality in UPPAAL. In: Margaria, T., Yi, W. (eds.)

TACAS 2001. LNCS, vol. 2031, pp. 174–188. Springer, Heidelberg (2001). https://doi.org/10.1007/3-540-45319-9_13

16. Bengtsson, J., Yi, W.: Timed automata: semantics, algorithms and tools. In: Desel, J., Reisig, W., Rozenberg, G. (eds.) ACPN 2003. LNCS, vol. 3098, pp. 87–124. Springer, Heidelberg (2004). https://doi.org/10.1007/978-3-540-27755-2_3

17. Bloem, R., et al.: Decidability of Parameterized Verification. Synthesis Lectures on Distributed Computing Theory, Morgan & Claypool Publishers, San Rafael (2015). https://doi.org/10.2200/S00658ED1V01Y201508DCT013

18. Bouajjani, A., Habermehl, P., Vojnar, T.: Verification of parametric concurrent systems with prioritised FIFO resource management. Formal Methods Syst. Des. 32(2), 129–172 (2008). https://doi.org/10.1007/s10703-008-0048-7

19. Bouyer, P., Gastin, P., Herbreteau, F., Sankur, O., Srivathsan, B.: Zone-based verification of timed automata: extrapolations, simulations and what next? In: Bogomolov, S., Parker, D. (eds.) FORMATS. LNCS, vol. 13465, pp. 16–42. Springer, Cham (2022). https://doi.org/10.1007/978-3-031-15839-1_2

20. Bruyère, V., Dall'Olio, E., Raskin, J.F.: Durations and parametric model-checking in timed automata. ACM Trans. Comput. Logic 9(2), 12:1–12:23 (2008). https://doi.org/10.1145/1342991.1342996

21. Clarke, E., Talupur, M., Touili, T., Veith, H.: Verification by network decomposition. In: Gardner, P., Yoshida, N. (eds.) CONCUR 2004. LNCS, vol. 3170, pp. 276–291. Springer, Heidelberg (2004). https://doi.org/10.1007/978-3-540-28644-8_18

22. Courcoubetis, C., Yannakakis, M.: Minimum and maximum delay problems in real-time systems. Formal Methods Syst. Des. 1(4), 385–415 (1992). https://doi.org/10.1007/BF00709157

23. Emerson, E.A., Kahlon, V.: Reducing model checking of the many to the few. In: McAllester, D. (ed.) CADE 2000. LNCS (LNAI), vol. 1831, pp. 236–254. Springer, Heidelberg (2000). https://doi.org/10.1007/10721959_19

24. Emerson, E.A., Kahlon, V.: Model checking guarded protocols. In: LICS, pp. 361–370. IEEE Computer Society (2003). https://doi.org/10.1109/LICS.2003.1210076

25. Emerson, E.A., Namjoshi, K.S.: On reasoning about rings. Int. J. Found. Comput. Sci. 14(4), 527–550 (2003). https://doi.org/10.1142/S0129054103001881

26. Esparza, J., Ganty, P., Majumdar, R., Weil-Kennedy, C.: Verification of immediate observation population protocols. In: Schewe, S., Zhang, L. (eds.) CONCUR. LIPIcs, vol. 118, pp. 31:1–31:16. Schloss Dagstuhl - Leibniz-Zentrum für Informatik (2018). https://doi.org/10.4230/LIPIcs.CONCUR.2018.31

27. Esparza, J., Jaax, S., Raskin, M.A., Weil-Kennedy, C.: The complexity of verifying population protocols. Distrib. Comput. 34(2), 133–177 (2021). https://doi.org/10.1007/s00446-021-00390-x

28. Esparza, J., Raskin, M., Weil-Kennedy, C.: Parameterized analysis of immediate observation petri nets. In: Donatelli, S., Haar, S. (eds.) PETRI NETS 2019. LNCS, vol. 11522, pp. 365–385. Springer, Cham (2019). https://doi.org/10.1007/978-3-030-21571-2_20

29. Hanna, Y., Samuelson, D., Basu, S., Rajan, H.: Automating cut-off for multi-parameterized systems. In: Dong, J.S., Zhu, H. (eds.) ICFEM 2010. LNCS, vol. 6447, pp. 338–354. Springer, Heidelberg (2010). https://doi.org/10.1007/978-3-642-16901-4_23

30. Herbreteau, F., Srivathsan, B.: efficient on-the-fly emptiness check for timed büchi automata. In: Bouajjani, A., Chin, W.-N. (eds.) ATVA 2010. LNCS, vol. 6252, pp. 218–232. Springer, Heidelberg (2010). https://doi.org/10.1007/978-3-642-15643-4_17

31. Herbreteau, F., Srivathsan, B., Walukiewicz, I.: Efficient emptiness check for timed Büchi automata. Formal Methods Syst. Des. **40**(2), 122–146 (2012). https://doi.org/10.1007/s10703-011-0133-1

32. Herbreteau, F., Srivathsan, B., Walukiewicz, I.: Lazy abstractions for timed automata. In: Sharygina, N., Veith, H. (eds.) CAV 2013. LNCS, vol. 8044, pp. 990–1005. Springer, Heidelberg (2013). https://doi.org/10.1007/978-3-642-39799-8_71

33. Jacobs, S., Sakr, M.: Analyzing guarded protocols: better cutoffs, more systems, more expressivity. In: VMCAI 2018. LNCS, vol. 10747, pp. 247–268. Springer, Cham (2018). https://doi.org/10.1007/978-3-319-73721-8_12

34. Jones, N.D., Landweber, L.H., Lien, Y.E.: Complexity of some problems in petri nets. Theoret. Comput. Sci. **4**(3), 277–299 (1977). https://doi.org/10.1016/0304-3975(77)90014-7

35. Merro, M., Ballardin, F., Sibilio, E.: A timed calculus for wireless systems. Theoret. Comput. Sci. **412**(47), 6585–6611 (2011). https://doi.org/10.1016/j.tcs.2011.07.016

36. Pagliarecci, F., Spalazzi, L., Spegni, F.: Model checking grid security. Futur. Gener. Comput. Syst. **29**(3), 811–827 (2013). https://doi.org/10.1016/j.future.2011.11.010

37. Raskin, M.A., Weil-Kennedy, C., Esparza, J.: Flatness and complexity of immediate observation Petri nets. In: Konnov, I., Kovács, L. (eds.) CONCUR. LIPIcs, vol. 171, pp. 45:1–45:19. Schloss Dagstuhl - Leibniz-Zentrum für Informatik (2020). https://doi.org/10.4230/LIPIcs.CONCUR.2020.45

38. Spalazzi, L., Spegni, F.: Parameterized model checking of networks of timed automata with boolean guards. Theoret. Comput. Sci. **813**, 248–269 (2020). https://doi.org/10.1016/j.tcs.2019.12.026

39. Suzuki, I.: Proving properties of a ring of finite-state machines. Inf. Process. Lett. **28**(4), 213–214 (1988). https://doi.org/10.1016/0020-0190(88)90211-6

40. Zhang, Z., Nielsen, B., Larsen, K.G.: Distributed algorithms for time optimal reachability analysis. In: Fränzle, M., Markey, N. (eds.) FORMATS 2016. LNCS, vol. 9884, pp. 157–173. Springer, Cham (2016). https://doi.org/10.1007/978-3-319-44878-7_10

Resilience and Home-Space for WSTS

Alain Finkel[1,2] and Mathieu Hilaire[1(✉)]

[1] Université Paris-Saclay, CNRS, ENS Paris-Saclay, LMF, Gif-sur-Yvette, France
`mathieu.hilaire@univ-lyon1.fr`
[2] Institut Universitaire de France, Paris, France

Abstract. Resilience of unperfect systems is a key property for improving safety by insuring that if a system could go into a bad state in **Bad** then it can also leave this bad state and reach a safe state in **Safe**. We consider six types of resilience (one of them is the home-space property) defined by an upward-closed set or a downward-closed set **Safe**, and by the existence of a bound on the length of minimal runs starting from a set **Bad** and reaching **Safe** (**Bad** is generally the complementary of **Safe**).

We first show that all resilience problems are undecidable for effective Well Structured Transition Systems (WSTS) with strong compatibility. We then show that resilience is decidable for Well Behaved Transition Systems (WBTS) and for WSTS with adapted effectiveness hypotheses. Most of the resilience properties are shown decidable for other classes like WSTS with the downward compatibility, VASS, lossy counter machines, reset-VASS, integer VASS and continuous VASS.

Keywords: Verification · Resilience · Home-Space · Well-structured transition systems · Vector addition system with states

1 Introduction

Context. Resilience is a key notion for improving safety of unperfect systems and resilience engineering is a paradigm for safety management that focuses on systems coping with complexity and balancing productivity with safety [23]. Some systems are subjects at frequent intervals to accidents, attacks or changes. In such cases, a question that arises is that of whether the system can return to its normal (safe) behavior after an accident or attack pushed it towards some kind of 'error state' and, if it can, whether it can perform the return in a satisfactory timeframe.

Resiliences. Given a transition system whose set of states is S and let $X, H \subseteq S$, we say that (X, H) satisfies the *home-space* property [20] if the reachability set from X is included in the set of predecessors of H. In 1986, Memmi and Vautherin introduced the notion of home-space [20] in Petri nets for X a singleton. In

This work was partly done while the authors were supported by the Agence Nationale de la Recherche grant BraVAS (ANR-17-CE40-0028).

1989, de Frutos Escrig and Johnen proved that the home-space problem (for X a singleton and H a finite union of linear sets with the same period) was decidable for VASS [13]. In 2023, Jancar and Leroux proved the decidability of the (complete) semilinear home-space problem (X and H are both semilinear) in VASS [17].

A transition system is *resilient for* (*Bad,Safe*) if (Bad, Safe) satisfies the home-space property. A transition system is *resilient for* Safe if it is resilient for (Bad, Safe) with Bad is the complementary to Safe. It is *state-resilient* if it is resilient for (Bad, Safe) where Bad contains an unique state. The *k-resilience* problem, for $k \geq 0$, is to decide whether from any state is it always possible to reach Safe with a run of length smaller than k and the *bounded resilience* problem is to decide whether there exists an k such that the system is k-resilient.

State of the Art

In 2016, Prasad and Zuck introduced in [25] intuitions, definitions (without any connection to the concept of the home-space) and results (without detailed proofs) about resilience in the framework of process algebra. They show that resilience is decidable for effective Well Structured Transition Systems (WSTS) with both upward and reflexive downward compatibilities and some other technical conditions (that are defined in Sect. 2). In 2021, Özkan and Würdemann [22] and Özkan [21], in 2022, proved the decidability of the bounded state-resilience problem and the k-state-resilience problem for WSTS with strong compatibility (see Sect. 2) and with the supplementary (strong) hypothesis that there exists an algorithm that computes a finite basis of the upward-closure of the reachability set from s when Safe is upward-closed.

Our Main Contributions

- Surprisingly, the general undecidability statements about resilience were not known neither proved. We show that resilience and state-resilience (i.e. X a singleton) problems are both undecidable for WSTS with strong compatibility and for Safe upward-closed or downward-closed.
- In the subsequent three figures, we present the majority of our results pertaining to various models and resilience categories. The figures depict the boundaries of decidability within the context of well-known VASS and reset-VASS models.
- In particular, the three resilience problems are decidable for post-ideal-effective (Definition 5) Well Behaved Transition Systems (a generalisation of WSTS where the quasi-ordering is not necessarily well founded) with strong (upward) compatibility and when Safe is upward-closed (Theorem 4). Resilience is also decidable, when Safe is upward-closed, for effective predbasis (Definition 5) WSTS with strong (upward) compatibility (Theorem 5).
- We clarify the different effectiveness conditions on WBTS and WSTS (Definition 5) that allow the decidability of the six resilience.
- We generalize with Theorem 8 the main theorem of [21,22] and we show that relaxing some hypothesis leads to undecidability (Proposition 5).
- We show that the three state-resilience problems are decidable for post-ideal-effective WBTS with downward and upward compatibilities and Safe

downward-closed (Theorem 9 extends [25, Theorem 1]). The two other types of resilience, k-state-resilience and bounded-state-resilience, are decidable for post-ideal-effective WBTS with strong downward compatibility.

- We study the resilience problems for VASS and variations of VASS where most of the resilience problems are shown decidable (Fig. 1).

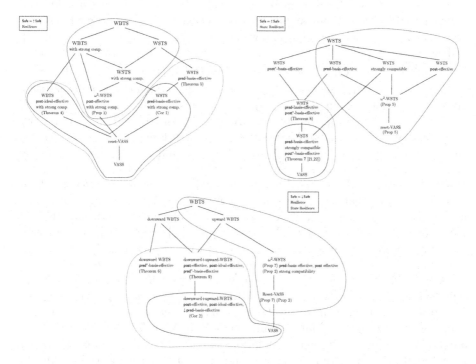

Fig. 1. Hasse diagram of some classes of transition systems, together with the decidability (in green) or undecidability (in red) of the resilience problems. Decidability of bounded resilience and $k-$resilience variants are indicated in blue. (Color figure online)

Plan of the Paper

Section 2 introduces the necessary definitions of WSTS and variations as well as the different notions of effectivity. Section 3 and 4 are concerned with the general framework of WSTS and contain our results on the three resilience problems and the three state-resilience problems respectively. Lastly, Sect. 5 concerns itself with the resilience problems for VASS and variations of VASS.

2 Well-Structured Transition Systems and VASS

A *transition system* is a pair $\mathscr{S} = (S, \rightarrow)$ where S is a set of *states* and $\rightarrow \subseteq S \times S$ is a binary relation on the set of states, denoted as the set of *transitions*. We write $s \to s'$ to denote $(s, s') \in \, \to$. We write $\to^k$, $\to^+$, $\to^=$, $\to^*$ for the k-step iteration of $\to$, its transitive closure, its reflexive closure, its reflexive and transitive closure. Let $X, Y \subseteq S$ and $k \in \mathbb{N}$; we denote $X \longrightarrow^* Y$ (resp. $X \longrightarrow^{\leq k} Y$) if from all states $x \in X$ there exists a path (resp. of length smaller than k) that reaches a state $y \in Y$. The set of *(immediate) successors* of a state $s \in S$ is defined as $\mathrm{post}_{\mathscr{S}}(s) = \{s' \in S \mid s \to s'\}$. The set of *(immediate) predecessors* of a state $s \in S$ is defined as $\mathrm{pred}_{\mathscr{S}}(s) = \{s' \in S \mid s' \to s\}$. To simplify the notations, we write without ambiguity $\mathrm{post}_{\mathscr{S}}(D)$ by $\mathrm{post}(D)$ and $\mathrm{pred}_{\mathscr{S}}(D)$ by $\mathrm{pred}(D)$. By iterating pred and post we obtain $\mathrm{post}^n(s) = \{s' \in S \mid s \to^n s'\}$ and $\mathrm{pred}^n(s) = \{s' \in S \mid s' \to^n s\}$. However, we are generally more interested in $\mathrm{post}^{\leq n}(s) = \bigcup_{1 \leq i \leq n} \mathrm{post}^i(s)$, $\mathrm{post}^*(s) = \bigcup_{1 \leq i} \mathrm{post}^i(s)$ and $\mathrm{pred}^{\leq n}(s) = \bigcup_{1 \leq i \leq n} \mathrm{pred}^i(s)$ and $\mathrm{pred}^*(s) = \bigcup_{1 \leq i} \mathrm{pred}^i(s)$. The *reachability problem* asks, given a transition system $\mathscr{S} = (S, \to)$, two states $s, t \in S$, whether $s \to^* t$.

A *quasi-ordering* (a qo) is any reflexive and transitive relation $\leq$ over some set X and we often write $(X, \leq)$. Given a quasi-ordering $(X, \leq)$, an *upward-closed set* is any set $U \subseteq X$ such that if $x \leq y$ and $x \in U$ then $y \in U$. A *downward-closed set* is any set $D \subseteq X$ such that if $y \leq x$ and $x \in D$ then $y \in D$. It is an *ideal* if it is also *directed*, i.e. it is nonempty and for every $a, b \in D$, there exists $c \in D$ such that $a \leq c$ and $b \leq c$. To any subset $A \subseteq X$, we may associate its *upward-closure*, $\uparrow A = \{x \in X \mid \exists a \in A \, a \leq x\}$ and its *downward-closure*, $\downarrow A = \{x \in X \mid \exists a \in A \, x \leq a\}$. We abbreviate $\uparrow\{x\}$ (resp. $\downarrow\{x\}$) as $\uparrow x$ (resp. $\downarrow x$). A *basis* of an upward-closed set U is a set U_b such that $U = \uparrow U_b$; similarly, a *basis* of a downward-closed set D is a set D_b such that $D = \downarrow D_b$.

A *well-quasi-ordering* (wqo) is any quasi-ordering $(X, \leq)$ such that, for any infinite sequence $x_0, x_1, x_2, \ldots$ in X, there exist indexes $i \leq j$ with $x_i \leq x_j$. This property is equivalent to the *finite decomposition* property: every upward-closed set $\emptyset \neq U \subseteq X$ admits a finite basis $B \subseteq X$ such that $U = \uparrow B$. Wqo admits many other equivalent formulations like: a qo $(X, \leq)$ is a wqo iff $(X, \leq)$ is well founded (i.e. there is no infinite strictly decreasing sequence of elements of X) and $(X, \leq)$ contains no infinite antichains (an antichain is a subset of mutually incomparable elements of X). See other equivalences in [15,27]. As an example, $(\mathbb{N}^d, \leq)$, the set of vectors of d natural numbers (where d is finite) with component-wise order is a wqo.

Quasi-orderings that have no infinite antichains enjoy a similar *finite decomposition* property than wqo: every downward-closed subset $D \subseteq X$ can be decomposed into a *finite* set of ideals $J_1, J_2, \ldots, J_n$ such that $D = J_1 \cup J_2 \cup \ldots \cup J_n$ See for example [7]. In what follows, a downward-closed set D is represented by its finite set of ideals (or by the minimal elements of its upward-closed complement), and an upward-closed set U is represented by its finite set of minimal elements. Let us now recall the (most general) definition of well-structured transition systems.

Definition 1 (Definition 3.10, [9]). *A Well-Structured Transition System (WSTS) $\mathscr{S} = (S, \to, \leq)$ is a transition system $(S, \to)$ equipped with a wqo $\leq \, \subseteq S \times S$ such that the transition relation $\to$ is (upward) compatible with*

$\leq$, *i.e., for all* $s_1, t_1, s_2 \in S$ *with* $s_1 \leq s_2$ *and* $s_1 \to t_1$, *there exists* $t_2 \in S$ *with* $t_1 \leq t_2$ *and* $s_2 \to^* t_2$.

An (upward) compatible transition relation $\to$ is *reflexive* when the sequence $s_2 \to^* t_2$ is not empty: formally, for all $s_1, t_1, s_2 \in S$ with $s_1 \leq s_2$ and $s_1 \to t_1$, there exists $t_2 \in S$ with $t_1 \leq t_2$ and $s_2 \to^+ t_2$. We say that a WSTS $\mathscr{S}$ has *strong (upward) compatibility* when moreover for all $s_1, t_1, s_2 \in S$ with $s_1 \leq s_2$ and $s_1 \to t_1$, there exists $t_2 \in S$ with $t_1 \leq t_2$ and $s_2 \to t_2$.

Several families of formal models of processes [12] give rise to WSTSs in a natural way with different compatibilities, e.g. compatible counter machines like VASS with d counters and Q a finite set of control-states (and equivalently Petri nets) data Petri nets, reset/transfer VASS are WSTS with strong compatibility for the usual ordering $= \times \leq^d$ on the set of states $S = Q \times \mathbb{N}^d$. Similarly, lossy channel systems with d channels and Q a finite set of control-states are WSTS (with a non-strong compatibility) for the ordering $= \times \sqsubseteq^d$ (where $\sqsubseteq$ is the subword ordering on Σ^*) on $S = Q \times (\Sigma^*)^d$.

But there is a more general class of (upward) compatible ordered transition systems than WSTS for which coverability is still decidable: recall that a Well Behaved Transition System (WBTS) [6] is an upward compatible ordered transition system $\mathscr{S} = (S, \to, \leq)$ where $(S, \leq)$ contains no infinite antichains (but it can be not well founded). The class of WBTS is strictly larger than WSTS: for example, $\mathbb{Z}^d$–VASS under the lexicographical ordering are WBTS but not WSTS [6].

By applying Proposition 4.3 from [10], we can reintroduce a straightforward definition of a specific proper subset of wqos that allow to construct coverability trees:

Definition 2. *A wqo* $(X, \leq)$ *is an* ω^2-*wqo if* $(Ideals(X), \subseteq)$ *forms a wqo.*

Although there exists wqo that are not ω^2-wqo (see the Rado ordering in [16]), all naturally occurring wqos are ω^2-wqo, perhaps to the notable exception of finite graphs well-quasi-ordered by the graph minor relation. The class of ω^2-wqo is robust because every datatype in the following list - natural number, finite set, finite product, finite sum, finite disjoint sum, finite words, finite multisets and finite trees - is an ω^2-wqo [10, Proposition 4.5]. Now we are able to define ω^2-WSTS:

Definition 3. *(Finkel and Goubault-Larrecq [10]) A WSTS* $\mathscr{S} = (S, \to, \leq)$ *is an* ω^2-*WSTS if* $(S, \leq)$ *is an* ω^2-*wqo.*

Since almost of usual wqo are ω^2-wqo, all naturally occurring WSTS are in fact ω^2-WSTS: for example, reset/transfer VASS are ω^2-WSTS.

Let us recall that the *completion* [7] of a WSTS $\mathscr{S} = (S, \to, \leq)$ is the associated ordered transition system $\hat{\mathscr{S}} = (Ideals(S), \to, \subseteq)$ where $Ideals(S)$ is the set of ideals of S and $I \to J$ if J belongs to the finite ideal decomposition of $\downarrow \mathrm{post}_{\mathscr{S}}(I)$. The completion is always *finitely branching* but it is not necessarily a WSTS since $\subseteq$ is not necessarily a wqo. It is proved in [7] that $\hat{\mathscr{S}}$ is WSTS iff $\mathscr{S} = (S, \to, \leq)$ is ω^2-WSTS.

Another type of WSTS exists that enjoys downward compatibility.

Definition 4. *A* downward (compatible) WBTS (short, downward-WBTS) $\mathscr{S} = (S, \rightarrow, \leq)$ *is an ordered transition system* $(S, \rightarrow, \leq)$ *such that* $\leq$ *is without infinite antichains and the transition relation* $\rightarrow$ *is downward compatible with* $\leq$, *i.e., for all* $s_1, t_1, s_2 \in S$ *with* $s_2 \leq s_1$ *and* $s_1 \rightarrow t_1$, *there exists* $t_2 \in S$ *with* $t_2 \leq t_1$ *and* $s_2 \rightarrow^* t_2$.

A downward (compatible) WSTS (short, downward-WSTS) *[12, Definition 5.1] is a downward-WBTS where* $\leq$ *is a wqo.*

There are fewer downward-WSTS than WSTS, but we can still mention FIFO automata with insertion errors and Basic Process Algebra (BPA).

Let us reformulate the downward compatibility property.

Lemma 1. *For an ordered transition system* $\mathscr{S} = (S, \rightarrow, \leq)$ *(not necessarly a WSTS), the two following properties are equivalent: (1)* $\mathscr{S} = (S, \rightarrow, \leq)$ *is downward compatible; (2) for every downward-closed set* $D \subseteq S$, *the set* $\mathsf{pred}^*(D)$ *is downward-closed.*

To obtain decidability results, we must introduce some notions of *effectiveness*. We use a mixture of the notions defined both by Blondin and al. in [6] and by Halfon in [15]. First, to simplify and w.l.o.g., we suppose that all classes of considered ordered transition systems $\mathscr{S} = (S, \rightarrow, \leq)$ satisfy the five following properties:

1. there exists an algorithm that decides whether $s \rightarrow t$ is true or not, for any states $s, t \in S$,
2. there is a computational representation for S for which membership in S is decidable,
3. $\leq$ is decidable,
4. there is a computational representation for $Ideals(S)$ for which membership in $Ideals(S)$ is decidable, and
5. inclusion of ideals is decidable.

Blondin and al. showed [6, Lemma 4.3] that under the previous hypotheses, one are able to enumerate *downward-closed sets* (by their finite decomposition in ideals), to decide inclusion between downward-closed sets and to decide if a state belongs to a given finite set of ideals. But the five properties are not sufficient to compute the complementary and the intersection of downward-closed sets. Halfon introduced *ideally-effective wqo* as wqo that essentially allow to compute representations of (principal) ideals $\downarrow s$, of the complementary of an ideal, of the complementary of filters (a filter is a set $\uparrow s$ for $s \in S$) and to compute representations of finite intersections of filters and finite intersections of ideals. Most well-known wqo are ideally-effective [15].

Starting now, we assume that all considered WBTS $\mathscr{S} = (S, \rightarrow, \leq)$ satisfy the five previous properties and that the wqo $\leq$ is ideally-effective. We refer to such WBTS as *effective*.

In order to construct algorithms, we also require effective hypotheses about the computations of certain sets of predecessors and successors. More precisely:

Definition 5. *We say that an ordered transition system* $\mathscr{S} = (S, \rightarrow, \leq)$ *is*

1. *post-effective if* $\mathscr{S}$ *is effective, and if there exists an algorithm that computes* $|post(s)| \in \mathbb{N} \cup \{\infty\}$ *on input* $s \in S$.
2. *post-ideal-effective if* $\mathscr{S}$ *is effective and there exists an algorithm accepting any ideal* $I \in Ideals(S)$ *and returning* $\downarrow post(I)$, *expressed as a finite union of (maximal) ideals.*
3. *pred-basis-effective [1, 12] if* $\mathscr{S}$ *is effective and there exists an algorithm accepting any state* $s \in S$ *and returning a finite basis of* $\uparrow pred(\uparrow s)$.
4. *post*-basis-effective [21, 22] if* $\mathscr{S}$ *is effective and there exists an algorithm accepting any state* $s \in S$ *and returning a finite basis of* $\uparrow post^*(s)$.
5. *pred*-basis-effective if* $\mathscr{S}$ *is effective and there exists an algorithm accepting any ideal* $I \in Ideals(S)$ *and returning a finite basis of* $\downarrow pred^*(I)$.

Counter machines are effective but don't enjoy any other effectivities among the list of five. Reset VASS (hence VASS) and (front) lossy fifo automata are post-effective, post-ideal-effective, and pred-basis-effective. There exist post-effective WSTS that are not post-ideal-effective [7, Proposition 35]; there exist post-ideal-effective WSTS that are not post-effective [7, Proposition 36].There also exist WSTS that are not pred-basis-effective [7, Proposition 45]. Reset VASS are not post*-basis-effective since post*-basis-effectiveness allows to compute the finite set of the set of minimal reachable states, hence it would allow to decide the zero-reachability problem that is undecidable for reset-VASS. However, VASS are post*-basis-effective [21, Proposition 2] and LCS (with d fifo channels) are also post*-basis-effective because we have $(q, w_1, w_2, ..., w_d) \in \uparrow post^*(q_0, u_1, u_2, ..., u_d)$ iff $(q, \epsilon, ..., \epsilon) \in \uparrow post^*(q_0, u_1, u_2, ..., u_d)$ iff $(q, \epsilon, ..., \epsilon)$ is reachable from $(q_0, u_1, u_2, ..., u_d)$ that is decidable in LCS. Reset-VASS are not pred*-basis-effective because computing a finite basis of $\downarrow pred^*(q, 0, ..., 0)$ would allow to decide whether $(q, 0, ..., 0)$ is reachable that is undecidable for reset-VASS. By using the decidability of reachability and [29, Theorem 3.11], we may prove that VASS are pred*-basis-effective. A more complete study of these properties will be done in the long version of this paper.

Recall the *coverability problem* for ordered transition systems.

COVERABILITY PROBLEM

INPUT: An ordered transition system $\mathscr{S} = (S, \rightarrow, \leq)$ and two states $s_0, s \in S$.

QUESTION: $s_0 \in pred^*(\uparrow s)$?

With the pred-basis-effective hypothesis, we obtain:

Theorem 1 (Theorem 3.6, [12], Theorem 4.1, [1]). *A finite basis of* $pred^*(U)$ *is computable for any pred-basis-effective WSTS* $\mathscr{S} = (S, \rightarrow, \leq)$ *and any upward-closed set* $U \subseteq S$ *given with its finite basis. Hence coverability is decidable.*

With the post-ideal-effective hypothesis, we obtain the decidability of coverability for WBTS (without the pred-basis-effective hypothesis).

Theorem 2 (Corollaire 4.4, [6]). *Coverability is decidable for any post-ideal-effective WBTS.*

Downward-WSTS enjoy a powerfull property.

Theorem 3 (Proposition 5.4, [12]). *Finitely branching downward-WSTS with reflexive compatibility and post-effective are post*-basis-effective.*

Let us recall the definition of vector addition system with (control-)states.

Definition 6. *A* vector addition system with (control-)states (VASS) *in dimension d (d-VASS for short) is a finite $\mathbb{Z}^d$-labeled directed graph $V = (Q, T)$, where Q is the set of* control-states, *and $T \subseteq Q \times \mathbb{Z}^d \times Q$ is the set of* control-transitions.

Subsetquently, $Q \times \mathbb{N}^d$ is the set of states of the transition system associated with a d-VASS V. For all states $p(\mathbf{u}), q(\mathbf{v}) \in Q \times \mathbb{N}^d$ and for every control-transition $t = (p, \mathbf{z}, q)$, we write $p(\mathbf{u}) \xrightarrow{t} q(\mathbf{v})$ whenever $\mathbf{v} = \mathbf{u} + \mathbf{z} \geq \mathbf{0}$. When in the context of a d-VASS, we denote 0^d by $\mathbf{0}$.

A *vector addition system (VAS)* in dimension d (d-VAS for short) is a d-VASS where the set of control-states is a singleton; hence one only needs T (Fig. 2).

VASS can be extended with resets.

Definition 7. *A* reset-VASS *in dimension d is a finite labeled directed graph $V = (Q, T)$, where Q is the set of* control-states, *$T \subseteq Q \times Op \times Q$ is the set of* control-transitions, *and $Op = \{add(\mathbf{z}) \mid \mathbf{z} \in \mathbb{Z}^d\} \cup \{reset(i) \mid i \in \{1, \dots, d\}\}$.*

For every states $p(\mathbf{u}), q(\mathbf{v}) \in Q \times \mathbb{N}^d$ and every control-transition t we write $p(\mathbf{u}) \xrightarrow{t} q(\mathbf{v})$ when

- $t = (q, add(\mathbf{z}), q') \in T$ and $\mathbf{u} + \mathbf{z} = \mathbf{v} \geq 0$,
- $t = (q, reset(i), q') \in T$ and $\mathbf{v}[i] = 0$, and $\mathbf{v}[i'] = \mathbf{u}[i']$ for all $i' \in \{1, \dots, d\} \setminus i$.

As an example, let us remark that $(\mathbb{N}^d, \leq)$ is an ideally-effective wqo.

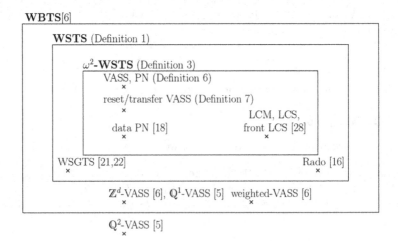

Fig. 2. A taxonomy of WSTS and variants.

3   Resilience for WSTS

In a (not necessarly ordered) transition system $\mathscr{S} = (S, \rightarrow)$, we consider a subset of states Safe $\subseteq S$, and its complement, Bad. The *resilience problem* (resp. the *k-resilience problem*) for $(\mathscr{S}, \text{Safe})$ is to decide whether from *any* state in S, *there exists* a path (resp. a path of length smaller than or equal to k) that reaches a state in Safe. Resilience is then akin to the Home-Space problem (defined in the introduction) for the set Safe. Resilience can also be viewed as a generalizeation of coverability, as it asks whether for *every* element of Bad it is possible to cover an element of the basis of Safe. We use the notation $S \longrightarrow^* \text{Safe}$ (resp. $S \longrightarrow^{\leq k} \text{Safe}$) for $\forall x \in S, \exists y \in \text{Safe}$ such that $x \longrightarrow^* y$ (resp. $\forall x \in S, \exists y \in \text{Safe}$ such that $x \longrightarrow^{\leq k} y$). In our framework, Safe $\subseteq S$ is possibly infinite but must admit a computable finite representation: for example, downward-closed sets and upward-closed sets in wqos and semilinear sets in $\mathbb{N}^d$ have finite representations.

Let us formalize three resilience problems.

RESILIENCE PROBLEMS

INPUT: A transition system $\mathscr{S} = (S, \rightarrow)$, and a set Safe $\subseteq S$.

QUESTION: (RESILIENCE PROBLEM (RP)) Bad $\longrightarrow^*$ Safe ?

(*k*-RESILIENCE PROBLEM (kRP)) Bad $\longrightarrow^{\leq k}$ Safe ?

(BOUNDED RESILIENCE PROBLEM (BRP)) $\exists k \geq 0$ $S \longrightarrow^{\leq k}$ Safe ?

These three resilience problems are decidable for finite transition systems but undecidable for (general) infinite-state transition systems. So we restrict our framework to the class of infinite-state WSTS. Since most of decidable properties in WSTS rely on the computation of upward or downward-closed sets [1,12], we consider upward-closed or downward-closed sets Safe. Since Safe $\subseteq \text{pred}^*(\text{Safe})$, one only needs to decide whether the complement of Safe is in $\text{pred}^*(\text{Safe})$. From now on, we use Bad to denote the complement of Safe.

Surprisingly, the general undecidability statement regarding resilience had neither been known nor proven (it is simply mentioned as a future work in the conclusion of [22]). We show that the resilience problem is undecidable for ω^2-WSTS with strong compatibility and natural effectiveness hypothesis.

Proposition 1. RESILIENCE *and* BOUNDED RESILIENCE *for* **post-effective** ω^2-*WSTS with strong compatibility, for upward closed sets* Safe *are undecidable.*

Proof. Indeed, consider the family $\{f_j : \mathbb{N}^2 \rightarrow \mathbb{N}^2\}$ of increasing recursive functions from [11] defined as

$$f_j(n, k) = \begin{cases} (n, 0) & \text{if k} = 0 \text{ and TM}_j \text{ runs for more than } n \text{ steps} \\ (n, n + k) & \text{otherwise,} \end{cases}$$

where TM_j is the j-th Turing machine (in a classical enumeration) which moreover begins by writing the integer j on its tape, and consider additionally the function $g : \mathbb{N}^2 \to \mathbb{N}^2$ defined by $g(n,k) = (n+1,k)$. The transition system $S_j = (\mathbb{N}^2, \{f_j, g\}, \leq)$ is a post-effective ω^2-WSTS with strong compatibility and has the property that it is $\uparrow(1,1)$-resilient iff TM_j halts on input j, since $(1,1)$ is coverable from $(0,0)$ iff TM_j halts on input j. Hence there is no Turing machine which correctly determines resilience and halts whenever its input is a WSTS. Hence RESILIENCE for ω^2-WSTS with strong compatibility is undecidable. Furthermore, strong compatibility implies the existence of $k \in \mathbb{N}$ such that $\text{pred}^*(\text{Safe}) = \text{pred}^{\leq k}(\text{Safe})$, hence BOUNDED-RESILIENCE for ω^2-WSTS with strong compatibility is undecidable as well. □

Let us remark that the transition system $S_j = (\mathbb{N}^2, \{f_j, g\}, \leq)$ in the previous proof is not post-ideal-effective nor pred-basis-effective. Termination and boundedness are decidable for increasing recursive (functional) WSTS [11]; the decidability of coverability and place-boundedness depends on the effectiveness hypothesis on recursive functions.

We now prove that RESILIENCE is undecidable for reset-VASS when Safe = ↓ Safe by a reduction from the undecidable [3] decision problem of *zero-reachability* in reset-VASS.

ZERO-REACHABILITY

INPUT: A reset-VASS $V = (Q, T)$ of dimension d, $q_f \in Q$, $p(\mathbf{u}) \in Q \times \mathbb{N}^d$.
QUESTION: $p(\mathbf{u}) \to^* q_f(\mathbf{0})$?

Proposition 2. RESILIENCE *is undecidable for reset-VASS, hence it is also undecidable for* post-effective *and* pred-basis-effective ω^2-*WSTS with strong compatibility and with* Safe = ↓ Safe.

Proof. Let V be a reset-VASS, $q_0(\mathbf{0})$ an initial state of V, and $q_f(\mathbf{0})$ a terminal state of V. We build V' from V by additionally allowing from any control-state to get back to $q_0(\mathbf{0})$ by a transition reseting every counter and changing the control-state to q_0. Remark that the set of states reachable from $q_0(\mathbf{0})$ is the same in V' and in V. Then $q_f(\mathbf{0})$ is reachable from $q_0(\mathbf{0})$ in V iff $q_f(\mathbf{0})$ is reachable from every state in V', which corresponds to RESILIENCE in V' for Safe = $\{q_f(\mathbf{0})\}$. Indeed, if $q_f(\mathbf{0})$ is reachable from $q_0(\mathbf{0})$ in V, then, since $q_0(\mathbf{0})$ is reachable from every state in V', $q_f(\mathbf{0})$ is reachable from every state in V'. In the other direction, if $q_f(\mathbf{0})$ is reachable from every state in V', then in particular $q_f(\mathbf{0})$ is reachable from $q_0(\mathbf{0})$ in V', hence $q_f(\mathbf{0})$ is reachable from $q_0(\mathbf{0})$ in V. □

3.1   Case: Safe = ↑ Safe

We start with the case Safe = ↑ Safe, hence Bad = ↓ Bad. Since Resilience is undecidable for ω^2-WSTS with strong compatibility (Theorem 1), we still consider WSTS, and even WBTS, with strong compatibility but by strengthening the assumptions of effectiveness (we now consider the post-ideal-effective hypothesis) and we demonstrate that the three resilience problems are now decidable.

Theorem 4. *Let* $\mathscr{S} = (S, \rightarrow, \leq)$ *be a* **post-ideal-effective** *WBTS with strong compatibility and a set* Safe $= \uparrow$ Safe. RESILIENCE, BOUNDED RESILIENCE *and* k-RESILIENCE *are decidable.*

Proof. Let us first recall two results in [7] that are stated for WSTS but are also true for WBTS since the proofs rely only on compatibility and not on the property of wqo. [7, Proposition 30] establishes a strong relation between the runs of a WSTS $\mathscr{S} = (S, \rightarrow, \leq)$ and the runs of its completion $\hat{\mathscr{S}}$. It states that if $x \xrightarrow{k} y$ in $\mathscr{S}$ then for every ideal $I \supseteq \downarrow x$, there exists an ideal $J \supseteq \downarrow y$ such that $I \xrightarrow{k} J$ in $\hat{\mathscr{S}}$. [7, Proposition 29] establishes that if $I \xrightarrow{k} J$ in $\hat{\mathscr{S}}$ then for every $y \in J$, there exists $x \in I$ and $y' \geq y$ such that $x \xrightarrow{k'} y'$ in $\mathscr{S}$. Moreover, if $\mathscr{S}$ has transitive compatibility then $k' \geq k$; if $\mathscr{S}$ has strong compatibility then $k' = k$.

Let $\{s_1, s_2, ..., s_m\}$ be the (unique) minimal basis of Safe and $\{J_1, J_2, ..., J_n\}$ be the ideal decomposition of Bad. The resilience problem can be reduced to the following infinite number of instances of the coverability problem in $\mathscr{S}$: for all $x \in$ Bad does there exist an j such that s_j is coverable from x. Let us show how this infinite set of coverability questions can be reduced to a *finite* set of coverability questions in the completion $\hat{\mathscr{S}} = (Ideals(S), \rightarrow, \subseteq)$ of $\mathscr{S} = (S, \rightarrow, \leq)$.

Let us prove that s_j is coverable from x in $\mathscr{S}$ if and only if $\downarrow s_j$ is coverable (for inclusion) from $\downarrow x$ in $\hat{\mathscr{S}}$. Suppose that s_j is coverable from x then there exists a run $x \xrightarrow{k} y \geq s_j$. From [7, Proposition 30], there exist an ideal J and a run $\downarrow x \xrightarrow{k} J$ where $J \supseteq \downarrow y \supseteq \downarrow s_j$ in $\hat{\mathscr{S}}$, hence $\downarrow s_j$ is covered from $\downarrow x$. Conversely, if $I \xrightarrow{k} J$ in $\hat{\mathscr{S}}$ with $\downarrow s_j \subseteq J$ then there exists $x \in I$ and $y' \geq s_j$ such that $x \xrightarrow{k} y' \geq s_j$ in $\mathscr{S}$ and then s_j is coverable from x in $\mathscr{S}$.

Hence we obtain: $\mathscr{S}$ is resilient iff for all $i = 1, .., n$ and $j = 1, ..m$, $\downarrow s_j$ is coverable from ideal J_i in $\hat{\mathscr{S}}$. Let us denote by $k_{i,j}$ the length of a covering sequence that covers $\downarrow s_j$ from J_i in $\hat{\mathscr{S}}$ and let $k_{i,j} \overset{\text{def}}{=} \infty$ if $\downarrow s_j$ is not coverable from J_i. Let us now define $K_{\mathscr{S}}(\mathsf{Safe}) = \max(k_{i,j} \mid i = 1, .., n$ and $j = 1, ..m)$.

We now have $\mathscr{S}$ is resilient iff $K_{\mathscr{S}}(\mathsf{Safe})$ is finite iff $\mathscr{S}$ is $K_{\mathscr{S}}(\mathsf{Safe})$-resilient with $K_{\mathscr{S}}(\mathsf{Safe})$ finite.

This implies that resilience and bounded resilience can be reduced to coverability. Since coverability is decidable for post-ideal-effective WBTS [7, Theorem 44], we deduce that both the resilience problem and the bounded resilience problem are decidable.

Let us now show that the k-resilience problem, with $k \in \mathbb{N}$, is also decidable. Let us denote by $k'_{i,j}$ the *minimal* length of a covering sequence that covers $\downarrow s_j$ from J_i in $\hat{\mathscr{S}}$ if it exists and let $k'_{i,j} \overset{\text{def}}{=} \infty$ if $\downarrow s_j$ is not coverable from J_i. If $\downarrow s_j$ is coverable from J_i, we first compute an $k_{i,j}$, and then we compute $k'_{i,j}$ by iteratively checking whether there exists a sequence of length $0, 1, ..., k_{i,j} - 1$ that covers $\downarrow s_j$ from J_i until we find the minimal one which is necessarily smaller (or equal to) than $k_{i,j}$.

Let us now define $K'_{\mathscr{S}}(\mathsf{Safe}) = \max(k'_{i,j} \mid i = 1,..,n$ and $j = 1,..m)$ and we deduce that $\mathscr{S}$ is k-resilient iff $k \geq K'_{\mathscr{S}}(\mathsf{Safe})$.   □

Theorem 5. *Let $\mathscr{S} = (S, \rightarrow, \leq)$ be a pred-basis-effective WSTS and a set* Safe $= \uparrow$ *Safe.* RESILIENCE *is decidable.*

Proof. The resilience problem can be reformulated as $\mathsf{Bad} \subseteq \mathsf{pred}^*(\mathsf{Safe})$ that is equivalent to $\mathsf{Bad} \cap (S \setminus \mathsf{pred}^*(\mathsf{Safe})) = \emptyset$. The set $\mathsf{pred}^*(\mathsf{Safe})$ is upward-closed, since Safe is upward-closed. Since $\mathscr{S} = (S, \rightarrow, \leq)$ is pred-basis-effective, we can compute a basis of $\mathsf{pred}^*(\mathsf{Safe})$. Since the wqo is ideally-effective, we can compute the intersection of Bad and $S \setminus \mathsf{pred}^*(\mathsf{Safe})$, which are both downward-closed, and test if this intersection is empty or not. Hence, the resilience problem is decidable.   □

Strong compatibility implies furthermore the decidability of k-RESILIENCE and BOUNDED-RESILIENCE.

Corollary 1. *Let $\mathscr{S} = (S, \rightarrow, \leq)$ be a pred-basis-effective WSTS with strong compatibility and a set* Safe $= \uparrow$ *Safe.* BOUNDED RESILIENCE *and k-RESILIENCE are decidable.*

Proof. For strongly compatible WSTS, $\mathsf{pred}^k(\mathsf{Safe}) = \uparrow \mathsf{pred}^k(\mathsf{Safe})$, and hence $\mathsf{pred}^{\leq k}(\mathsf{Safe}) = \uparrow \mathsf{pred}^{\leq k}(\mathsf{Safe})$, when Safe $= \uparrow$ Safe. Since $\mathscr{S} = (S, \rightarrow, \leq)$ is pred-basis-effective, we can compute a basis of $\uparrow \mathsf{pred}^{\leq k}(\mathsf{Safe}) = \mathsf{pred}^{\leq k}(\mathsf{Safe})$. Like above, we can test if the intersection of Bad and $S \setminus \mathsf{pred}^{\leq k}(\mathsf{Safe})$ is empty or not, hence k-RESILIENCE is decidable. To decide BOUNDED RESILIENCE, we check k-RESILIENCE starting with $k = 0$ until we find some k_0 such that either k_0-RESILIENCE holds, either $\mathsf{pred}^{\leq k_0}(\mathsf{Safe}) = \mathsf{pred}^{\leq k_0 + 1}(\mathsf{Safe})$, whichever comes first. The convergence of $(\mathsf{pred}^{\leq k}(\mathsf{Safe}))_{k \in \mathbb{N}}$ guarantees the latter eventually happens.   □

Remark that the above proofs do not make use of the property that Bad is the complement of Safe, simply using $\mathsf{Bad} = \downarrow \mathsf{Bad}$ and $\mathsf{Safe} = \uparrow \mathsf{Safe}$, thus the above results still hold in the more general case where Bad and Safe are not complements of each others. Remark furthermore that reset-VASS are pred-basis-effective hence satisfy the conditions from Theorem 5 which thus implies decidability of RESILIENCE, BOUNDED RESILIENCE and k-RESILIENCE in reset-VASS in the case Safe $= \uparrow$ Safe.

3.2   Case: Safe $= \downarrow$ Safe

Let us now consider the case Safe $= \downarrow$ Safe hence Bad $= \uparrow$ Bad. It is of interest to note this case can be linked to the problem of mutual exclusion. Indeed the well-known mutual exclusion property can be modeled, in a d-VASS with d counters, by the property that a special counter c_{mutex} must be bounded by $k \geq 1$ which counts the (maximal) number of processes that are allowed to be simultaneously in the critical section. Then, the set Safe $= \{c_{mutex} \leq k\} \times \mathbb{N}^{d-1}$ is downward-closed and Bad $= \{c_{mutex} \geq k+1\} \times \mathbb{N}^{d-1}$ is the upward-closed complementary of Safe.

Theorem 6. RESILIENCE *is decidable for* pred**-basis-effective downward-WBTS with* Safe $= \downarrow$ Safe.

Proof. Let $\mathscr{S}$ be a pred*-basis-effective downward-WBTS with Safe $= \downarrow$ Safe. By Lemma 1, pred*(Safe) is downward-closed; moreover, pred*(Safe) is computable because $\mathscr{S}$ is pred*-basis-effective. The resilience problem can be reformulated as Bad $\cap (S \setminus$ pred*(Safe)) $= \emptyset$. Since $\le$ is ideally-effective, we can compute complementaries and intersections of upward and downward-closed subsets. Hence we can compute $S \setminus$ pred*(Safe) and then the intersection of Bad and $S \setminus$ pred*(Safe), which are both upward-closed. We may decide whether this intersection is empty, hence the resilience problem is decidable. $\square$

In the case of a pred*-basis-effective WBTS, not necessarily downward compatible, the above construction can provide a proof of non-resilience i.e. when Bad $\cap (S \setminus \downarrow$ pred*(Safe)) $\ne \emptyset$ then Bad $\not\subseteq \downarrow$ pred*(Safe) and hence Bad $\not\subseteq$ pred*(Safe). When Bad $\cap (S \setminus \downarrow$ pred*(Safe)) $= \emptyset$ however it is not enough to conclude.

Remark we did not make use of the property Bad complement of Safe, simply Bad $= \uparrow$ Bad and Safe $= \downarrow$ Safe, thus the above results still hold in the more general case where Bad and Safe are not complements of each others.

4   State-Resilience

Resilience is a strong property that implies that from every element there must exist a path to Safe. However, when one considers a system with an initial state s_0, it could be sufficient to only ask that from post*(s_0), there must exist a path to Safe. The three previous problems become:

STATE RESILIENCE PROBLEMS

INPUT: A transition system $\mathscr{S} = (S, \rightarrow)$, $s \in S$, and a set Safe $\subseteq S$.

QUESTION: (STATE-RESILIENCE PROBLEM (SRP)) post*(s) $\longrightarrow^*$ Safe ?
(k-STATE-RESILIENCE PROBLEM (KSRP)) post*(s) $\longrightarrow^{\le k}$ Safe ?
(BOUNDED-STATE-RESILIENCE PROBLEM (BSRP)) $\exists k \ge 0$ such that post*(s) $\longrightarrow^{\le k}$ Safe?

Remark that STATE-RESILIENCE is HOME-SPACE with input set Safe. Since these problems are undecidable for general infinite-state transition systems, we still restrict our study to WSTS. As in the Sect. 3, we study decidability results for Safe downward-closed and upward-closed.

4.1   Case: Safe $= \uparrow$ Safe

We start with the case Safe $= \uparrow$ Safe. Unfortunately, in this case STATE-RESILIENCE is undecidable for (general) WSTS even with strong upward-compatibility. This stems from the fact that it is undecidable in the particular case of reset-VASS, where t-liveness is both undecidable and reducible to STATE-RESILIENCE. This undecidability result furthermore implies the undecidability of the other two state resilience problems by straightforward reductions.

A control-transition t of a reset-VASS is *live* in a state $r(\mathbf{w})$ if for each $q(\mathbf{v}) \in$ post$^*(r(\mathbf{w}))$ there exists two states $p(\mathbf{u}), p'(\mathbf{u'})$ such that $q(\mathbf{v}) \to^* p(\mathbf{u}) \xrightarrow{t} p'(\mathbf{u'})$. We say that the whole reset-VASS is live if all its control-transitions are live. This leads to the following problem.

t-LIVENESS

INPUT: A reset-VASS $V = (Q, T)$ of dimension d, a transition $t \in T$, an initial state $s_0 \in Q \times \mathbb{N}^d$

QUESTION: Is t live in s_0 ?

Let us define the set of states, pre(t), from which a control-transition t is enabled: pre$(t) = \{p(\mathbf{u}) \in Q \times \mathbb{N}^d \mid \exists q(\mathbf{v}) \in Q \times \mathbb{N}^d$ such that $p(\mathbf{u}) \xrightarrow{t} q(\mathbf{v})\}$. Remark that pre$(t)$ is upward-closed.

Proposition 3. t-LIVENESS *is reducible to* STATE-RESILIENCE *in reset-VASS.*

Proof. We reformulate t-liveness in a reset-VASS (Q, T) with initial state s_0 as the following formula

$$\forall p(\mathbf{u}) \in Q \times \mathbb{N}^d, \ s_0 \to^* p(\mathbf{u}) \implies \exists q(\mathbf{v}) \in \mathsf{pre}(t), \ p(\mathbf{u}) \to^* q(\mathbf{v})$$

The previous formula reduces itself to STATE-RESILIENCE where Safe = pre(t).
$\square$

It now remains to argue why t-LIVENESS is undecidable. Recall ZERO-REACHABILITY in reset-VASS is undecidable [3]. The following Proposition is a variation to reset-VASS of [24, Theorem 5.5] originally stated for Petri nets, whose proof can be seen in the full version of the paper.

Proposition 4. ZERO-REACHABILITY *can be reduced to* t-LIVENESS.

Since ZERO-REACHABILITY for reset-VASS is undecidable [8], the reduction implies t-LIVENESS is undecidable. We deduce undecidability by reduction to STATE-RESILIENCE, and straightforward reductions to the other two state resilience problems.

Proposition 5. STATE-RESILIENCE, BOUNDED-STATE-RESILIENCE *and* k-STATE-RESILIENCE *are undecidable for reset-VASS, hence for strongly compatible, post-effective, pred-basis-effective* ω^2-*WSTS when* Safe = $\uparrow$ Safe.

Recall that on the other hand RESILIENCE is decidable for reset-VASS when Safe = $\uparrow$ Safe by Theorem 5. This suggests that the undecidability of STATE-RESILIENCE comes more from the fact $\uparrow$ post$^*(s_0)$ is not constructible in reset-VASS - a consequence of the undecidability of reachability - rather than from the difficulties inherent in searching for paths from post$^*(s_0)$ to Safe.

On the positive side, let us recall a result about BOUNDED-STATE-RESILIENCE (called resilience in [21, 22]).

Theorem 7 (Theorem 1, [21], Theorem 1, [22]). BOUNDED-STATE-RESILIENCE *and* k-STATE-RESILIENCE *are decidable for* **pred-basis-effective**, **post****-basis-effective WSTS with strong compatibility when* Safe = ↑ Safe.

The proof of Theorem 7 rely on the computability of ↑ post*(s) and on the following lemma.

Lemma 2. *Let* $A \subseteq S$, $D \subseteq S$ *be a downward-closed set and* $U \subseteq S$ *be an upward-closed set. Then* $A \cap D \subseteq U$ *iff* $(\uparrow A) \cap D \subseteq U$.

Özkan [21] argues that it is precisely the WSTS for which the following problem is decidable.

DOWNWARD-REACHABILITY PROBLEM
INPUT: A transition system $\mathscr{S} = (S, \rightarrow)$, $s \in S$ and a downward-closed set $D \subseteq S$.
QUESTION: $s \rightarrow^* D$?

Proposition 6 (Proposition 1, [21]). *For* **post-effective**, **pred-basis-effective** *finite-branching WSTS, a finite basis of* ↑ post*(s) *is computable for every state* s *iff the downward-reachability problem is decidable.*

The idea behind the proof is the following. For deciding whether a downward-closed set D is reachable from s, one checks whether $B_{\uparrow \, \text{post}^*(s)} \cap D = \emptyset$, where $B_{\uparrow \, \text{post}^*(s)}$ is a basis of ↑ post*(s), that is equivalent to post*(s) ∩ D = ∅ by Lemma 2. For the converse direction, one computes the sequence of upward-closed sets $U_n = \uparrow \text{post}^{\leq n}(s)$ until it becomes stationary. Decidability of downward-reachability leads to the decidability of the following stop condition: asking whether $S \setminus U_n$ is reachable from s.

For instance, VASS are post*-basis-effective WSTS [22]. It is well-known that VASS are WSTS with strong compatibility and since there is an algorithm that computes a finite basis of ↑ post*(s), [21] deduced that BOUNDED-STATE-RESILIENCE is decidable for VASS. Hence STATE-RESILIENCE is decidable for VASS.

However, the hypothesis that ↑ post* is computable cannot be tested in the general WSTS framework. Moreover there exist classes of WSTSs with strong compatibility for which there doesn't exist an algorithm computing a basis of ↑ post*, such as reset-VASS.

Keeping the ↑ post* effectiveness hypothesis but *loosening* the strong compatibility one still yields some decidability result for the general STATE-RESILIENCE. Using the same proof structure as Theorem 1 from [22] we obtain:

Theorem 8. STATE-RESILIENCE *is decidable for* **pred-basis-effective**, **post****-basis-effective WSTS when* Safe = ↑ Safe.

Proof. Let $\mathscr{S}$ be a pred-basis-effective, post*-basis-effective WSTS. Since $\mathscr{S}$ is post*-basis-effective, let $B_{\uparrow \, \text{post}^*(s)}$ be a finite basis of ↑ post*(s), hence

$\uparrow \mathsf{post}^*(s) = \uparrow B_{\uparrow \mathsf{post}^*(s)}$. Let B_{Safe} be a finite basis of Safe, hence $\mathsf{Safe} = . \uparrow B_{\mathsf{Safe}}$. Since $\mathscr{S}$ is pred-basis-effective, we may compute a finite basis $B_{\mathsf{pred}^*(\mathsf{Safe})}$ of $\mathsf{pred}^*(\mathsf{Safe})$ from B_{Safe}. By applying Lemma 2 twice, we obtain that

$$\mathsf{post}^*(s) \subseteq \mathsf{pred}^*(\mathsf{Safe}) \text{ iff } \uparrow \mathsf{post}^*(s) \subseteq \mathsf{pred}^*(\mathsf{Safe})$$

Now the last inclusion is equivalent to:

$$\uparrow B_{\uparrow \mathsf{post}^*(s)} \subseteq \uparrow B_{\mathsf{pred}^*(\mathsf{Safe})} \text{ iff } \forall b \in B_{\uparrow \mathsf{post}^*(s)} \exists b' \in B_{\mathsf{pred}^*(\mathsf{Safe})} \text{ such that } b' \le b$$

Since both $B_{\uparrow \mathsf{post}^*(s)}$ and $B_{\mathsf{pred}^*(\mathsf{Safe})}$ are finite, STATE-RESILIENCE is decidable.

□

However when removing strong compatibility, some precision is lost. Since $\mathsf{pred}(\uparrow \mathsf{Safe})$ is not necessarily upward-closed, it is possible to have $\uparrow \mathsf{post}^*(s) \cap S \not\subseteq \mathsf{pred}(\mathsf{Safe})$, despite having $\mathsf{post}^*(s) \cap S \subseteq \mathsf{pred}(\mathsf{Safe})$. In such a case, the algorithm in [21] would deduce that 1-STATE-RESILIENCE does not hold, which is incorrect.

Thus in case of post^*-basis-effective WSTS (when the compatibility is not strong), we don't know the decidability status of k-STATE-RESILIENCE and BOUNDED-STATE-RESILIENCE.

4.2   Case: Safe = ↓ Safe

We now consider the case $\mathsf{Safe} = {\downarrow} \mathsf{Safe}$. Unfortunately STATE-RESILIENCE is undecidable in this case. This stems from undecidability of ZERO-REACHABILITY in reset-VASS, as seen in the reduction proof of Proposition 2.

Proposition 7. STATE-RESILIENCE *is undecidable for reset-VASS when* $\mathsf{Safe} = {\downarrow} \mathsf{Safe}$, *hence also for* **post**-*effective,* **pred**-*basis-effective* ω^2-*WSTS with strong compatibility.*

Despite this, it is possible to yield positive results. Indeed, in many ways the case where $\mathsf{Safe} = {\downarrow} \mathsf{Safe}$ is symmetrical to the case $\mathsf{Safe} = {\uparrow} \mathsf{Safe}$. We will need the following lemma in order to obtain symmetrical results.

Lemma 3. *(Symmetrical from Lemma 2) Let* $A \subseteq S$, $D \subseteq S$ *be a downward-closed set and* $U \subseteq S$ *be an upward-closed set. Then* $A \cap U \subseteq D$ *iff* $({\downarrow} A) \cap U \subseteq D$.

In the case of a WBTS with *downward* compatibility, not necessarily strong, then Safe downward-closed implies $\mathsf{pred}^*(\mathsf{Safe})$ downward-closed and Lemma 3 can be used to show that if $\mathsf{Safe} = {\downarrow} \mathsf{Safe}$, then $\mathsf{post}^*(s) \subseteq \mathsf{pred}^*(\mathsf{Safe})$ iff $({\downarrow} \mathsf{post}^*(s)) \subseteq \mathsf{pred}^*(\mathsf{Safe})$.

Theorem 9. STATE-RESILIENCE *is decidable for* **post**-*effective,* **post**-*ideal-effective and* **pred***-*basis-effective WBTS with downward and upward compatibilities and* $\mathsf{Safe} = {\downarrow} \mathsf{Safe}$.

Proof. Since $\mathscr{S}$ is pred^*-basis-effective, $\mathsf{Safe} = \downarrow\mathsf{Safe}$ and $\mathsf{pred}^*(\mathsf{Safe}) = \downarrow\mathsf{pred}^*(\mathsf{Safe})$, we may compute the finite decomposition in ideals of $\mathsf{pred}^*(\mathsf{Safe})$.

In order to decide whether $\mathsf{post}^*(s) \subseteq \mathsf{pred}^*(\mathsf{Safe})$, we execute two procedures in parallel, one looking for a resilience certificate and one looking for a non-resilience certificate.

Procedure 1 enumerates every downward-closed subsets in some fixed order D_1, D_2, ... by their ideal decomposition. The computability of the enumeration comes from the hypothesis that $\leq$ is ideally-effective (Lemma 4.3. from [6]). The procedure then checks for every downward-closed subset D_i whether $\downarrow\mathsf{post}(D_i) \subseteq D_i$ (this inclusion is decidable because $\mathscr{S}$ is post-ideal-effective). Every subset D_i such that $\downarrow\mathsf{post}(D_i) \subseteq D_i$ is an "over-approximation" of $\downarrow\mathsf{post}^*(s)$ if it contains s. Notice that, by upward compatibility, $\downarrow\mathsf{post}^*(s)$ is such a subset and may eventually be found.

Procedure 1 stops when it finds a downward-closed subset D such that $\downarrow\mathsf{post}(D) \subseteq D$, $s \in D$ and $D \subseteq \mathsf{pred}^*(\mathsf{Safe})$.

Indeed $D \subseteq \mathsf{pred}^*(\mathsf{Safe})$ implies $\downarrow\mathsf{post}^*(s) \subseteq \mathsf{pred}^*(\mathsf{Safe})$ since $\downarrow\mathsf{post}^*(s) \subseteq D$.

The second procedure iteratively computes $\mathsf{post}^{\leq n}(s)$ (this is effective because $\mathscr{S}$ is post-effective) until it finds an element not in $\mathsf{pred}^*(\mathsf{Safe})$.

If resilience hold, then procedure 1 terminates since it eventually finds $D = \downarrow\mathsf{post}^*(s)$ such that $D \subseteq \mathsf{pred}^*(\mathsf{Safe})$. If resilience does not hold, then procedure 2 terminates since there exists a witness that resilience does not hold which can eventually be found. $\square$

In the case of strong downward compatibility, $\mathsf{pred}^{\leq k}(\mathsf{Safe})$ is downward-closed when Safe is. Hence, assuming a pred^*-basis-effective variation consisting in an algorithm accepting any ideal I and returning a finite basis of $\downarrow\mathsf{pred}(I)$, it becomes possible to perform the same procedures as above except with comparisons against $\mathsf{pred}^{\leq k}(\mathsf{Safe})$ rather than $\mathsf{pred}^*(\mathsf{Safe})$, and hence k-STATE-RESILIENCE and BOUNDED-STATE-RESILIENCE are decidable in this case.

Corollary 2. k-STATE-RESILIENCE *and* BOUNDED-STATE-RESILIENCE *are decidable for* **post**-effective, **post**-ideal-effective *WBTS, strong downward compatibility and upward compatibility when* $\mathsf{Safe} = \downarrow\mathsf{Safe}$ *and there exists an algorithm accepting any ideal I and returning a finite basis of $\downarrow\mathsf{pred}(I)$.*

Note that post-effective, post-ideal-effective WSTS, strong downward compatibility and upward compatibility include for instance Lossy Channel Systems with insertions [2].

5 Resilience for VASS and Variations

In this section we study VASS. Since they enjoy many properties of effectivity ($\leq$ is ideally-effective, VASS are post-effective, post-ideal-effective, pred-basis-effective, post*-basis-effective), they inherit the decidability results for WSTS in the case $\mathsf{Safe} = \uparrow\mathsf{Safe}$. Lacking downward compatibility or a more relaxed hypothesis that for all downward-closed set D, the set $\mathsf{pred}^*(D)$ is downward-closed, VASS do not inherit the decidability results for WSTS in the case $\mathsf{Safe} = \downarrow \mathsf{Safe}$. In this section, we work to re-establish decidability results for VASS when Safe is downward-closed and $\mathsf{Bad} = S \setminus \mathsf{Safe}$. We also extend decidability to resilience for semilinear sets rather than simply upward and downward-closed ones.

Surprisingly, when Safe is downward-closed, we have the following result for VAS (not VASS):

Proposition 8. BOUNDED RESILIENCE *and k-*RESILIENCE *never hold for d-VAS when $\mathsf{Safe} = \downarrow \mathsf{Safe}$ and $\mathsf{Safe} \neq \mathbb{N}^d$.*

Proof. Consider a given $k \in \mathbb{N}$, $\mathsf{Safe} \subseteq \mathbb{N}^d$ is downward-closed such that its upward-closed complement $\mathsf{Bad} \subseteq \mathbb{N}^d$ is nonempty, and consider a given VAS V. Let us call $c_{\max}$ the maximal absolute value of a constant appearing in a coordinate of a transition of V. The set $\mathsf{Bad} \neq \emptyset$ admits a finite basis B_{Bad}. Consider the vector $\mathbf{v}_{\mathsf{Bad}}$ obtained by summing all members of the basis of Bad and then consider the vector $\mathbf{u}_k = \mathbf{v}_{\mathsf{Bad}} + (k+1) \cdot (c_{\max}, c_{\max}, \ldots, c_{\max})$.

All states reachable from $\mathbf{u}_k$ in k or less steps are above $\mathbf{v}_{\mathsf{Bad}}$ and thus, are in Bad, by upward-closedness. Hence Safe is not reachable from $\mathbf{u}_k$ in k or less steps and k-resilience does not hold. Since the reasoning hold for all $k \in \mathbb{N}$, BOUNDED-RESILIENCE does not hold either.

$\square$

This all changes when one considers VASS (having control-states). For a VASS $V = (Q, T)$, a set $\mathsf{Bad} = \uparrow \mathsf{Bad}$, for all $q \in Q$, either there is an element of the basis of Bad with control-state q—then $q(\mathbf{v})$ with $\mathbf{v} > \mathbf{v}_{\mathsf{Bad}}$ is necessarily in Bad by upward closure; if this hold for all $q \in Q$, then k-resilience does not hold—either there is none. If there is none, then $\uparrow q(\mathbf{0})$ is in the complement of Bad, i.e. Safe. Based on this, we partition Safe into two subsets: an upward-closed union of sets of the form $\uparrow q(\mathbf{0})$, and a remaining downward-closed subset. The key now is that there is only a finite number of elements of Bad of the form $p(\mathbf{v})$ with $\mathbf{v} \leq \mathbf{u}_k$ for all of which we will check reachability of Safe in k or less steps, while elements of Bad of the form $p(\mathbf{v})$ with $\mathbf{v} > \mathbf{u}_k$ can only potentially reach the upward-closed subset of Safe in k or less steps. Only being concerned by the upward-closed subset of Safe in the latter case enable us to reuse techniques seen when Safe is upward-closed. Indeed we can compute the basis for the sets of elements from which our upward-closed subset of Safe is reachable in at most k steps. Substracting these predecessor from Bad yields either a finite number of elements from which one has to check Safe is reachable in at most k steps, either an infinite number of elements of which there is one which cannot reach Safe in at most k steps - for much the same reasons BOUNDED RESILIENCE and k-RESILIENCE never hold for VAS when $\mathsf{Safe} = \downarrow \mathsf{Safe}$ and $\mathsf{Bad} = \uparrow \mathsf{Bad}$. We check the finiteness of the set by comparing the elements from the finite basis of both upward-closed sets. This leads to a decision procedure detailed in the full version of the paper, which lead to the following decidability result:

Theorem 10. k-RESILIENCE *and* BOUNDED RESILIENCE *are decidable for VASS when* Safe $= \downarrow$ Safe.

Let us now extend decidability of RESILIENCE for semilinear sets and VASS variants. $\mathbb{Z}^d$–VASS (resp. $\mathbb{Z}$–VASS) [14] are d-VASS (resp. VASS) that are allowed to take values from the integers.

Theorem 11. RESILIENCE *is decidable for VASS, and* $\mathbb{Z}$–*VASS, when* Safe *is a semilinear set.*

Proof. We consider the case where Safe is semilinear and Bad $= S \setminus$ Safe is semilinear too. We want to use the fact that, for VASS, it is decidable whether $\mathsf{post}^*(X) \subseteq \mathsf{pred}^*(Y)$ when X and Y are both semilinear sets [17]. RESILIENCE asks whether Bad $\subseteq \mathsf{pred}^*($Safe$)$. We have $\mathsf{post}^*($Bad$) \setminus$ Safe $=$ Bad by extension of $S \setminus$ Safe $=$ Bad. Thus $\mathsf{post}^*($Bad$) =$ Bad $\cup$ (Safe $\cap \mathsf{post}^*($Bad$))$. Since (Safe $\cap \mathsf{post}^*($Bad$)) \subseteq \mathsf{pred}^*($Safe$)$, we have

$$\mathsf{post}^*(\mathsf{Bad}) \subseteq \mathsf{pred}^*(\mathsf{Safe}) \quad \text{iff} \quad \mathsf{Bad} \subseteq \mathsf{pred}^*(\mathsf{Safe}).$$

and hence, RESILIENCE is decidable. The reachability relation of $\mathbb{Z}^d$–VASS is definable by a well-formed formula with no free variables in the first-order theory of the integers with addition and orders (Presburger arithmetic) in $\mathbb{Z}^{2d}$, hence Bad $\subseteq \mathsf{pred}^*($Safe$)$ is decidable when Safe and Bad are semilinear sets. $\square$

RESILIENCE is also decidable for other classes of counter machines for which the reachability relation can be expressed in a decidable logic. Recall that lossy counter machines (LCM) [28] are counter machines that may loose tokens in each control-state.

Resilience and the home-space problem are also linked to the model-checking of basic reachability and safety formulae. In particular [28] shows that the "from-all" formula $\forall s \in X \ \exists t \in Y \ s \rightarrow^* t$ is decidable for lossy counter machine when X and Y are semi-linear sets. Other decidable formulae include "one-to-one" ($\exists s \in X \ \exists t \in Y \ s \rightarrow^* t$), and "all-to-same" ($\exists t \in Y \ \forall s \in X \ s \rightarrow^* t$), whereas "one-to-all" ($\exists s \in X \ \forall t \in Y \ s \rightarrow^* t$), "all-to-all" ($\forall s \in X \ \forall t \in Y \ s \rightarrow^* t$) and "to-all" ($\forall t \in Y \ \exists s \in X \ s \rightarrow^* t$) are undecidable (again, for LCM).

Theorem 12. RESILIENCE *is decidable for lossy counter machines when* Safe *is a semilinear set.*

Proof. We deduce from [28, Theorem 3.6] that $\mathsf{pred}^*($Safe$)$ is a computable semilinear set if Safe is semilinear. Hence since the inclusion between two semilinear sets is decidable, we deduce that Bad $\subseteq \mathsf{pred}^*($Safe$)$ is decidable if Bad is semilinear. $\square$

$\mathbb{Q}^d$-VASS [5] (or continuous VASS) are a relaxation of classical (discrete) d-VASS in which transitions can be fired a fractional number of times, and consequently counters may contain a fractional number of tokens.

Theorem 13. RESILIENCE *is decidable for continuous VASS when* Safe *is definable in the existential theory of the rationals with addition and order.*

Proof. The reachability relation of continuous VASS is definable by a sentence of linear size in the existential theory of the rationals with addition and order whose complexity is EXPSPACE [5]. Hence, Bad $\subseteq$ pred*(Safe) is decidable (and also in EXPSPACE).

$\square$

Remark that Theorem 12 (resp. Theorem 13) did not make use of the hypothesis that Bad is the complement of Safe, simply relying on the semilinearity of the set (resp. its definability in the existential theory of the rationals with addition and order). Thus the mentionned theorems still hold in the more general case where Bad and Safe are not complements of each others.

6   Conclusion and Perspectives

We complemented previous results (decidability of two state-resilience problems for a restricted class of WSTS in [21,22,25]) by providing some undecidability proofs for resilience and state-resilience in general. We exhibited classes of WBTS, WSTS, VASS and extensions of VASS with decidable resilience.

Several questions still remain. For instance, we have been concerned with decidability only, and a detailed complexity analysis of the different resilience problems still remains to be done for concrete models. Another question could be to analyse the resilience in the framework of a controller and its environment. One could also extend upon the classes of set Safe considered. As with semilinear sets for VASS, one could study resilience for sets defined in a boolean logic on upward and downward-closed subsets [4]. Finally, while we mention VASS, a more detailed analysis of the resilience problems could be also done for other computational models such as pushdown automata, one-counter automata or timed automata.

Acknowledgements. We express our thanks to the reviewers of the VMCAI 2024 Conference for their numerous and relevant comments and improvement suggestions.

References

1. Abdulla, P.A., Cerans, K., Jonsson, B., Tsay, Y.: Algorithmic analysis of programs with well quasi-ordered domains. Inf. Comput. **160**(1–2), 109–127 (2000). https://doi.org/10.1006/inco.1999.2843
2. Abdulla, P.A., Jonsson, B.: Verifying programs with unreliable channels. Inf. Comput. **127**(2), 91–101 (1996). https://doi.org/10.1006/inco.1996.0053
3. Araki, T., Kasami, T.: Some decision problems related to the reachability problem for petri nets. Theoret. Comput. Sci. **3**(1), 85–104 (1976)
4. Bertrand, N., Schnoebelen, P.: Computable fixpoints in well-structured symbolic model checking. Formal Methods Syst. Des. **43**(2), 233–267 (2013). https://doi.org/10.1007/s10703-012-0168-y

5. Blondin, M., Finkel, A., Haase, C., Haddad, S.: The logical view on continuous petri nets. ACM Trans. Comput. Log. **18**(3), 24:1–24:28 (2017). https://doi.org/10.1145/3105908

6. Blondin, M., Finkel, A., McKenzie, P.: Well behaved transition systems. Log. Methods Comput. Sci. **13**(3) (2017). https://doi.org/10.23638/LMCS-13(3:24)2017

7. Blondin, M., Finkel, A., McKenzie, P.: Handling infinitely branching well-structured transition systems. Inf. Comput. **258**, 28–49 (2018). https://doi.org/10.1016/j.ic.2017.11.001

8. Dufourd, C., Finkel, A., Schnoebelen, P.: Reset nets between decidability and undecidability. In: Larsen, K.G., Skyum, S., Winskel, G. (eds.) Automata, Languages, and Programming. LNCS, vol. 1443, pp. 103–115. Springer, Heidelberg (1998). https://doi.org/10.1007/BFb0055044

9. Finkel, A.: Reduction and covering of infinite reachability trees. Inf. Comput. **89**(2), 144–179 (1990). https://doi.org/10.1016/0890-5401(90)90009-7

10. Finkel, A., Goubault-Larrecq, J.: Forward analysis for WSTS, part II: complete WSTS. Log. Methods Comput. Sci. **8**(3) (2012). https://doi.org/10.2168/LMCS-8(3:28)2012

11. Finkel, A., McKenzie, P., Picaronny, C.: A well-structured framework for analysing petri net extensions. Inf. Comput. **195**(1–2), 1–29 (2004)

12. Finkel, A., Schnoebelen, P.: Well-structured transition systems everywhere! Theor. Comput. Sci. **256**(1–2), 63–92 (2001). https://doi.org/10.1016/S0304-3975(00)00102-X

13. de Frutos Escrig, D., Johnen, C.: Decidability of home space property. Université de Paris-Sud. Centre d'Orsay. Laboratoire de Recherche en Informatique, Rapport de recherche n°503 (1989)

14. Haase, C., Halfon, S.: Integer vector addition systems with states. In: Ouaknine, J., Potapov, I., Worrell, J. (eds.) RP 2014. LNCS, vol. 8762, pp. 112–124. Springer, Cham (2014). https://doi.org/10.1007/978-3-319-11439-2_9

15. Halfon, S.: On Effective Representations of Well Quasi-Orderings. (Représentations Effectives des Beaux Pré-Ordres). Ph.D. thesis, University of Paris-Saclay, France (2018). https://tel.archives-ouvertes.fr/tel-01945232

16. Jancar, P.: A note on well quasi-orderings for powersets. Inf. Process. Lett. **72**(5–6), 155–160 (1999). https://doi.org/10.1016/S0020-0190(99)00149-0

17. Jancar, P., Leroux, J.: Semilinear home-space is decidable for petri nets. CoRR abs/2207.02697 (2022). https://doi.org/10.48550/arXiv.2207.02697

18. Lazic, R., Newcomb, T.C., Ouaknine, J., Roscoe, A.W., Worrell, J.: Nets with tokens which carry data. Fundam. Informaticae **88**(3), 251–274 (2008). http://content.iospress.com/articles/fundamenta-informaticae/fi88-3-03

19. Mayr, R.: Lossy counter machines. Technical report TUM-I9827, Institut für Informatik (1998)

20. Memmi, G., Vautherin, J.: Analysing nets by the invariant method. In: Brauer, W., Reisig, W., Rozenberg, G. (eds.) ACPN 1986. LNCS, vol. 254, pp. 300–336. Springer, Heidelberg (1986). https://doi.org/10.1007/BFb0046843

21. Özkan, O.: Decidability of resilience for well-structured graph transformation systems. In: Behr, N., Strüber, D. (eds.) ICGT 2022. LNCS, vol. 13349, pp. 38–57. Springer, Cham (2022). https://doi.org/10.1007/978-3-031-09843-7_3

22. Özkan, O., Würdemann, N.: Resilience of well-structured graph transformation systems. In: Hoffmann, B., Minas, M. (eds.) Proceedings Twelfth International Workshop on Graph Computational Models, GCM@STAF 2021, Online, 22nd June 2021. EPTCS, vol. 350, pp. 69–88 (2021). https://doi.org/10.4204/EPTCS.350.5

23. Patriarca, R., Bergström, J., Di Gravio, G., Costantino, F.: Resilience engineering: current status of the research and future challenges. Saf. Sci. **102**, 79–100 (2018). https://doi.org/10.1016/j.ssci.2017.10.005

24. Peterson, J.L.: Petri Net Theory and the Modeling of Systems. Prentice Hall PTR, Hoboken (1981)

25. Prasad, S., Zuck, L.D.: Self-similarity breeds resilience. In: Gebler, D., Peters, K. (eds.) Proceedings Combined 23rd International Workshop on Expressiveness in Concurrency and 13th Workshop on Structural Operational Semantics, EXPRESS/SOS 2016, Québec City, Canada, 22nd August 2016. EPTCS, vol. 222, pp. 30–44 (2016). https://doi.org/10.4204/EPTCS.222.3

26. Schmitz, S.: The complexity of reachability in vector addition systems. ACM SIGLOG News **3**(1), 4–21 (2016). https://doi.org/10.1145/2893582.2893585

27. Schmitz, S., Schnoebelen, P.: Algorithmic Aspects of WQO Theory (2012). https://cel.hal.science/cel-00727025

28. Schnoebelen, P.: Lossy counter machines decidability cheat sheet. In: Kučera, A., Potapov, I. (eds.) RP 2010. LNCS, vol. 6227, pp. 51–75. Springer, Heidelberg (2010). https://doi.org/10.1007/978-3-642-15349-5_4

29. Valk, R., Jantzen, M.: The residue of vector sets with applications to decidability problems in petri nets. Acta Informatica **21**, 643–674 (1985). https://doi.org/10.1007/BF00289715

Model Checking and Synthesis

Generic Model Checking for Modal Fixpoint Logics in COOL-MC

Daniel Hausmann[1]([✉])[iD], Merlin Humml[2][iD], Simon Prucker[2][iD],
Lutz Schröder[2][iD], and Aaron Strahlberger[2]

[1] University of Gothenburg, Gothenburg, Sweden
`hausmann@chalmers.se`
[2] Friedrich-Alexander-Universität Erlangen-Nürnberg,
Erlangen, Germany
`merlin.humml@fau.de`

Abstract. We report on COOL-MC, a model checking tool for fixpoint logics that is parametric in the branching type of models (nondeterministic, game-based, probabilistic etc.) and in the next-step modalities used in formulae. The tool implements generic model checking algorithms developed in coalgebraic logic that are easily adapted to concrete instance logics. Apart from the standard modal μ-calculus, COOL-MC currently supports alternating-time, graded, probabilistic and monotone variants of the μ-calculus, but is also effortlessly extensible with new instance logics. The model checking process is realized by polynomial reductions to parity game solving, or, alternatively, by a *local* model checking algorithm that directly computes the extensions of formulae in a lazy fashion, thereby potentially avoiding the construction of the full parity game. We evaluate COOL-MC on informative benchmark sets.

Keywords: Model checking · parity games · μ-calculus · lazy evaluation

1 Introduction

The μ-calculus [25] is one of the most expressive logics for the temporal verification of concurrent systems. Model checking the μ-calculus is equivalent to parity game solving, and as such enjoys diversified tool support in the shape of both well-developed parity game solving suites such as PGSolver [11,38] or Oink [7] and dedicated model checking tools such as mCRL2 [3]. While the μ-calculus is standardly interpreted over relational transition systems, a wide range of alternative flavours have emerged whose semantics is variously based on concurrent

D. Hausmann—Funded by the ERC Consolidator grant D-SynMA (No. 772459).
M. Humml—Funded by the Deutsche Forschungsgemeinschaft (DFG, German Research Foundation) – project number 377333057 and 393541319/GRK2475/1-2019.
L. Schröder—Funded by the Deutsche Forschungsgemeinschaft (DFG, German Research Foundation) – project number 419850228.

R. Dimitrova et al. (Eds.): VMCAI 2024, LNCS 14499, pp. 171–185, 2024.
https://doi.org/10.1007/978-3-031-50524-9_8

games as in the alternating-time μ-calculus [2]; on probabilistic transition systems as in the (two-valued) probabilistic μ-calculus [4,5,30]; on counting successors as in the graded μ-calculus [26]; or on neighbourhood structures as in the monotone μ-calculus, the ambient fixpoint logic of game logic [8,32,34]. Model checking tools for such μ-calculi are essentially non-existent or limited to fragments (see additional comments under 'related work'). We present the generic model checker COOL-MC, which implements generic model checking algorithms for the *coalgebraic μ-calculus* [5] developed in previous work [23]. The coalgebraic μ-calculus is based on the semantic framework of *coalgebraic logic*, which treats systems generically as coalgebras for a set functor encapsulating the system type, following the paradigm of *universal coalgebra* [35], and parametrizes the semantics of modalities using so-called *predicate liftings* [33,36]. By fairly simple instantiation to concrete logics, COOL-MC thus serves as the first available model checker for the probabilistic μ-calculus, the graded μ-calculus, and the full alternating-time μ-calculus AMC (model checking tools for alternating-time temporal logic ATL, a fragment of the AMC, do exist, as discussed further below). Besides presenting the tool itself and discussing implementation issues, we conduct an experimental evaluation of COOL-MC on benchmark series of parity games [11,38] that we adapt to the generalized coalgebraic setting. We thus show that COOL-MC scales even on series of problems designed to be hard in the relational base case.

Related Work. As mentioned above, COOL-MC is the only currently available model checker for most of the logics that it supports, other than the standard modal μ-calculus (and the main point of its genericity is that support for further logics can be added easily). We refrain from benchmarking COOL-MC against modal μ-calculus model checkers (e.g. mCRL2 [3]) as this would essentially amount to comparing the respective backend parity game solvers. Model checking tools for alternating-time temporal logic ATL [2] do exist, such as MOCHA [1], MCMAS [31], and UMC4ATL [24], out of which MCMAS appears to be the fastest one currently available [24]. We do compare COOL-MC to MCMAS on two benchmarks, confirming that MCMAS is faster on ATL. Note however that ATL model checking works along essentially the same lines as for CTL, and as such is much simpler than model checking the alternating-time μ-calculus AMC (e.g., it does not require parity conditions, and unlike AMC model checking it is known to be in PTIME [2]), so it is expected that dedicated ATL model checkers will be faster than an AMC model checker like COOL-MC on ATL. Local solving has been shown to be advantagous in model checking for the relational μ-calculus [37] and for standard parity games [12].

COOL-MC uses the basic infrastructure, such as parsers and data structures for formulae, of the *Coalgebraic Ontology Logic Solver (COOL/COOL 2)* [13,14], a generic reasoner aimed at satisfiability checking rather than model checking. The algorithms we use here [23] improve on either the theoretical complexity or the complexity analysis of previous model checking algorithms for concrete instance logics including the alternating-time [2], graded [10], and monotone [16] μ-calculi, as well as of a previous generic model checking algorithm for the coalgebraic μ-calculus [18]; see [23] for details.

2   Model Checking for the Coalgebraic μ-Calculus

We briefly recall the syntax and semantics of the underlying generic logic of COOL-MC, the *coalgebraic μ-calculus* [5], and subsequently sketch two different model checking algorithms implemented for this logic in COOL-MC, a *local* algorithm that directly computes extensions of formulae, and a more global algorithm that reduces instances of the model checking problem to parity games [23].

Syntax. Formulae of the *coalgebraic μ-calculus* are given by the following grammar, parametrized over a choice of countable sets Λ and V of *modalities* and *(fixpoint) variables*, respectively.

$$\varphi, \psi ::= \top \mid \bot \mid \varphi \wedge \psi \mid \varphi \vee \psi \mid \heartsuit\varphi \mid X \mid \nu X.\, \varphi \mid \mu X.\, \varphi$$

where $X \in V$; we assume that Λ contains, for each modality $\heartsuit \in \Lambda$, also the dual $\overline{\heartsuit}$ (with $\overline{\overline{\heartsuit}} = \heartsuit$). The logic generalizes the standard μ-calculus by supporting arbitrary monotone modalities $\heartsuit$ in place of $\Diamond$ and $\Box$, assuming that the semantics of $\heartsuit$ can be defined in the framework of coalgebraic logic, recalled below. To ensure monotonicity, the logic does not contain negation as an explicit operator; however, negation of closed formulae can, as usual, be defined via negation normal forms. Given a formula φ, we let $|\varphi|$ denote its syntactic size. The algorithms we use work on the (Fischer-Ladner) *closure* $\mathsf{cl}(\varphi)$ of φ, a succinct graph representation of the respective formula, intuitively obtained from its syntax tree by identifying occurrences of fixpoint variables with their binding fixpoint operators; we have $|\mathsf{cl}(\varphi)| \leq |\varphi|$ [25]. The *alternation-depth* $\mathsf{ad}(\varphi)$ of fixpoint formulae $\varphi = \eta X.\, \psi$ is defined in the usual way as the number of dependent alternations between least and greatest fixpoints in φ; for a detailed account, see [27].

Semantics. The semantics of the coalgebraic μ-calculus is parametrized over the choice of a set functor $\mathcal{F}$ that encapsulates the branching type of systems, e.g. nondeterministic ($\mathcal{F}X = \mathcal{P}X$, the powerset of X) or probabilistic ($\mathcal{F}X = \mathcal{D}X$, the set of discrete probability distributions on X). Formulae are then evaluated over *coalgebras* $(C, \xi : C \to \mathcal{F}C)$ for $\mathcal{F}$, that is, over generalized transition systems consisting of a set C of states and *transition function* ξ that associates to each state c a collection $\xi(c) \in \mathcal{F}C$ of observations and successors, structured according to $\mathcal{F}$. For the most basic case, we can pick $\mathcal{F} = \mathcal{P}$ to be the powerset functor, so that $\mathcal{F}$-coalgebras are standard transition systems, with $\xi(c) \in \mathcal{P}C$ being the set of successor states of c.

The semantics of modalities $\heartsuit \in \Lambda$ is defined in terms of so-called *predicate liftings*, that is, functions $[\![\heartsuit]\!]$ that lift predicates $D \subseteq C$ on C to predicates $[\![\heartsuit]\!](D) \in \mathcal{P}(\mathcal{F}C)$ on $\mathcal{F}C$. A state $c \in C$ in a coalgebra (C, ξ) then satisfies a formula $\heartsuit\psi$ if $\xi(c) \in [\![\heartsuit]\!]([\![\psi]\!])$ where $[\![\psi]\!]$ is the set of states that satisfy ψ.

This concept instantiates to the standard modalities $\Diamond$ and $\Box$ over transition systems (that is, over coalgebras $(C, \xi : C \to \mathcal{P}C)$ for the functor $\mathcal{P}$) by taking predicate liftings

$$[\![\Diamond]\!](D) = \{E \in \mathcal{P}C \mid E \cap D \neq \emptyset\} \qquad [\![\Box]\!](D) = \{E \in \mathcal{P}C \mid E \subseteq D\}.$$

For another example, consider *graded* modalities of the shape $\langle n \rangle$ and $[n]$ (for $n \in \mathbb{N}$), expressing that more than n successors or all but at most n successors, respectively, satisfy the argument formula. We interpret such modalities over *graded transition systems*, in which every transition from one state to another is equipped with a non-negative integer *multiplicity*; these are coalgebras $(C, \xi : C \to \mathcal{G}C)$ for the *multiset functor* $\mathcal{G}$ that maps a set X to the set $\mathcal{G}X$ of finite multisets over X, represented as maps $X \to \mathbb{N}$ with finite support [6]. For $\theta : C \to \mathbb{N}$ and $D \subseteq C$, we put $\theta(D) = \Sigma_{d \in D} \theta(d)$, and interpret $\langle n \rangle$, $[n]$ as the predicate liftings

$$[\![\langle n \rangle]\!](D) = \{\theta \in \mathcal{G}C \mid \theta(D) > n\} \quad [\![[n]]\!](D) = \{\theta \in \mathcal{G}C \mid \theta(C \setminus D) \leq n\}.$$

Having defined the semantics of single modal steps, we now extend the semantics to the full logic, introducing the *game-based semantics* of the coalgebraic μ-calculus (which is equivalent to a recursively defined algebraic semantics [22, 23, 39]). To treat least and greatest fixpoints correctly, this semantics uses *parity games*, which are infinite-duration games played by two players $\exists$ and $\forall$. A parity game $G = (V, V_\exists, E, \Omega)$ consists of a set V of positions, with positions $V_\exists \subseteq V$ owned by $\exists$ and the others by $\forall$, a *move* relation $E \subseteq V \times V$, and a priority function $\Omega : V \to \mathbb{N}$ that assigns a natural number $\Omega(v)$ to each position $v \in V$. A *play* is a path in the directed graph (V, E) that is either infinite or ends in a node $v \in V$ with no outgoing moves. Finite plays $v_0 v_1 \ldots v_n$ are won by $\exists$ if and only if $v_n \in V_\forall$ (i.e. if $\forall$ is stuck); infinite plays are won by $\exists$ if and only if the maximal priority that is visited infinitely often is even. A *(history-free) $\exists$-strategy* is a partial function $s : V_\exists \rightharpoonup V$ that assigns moves to $\exists$-nodes. A play *follows* a strategy s if for all $i \geq 0$ such that $v_i \in V_\exists$, $v_{i+1} = s(v_i)$. An $\exists$-strategy *wins* a node $v \in V$ if $\exists$ wins all plays that start at v and follow s.

For the remainder of the paper, we fix a functor $\mathcal{F}$, an $\mathcal{F}$-coalgebra (C, ξ), a set Λ of modalities with associated monotone predicate liftings, and a formula χ (that uses modalities from Λ); further we let $\mathsf{cl} = \mathsf{cl}(\chi)$ denote the closure of χ, and put $n := |\mathsf{cl}|$ and $k := \mathsf{ad}(\chi)$.

Definition 1. The *model checking game* $G_{(C, \xi), \chi} = (V, V_\exists, E, \Omega)$ is the parity game defined by the following table, where game nodes $v \in V = V_\exists \cup V_\forall$ are of the shape $v = (c, \psi) \in C \times \mathsf{cl}$ or $v = (D, \psi) \in \mathcal{P}(C) \times \mathsf{cl}$.

node	owner	set of allowed moves
$(c, \top)$	$\forall$	$\emptyset$
$(c, \bot)$	$\exists$	$\emptyset$
$(c, \varphi \wedge \psi)$	$\forall$	$\{(c, \varphi), (c, \psi)\}$
$(c, \varphi \vee \psi)$	$\exists$	$\{(c, \varphi), (c, \psi)\}$
$(c, \eta X. \psi)$	$\exists$	$\{(c, \psi[\eta X. \psi/X])\}$
$(c, \heartsuit \psi)$	$\exists$	$\{(D, \psi) \mid \xi(c) \in [\![\heartsuit]\!](D)\}$
(D, ψ)	$\forall$	$\{(d, \psi) \mid d \in D\}$

In order to show satisfaction of $\heartsuit \psi$ at $c \in C$, player $\exists$ thus has to claim satisfaction of ψ at a sufficiently large set $D \subseteq C$ of states; player $\forall$ in turn can challenge the satisfaction of ψ at any node $d \in D$.

As usual in μ-calculi, the priority function Ω serves to detect that the outermost fixpoint that is unfolded infinitely often is a greatest fixpoint. It is thus defined ensuring that for nodes (c, φ), $\Omega(c, \varphi)$ is even if $\varphi = \nu X. \psi$, odd if $\varphi = \mu X. \psi$, and $\Omega(c, \varphi) = 0$ otherwise, and moreover that larger numbers are assigned to outer fixpoints, using the alternation depth of fixpoints. The formal definition of Ω follows the standard method, see e.g. [27].

We say that $c \in C$ *satisfies* χ (denoted $(C, \xi), c \models \chi$) if and only if player $\exists$ wins the position (c, χ) in $G_{(C,\xi),\chi}$. The *model checking problem* for the coalgebraic μ-calculus consists in deciding, for state $c \in C$ in a coalgebra (C, ξ), and formula χ of the coalgebraic μ-calculus, whether $(C, \xi), c \models \chi$.

We point out that $G_{(C,\xi),\chi}$ is a parity game with k priorities that contains up to $n \cdot 2^{|C|}$ positions of the form (D, ψ) for $D \subseteq C$. Therefore it is not feasible to perform model checking by explicitly constructing and solving this parity game. In previous work [21,23], we have shown that the model checking problem for the coalgebraic μ-calculus is in NP∩co-NP and in QP (under mild assumptions on the complexity of evaluating single modal steps using the predicate liftings), providing two methods to circumvent the explicit construction of the full game:

1. Compute the winning region in $G_{(C,\xi),\chi}$ as a nested fixpoint over the set of positions of the shape (c, ψ); intuitively, this avoids the explicit construction of the intermediate positions of the shape (D, ψ) by directly computing the extension of subformulae over C. This solution is generic in the sense that it works for any instance of the coalgebraic μ-calculus.
2. Provide a polynomial-sized game-characterization of the modalities of the concrete logic at hand, enabling a polynomial reduction of the model checking problem to solving parity games. This makes it possible to use parity game solvers, but relies on the logic-specific game characterization of the modalities.

As part of this work, we have implemented and evaluated both methods as an extension of the reasoner COOL, as described next.

3   Implementation – Model Checking in COOL-MC

We report on the implementation of model checking for the coalgebraic μ-calculus within the framework provided by the COalgebraic Ontology Logic solver (COOL), a coalgebraic reasoner for modal fixpoint logics [13], implemented in OCaml. The satisfiability-checking capacities of COOL have been reported elsewhere [14]. Our tool COOL-MC extends this framework with comprehensive functionality for model checking, along the lines of Sect. 2. To this end, we use existing infrastructure and data structures of COOL for parsing and representing (the closure of) input formulae χ for an extensible selection of logics, induced by the choice of a set functor $\mathcal{F}$; a newly added parser reads input models $(C, \xi : C \to \mathcal{F}(C))$ in the form of coalgebras for the selected functor (more details on the introduced specification format for coalgebras can be found in the artifact [19]). We thus obtain model checking support for

– the standard modal μ-calculus (including its fragment CTL) [25],

- the monotone μ-calculus (including its fragment game logic) [8,32,34],
- the alternating-time μ-calculus (including its fragment ATL) [2],
- the graded μ-calculus [26],
- the probabilistic μ-calculus [4,5,30].

By the relation between μ-calculus model checking and the solution of games with parity conditions, made more precise in Sect. 4 below, COOL-MC can also be seen as a generic qualitative solver for (standard, monotone, alternating-time, graded and probabilistic) parity games.

The core model checking functionality is provided by implementations of the two approaches described in Sect. 2: On the one hand, we implement the direct evaluation of formulae in the form of a generic *local model checking algorithm*; on the other hand, we also implement a *polynomial reduction to standard parity games* for each of the logics currently supported. Below, we provide intuitive explanations of the two algorithms, pointing out concrete implementational details only where the implementation is not straight-forward.

Local Model Checking. The local model checking algorithm follows the ideas of [23] by directly encoding the one-step evaluation of formulae $\psi \in$ cl by means of functions $\mathsf{eval}_\psi : \mathcal{P}(C \times \mathsf{cl}) \to \mathcal{P}(C \times \mathsf{cl})$, corresponding to all moves in the model checking game that evaluate ψ at some state. For instance, we have

$$\mathsf{eval}_{\varphi \vee \psi}(X) = \{(c, \varphi \vee \psi) \mid (c, \varphi) \in X \text{ or } (c, \psi) \in X\}$$
$$\mathsf{eval}_{\heartsuit\psi}(X) = \{(c, \heartsuit\psi) \mid \xi(c) \in \llbracket \heartsuit \rrbracket (\{d \mid (d, \psi) \in X\})\}$$

for $X \subseteq C \times \mathsf{cl}$, and similar functions for the remaining operators; intuitively, $\mathsf{eval}_\psi(X)$ computes the set of positions in the model checking game that have formula component ψ and are won by player $\exists$, assuming that it is already known that $\exists$ wins all positions in X. Crucially, the evaluation function for modal operators skips the exploration of the intermediate nodes (D, ψ) in the model checking game by directly evaluating the predicate lifting over the set X. Then we can compute the winning regions in the model checking game as nested fixpoints of the one-step solving function:

$$\mathsf{win}_\exists = \mu X_k. \nu X_{k-1}. \ldots . \nu X_0. \cup_{\psi \in \mathsf{cl}} \mathsf{eval}_\psi(X_{\Omega(\psi)})$$
$$\mathsf{win}_\forall = \nu X_k. \nu X_{k-1}. \ldots . \mu X_0. \cup_{\psi \in \mathsf{cl}} \mathsf{eval}_{\neg\psi}(X_{\Omega(\psi)}),$$

assuming w.l.o.g. that k is odd, and denoting by $\Omega(\psi)$ the priority of all game nodes of the shape (c, ψ); thus the functions below the fixpoints directly correspond to the functions f, g from [23], Definition 5, noting that $\Omega(\psi) = 0$ whenever ψ is not a fixpoint formula. We implement this game solving procedure by a higher order function that receives the semantic function for modalities as an argument, and then computes the relevant fixpoints by Kleene fixpoint iteration.

The overall local model checking implementation then builds the model checking game step by step, starting from the initial position (c, φ) and adding nodes to which the respective player can move; crucially, the evaluation functions for modalities allow us to skip all nodes of the form (D, ψ) during the exploration of the game arena. At any point during the game construction, the algorithm can attempt to solve the partially constructed game by computing the fixpoints defined above, allowing it (in some cases) to finish *early*, that is, before the whole search space has been explored; this constitutes the *local* nature of the algorithm in the sense that satisfaction of a formula may be proved or refuted without traversing the whole model.

Parity Game Model Checking. Relying on polynomial reductions of modality evaluation to game fragments [23], we implement the generation of model checking parity games in COOL-MC by a higher order function which traverses the input model and formula and translates all connectives into game nodes as described in Sect. 2, interpreting modal operators using a function it receives as an argument. The parity game thus constructed then can be solved using any parity game solver (including an unoptimized native solver provided by the COOL-MC framework); the current version of COOL-MC uses PGSolver as external parity game solver (support for Oink is planned).

The subgames that evaluate individual modalities in this construction are specific to the logic at hand. Due to space restrictions, we provide sketches of the reductions for two central logics here and refer to [23], Example 15 for full details. For the standard μ-calculus, we have modal positions $(c, \Diamond \psi)$ (or $(c, \Box \psi)$), for which the one-step evaluation games just consist of that single position controlled by player $\exists$ (or $\forall$), with moves to all positions (d, ψ) such that $d \in \xi(c)$. The evaluation games for, e.g., graded modalities are significantly more involved: For instance, from a position $(c, \langle n \rangle \psi)$, the game proceeds in layers, with one layer for each $d \in C$ to which c has an edge with multiplicity at least 1. In each layer, player $\exists$ decides whether or not to include d in the set of states that she claims to satisfy ψ; all game positions also contain a counter that keeps track of the joint multiplicities of all successors included so far. Player $\exists$ wins the subgame as soon as this counter exceeds n but loses when the subgame exits the final layer while the counter is still below n. Additionally, player $\forall$ can choose, for any state d that player $\exists$ decides to use, to either challenge the satisfaction of ψ at d by continuing the model checking game a position (d, ψ), or accept the choice of d and proceed to the next layer of the local game, increasing the counter by the multiplicity of d as a successor of c.

4   Experimental Evaluation of the Implementation

We experimentally evaluate the performance of our two generic model checking implementations for all logics currently supported. The main interest in COOL-MC lies in its genericity, which enables it to cover a wide range of logics not supported by other tools, so comparison to other tools is mostly omitted for lack of competitors; additional discussion is provided below.

Generalized Parity Games. As we have seen above, model checking for (coalgebraic) μ-calculi reduces to solving parity games. Conversely, parity games can also be solved by model checking: It is well known that player $\exists$ wins a node $v \in V$ in a parity game with k priorities if and only if v satisfies the formula

$$\chi_k := \mu X_k. \nu X_{k-1} \ldots \mu X_1. \nu X_0. \bigvee_{0 \leq i \leq k} \Omega^-(i) \wedge ((V_\exists \wedge \heartsuit X_i) \vee (V_\forall \wedge \overline{\heartsuit} X_i))$$

where $\Omega^-(i) = \{v \in V \mid \Omega(v) = i\}$, $\heartsuit = \Diamond$ and $\overline{\heartsuit} = \Box$ and k is w.l.o.g. assumed to be odd. We exploit this characterization to lift benchmarking problems for standard parity games to a coalgebraic level of generality: A parity game is essentially a Kripke structure with propositional atoms for priorities and player ownership, that is, a coalgebraic model based (for transitions) on the powerset functor $\mathcal{P}$. We generalize this situation by replacing $\mathcal{P}$ with other functors $\mathcal{F}$, and $\Diamond, \Box$ with suitable pairs $\heartsuit, \overline{\heartsuit}$ of dual modalities. In order to win the resulting generalized game, player $\exists$ then requires a strategy that picks, at each game node v, a set of successors that satisfies $\heartsuit$ if $v \in V_\exists$ or $\overline{\heartsuit}$ if $v \in V_\forall$; e.g. in the case of standard games player $\exists$ has to pick a single successor at their nodes ($\heartsuit = \Diamond$), while they have to allow all successors at nodes belonging to $\forall$ ($\overline{\heartsuit} = \Box$). Furthermore, all plays adhering to such a strategy have to satisfy the parity condition. We then systematically enrich given standard parity games to supply additional functor-specific transition structure in a deterministic way to strike a balance between making the games much harder or much easier than the original game while still making use of the added structure; in our leading examples, we proceed as follows:

- For the monotone μ-calculus, we construct *monotone parity games*; concretely, we build *monotone neighbourhood structures* N (e.g. [32]; these are coalgebras for the *monotone neighbourhood functor* [15]), in which two consecutive steps in the original parity game G are merged so that a single step in N corresponds to the evaluation of two-step strategies in G, that is, we define $\xi(v)$ to be the set $\{D_1, \ldots, D_m\}$ of (minimal) neighbourhoods $D_i \subseteq V$ such that the owner of v has a strategy in G to ensure that starting from v and playing two steps, some node from D_i is reached. Then, $\Diamond\varphi$ essentially says that $\exists$ can enforce φ (in two steps), while $\Box\varphi$ says that $\exists$ cannot prevent φ.
- For the graded μ-calculus (Sect. 2), we construct *graded parity games* by equipping moves in G with multiplicities summing up to at least 10 at each node, that is, we assign multiplicity $(\xi(v))(u) = \lceil 10 \div E(v) \rceil$ to each successor $u \in E(v)$ of v in G. Then we take $\heartsuit = \langle 5 \rangle, \overline{\heartsuit} = [5]$, so to win in the graded parity game, player $\exists$ requires a strategy that picks more than five moves, counting multiplicities, at $\exists$ nodes, and all but at most five moves at $\forall$ nodes.
- For the two-valued probablistic μ-calculus, we construct *qualitive stochastic parity games* by imposing a uniform distribution on the moves, thus obtaining *probabilistic transition systems*, which are coalgebras for the *distribution functor* that assigns to a set X the set of (discrete) probability distributions on X. Then we take $\heartsuit = \langle \frac{1}{2} \rangle, \overline{\heartsuit} = [\frac{1}{2}]$ where $\langle p \rangle$ is read "with probability more than p", so player $\exists$ wins the resulting stochastic parity game if they have a strategy that in each $\exists$-move stays within the winning region with probability more than $\frac{1}{2}$, and forces $\forall$ to stay within $\exists$'s winning region with probability at least $\frac{1}{2}$.

We apply the above constructions to various established parity game benchmarking series, and in each case evaluate the respective variant of the formula χ_k, thereby solving the respective monotone, graded, or probabilistic variants of the game. Specifically, we use series of *clique games*, *ladder games*, *Jurdzinski games*, *Towers of Hanoi games*, and *language inclusion games* generated by the parity game solver PGSolver [11, 38].

Lazy Games. To illustrate the potential advantages of local model checking, we also devise an experiment in which each game from a series of generalized parity games (as detailed above) is prepended with a node owned by player $\exists$ which has one move that leads to the original game, but also a move to an additional self-looping node with priority 0. The resulting games all have very small solutions that can be found by the local solver, while global solving becomes more and more expensive as the parameters of the game grow.

Modulo Game. To evaluate the alternating-time μ-calculus [2] instance of the model checking implementation in COOL-MC, we devise a series of games, parameterized by a number of agents and a number m of moves per agent, but with a fixed number of positions $p_0, \ldots, p_9$ marked by propositional atoms of the same name. At p_j, the agents concurrently each pick a number from the set $\{1, \ldots, m\}$, causing the game to proceed to position $p_{(h+j) \bmod 10}$ where h is the sum of the numbers played. Given a set C of agents, we evaluate the formulae $\varphi_1 = \bigwedge_{0 \le i < 10} \mu X. \, p_i \vee [C] X$ and $\varphi_2 = \nu X. \, \mu Y. \, (X \wedge (p_0 \vee [C] Y) \wedge (p_5 \vee [C] Y))$ over the modulo game. Formula φ_1 says that the coalition C has a joint strategy to reach any given state eventually, while φ_2 expresses the Büchi property that C can enforce that both p_0 and p_5 are visited infinitely often.

Evaluation Setup. Our main aim in the evaluation is to show that COOL-MC scales even on the benchmark series we use, which are designed to be hard. In the process, we compare the local model checking method with the reduction to parity games (Sect. 2). To solve the parity games obtained, we use PGSolver's [11, 38] implementation of Zielonka's recursive algorithm; we expect that practical performance can be further improved by instead using Oink [7] as a back-end parity game solver, but leave this issue as future work.

For the standard and monotone μ-calculi, the reduction to parity games is straightforward, blurring the difference between model checking and parity game solving. For these logics, we thus refrain from a comparison between COOL-MC and other existing model checking tools [3,29], which would essentially boil down to a comparison of the respective backend parity game solvers. On the alternating-time μ-calculus (AMC), we do conduct a brief comparison with the model checker MCMAS [31] (further comparison between COOL-MC and MCMAS can be found in the extended version of this paper [20]). We emphasize that the meaningfulness of such a comparison is limited, as on the one hand, MCMAS represents models symbolically while COOL-MC uses an explicit-state representation, and on the other hand, MCMAS only supports alternating-time temporal logic ATL (for which parity-game-based model checking is overkill) while COOL-MC supports the full AMC. For graded and probabilistic μ-calculi, COOL-MC appears to be the only existing model checker, so for these logics we evaluate only the two variants of model checking in COOL-MC; we note that the Probabilistic Symbolic Model Checker (PRISM) [28] uses a specification language based largely on PCTL [17], which is incomparable to the two-valued probabilistic μ-calculus [4].

Below, we refer to the different instantiations of COOL-MC by indexing a logic name with either l (for local model checking) or g (for model checking by game reduction); for instance "graded$_g$" refers to the variant of COOL-MC that reduces model checking for the graded μ-calculus to parity game solving.

We measure runtimes as well as the sizes of the graph structures and games constructed, averaging the values measured in our experiments over at least five executions, with a timeout of 60 s. All experiments have been executed on a machine with an AMD Ryzen 7 2700 CPU and 32GB of RAM. An artifact containing the source code, evaluation scripts, and benchmarking sets for all experiments described above is available online [19].

Results and Interpretation. The runtime results on the generalized parity games experiment are shown in Figs. 1, 2 and 3. The trends for the different logics and variants of generalized games are similar. For readability, we show the measurements for just three logics in each case; additional results can be found in the extended version of this paper [20] and in the artifact.

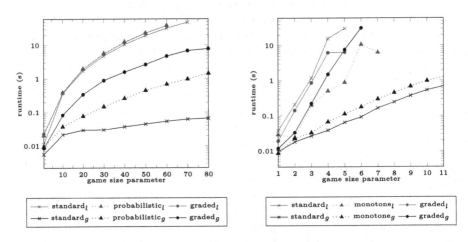

Fig. 1. Ladder games runtimes   **Fig. 2.** Language inclusion game runtime

It appears that the concrete choice of the logic does not strongly effect the runtimes of the local solvers (the blue plots in Figs. 1, 2 and 3). For game-based solving (the black plots), we observe a considerable impact of the choice of logic on the runtimes, in particular solving the graded and probabilistic parity games through PGSolver takes much longer than for the standard variants. This is in line with expectations: As mentioned in the end of Sect. 3, the game characterization of the standard (or monotone) modalities $\Diamond, \Box$ is straightforward, but the encoding of graded and probabilistic modalities leads to quadratic blow-up in the resulting games. The local solver however, directly evaluates modalities and thereby avoids this blow-up so that the performance of the local solver is hardly affected by the concrete choice of modalities.

On the other hand, game-based solving typically is faster than local solving. We note that the native fixpoint computation that COOL-MC uses for local solving is completely unoptimized and performs naive Kleene fixpoint iteration, while PGSolver is an optimized tool, and in particular its recursive algorithm shows good performance in practice.

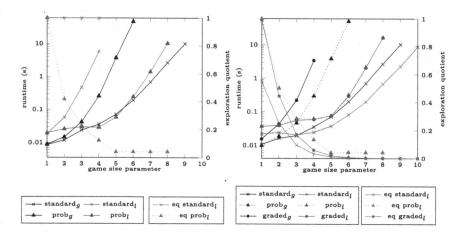

Fig. 3. Towers of Hanoi runtimes   **Fig. 4.** *Lazy* Towers of Hanoi runtimes

Also, the generalized games used in the benchmarks are constructed from parity games designed to be hard to solve; in particular, we observe that with the notable exceptions of the language inclusion games (Figs. 2 and 5) and the probabilistic variant of the Towers of Hanoi games (Fig. 3), these games typically do not have small solutions so that the local solver cannot play out the strength of on-the-fly model checking.

This line of argumentation is substantiated by the lazy games experiment conducted on games built from the Towers of Hanoi series, shown in Fig. 4 (the sizes of the constructed graphs and games are listed in Fig. 5). These results are representative for the lazy modifications of the other parity game series as well. Here, the local solver significantly outperforms the algorithm that first constructs the full game. It appears that the local solver does indeed manage to detect the existence of small winning strategies in these games, thereby avoiding the full exploration of the search space. In each case, the extent to which the local solver explores the full game is shown in Fig. 4 with a red plot that depicts the *exploration quotient*, i.e. the percentage of the total number of nodes that

Experiment series	parameter	worlds	full graph	lazy graph	game size
Language incl., monotone	1	3	93	59	126
	7	313	9,703	937	13,146
	30	†	†	†	1,099,896
Lazy Hanoi, standard	1	5	103	57	133
	5	245	5,143	57	6,613
	9	19,685	413,383	53	531,493
	10	59,051	1,240,069	53	†
Lazy Hanoi, graded	1	5	103	102	523
	2	11	229	102	2,345
	4	83	1,741	102	126,222
	10	59,051	1,240,069	102	†

Fig. 5. Sizes of (full and lazy) graphs and constructed parity games

are actually explored. This effect is observed for all logics currently supported, including the graded and probabilistic variants.

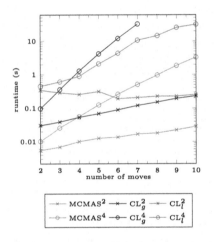

Fig. 6. Modulo game runtimes (φ_1)

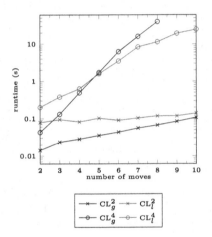

Fig. 7. Modulo game runtimes (φ_2)

Figures 6 and 7 show the runtimes for φ_1 and φ_2 on the modulo game with 2 and 4 agents, respectively. We include runtime plots for MCMAS on φ_1, which is expressible in ATL, while ϕ_2 is goes beyond ATL and is thus not handled by MCMAS. As expected, MCMAS is faster on the fragment that it supports; presumably, this is due partly to the fact that ATL allows for dedicated model checking algorithms that avoid parity games and in fact run in polynomial time [2].

5   Conclusions and Future Work

We have presented and evaluated the generic model checker COOL-MC, which implements generic model checking algorithms for the coalgebraic μ-calculus [23], and has been instantiated to a range of instance logics. In particular, COOL-MC thus constitutes the first available model checker for the two-valued probabilistic μ-calculus [4,5,30], the graded μ-calculus [26], and the full alternating-time μ-calculus [2] (model checkers for alternating-time temporal logic exist [1,24,31]). The benchmarking results suggest the direct evaluation of modalities in combination with lazy solving as a setup for coalgebraic model checking that scales well in practice. An important issue for future work is to develop and implement symbolic model checking algorithms for the coalgebraic μ-calculus.

Data Availability Statement. All data to reproduce the findings in this paper are available online. The COOL-MC source code used to compile the artifact is available at tag `VMCAI-2024` of the COOL git repository [9]. Pre-compiled Linux executables as well as a docker container to reproduce the measurements displayed in the figures and tables of this paper are available online [19].

References

1. Alur, R., et al.: JMOCHA: a model checking tool that exploits design structure. In: International Conference on Software Engineering, ICSE 2001, pp. 835–836. IEEE Computer Society (2001). https://doi.org/10.1109/ICSE.2001.919196
2. Alur, R., Henzinger, T.A., Kupferman, O.: Alternating-time temporal logic. J. ACM **49**, 672–713 (2002). https://doi.org/10.1145/585265.585270
3. Atif, M., Groote, J.F.: Understanding behaviour of distributed systems using mCRL2. Springer (2023). https://doi.org/10.1007/978-3-031-23008-0
4. Chakraborty, S., Katoen, J.: On the satisfiability of some simple probabilistic logics. In: Logic in Computer Science, LICS 2016, pp. 56–65. ACM (2016). https://doi.org/10.1145/2933575.2934526
5. Cîrstea, C., Kupke, C., Pattinson, D.: EXPTIME tableaux for the coalgebraic mu-calculus. Log. Methods Comput. Sci. **7**(3) (2011). https://doi.org/10.2168/LMCS-7(3:3)2011
6. D'Agostino, G., Visser, A.: Finality regained: a coalgebraic study of Scott-sets and multisets. Arch. Math. Logic **41**, 267–298 (2002). https://doi.org/10.1007/S001530100110
7. Dijk, T.: Oink: an implementation and evaluation of modern parity game solvers. In: Beyer, D., Huisman, M. (eds.) TACAS 2018. LNCS, vol. 10805, pp. 291–308. Springer, Cham (2018). https://doi.org/10.1007/978-3-319-89960-2_16
8. Enqvist, S., Hansen, H.H., Kupke, C., Marti, J., Venema, Y.: Completeness for game logic. In: Logic in Computer Science, LICS 2019, pp. 1–13. IEEE (2019). https://doi.org/10.1109/LICS.2019.8785676
9. fauprojects: COOL - The Coalgebraic Ontology Logic Reasoner (git repository). https://git8.cs.fau.de/software/cool/-/tree/VMCAI-2024
10. Ferrante, A., Murano, A., Parente, M.: Enriched μ-calculi module checking. Log. Methods Comput. Sci. **4**(3) (2008). https://doi.org/10.2168/LMCS-4(3:1)2008
11. Friedmann, O., Lange, M.: The PGSolver collection of parity game solvers. Technical report, University of Munich (2009)
12. Friedmann, O., Lange, M.: Local strategy improvement for parity game solving. In: Proceedings First Symposium on Games, Automata, Logic, and Formal Verification, GANDALF 2010. EPTCS, vol. 25, pp. 118–131 (2010). https://doi.org/10.4204/EPTCS.25.13

13. Gorín, D., Pattinson, D., Schröder, L., Widmann, F., Wißmann, T.: COOL – a generic reasoner for coalgebraic hybrid logics (system description). In: Demri, S., Kapur, D., Weidenbach, C. (eds.) IJCAR 2014. LNCS (LNAI), vol. 8562, pp. 396–402. Springer, Cham (2014). https://doi.org/10.1007/978-3-319-08587-6_31

14. Görlitz, O., Hausmann, D., Humml, M., Pattinson, D., Prucker, S., Schröder, L.: COOL 2 - a generic reasoner for modal fixpoint logics (system description). In: Pientka, B., Tinelli, C. (eds.) CADE 2023. LNCS, vol. 14132, pp. 234–247. Springer, Cham (2023). https://doi.org/10.1007/978-3-031-38499-8_14

15. Hansen, H.H., Kupke, C.: A coalgebraic perspective on monotone modal logic. In: Coalgebraic Methods in Computer Science, CMCS 2004. ENTCS, vol. 106, pp. 121–143. Elsevier (2004). https://doi.org/10.1016/j.entcs.2004.02.028

16. Hansen, H.H., Kupke, C., Marti, J., Venema, Y.: Parity games and automata for game logic. In: Madeira, A., Benevides, M. (eds.) DALI 2017. LNCS, vol. 10669, pp. 115–132. Springer, Cham (2018). https://doi.org/10.1007/978-3-319-73579-5_8

17. Hansson, H., Jonsson, B.: A logic for reasoning about time and reliability. Formal Aspects Comput. **6**(5), 512–535 (1994). https://doi.org/10.1007/BF01211866

18. Hasuo, I., Shimizu, S., Cîrstea, C.: Lattice-theoretic progress measures and coalgebraic model checking. In: Principles of Programming Languages, POPL 2016, pp. 718–732. ACM (2016). https://doi.org/10.1145/2837614.2837673

19. Hausmann, D., Humml, M., Prucker, S., Schröder, L., Strahlberger, A.: Generic model checking for modal fixpoint logics in COOL-MC (artifact). Zenodo (2023). https://doi.org/10.5281/zenodo.8332511

20. Hausmann, D., Humml, M., Prucker, S., Schröder, L., Strahlberger, A.: Generic model checking for modal fixpoint logics in COOL-MC (extended version). CoRR abs/2311.01315 (2023). https://doi.org/10.48550/arXiv.2311.01315

21. Hausmann, D., Schröder, L.: Quasipolynomial computation of nested fixpoints. In: TACAS 2021. LNCS, vol. 12651, pp. 38–56. Springer, Cham (2021). https://doi.org/10.1007/978-3-030-72016-2_3

22. Hausmann, D., Schröder, L.: Coalgebraic satisfiability checking for arithmetic μ-calculi. CoRR abs/2212.11055 (2022). https://doi.org/10.48550/arXiv.2212.11055

23. Hausmann, D., Schröder, L.: Game-based local model checking for the coalgebraic mu-calculus. In: 30th International Conference on Concurrency Theory, CONCUR 2019. LIPIcs, vol. 140, pp. 35:1–35:16. Schloss Dagstuhl - Leibniz-Zentrum für Informatik (2019). https://doi.org/10.4230/LIPIcs.CONCUR.2019.35

24. Kański, M., Niewiadomski, A., Kacprzak, M., Penczek, W., Nabiałek, W.: Unbounded model checking for ATL. Studia Informatica **25**(1–2) (2021). https://doi.org/10.34739/si.2021.25.01

25. Kozen, D.: Results on the propositional μ-calculus. Theor. Comput. Sci. **27**, 333–354 (1983). https://doi.org/10.1016/0304-3975(82)90125-6

26. Kupferman, O., Sattler, U., Vardi, M.Y.: The complexity of the graded μ-calculus. In: Voronkov, A. (ed.) CADE 2002. LNCS (LNAI), vol. 2392, pp. 423–437. Springer, Heidelberg (2002). https://doi.org/10.1007/3-540-45620-1_34

27. Kupke, C., Marti, J., Venema, Y.: Size measures and alphabetic equivalence in the μ-calculus. In: Logic in Computer Science, LICS 2022, pp. 18:1–18:13. ACM (2022). https://doi.org/10.1145/3531130.3533339

28. Kwiatkowska, M., Norman, G., Parker, D.: PRISM 4.0: verification of probabilistic real-time systems. In: Gopalakrishnan, G., Qadeer, S. (eds.) CAV 2011. LNCS, vol. 6806, pp. 585–591. Springer, Heidelberg (2011). https://doi.org/10.1007/978-3-642-22110-1_47

29. Landsaat, E.: A model checker for game logic via parity games. BSc thesis, University of Groningen (2022). https://fse.studenttheses.ub.rug.nl/28126/

30. Liu, W., Song, L., Wang, J., Zhang, L.: A simple probabilistic extension of modal mu-calculus. In: International Joint Conference on Artificial Intelligence, IJCAI 2015, pp. 882–888. AAAI Press (2015). http://ijcai.org/proceedings/2015

31. Lomuscio, A., Qu, H., Raimondi, F.: MCMAS: an open-source model checker for the verification of multi-agent systems. Int. J. Softw. Tools Technol. Transf. **19**(1), 9–30 (2017). https://doi.org/10.1007/s10009-015-0378-x

32. Parikh, R.: The logic of games and its applications. Ann. Discr. Math. **24**, 111–140 (1985). https://doi.org/10.1016/S0304-0208(08)73078-0

33. Pattinson, D.: Expressive logics for coalgebras via terminal sequence induction. Notre Dame J. Formal Log. **45**(1), 19–33 (2004). https://doi.org/10.1305/ndjfl/1094155277

34. Pauly, M.: Logic for Social Software. Ph.D. thesis, Universiteit van Amsterdam (2001)

35. Rutten, J.J.M.M.: Universal coalgebra: a theory of systems. Theor. Comput. Sci. **249**(1), 3–80 (2000). https://doi.org/10.1016/S0304-3975(00)00056-6

36. Schröder, L.: Expressivity of coalgebraic modal logic: the limits and beyond. Theor. Comput. Sci. **390**(2–3), 230–247 (2008). https://doi.org/10.1016/j.tcs.2007.09.023

37. Stevens, P., Stirling, C.: Practical model-checking using games. In: Steffen, B. (ed.) TACAS 1998. LNCS, vol. 1384, pp. 85–101. Springer, Heidelberg (1998). https://doi.org/10.1007/BFb0054166

38. tcsprojects: PGSolver (git repository). https://github.com/tcsprojects/pgsolver

39. Venema, Y.: Automata and fixed point logic: a coalgebraic perspective. Inf. Comput. **204**(4), 637–678 (2006). https://doi.org/10.1016/j.ic.2005.06.003

Model-Guided Synthesis for LTL over Finite Traces

Shengping Xiao[1], Yongkang Li[1], Xinyue Huang[1], Yicong Xu[1], Jianwen Li[1(✉)], Geguang Pu[1,2], Ofer Strichman[3], and Moshe Y. Vardi[4]

[1] East China Normal University, Shanghai, China
{spxiao,51265902012,52265902016,51215902150}@stu.ecnu.edu.cn,
{jwli,ggpu}@sei.ecnu.edu.cn
[2] Shanghai Trusted Industrial Control Platform Co., Ltd., Shanghai, China
[3] Technion, Haifa, Israel
ofers@technion.ac.il
[4] Rice University, Houston, USA
vardi@cs.rice.edu

Abstract. Satisfiability and synthesis are two fundamental problems for Linear Temporal Logic, both of which can be solved on the automaton constructed from the input formula. In general, satisfiability is easier than synthesis in both theory and practice, as satisfiability needs only to find a satisfying trace, while synthesis has to find a winning strategy.

This paper presents a novel technique called MoGuS, which improves the performance of synthesis for LTL_f, a variant of LTL interpreted over finite traces, by repeatedly invoking an LTL_f satisfiability checker to guide its search for a winning strategy. Satiisfiabiity checkers have not been used before in the context of LTL_f synthesis. MoGuS computes a satisfying trace of the input formula, and then uses the formula-progression technique to compute the states on the fly in the automaton run. It then checks whether there exists a winning strategy from each of the states. If not, the current state is marked as a 'failure' state (as it can never produce a winning strategy), the checking rolls back to its predecessor state, and the process repeats. MoGuS returns 'Realizable' if the initial state turns out to be winning, and 'Unrealizable' otherwise. We conducted an extensive experimental evaluation of MoGuS by comparing it to different state-of-the-art LTL_f synthesis algorithms on a large set of benchmarks. The results show that MoGuS has the most stable and the best overall performance on the tested benchmarks.

1 Introduction

Temporal *synthesis* is the automated construction of a reactive system from a given temporal logic formula (specification), e.g., LTL [38], such that the interactive behaviors between the system and the external environment are guaranteed to satisfy the specification [18,39]. The problem of determining whether such a system exists is called *realizability*. LTL realizability and synthesis are major research topics in formal methods, and fruitful works have been accomplished on both the theoretical and practical aspects, e.g., [8,36,45], to name a few. In recent years, an annual synthesis competition [1] has played an important role in motivating tool development in this area. Nevertheless, LTL synthesis is still considered a very challenging problem, as generating deterministic automata from LTL specifications, which is a critical part of the algorithmic solution, involves a doubly-exponential blow-up [2].

An LTL formula is interpreted over infinite traces, so the constructed automaton from the formula has to accept infinite traces as well. Such automata with infinite accepting conditions, e.g., Büchi [11], are notorious for their challenging determinization, e.g., Safra Construction [41], which is a barrier to effective LTL realizability/synthesis. A recent argument, however, has been made that synthesis of the system with finite behaviors is sufficient in practice [20,28].

LTL_f, which is defined over finite traces, has emerged as a popular logic in AI-related domains since its invention [20]. Given an LTL_f formula φ, there exists a non-deterministic finite automaton (NFA) that represents φ's language. Determinization of an NFA can be performed via the classical subset construction [28]. Although the worst-case complexity remains the same (2EXPTIME), this leads to a much simpler synthesis procedure [45], as we will discuss below. Indeed, LTL_f has emerged as a popular temporal description language in AI-related domains, especially for specifying *motion planning* problems [3,4,14–16,22,27,46]. LTL_f synthesis is then used to build a model that satisfies these specifications.

Recently, several works have been conducted to study the theory and practice of fundamental problems related to LTL_f, e.g., translation to automata [21,42], satisfiability checking [33–35], and synthesis [28,43,45]. The asymptotic complexity of both LTL_f satisfiability and LTL_f synthesis is the same as in LTL, namely PSPACE-complete and 2EXPTIME-complete, respectively [20,28]. The focus of this paper is on using LTL_f satisfiability checking to speed up LTL_f synthesis and realizability.

There are so far two kinds of approaches to solving LTL_f synthesis. The first one is *bottom-up* [7,45], which first constructs the whole (minimal) DFA for the input formula, using the efficient DFA-construction tool MONA [30], and then computes all winning states in the DFA back from the accepting states. Representative LTL_f synthesizers based on this approach include Lisa [7] and Lydia [21]. The second one is *top-down* [29,43], which takes the input formula as the initial state, and then computes the remaining states of the DFA *on-the-fly*, while rolling back once a winning/failure state is identified. Representative LTL_f synthesizers based on this approach include OLFS [43] and Cynthia [29]. Both solutions return 'Realizable' iff the initial state is winning. According to previous studies [29,43], the top-down solution performs better in some particular benchmarks than the bottom-up one, while in general the latter approach gains a better overall performance in the selected benchmarks where synthesis from the formulas is not challenging enough.

The on-the-fly top-down solution has been conducted by using either SAT [24] or Sentential Decision Diagram (SDD) [19] techniques. The former utilizes SAT solvers to compute exactly one (deterministic) state at a time, making the whole framework very flexible. Using SAT solvers for state enumeration is, however, not quite efficient, as the SAT solver cannot distinguish between the system and environment variables. As a result, the satisfying assignment that it finds is rather arbitrary from the perspective of the synthesis process. Meanwhile, the latter solution leverages SDD to encode the variables' information in order, such that the enumerated DFA transitions (with states) can be more compact. As shown in [29], the SDD-based approach is able to outperform the SAT-based one on a considerable number of test instances. Nevertheless, the drawback of this approach is that one SDD computation has to generate all of the one-step successors, which is a much more computationally expensive operation than a single SAT call. So the question is whether there is a way to compute states in a light way, and enumerate the transitions and states in a more 'targeted' way, which will lead to faster convergence.

This paper tries to address this question and proposes to conduct LTL_f synthesis via an algorithm that is based on multiple LTL_f satisfiability checks, hence leveraging the relative efficiency of those tools. We call this approach MoGuS (**Mo**del-**Gu**ided **S**ynthesis). Instead of computing only one state (and transition), MoGuS utilizes an LTL_f satisfiability solver to generate one satisfying trace at a time, which corresponds to a sequence of states and transitions. The insight is that a satisfying trace is more likely to be compatible with a winning strategy, which can potentially make the state search more precise. MoGuS then progresses backward on the satisfying state sequence, trying each time to prove that the state is winning. If it is not, it asks the LTL_f satisfiability solver for a new assignment that does not go through that state (which we call a 'failure' state), by blocking it. This process is repeated until the initial state becomes winning, or the formula becomes unsatisfiable. In the former case, the formula is declared to be 'Realizable', and in the latter, 'Unrealizable'. The correctness of our procedure relies on the fact that if a satisfying trace runs across a failure state, it cannot be produced by a winning strategy and hence can be blocked.

Generally speaking, MoGuS is a top-down solution for LTL_f synthesis. Compared to the SAT-based approach, it can enumerate states and transitions more precisely, as the search inside is guided by a satisfying trace that is more likely to target a winning state. Compared to the SDD-based approach, MoGuS is more flexible in computing states and transitions, as the state-of-the-art satisfiability solvers, e.g., aaltaf [33], provide a way to compute one state (and transition) at a time.

We implemented MoGuS inside the tool MoGuSer and evaluated its performance by comparing it to the state-of-the-art LTL_f synthesis solvers Cynthia, Lisa, and Lydia on the collected benchmarks from [29] (1454 in total), as well as the *Ascending* benchmarks (1800 in total) generated by Spot [23] for the purpose of scalability testing[1]. For the collected benchmarks, MoGuSer solves a total of 1287 (out of 1454) instances, which is twice that solved by Cynthia (605) but

---

[1] From the preliminary evaluations, our previous synthesizer OLFS [43] performs much worse than other tested tools, so it is excluded in the comparison.

slightly less than that solved by Lisa (1316) and Lydia (1339). For the *Ascending* benchmarks, MoGuSer solves a total of 1559 (out of 1800) cases, which is better than that solved by all other solvers, i.e., Cynthia (1430), Lisa (1101), Lydia (916). Our tool MoGuSer has the most stable and the best overall performance in these two evaluations.

This paper is organized as follows. The next section introduces preliminaries. Section 3 presents the construction from an LTL_f formula to its corresponding TDFA, by leveraging the formula progression technique. Section 4 describes the details of MoGuS and its correctness guarantee. Section 5 shows the experimental results. And we discuss a brief history of LTL synthesis in Sect. 6. Finally, Sect. 7 summarizes the contributions and discusses future work.

2 Preliminaries

2.1 LTL over Finite Traces (LTL$_f$)

Linear Temporal Logic over finite traces, or LTL_f [20], extends propositional logic with finite-horizon temporal connectives. Generally speaking, LTL_f is a variant of Linear Temporal Logic (LTL) [38] that is interpreted over finite traces. Given a set of atomic propositions $\mathcal{P}$, the syntax of LTL_f is identical to LTL, and defined as:

$$\varphi ::= tt \mid p \mid \neg\varphi \mid \varphi \wedge \varphi \mid \bigcirc\varphi \mid \varphi\,\mathcal{U}\,\varphi$$

where tt represents the *true* formula, $p \in \mathcal{P}$ is an atomic proposition, $\neg$ represents *negation*, $\wedge$ represents *and*, $\bigcirc$ represents the *strong Next* operator and $\mathcal{U}$ represents the *Until* operator. We also have the corresponding dual operators $f\!f$ (*false*) for tt, $\vee$ (or) for $\wedge$, $\bullet$ (weak Next) for $\bigcirc$ and $\mathcal{R}$ (Release) for $\mathcal{U}$. Moreover, we use the notation $\mathcal{G}\varphi$ (Global) and $\mathcal{F}\varphi$ (Future) to represent $f\!f\,\mathcal{R}\,\varphi$ and $tt\,\mathcal{U}\,\varphi$, respectively. Notably, $\bigcirc$ is the standard *Next* operator, while $\bullet$ is *weak Next*; $\bigcirc$ requires the existence of a successor instance, while $\bullet$ does not. Thus $\bullet\phi$ is always true in the last instance of a finite trace, since no successor exists there.

A finite *trace* $\rho = \rho[0], \rho[1], \cdots, \rho[n]$ is a sequence of propositional interpretations (sets), in which $\rho[m] \in 2^{\mathcal{P}}$ $(0 \leq m < |\rho|)$ is the m-th interpretation of ρ, and $|\rho| = n + 1$ represents the length of ρ. Intuitively, $\rho[m]$ is interpreted as the set of propositions which are *true* at instance m. We denote ρ^i to represent $\rho[0], \rho[1], \ldots, \rho[i-1]$ $(i \geq 1)$, which is the prefix of ρ to position i (not including i), and ρ_i to represent $\rho[i], \rho[i+1], \cdots, \rho[n]$, which is the suffix of ρ from position i (including i). Two finite traces, ρ_1 and ρ_2, can be concatenated to one trace ρ, denoted by $\rho = \rho_1 \cdot \rho_2$.

LTL_f formulas are interpreted over finite traces. For a finite trace ρ and an LTL_f formula φ, we define the satisfaction relation $\rho \models \varphi$ (i.e., ρ is a model of φ) as follows:

- $\rho \models tt$;
- $\rho \models p$ iff $p \in \rho[0]$, where p is an atomic proposition;

- $\rho \models \neg\varphi$ iff $\rho \not\models \varphi$;
- $\rho \models \varphi_1 \wedge \varphi_2$ iff $\rho \models \varphi_1$ and $\rho \models \varphi_2$;
- $\rho \models \bigcirc\varphi$ iff $|\rho| > 1$ and $\rho_1 \models \varphi$;
- $\rho \models \varphi_1 \mathcal{U} \varphi_2$ iff there exists i with $0 \leq i < |\rho|$ such that $\rho_i \models \varphi_2$, and for every j with $0 \leq j < i$ it holds that $\rho_j \models \varphi_1$.

The set of finite traces that satisfy LTL_f formula φ is the language of φ, denoted as $\mathcal{L}(\varphi) = \{\rho \in (2^{\mathcal{P}})^+ \mid \rho \models \varphi\}$. Two LTL_f formulas φ_1 and φ_2 are semantically equivalent, denoted as $\varphi_1 \equiv \varphi_2$, iff for every finite trace ρ, $\rho \models \varphi_1$ iff $\rho \models \varphi_2$. A *literal* is an atom $p \in \mathcal{P}$ or its negation ($\neg p$). We say an LTL_f formula is in *Negation Normal Form* (*NNF*), if the negation operator appears only in front of an atom. Every LTL_f formula can be converted into its *NNF* in linear time. We assume that all LTL_f formulas are in *NNF* in this paper.

2.2 Transition-Based DFA

The Transition-based Deterministic Finite Automaton (TDFA) is a variant of the Deterministic Finite Automaton (DFA) [42].

Definition 1 (Transition-based DFA). *A transition-based* DFA *(TDFA) is a tuple $\mathcal{A} = (2^{\mathcal{P}}, S, s_0, \delta, T)$ where*

- $2^{\mathcal{P}}$ *is the alphabet;*
- S *is the set of states;*
- $s_0 \in S$ *is the initial state;*
- $\delta : S \times 2^{\mathcal{P}} \to S$ *is the transition function;*
- $T \subseteq \delta$ *is the set of accepting transitions.*

For simplicity, we use the notation $s_1 \xrightarrow{\omega} s_2$ to denote $\delta(s_1, \omega) = s_2$. The run r of a TDFA $\mathcal{A}$ on a finite trace $\rho = \rho[0], \rho[1], \cdots, \rho[n] \in (2^{\mathcal{P}})^+$ is a finite state sequence $r = s_0, s_1, \cdots, s_n$ such that s_0 is the initial state, $s_i \xrightarrow{\rho[i]} s_{i+1}$ is true for $0 \leq i < n$. Note that runs of TDFA do not need to include the destination state of the last transition, which is implicitly $s_{n+1} = \delta(s_n, \rho[n])$, since the starting state (s_n) together with the labels of the transition ($\rho[n]$) are sufficient to determine the destination. r is called *acyclic* iff $(s_i = s_j) \Leftrightarrow (i = j)$ for $0 \leq i, j < n$. Also, we say that ρ *runs across* s_i iff s_i is in the corresponding run r. The trace ρ is accepted by $\mathcal{A}$ iff the corresponding run r ends with an accepting transition, i.e., $\delta(s_n, \rho[n]) \in T$. The set of finite traces accepted by a TDFA $\mathcal{A}$ is the language of $\mathcal{A}$, denoted as $\mathcal{L}(\mathcal{A})$.

According to [42], TDFA has the same expressiveness as the normal DFA, and for an LTL_f formula φ, there is a TDFA $\mathcal{A}_\varphi$ such that $\mathcal{L}(\varphi) = \mathcal{L}(\mathcal{A}_\varphi)$. As a result, the LTL_f satisfiability-checking problem can be solved on the corresponding TDFA. That is, an LTL_f formula φ is satisfiable iff there is a finite trace accepted by its corresponding TDFA $\mathcal{A}_\varphi$ [42].

2.3 LTL$_f$ Realizability, Synthesis, and TDFA Games

Definition 2 (LTL$_f$ **realizability**). *Let φ be an LTL$_f$ formula whose alphabet is $\mathcal{P}$ and $\mathcal{X}, \mathcal{Y}$ be two subsets of $\mathcal{P}$ such that $\mathcal{X} \cap \mathcal{Y} = \emptyset$ and $\mathcal{X} \cup \mathcal{Y} = \mathcal{P}$. $\mathcal{X}$ is the set of input variables controlled by the environment and $\mathcal{Y}$ is the set of output variables controlled by the system. φ is realizable with $\langle \mathcal{X}, \mathcal{Y} \rangle$ if there exists a strategy $g : (2^{\mathcal{X}})^* \to 2^{\mathcal{Y}}$ such that for an arbitrary infinite sequence $\lambda = X_0, X_1, \cdots \in (2^{\mathcal{X}})^{\omega}$ of propositional interpretations over $\mathcal{X}$, there is $k > 0$ such that $\rho \models \phi$ holds, where $\rho = (X_0 \cup g(\epsilon)), (X_1 \cup g(X_0)), \cdots, (X_k \cup g(X_0, \cdots, X_{k-1}))$. ($\epsilon$ means the empty trace.)*

Less formally, an LTL$_f$ formula φ is realizable if there exists a winning strategy g for the outputs, namely that for every sequence of inputs, its combination with g's outputs up to some k, satisfies φ (it can be a different value of k for different input sequences). Notably, the synthesis defined above is called *system-first synthesis*, which means that at each point the value of Y depends on the value history of X. This paper focuses on the system-first synthesis.

The synthesis problem can be reduced to TDFA *games* [28,43] specified by $\mathcal{A}_{\varphi}$, with the help of definitions on the *winning/failure* states.

Definition 3 (System Winning/Failure State [43]). *For a TDFA game specified by $\mathcal{A} = (2^{\mathcal{Y} \cup \mathcal{X}}, S, s_0, \delta, T)$, $s \in S$ is a system winning state iff there is $Y \in 2^{\mathcal{Y}}$ such that for every $X \in 2^{\mathcal{X}}$, either $\delta(s, Y \cup X) = s'$ is an accepting transition or s' is a system winning state. State s is a system failure state iff s is not a system winning state.*

Theorem 1. *Given an LTL$_f$ formula φ with $\langle \mathcal{X}, \mathcal{Y} \rangle$, a TDFA $\mathcal{A}_{\varphi} = (2^{\mathcal{Y}} \times 2^{\mathcal{X}}, S, s_0, \delta, T)$ such that $\mathcal{L}(\mathcal{A}_{\varphi}) = \mathcal{L}(\varphi)$, and the TDFA game specified by $\mathcal{A}_{\varphi}$, the following are equivalent:*

- *φ with $\langle \mathcal{X}, \mathcal{Y} \rangle$ is realizable;*
- *the system wins the game specified by $\mathcal{A}_{\varphi}$;*
- *s_0 is a system-winning state of the game specified by $\mathcal{A}_{\varphi}$.*

3 LTL$_f$-to-TDFA via Progression

In our new framework MoGuS, we translate the formula to its TDFA via *formula progression*. The progression technique originates in [5] for goal planning with temporal logics, and a definition of LTL$_f$ progression has been used in [29]. Here we adapt the progression to finite traces instead of single propositions and adjust the translation process for TDFA.

Definition 4 (Formula Progression for LTL$_f$). *Given an LTL$_f$ formula φ and a non-empty finite trace ρ, the progression formula $\mathsf{fp}(\varphi, \rho)$ is recursively defined as follows:*

- *$\mathsf{fp}(tt, \rho) = tt$ and $\mathsf{fp}(ff, \rho) = ff$;*
- *$\mathsf{fp}(p, \rho) = tt$ if $p \in \rho[0]$; $\mathsf{fp}(p, \rho) = ff$ if $p \notin \rho[0]$;*

- $\mathsf{fp}(\neg\varphi, \rho) = \neg\mathsf{fp}(\varphi, \rho)$;
- $\mathsf{fp}(\varphi_1 \wedge \varphi_2, \rho) = \mathsf{fp}(\varphi_1, \rho) \wedge \mathsf{fp}(\varphi_2, \rho)$;
- $\mathsf{fp}(\varphi_1 \vee \varphi_2, \rho) = \mathsf{fp}(\varphi_1, \rho) \vee \mathsf{fp}(\varphi_2, \rho)$;
- $\mathsf{fp}(\bigcirc\varphi, \rho) = \varphi$ if $|\rho| = 1$; Else $\mathsf{fp}(\bigcirc\varphi, \rho) = \mathsf{fp}(\varphi, \rho_1)$;
- $\mathsf{fp}(\bullet\varphi, \rho) = \varphi$ if $|\rho| = 1$; Else $\mathsf{fp}(\bullet\varphi, \rho) = \mathsf{fp}(\varphi, \rho_1)$;
- $\mathsf{fp}(\varphi_1 \,\mathcal{U}\, \varphi_2, \rho) = \mathsf{fp}(\varphi_2, \rho) \vee (\mathsf{fp}(\varphi_1, \rho) \wedge \mathsf{fp}(\bigcirc(\varphi_1 \,\mathcal{U}\, \varphi_2), \rho))$;
- $\mathsf{fp}(\varphi_1 \,\mathcal{R}\, \varphi_2, \rho) = \mathsf{fp}(\varphi_2, \rho) \wedge (\mathsf{fp}(\varphi_1, \rho) \vee \mathsf{fp}(\bullet(\varphi_1 \,\mathcal{R}\, \varphi_2), \rho))$.

The following lemmas are not hard to obtain based on Definition 4, whose proofs are omitted here.

Lemma 1. *Given an* LTL_f *formula* φ *and two non-empty finite traces* ρ_1 *and* ρ_2, $\rho_2 \models \mathsf{fp}(\varphi, \rho_1)$ *implies* $\rho_1 \cdot \rho_2 \models \varphi$.

Lemma 2. *Given an* LTL_f *formula* φ *and two non-empty finite traces* ρ_1 *and* ρ_2, *it holds that* $\mathsf{fp}(\mathsf{fp}(\varphi, \rho_1), \rho_2) = \mathsf{fp}(\varphi, \rho_1 \cdot \rho_2)$.

Lemma 3. *Given an* LTL_f *formula and a non-empty finite trace* ρ, $\rho \models \varphi$ *implies* $\rho_i \models \mathsf{fp}(\varphi, \rho^i)$ *for every* $0 \le i < |\rho|$.

Now we re-construct the TDFA for an LTL_f formula.

Definition 5 (LTL$_f$ to TDFA). *Given an* LTL_f *formula* φ, *the* TDFA $\mathcal{A}_\varphi$ *is a tuple* $(2^{\mathcal{P}}, S, \delta, s_0, T)$ *such that*

- $2^{\mathcal{P}}$ *is the alphabet, where* $\mathcal{P}$ *is the set of atoms of* φ;
- $S = \{\varphi\} \cup \{\mathsf{fp}(\varphi, \rho) \mid \forall \rho \in (2^{\mathcal{P}})^+\}$ *is the set of states;*
- $s_0 = \varphi$ *is the initial state;*
- $\delta : S \times 2^{\mathcal{P}} \to S$ *is the transition function such that* $\delta(s, \sigma) = \mathsf{fp}(s, \sigma)$ *for* $s \in S$ *and* $\sigma \in 2^{\mathcal{P}}$ *(Here* σ *is considered a trace with length 1);*
- $T = \{s_1 \xrightarrow{\sigma} s_2 \in \delta \mid \sigma \models s_1\}$ *is the set of accepting transitions.*

Theorem 2. *Given an* LTL_f *formula* φ *and the* TDFA $\mathcal{A}_\varphi$ *constructed by Definition 5, it holds that* $\mathcal{L}(\varphi) = \mathcal{L}(\mathcal{A}_\varphi)$.

Proof. Let $|\rho| = n + 1$ $(n \ge 0)$ and the corresponding run r of $\mathcal{A}_\varphi$ on ρ is $s_0, s_1, \ldots, s_n$, where $s_0 = \varphi$.

($\Leftarrow$) According to Definition 5, ρ is accepted by $\mathcal{A}_\varphi$ implies $(\rho_n = \rho[n]) \models s_n$ and $s_n = \mathsf{fp}(\varphi, \rho^n)$. Then from Lemma 1, we have $(\rho^n \cdot \rho_n = \rho) \models (s_0 = \varphi)$.

($\Rightarrow$) First from Definition 5, every $\mathsf{fp}(\varphi, \rho^i)$ for $0 \le i \le n$ is a state of $\mathcal{A}_\varphi$. Secondly, $\delta(\mathsf{fp}(\varphi, \rho^i), \rho[i]) = \mathsf{fp}(\varphi, \rho^{i+1})$ is true for $0 \le i \le n$, because $\mathsf{fp}(\varphi, \rho^{i+1}) = \mathsf{fp}(\mathsf{fp}(\varphi, \rho^i), \rho[i])$ is true (Lemma 2). Therefore, let $s_i = \mathsf{fp}(\varphi, \rho^i)$ $(0 \le i \le n)$ and the state sequence $r = s_0, s_1, \ldots, s_n$ is a run of $\mathcal{A}_\varphi$ on ρ. Finally, $\rho \models \varphi$ implies that $\rho_n \models (s_n = \mathsf{fp}(\varphi, \rho^n))$ is true because of Lemma 3. So ρ is accepted by $\mathcal{A}_\varphi$. $\qquad\square$

Theorem 3 (Complexity). *Given an* LTL_f *formula* φ *and the* TDFA $\mathcal{A}_\varphi$ *constructed by Definition 5, the size of* $\mathcal{A}_\varphi$ *is at most* $2^{2^{|cl(\varphi)|}}$, *where* $cl(\varphi)$ *is the set of all subformulas of* φ.

Proof. From Definition 5, every state in $\mathcal{A}_\varphi$ excluding the initial one is computed via progression. It is not hard to prove that every formula from progression can be converted into the form of $\bigvee \bigwedge \psi$ where $\psi \in cl(\varphi)$. Since there are at most $2^{2^{|cl(\varphi)|}}$ formulas with the form $\bigvee \bigwedge \psi$ (including φ), the size of $\mathcal{A}_\varphi$ is also at most $2^{2^{|cl(\varphi)|}}$.   □

4   Guided LTL $_f$ Synthesis with Satisfiable Traces

The previous forward synthesis approaches [29,43] require determining whether the state in the corresponding TDFA game is a winning or a failure state. In this process, the edges from the state are explored enumeratively, which is performed randomly without direction. By Definition 3, winning states are recursively defined with its base case falling on the accepting edges of TDFA. Therefore, we can intuitively infer that edges associated with some satisfiable traces are more likely to make the current state determined as winning. Inspired by that, we propose our new synthesis algorithm MoGuS (**Mo**del-**Gu**ided **S**ynthesis). In the following, we first illustrate our new synthesis approach at a high level with an example (Sect. 4.1), introduce the details of the approach (Sect. 4.2), and then run the algorithmic details using the example again (Sect. 4.3).

4.1   An Example

We will use the formula

$$\varphi = \bigcirc\mathcal{F}(\bigcirc a \wedge \mathcal{G}b) \tag{1}$$

with $\mathcal{X} = \{a\}$ and $\mathcal{Y} = \{b\}$ as a running example. It is clearly unrealizable, because no system can guarantee $\bigcirc a$, since a is an input. In the beginning, we cannot determine whether s_0 is winning or failure and a finite trace ρ that satisfies φ is computed by an LTL$_f$ satisfiability solver (Fig. 1(a)). Suppose that $\rho = a \wedge b, a \wedge b, a \wedge b$. The corresponding TDFA run on ρ is $r = s_0, s_1, s_2$, where

$$s_1 = \mathcal{F}(\bigcirc a \wedge \mathcal{G}b), \tag{2}$$

and

$$s_2 = (a \wedge \mathcal{G}b) \vee (\bigcirc\mathcal{F}(\bigcirc a \wedge \mathcal{G}b)). \tag{3}$$

We now check whether s_2 is a winning state. As illustrated in Fig. 1(b), for every $Y \in 2^{\mathcal{Y}}$ there exists $X \in 2^{\mathcal{X}}$ such that it forms a loop from s_2. So it concludes that s_2 is a failure state and rolls back to s_1. For $Y = b$, it leads to a known failure state s_2, and for $Y = \neg b$, it detects a loop (Fig. 1(c)). This implies that s_1 is a failure state. Similarly, the initial state s_0 is found to be a failure state, and then MoGuS returns 'Unrealizable'.

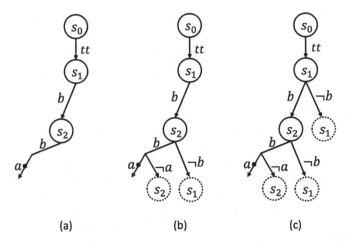

Fig. 1. Given the input formula $\bigcirc\mathcal{F}(\bigcirc a \wedge \mathcal{G}b)$ (recall that $\bigcirc$ denotes *strong Next*) with $\mathcal{X} = \{a\}$ and $\mathcal{Y} = \{b\}$, these diagrams show the progression of $\mathsf{MoGuS}(\varphi, \langle \mathcal{X}, \mathcal{Y} \rangle)$ in its satisfiability-based search for a winning strategy. Dashed circles represent loops, and points on the arrows represent accepting transitions. For simplicity, we merge edges with the same successor, i.e., tt, b, and $\neg b$ represent 4, 2, and 2 edges respectively. And here $s_0 = \bigcirc\mathcal{F}(\bigcirc a \wedge \mathcal{G}b)$, $s_1 = \mathcal{F}(\bigcirc a \wedge \mathcal{G}b)$, and $s_2 = (a \wedge \mathcal{G}b) \vee (\bigcirc\mathcal{F}(\bigcirc a \wedge \mathcal{G}b))$.

4.2   The Synthesis Algorithm MoGuS

Given a formula φ with inputs $\mathcal{X}$ and outputs $\mathcal{Y}$, we explore the TDFA by on-the-fly construction and perform a top-down traversal of the search space. Every time a TDFA state is visited, we first try to determine based on known information whether it is winning, failure, or forming a loop. If one of these three cases occurs, we backtrack to the previous state. Otherwise, we invoke the LTL_f satisfiability solver to find a satisfiable trace, subsequently exploring the states in the corresponding TDFA run. This is the primary distinction from the approach in [43]. Our search direction is guided by a satisfiable trace, rather than randomly selected by the (Boolean) SAT solver.

Algorithm 1 shows the implementation of MoGuS. It first declares four global sets: *winning* and *failure* to store the known winning and failure states respectively, *to_win* to collect winning state-edge pairs $\langle s, X \cup Y \rangle$ such that $X \cup Y \models s$ or $\mathsf{fp}(s, X \cup Y)$ is a winning state, and *to_fail* to collect state-edge pairs $\langle s, X \cup Y \rangle$ such that $X \cup Y \not\models s$ and $\mathsf{fp}(s, X \cup Y)$ is a failure state. *to_win* and *to_fail* are maintained to compute the edge constraint (Line 8). At the main entry of the algorithm, the parameter *path* refers to the state sequence that leads from the initial state to the current state ψ.

At Line 5, MoGuS checks whether ψ is winning or failure currently based on the state information collected so far, and Algorithm 2 presents the implementation of `currentWinning`. `currentWinning(`ψ`)` returns 'Winning' if (1) ψ is already in *winning*, or (2) based on Definition 3, there is $Y \in 2^{\mathcal{Y}}$ such that $X \cup Y \models \psi$ holds or $\mathsf{fp}(\psi, X \cup Y)$ is in *winning*, for every $X \in 2^{\mathcal{X}}$. During the process, those transitions accepting or leading to a winning state are added into to_win. The analogous process is performed to check whether ψ is a failure state currently.

An edge constraint is computed for current ψ at Line 8, which blocks all edges not requiring further exploration. Formally, $\texttt{edgeConstraint}(\psi)$ assigns $edge\_constraint$ as:

$$\bigwedge_{Y \in Y_f} \neg Y \ \wedge \ \bigwedge_{Y \in Y_u} \left(Y \to \bigwedge_{X \in X_w(Y)} \neg X \right). \tag{4}$$

Y_f (subscripts 'f'/'u' stand for 'failure'/'unknown') denotes a set of values for output variables, with which some assignments for input variables can lead the system to fail in the TDFA game.

$$Y_f = \left\{ Y \in 2^{\mathcal{Y}} \mid \exists X \in 2^{\mathcal{X}}. \langle \psi, X \cup Y \rangle \in to\_fail \right\} \tag{5}$$

For some output values collected in Y_u, the system retains the potential for winning. It no longer needs to explore inputs values that are known to lead the system winning, which are denoted by $X_w(Y)$ (subscript 'w' stands for 'winning'). And the right part of Eq. 4 addresses this scenario.

$$Y_u = \left\{ Y \in 2^{\mathcal{Y}} \mid Y \notin Y_f \text{ and } \exists X \in 2^{\mathcal{X}}. \langle \psi, X \cup Y \rangle \notin to\_win \right\} \tag{6}$$

$$X_w(Y) = \left\{ X \in 2^{\mathcal{X}} \mid \langle \psi, X \cup Y \rangle \in to\_win \right\} \tag{7}$$

It checks the satisfiability of the current state ψ under the edge constraint at Line 9. If 'sat' is returned, the solver would compute a satisfiable trace ρ (Line 10). The first for-loop at Lines 12–18 generates each state in the run on ρ by formula progression and checks whether the new states are winning, failure, or forming a loop. Then in the second for-loop (Lines 19–24), it recursively checks whether each state $r[i]$ is winning or failure and add $r[i]$ to $winning$ or $failure$ respectively. Notably, the check order has to be reverse, i.e., from $|r| - 1$ to 0, as MoGuS performs a Depth-First Search (DFS). At last, it recursively checks the current state ψ (Line 25) and returns the results.

While giving the search direction, the satisfiability solver also helps prevent the exploration of states that do not appear in any accepting run. The following lemma indicates that these states are system failure states.

Lemma 4. *Given a* TDFA *game over* $\mathcal{A}$, s *is a system failure state if* s *is not in any accepting run.*

And then, ψ can be determined as a failure state when $\psi \wedge edge\_constraint$ is unsatisfiable (Lines 31–32).

Lemma 5. *When the Algorithm 1 reaches Line 9, state* ψ *is a failure state if* $\psi \wedge edge\_constraint$ *is unsatisfiable, where* $edge\_constraint$ *is computed from Eqs. 4–7.*

Algorithm 1: MoGuS: Model-Guided Synthesis

Input: LTL_f formula φ with inputs $\mathcal{X}$ and outputs $\mathcal{Y}$
Output: Realizable or Unrealizable

1 $winning, \; failure, \; to\_win, \; to\_fail := \emptyset$
2 **return** isWinning$(\varphi, [\varphi])$? Realizable : Unrealizable

3

4 **Function** isWinning$(\psi, path)$
5     $peek := $ currentWinning$(\psi, path)$
6   **if** $peek \neq$ Unknown **then**
7   **return** $peek =$Winning

8     $edge\_constraint :=$edgeConstraint(ψ)
9   **if** ltlfSat$(\psi \wedge edge\_constraint)$=sat **then**
10        $\rho :=$ getModel$()$
11   Initialize r as an empty state sequence
12   **for** i *from* 0 *to* $|\rho| - 2$ **do**
13           $s :=$ fp$(\psi, \rho[0:i])$
14   **if** $s \in winning \cup failure \cup path$ **then**
15   **break**
16   **else**
17   r.pushBack(s)
18              $path$.pushBack(s)

19   **for** i *from* $|r| - 1$ *to* 0 **do**
20   **if** isWinning$(r[i], path)$ **then**
21              $winning := winning \cup \{r[i]\}$
22   **else**
23              $failure := failure \cup \{r[i]\}$
24           $path$.popBack$()$

25   **if** isWinning$(\psi, path)$ **then**
26           $winning := winning \cup \{r[i]\}$
27   **return** true
28   **else**
29           $failure := failure \cup \{r[i]\}$
30   **return** false

31   **else**
32   **return** false

Proof. At Line 9, ψ has not been determined as winning, which implies $Y_f \cup Y_u = 2^{\mathcal{Y}}$. Therefore, there are two cases for $Y \in 2^{\mathcal{Y}}$:

- $Y \in Y_f$. By Eq. 5, there exists $X \in 2^{\mathcal{X}}$ such that fp$(\psi, X \cup Y)$ is a failure state or forming a loop.
- $Y \in Y_u$. Here we consider $X \in 2^{\mathcal{X}}$ such that $X \notin X_w(Y)$. By Eqs. 4, it holds that $X \cup Y \models edge\_constrain$. Therefore, $\psi \wedge edge\_constraint$ is

---

Algorithm 2: Implementation of `currentWinning`

Input: A TDFA state ψ and a state sequence *path* storing the states visited from the initial state to ψ.

Output: Winning, Failure, or Unknown

```
1  if isCurrentWinning(ψ) then
2  │   return Winning

3  if isCurrentFailure(ψ) then
4  │   return Failure

5  return Unknown

6

7  function isCurrentWinning(ψ)
8  │   if ψ ∈ winning then
9  │   │   return true
10 │   for each Y ∈ 2^𝒴 do
11 │   │   all_win :=true
12 │   │   for each X ∈ 2^𝒳 do
13 │   │   │   if X ∪ Y ⊭ ψ and fp(ψ, X ∪ Y) ∉ winning then
14 │   │   │   │   all_win := false
15 │   │   │   │   break
16 │   │   │   else
17 │   │   │   │   to_win := to_win ∪ {⟨ψ, X ∪ Y⟩}
18 │   │   if all_win then
19 │   │   │   return true
20 │   return false

21 function isCurrentFailure(ψ)
22 │   if ψ ∈ failure then
23 │   │   return failure
24 │   for each Y ∈ 2^𝒴 do
25 │   │   Let exist_fail :=false
26 │   │   for each X ∈ 2^𝒳 do
27 │   │   │   if X ∪ Y ⊭ ψ and fp(ψ, X ∪ Y) ∈ failure ∪ path then
28 │   │   │   │   exist_fail :=true
29 │   │   │   │   to_fail := to_fail ∪ {⟨ψ, X ∪ Y⟩}
30 │   │   │   │   break
31 │   │   if ¬exist_fail then
32 │   │   │   return false
33 │   return true
```

---

unsatisfiable implying that $\mathsf{fp}(\psi, X \cup Y)$ cannot appear in any accepting run. By Lemma 4, $\mathsf{fp}(\psi, X \cup Y)$ is a failure state.

Based on the above cases, ψ is a failure state by Definition 3.   □

With Lemma 5 and Definition 3, we explicitly state this corollary in the context of Algorithm 1 for better understanding.

Corollary 1. *Given a* TDFA *state* ψ,

- ψ *is a winning state, if there exists* $Y \in 2^{\mathcal{Y}}$ *such that for every* $X \in 2^{\mathcal{X}}$, $X \cup Y \models \psi$ *or* $\mathsf{fp}(\psi, X \cup Y) \in winning$;
- ψ *is a failure state, if either* $\psi \wedge edge\_constrainr$ *is unsatisfiable, or for every* $Y \in 2^{\mathcal{Y}}$ *there exists* $X \in 2^{\mathcal{X}}$, $X \cup Y \not\models \psi$ *and* $\mathsf{fp}(\psi, X \cup Y) \in failure \cup path$.

Theorem 4. *Algorithm 1 is complete and sound. That is, given an* LTL$_f$ *formula* φ *with inputs* $\mathcal{X}$ *and outputs* $\mathcal{Y}$,

- *Algorithm 1 can terminate within time of* $O(2^{|\mathcal{X} \cup \mathcal{Y}|} \cdot 2^{2^{|cl(\varphi)|}})$;
- φ *with* $\langle \mathcal{X}, \mathcal{Y} \rangle$ *is realizable iff Algorithm 1 returns 'Realizable'.*

Proof. **Completeness.** The number of states related to recursive calls to isWinning is bounded by the worst case doubly-exponential number of states in the constructed TDFA (Theorem 3). Before every newly computed state is checked recursively, MoGuS first checks for winning, failure, and loop (Line 14). Then *edge_constraint* helps enumerate the edges and successors of each state, which guarantees that each state is visited at most $2^{|\mathcal{X} \cup \mathcal{Y}|}$ times. Hence, the time complexity of Algorithm 1 is $O(2^{|\mathcal{X} \cup \mathcal{Y}|} \cdot 2^{2^{|cl(\varphi)|}})$.

Soundness. ($\Leftarrow$) In Algorithm 1, there are two cases in which MoGuS returns 'Realizable'. The first one (Line 5) is when currentWinning(ψ, *path*) returns 'Winning', which implements exactly Definition 3. This indicates that φ is already a winning state and according to Theorem 1, φ with $\langle \mathcal{X}, \mathcal{Y} \rangle$ is realizable. The second case is recursively calling isWinning (Line 25), which finally falls into the base case, i.e., currentWinning(ψ, *path*) returns 'Winning'. This situation has been discussed before.

($\Rightarrow$) We perform this part of proof by contraposition. If Algorithm 1 returns 'Unrealizable', then φ with $\langle \mathcal{X}, \mathcal{Y} \rangle$ is unrealizable. There are three cases leading MoGuS to return 'Unealizable'. The first and second cases (Lines 5, 25) are analogous to those of ($\Leftarrow$). And the third case at Line 32 exactly corresponds to the conclusion of Lemma 5. □

4.3 Back to Our Example

Figure 2 illustrates how MoGuS works in detail, based on the example described in Subsect. 4.1. We now describe the main steps of Algorithm 1 when applied to this example. At Line 5 of Algorithm 1, currentWinning finds a transition $\langle s_0, a \wedge \neg b \rangle$, adds it to *to_fail*, and returns 'Unknown'. Then *edge_constraint* is assigned as b at Line 8. And ltlfSat computes a model of φ (Line 9), which in the figure returns $\rho = \{a \wedge b\}, \{a \wedge b\}, \{a \wedge b\}$. The corresponding run of ρ is $r = s_0, s_1, s_2$, of which states are checked whether winning, failure, or loop and added to *path* at Lines 14–18. Next, MoGuS recursively checks whether the

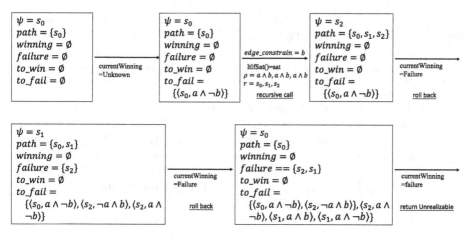

Fig. 2. The main steps of $\mathsf{MoGuS}(\varphi, \langle \mathcal{X}, \mathcal{Y} \rangle)$ when checking the realizability of Equation (1) with $\mathcal{X} = \{a\}$ and $\mathcal{Y} = \{b\}$, where we have $\varphi = s_0 = \mathcal{OF}(\mathcal{O}a \wedge \mathcal{G}b)$, $s_1 = \mathcal{F}(\mathcal{O}a \wedge \mathcal{G}b)$ and $s_2 = (a \wedge \mathcal{G}b) \vee (\mathcal{OF}(\mathcal{O}a \wedge \mathcal{G}b))$.

new states in r (with a reverse order) can be winning (Line 20). As shown in the figure, s_2 turns out to be a failure state by currentWinning and is added to $failure$, since for every $Y \in 2^{\mathcal{Y}}$ there is a chance of forming a loop through s_2 (see Fig. 1). Then, the algorithm rolls back state by state and sequentially updates the failure states and transitions into $failure$ and to_fail respectively. The recursive process finally goes back to s_0 and concludes that φ with $\langle \mathcal{X}, \mathcal{Y} \rangle$ is unrealizable.

5 Experimental Evaluation

5.1 Experimental Set-Up

Tools. We implemented MoGuS in a tool called MoGuSer using C++, and integrated aaltaf [33] as the engine for LTL$_f$ satisfiability checking. We compared the results with three state-of-the-art synthesis tools, Lisa [7], Lydia [21], and Cynthia [29]. The first two tools are based on the bottom-up approach, and Cynthia is SDD-based and performs forward synthesis. All three tools were run with their default parameters.

Benchmarks. We ran the experiment with the collected benchmarks in [29], which are in total 1454 instances, including 1400 *Random* instances, and 54 *Two-player-Games* instances. Based on our preliminary experimental results, the majority of existing cases can be solved by the evaluated tools. To better compare the scalability of different tools, we also created a new set of benchmarks, which we call *Ascending*. We employed the randltl command in Spot [23] to generate a batch of random LTL formulas, which are treated as LTL$_f$ formulas (since LTL$_f$ formulas share the same syntax as LTL), and then randomly divide the atomic variables from these formulas into input and output variables.

The `--tree-size`[2] option of `randltl` specifies the tree size of the generated formulas, and we generated 200 test cases for each size ranging from 100 to 900 (1800 in total). Although the *Ascending* benchmark consists of random formulas as well, they have much larger sizes than those in [29] and are more suitable to evaluate the tools' scalability.

Platform. We ran the experiments on a CentOS 7.4 cluster, where each instance had exclusive access to a processor core of the Intel Xeon 6230 CPU running at 2.1 GHz, with 8 GB of memory and a 30-minute time limit. We measured execution time with the Unix command `time`. When collecting the data, we recorded a running time of 1801 s for all instances that could not be solved by the tested tool within this time limit. We verified that the results emitted by the four tools are consistent (excluding those that timed out).

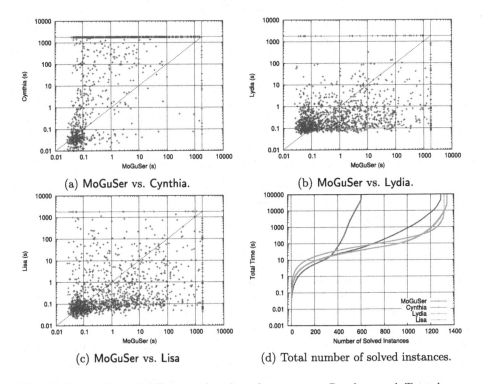

(a) MoGuSer vs. Cynthia.   (b) MoGuSer vs. Lydia.

(c) MoGuSer vs. Lisa   (d) Total number of solved instances.

Fig. 3. Comparison of different solvers' performance on Random and Two-player-Games benchmarks together. Points located on the red line represent instances that fail to be solved within the given time and memory resources. (Color figure online)

---

[2] See https://spot.lre.epita.fr/man/randltl.1.html.

5.2   Results and Analysis I: Random and Two-player-Games Benchmarks

We first evaluated the four tools on benchmarks collected from previous literature, i.e., Random and Two-player-Games benchmarks. The pairwise comparison
of MoGuSer against Cynthia, Lydia, and Lisa is shown in Figs. 3a-3c, respectively.
Figure 3d illustrates the total number of successfully solved cases accumulated
over time. Tables 1 and 2 present quantitative summaries of results on these
benchmarks.

Comparing MoGuSer with the other top-down tool Cynthia, we can observe
that MoGuS has significantly improved the top-down solving capability for LTL_f
synthesis problems. MoGuSer could solve many more instances than Cynthia,
and when both tools can solve an instance, MoGuSer demonstrates faster performance (Fig. 3a). As shown in Table 1, MoGuSer solved 918 unrealizable Random
instances, while Cynthia only solved 254. These facts indicate that the targeted
search strategy of MoGuS can avoid unnecessary searches (especially in unrealizable cases), resulting in faster convergence. Besides, MoGuSer achieves better
results on both *Counter(s)* instances than Cynthia, while Cynthia still keeps a
distinct advantage over other tools on the *Nim* test cases.

Table 1. Results on Random instances.

Tools	Realizable		Unrealizable		Total	
	Solved	Uniquely solved	Solved	Uniquely solved	Solved	Uniquely solved
MoGuSer	347	0	918	6	1265	6
Cynthia	324	0	254	0	578	0
Lydia	350	0	932	0	1282	0
Lisa	351	0	960	11	1311	11

Table 2. Results on Two-player-Games instances.

Tools	Single-counter		Double-counters		Nim	
	Solved	Uniquely solved	Solved	Uniquely solved	Solved	Uniquely solved
MoGuSer	8	0	6	1	8	0
Cynthia	4	0	2	0	21	4
Lydia	11	3	6	0	17	0
Lisa	8	0	7	0	13	1

As for the two bottom-up tools, Lydia and Lisa achieve similar performance
in most aspects and slightly outperform MoGuSer. The distribution pattern of
points in Fig. 3b closely resembles that in Fig. 3c. We can observe that each
tool has its own merits in solving speed, MoGuSer performs almost the same
as Lydia/Lisa among instances that can be solved by both tools. From the endpoints of the curves in Fig. 3d, Tables 1 and 2, the number of instances solved

202 S. Xiao et al.

by MoGuSer is slightly lower than that of Lydia and Lisa. Thus we conclude that the MoGuS has improved the top-down solving capability for LTL$_f$ synthesis problems to a level nearly equivalent to that of bottom-up tools (Lydia and Lisa).

Taking a comprehensive view of Tables 1 and 2, the bottom-up approach still holds a slight advantage over the top-down approach. However, no LTL$_f$ synthesis tool dominates all other tools, since each of them can uniquely solve some instances. Similarly to hardware model checking [44], this emphasizes the need for maintaining a portfolio consisting of diverse approaches for LTL$_f$ synthesis.

5.3 Results and Analysis II: Ascending Benchmarks

We further evaluated the scalability of the four tools by a collection of test cases sorted in ascending order based on the tree sizes of formulas. Figure 4 depicts the cumulative number of instances solved by each tool over time on benchmarks of varying sizes. Meanwhile, Fig. 5 illustrates the number of solved instances as a function of the formula's tree size.

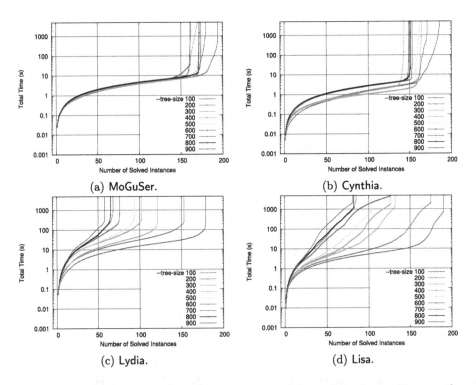

Fig. 4. The cumulative number of instances solved by each tool over time on the *Ascending* benchmarks.

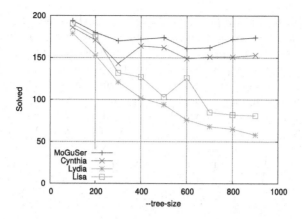

Fig. 5. The number of solved instances as a function of the formula's tree size.

Both MoGuSer and Cynthia demonstrate robust scalability, but MoGuSer performs better than Cynthia. The endpoints of the curves in Figs. 4a and 4b are not clearly located by the tree sizes of the formulas, and the curves corresponding to MoGuSer and Cynthia in Fig. 5 fluctuate but are generally stable. These suggest that the solving capabilities of MoGuSer and Cynthia are not significantly limited by the problem size. Besides, MoGuSer solved more cases than Cynthia, which reflects the efficiency of the top-down methodology by MoGuSer. Besides, Cynthia achieves relatively satisfactory results on the *Ascending* benchmarks. We infer that this can be attributed to the fact that realizable instances account for a large portion of the *Ascending* benchmarks, since Cynthia's ability to solve realizable instances is comparable to the other three tools (see Table 1). Excluding 101 instances that cannot be solved by any tool, there are 1475 realizable instances among the remaining 1699 instances.

On the other hand, the performance of the two tools (Lisa and Lydia) based on the bottom-up method drops significantly as the LTL_f formula size grows. The curves in Figs. 4c and 4d indicate that the solving speed of Lisa and Lydia gradually slows down as the tree sizes of the formulas increase. From the positions of the endpoints of the curves in Figs. 4c, 4d and the trend of the lower two curves in Fig. 5, the number of instances successfully solved by Lydia and Lisa also change significantly. When `-tree-size` is set to 100, Lydia and Lisa solve 179 and 190 respectively, and when –tree-size increases to 900, only 58 and 81 are solved respectively. We speculate that the reason is that both Lydia and Lisa require the construction of complete DFA before synthesis, which relies heavily on BDDs [10]. BDDs can require exponential space in the number of variables, which limits the capability of synthesis tools.

6   LTL Synthesis – Related Work

We have discussed various works related to LTL_f synthesis in the introduction. Let us mention here some comments about related work. Temporal synthesis is a classical problem, first proposed by A. Church in the 1960s [17]. The original logic specification to be synthesized was expressed by an S1S formula, for which the complexity to solve the problem is non-elementary [12,40]. The first work to consider LTL synthesis is [39], which solves the synthesis problem by reducing it to a Rabin game [25]. This approach constructs a non-deterministic Büchi automaton from the input LTL formula, and then determinizes it to its equivalent Rabin automaton, a process which takes worst-case double-exponential time. The complexity of solving a Rabin game is NP-Complete [25]. Nowadays, the standard approach is to reduce LTL synthesis to the parity game [26], because a parity game can be solved in quasi-polynomial time [13], even though the doubly-exponential process to obtain a deterministic parity automaton cannot be avoided. LTL synthesis tools like ltlsynt [37] and Strix [36], are built using the parity-game approach. Because of the challenge to determinize an ω automaton, researchers also consider other possibilities, e.g., by reducing LTL synthesis to the bounded safety game [31]. Acacia+ [9] is a representative tool following the safety-game approach. The annual reactive synthesis competition [1] drives progress in this field, yet the scalability issue is still a major problem.

7   Concluding Remarks

We have presented a new approach called MoGuS for synthesizing LTL_f formulas. By invoking an LTL_f satisfiability checker, MoGuS performs the search for a winning strategy in a more targeted way compared to previous top-down approaches. An empirical comparison of this method to state-of-the-art LTL_f synthesizers suggests that it can achieve the best overall performance. Several future works are being considered. Firstly, MoGuS relies heavily on an LTL_f satisfiability checker, and hence it can benefit from performance improvements of this stage. We can explore incrementally invoking the LTL_f satisfiability checker. Secondly, both the bottom-up tools Lydia and Lisa integrated with composition techniques [6,7,21], which decomposes the formula on conjunctions, perform synthesis of each conjunct, and then combine the results in order to solve the original problem. Similar composition ideas could also be applied to top-down synthesis approaches. Finally, it is interesting to check whether the optimizations described in this article can accelerate LTL synthesis of safety properties [32,47].

Acknowledgements. This work is supported by National Natural Science Foundation of China (Grant #U21B2015 and #62372178), "Digital Silk Road" Shanghai International Joint Lab of Trustworthy Intelligent Software under Grant 22510750100, Shanghai Collaborative Innovation Center of Trusted Industry Internet Software, by US NSF grants IIS-1527668, CCF-1704883, IIS-1830549, CNS-2016656, and by US DoD MURI grant N00014-20-1-2787.

Data Availability Statement. To support the experimental results, the source code of MoGuSer and benchmarks is available at https://drive.google.com/file/d/1ohOa4Kl4R4br095k-kVJcWV87U5XON5q/view?usp=sharing.

References

1. The reactive synthesis competition. http://www.syntcomp.org/
2. Althoff, C.S., Thomas, W., Wallmeier, N.: Observations on determinization of Büchi automata. In: Farré, J., Litovsky, I., Schmitz, S. (eds.) CIAA 2005. LNCS, vol. 3845, pp. 262–272. Springer, Heidelberg (2006). https://doi.org/10.1007/11605157_22
3. Aminof, B., De Giacomo, G., Murano, A., Rubin, S.: Synthesis under assumptions. In: Sixteenth International Conference on Principles of Knowledge Representation and Reasoning, pp. 615–616. AAAI Press (2018)
4. Aminof, B., De Giacomo, G., Murano, A., Rubin, S.: Planning under LTL environment specifications. In: Proceedings of the Twenty-Ninth International Conference on Automated Planning and Scheduling, pp. 31–39. AAAI Press (2019)
5. Bacchus, F., Kabanza, F.: Planning for temporally extended goals. Ann. Math. Artif. Intell. **22**, 5–27 (1998)
6. Bansal, S., Giacomo, G.D., Stasio, A.D., Li, Y., Vardi, M.Y., Zhu, S.: Compositional safety LTL synthesis. In: Verified Software: Theories, Tools, and Experiments (VSTTE) (2022)
7. Bansal, S., Li, Y., Tabajara, L., Vardi, M.: Hybrid compositional reasoning for reactive synthesis from finite-horizon specifications. In: The Thirty-Fourth AAAI Conference on Artificial Intelligence, vol. 34, pp. 9766–9774. AAAI Press (2020)
8. Bloem, R., Jobstmann, B., Piterman, N., Pnueli, A., Saar, Y.: Synthesis of reactive(1) designs. J. Comput. Syst. Sci. **78**(3), 911–938 (2012)
9. Bohy, A., Bruyère, V., Filiot, E., Jin, N., Raskin, J.-F.: Acacia+, a tool for LTL synthesis. In: Madhusudan, P., Seshia, S.A. (eds.) CAV 2012. LNCS, vol. 7358, pp. 652–657. Springer, Heidelberg (2012). https://doi.org/10.1007/978-3-642-31424-7_45
10. Bryant, R.: Graph-based algorithms for Boolean-function manipulation. IEEE Trans. Comput. **C-35**(8), 677–691 (1986)
11. Büchi, J.: On a decision method in restricted second order arithmetic. In: Proceedings of International Congress on Logic, Method, and Philosophy of Science. 1960, pp. 1–12. Stanford University Press (1962)
12. Büchi, J., Landweber, L.: Solving sequential conditions by finite-state strategies. Trans. AMS **138**, 295–311 (1969)
13. Calude, C.S., Jain, S., Khoussainov, B., Li, W., Stephan, F.: Deciding parity games in quasipolynomial time. In: Proceedings of the 49th Annual ACM SIGACT Symposium on Theory of Computing, STOC 2017, pp. 252–263. Association for Computing Machinery, New York (2017)
14. Camacho, A., Bienvenu, M., McIlraith, S.A.: Finite LTL synthesis with environment assumptions and quality measures. In: Sixteenth International Conference on Principles of Knowledge Representation and Reasoning, pp. 454–463. AAAI Press (2018)
15. Camacho, A., McIlraith, S.A.: Strong fully observable non-deterministic planning with LTL and LTLf goals. In: Proceedings of the Twenty-Eighth International Joint Conference on Artificial Intelligence, IJCAI 2019, pp. 5523–5531 (2019)
16. Camacho, A., Triantafillou, E., Muise, C.J., Baier, J.A., McIlraith, S.A.: Non-deterministic planning with temporally extended goals: LTL over finite and infinite traces. In: Proceedings of the Thirty-First AAAI Conference on Artificial Intelligence, pp. 3716–3724. AAAI Press (2017)

17. Church, A.: Logic, arithmetics, and automata. In: Proceedings of International Congress of Mathematicians, 1962, pp. 23–35. Institut Mittag-Leffler (1963)
18. Church, A.: Application of recursive arithmetic to the problem of circuit synthesis. J. Symb. Log. **28**(4), 289–290 (1963)
19. Darwiche, A.: SDD: a new canonical representation of propositional knowledge bases. In: Proceedings of the Twenty-Second International Joint Conference on Artificial Intelligence, pp. 819–826. AAAI Press (2011)
20. De Giacomo, G., Vardi, M.: Linear temporal logic and linear dynamic logic on finite traces. In: Proceedings of the Twenty-Third International Joint Conference on Artificial Intelligence, pp. 854–860. AAAI Press (2013)
21. De Giacomo, G., Favorito, M.: Compositional approach to translate LTLf/LDLf into deterministic finite automata. In: Proceedings of the International Conference on Automated Planning and Scheduling, vol. 31, pp. 122–130 (2021)
22. De Giacomo, G., Rubin, S.: Automata-theoretic foundations of fond planning for LTLf and LDLf goals. In: Proceedings of the 27th International Joint Conference on Artificial Intelligence, pp. 4729–4735. AAAI Press (2018)
23. Duret-Lutz, A., et al.: From spot 2.0 to spot 2.10: what's new? In: Shoham, S., Vizel, Y. (eds.) CAV 2022. LNCS, vol. 13372, pp. 174–187. Springer, Cham (2022). https://doi.org/10.1007/978-3-031-13188-2_9
24. Eén, N., Sörensson, N.: An extensible SAT-solver. In: Giunchiglia, E., Tacchella, A. (eds.) SAT 2003. LNCS, vol. 2919, pp. 502–518. Springer, Heidelberg (2004). https://doi.org/10.1007/978-3-540-24605-3_37
25. Emerson, E., Jutla, C.: The complexity of tree automata and logics of programs. In: Proceedings of 29th IEEE Symposium on Foundations of Computer Science, pp. 328–337 (1988)
26. Emerson, E., Jutla, C.: Tree automata, μ-calculus and determinacy. In: Proceedings of 32nd IEEE Symposium on Foundations of Computer Science, pp. 368–377 (1991)
27. Fuggitti, F.: FOND planning for LTLf and PLTLf goals (2020). https://doi.org/10.48550/ARXIV.2004.07027
28. Giacomo, G.D., Vardi, M.Y.: Synthesis for LTL and LDL on finite traces. In: Proceedings of the 24th International Conference on Artificial Intelligence, pp. 1558–1564. AAAI Press (2015)
29. Giacomo, G.D., Favorito, M., Li, J., Vardi, M.Y., Xiao, S., Zhu, S.: LTLf synthesis as and-or graph search: knowledge compilation at work. In: Proceedings of the Thirty-First International Joint Conference on Artificial Intelligence, pp. 3292–3298. AAAI Press (2022)
30. Henriksen, J.G., et al.: Mona: monadic second-order logic in practice. In: Brinksma, E., Cleaveland, W.R., Larsen, K.G., Margaria, T., Steffen, B. (eds.) TACAS 1995. LNCS, vol. 1019, pp. 89–110. Springer, Heidelberg (1995). https://doi.org/10.1007/3-540-60630-0_5
31. Kupferman, O.: Avoiding determinization. In: Proceedings of 21st IEEE Symposium on Logic in Computer Science, pp. 243–254 (2006)
32. Kupferman, O., Vardi, M.Y.: Model checking of safety properties. In: Halbwachs, N., Peled, D. (eds.) CAV 1999. LNCS, vol. 1633, pp. 172–183. Springer, Heidelberg (1999). https://doi.org/10.1007/3-540-48683-6_17
33. Li, J., Rozier, K.Y., Pu, G., Zhang, Y., Vardi, M.Y.: SAT-based explicit LTLf satisfiability checking. In: The Thirty-Third AAAI Conference on Artificial Intelligence, pp. 2946–2953. AAAI Press (2019)
34. Li, J., Zhang, L., Pu, G., Vardi, M.Y., He, J.: LTL$_f$ satisfibility checking. In: Proceedings of the Twenty-First European Conference on Artificial Intelligence, pp. 513–518. IOS Press (2014)

35. Luo, W., Wan, H., Du, J., Li, X., Fu, Y., Ye, R., Zhang, D.: Teaching LTLf satisfiability checking to neural network. In: Proceedings of the Thirty-First International Joint Conference on Artificial Intelligence, pp. 3292–3298. AAAI Press (2022)

36. Meyer, P.J., Sickert, S., Luttenberger, M.: Strix: explicit reactive synthesis strikes back! In: Chockler, H., Weissenbacher, G. (eds.) CAV 2018. LNCS, vol. 10981, pp. 578–586. Springer, Cham (2018). https://doi.org/10.1007/978-3-319-96145-3_31

37. Michaud, T., Colange, M.: Reactive synthesis from LTL specification with spot. In: Proceedings Seventh Workshop on Synthesis, SYNT@CAV 2018. Electronic Proceedings in Theoretical Computer Science (2018)

38. Pnueli, A.: The temporal logic of programs. In: 18th Annual Symposium on Foundations of Computer Science, pp. 46–57. IEEE (1977). https://doi.org/10.1109/SFCS.1977.32

39. Pnueli, A., Rosner, R.: On the synthesis of an asynchronous reactive module. In: Ausiello, G., Dezani-Ciancaglini, M., Della Rocca, S.R. (eds.) ICALP 1989. LNCS, vol. 372, pp. 652–671. Springer, Heidelberg (1989). https://doi.org/10.1007/BFb0035790

40. Rabin, M.: Automata on infinite objects and Church's problem. American Mathematical Society (1972)

41. Safra, S.: On the complexity of ω-automata. In: Proceedings of 29th IEEE Symposium on Foundations of Computer Science, pp. 319–327 (1988)

42. Shi, Y., Xiao, S., Li, J., Guo, J., Pu, G.: SAT-based automata construction for LTL over finite traces. In: 27th Asia-Pacific Software Engineering Conference (APSEC), pp. 1–10. IEEE (2020). https://doi.org/10.1109/APSEC51365.2020.00008

43. Xiao, S., Li, J., Zhu, S., Shi, Y., Pu, G., Vardi, M.Y.: On the fly synthesis for LTL over finite traces. In: The Thirty-Fourth AAAI Conference on Artificial Intelligence, pp. 6530–6537. AAAI Press (2021)

44. Zhang, X., Xiao, S., Xia, Y., Li, J., Chen, M., Pu, G.: Accelerate safety model checking based on complementary approximate reachability. IEEE Trans. Comput. Aided Des. Integr. Circuits Syst. **42**(9), 3105–3117 (2023). https://doi.org/10.1109/TCAD.2023.3236272

45. Zhu, S., Tabajara, L., Li, J., Pu, G., Vardi, M.: Symbolic LTLf synthesis. In: Proceedings of the 26th International Joint Conference on Artificial Intelligence, pp. 1362–1369. AAAI Press (2017)

46. Zhu, S., Giacomo, G.D., Pu, G., Vardi, M.Y.: LTLf synthesis with fairness and stability assumptions. In: The Thirty-Fourth AAAI Conference on Artificial Intelligence, pp. 3088–3095. AAAI Press (2020)

47. Zhu, S., Tabajara, L.M., Li, J., Pu, G., Vardi, M.Y.: A symbolic approach to safety LTL synthesis. In: HVC 2017. LNCS, vol. 10629, pp. 147–162. Springer, Cham (2017). https://doi.org/10.1007/978-3-319-70389-3_10

Solving Two-Player Games Under Progress Assumptions

Anne-Kathrin Schmuck[1], K. S. Thejaswini[2], Irmak Sağlam[1(✉)],
and Satya Prakash Nayak[1]

[1] Max Planck Institute for Software Systems (MPI-SWS), Kaiserslautern, Germany
{akschmuck,isaglam,sanayak}@mpi-sws.org
[2] Department of Computer Science, University of Warwick, Coventry, UK
thejaswini.raghavan.1@warwick.ac.uk

Abstract. This paper considers the problem of solving infinite two-player games over finite graphs under various classes of *progress assumptions* motivated by applications in cyber-physical system (CPS) design. Formally, we consider a game graph G, a temporal specification Φ and a temporal assumption ψ, where both Φ and ψ are given as linear temporal logic (LTL) formulas over the vertex set of G. We call the tuple (G, Φ, ψ) an *augmented game* and interpret it in the classical way, i.e., winning the augmented game (G, Φ, ψ) is equivalent to winning the (standard) game $(G, \psi \Rightarrow \Phi)$. Given a reachability or parity game $\mathcal{G} = (G, \Phi)$ and some progress assumption ψ, this paper establishes whether solving the augmented game $\mathfrak{G} = (G, \Phi, \psi)$ lies in the same complexity class as solving $\mathcal{G}$. While the answer to this question is negative for arbitrary combinations of Φ and ψ, a positive answer results in more efficient algorithms, in particular for large game graphs.

We therefore restrict our attention to particular classes of CPS-motivated progress assumptions and establish the worst-case time complexity of the resulting augmented games. Thereby, we pave the way towards a better understanding of assumption classes that can enable the development of efficient solution algorithms in augmented two-player games.

Keywords: Synthesis · Graph Games · Augmented Games · Progress Assumptions

1 Introduction

The automated and correct-by-design synthesis of control software for cyber-physical systems (CPS) has gained considerable attention in the last decades. Besides the numerous practical challenges that arise in CPS control, there are

The randomization record is publicly available at www.aeaweb.org/journals/policies/random-author-order/search. S. P. Nayak, I. Saglam and A.-K. Schmuck are supported by the DFG projects 389792660 TRR 248-CPEC and SCHM 3541/1-1.

also unique theoretical challenges due to the interplay of low-level physical control loops and higher-level logical decisions. As a motivating example, consider a fully automated air-traffic control at an airport. While each modern airplane is equipped with feedback-controllers automatically regulating flight dynamics, the problem of assigning landing and starting spots along with non-intersecting flight corridors to each airplane is mostly solved manually by *humans* these days.

The problem of designing a logical controller which automates these (higher-layer) logical decisions shares strong algorithmic similarities with the problem of *reactive synthesis* from the formal methods community. In reactive synthesis, the interplay of available logical decisions with the external environment's reactions is modelled as a two-player game over a finite graph. Given such a game graph, one automatically computes a *winning strategy* which takes logical decisions in response to any environment behavior such that a predefined specification always holds. While reactive synthesis is a mature field in computer science with rich tool support for solving games of various flavors, the application of these techniques to, e.g., the problem of air-traffic control, requires the construction of a two-player game graph, which reflects all possible interactions between the controller and the environment in an abstract manner.

Building a game graph with the 'right' level of abstraction correctly modelling the interplay of low-level continuous (physical) dynamics, external interactions and logical decisions is a known severe challenge in CPS design. This challenge has been extensively addressed in the past decades under the term abstraction-based controller synthesis (for an overview see e.g. [7,12,36,39]). From a reactive synthesis perspective, the challenge amounts to finding (i) the 'right' granularity of the abstract state space, i.e., the vertex set of the graph, and (ii) the 'right' power of the environment player in the resulting abstract game. While the first one determines the size of the resulting graph, the second one ensures that a winning strategy for the controller does not fail to exist due to an unnecessary conservative overapproximation of environment uncertainty.

In this context, the notion of *augmented* games appeared in different CPS control applications [25,26,28,31–33,38]. Here, the environment player is augmented with a restriction on its choice of moves, which is motivated by the physical laws governing the underlying CPS. Intuitively, these augmentations can for example abstractly model the fact that high disturbance spikes (e.g., strong wind) only occur sporadically, certain control actions (e.g., triggering a plane landing maneuver) will eventually result in a distinct system state (e.g., landing on the ground), or external logical decisions (e.g., resource allocations by a scheduler) have particular temporal poperies (e.g., are known to be fair). Here, a typical augmentation for modelling 'eventual success' of a triggered control action asserts that whenever the source vertex is visited infinitely often, a dedicated transition (e.g., the one that is not-self looping at this vertex) is eventually taken by the environment player. These assumptions are also known as *strong transition fairness* [5,20,35] and are illustrated in Fig. 1.

Formally, an *augmented game* is a tuple (G, Φ, ψ), consisting of a finite two-player game graph G, and temporal specifications Φ and assumption ψ, given as linear temporal logic (LTL) formulas over the vertex set of G. An augmented game is typically interpreted in the classical way, i.e., winning the augmented

Fig. 1. Examples illustrating how physical phenomena can be abstracted by *strong transition fairness assumptions* modelled by live edges (dashed) which need to be taken infinitely often if their source vertex is visited infinitely often.

game (G, Φ, ψ) is equivalent to winning the (standard) game $(G, \psi \Rightarrow \Phi)$ with modified specification $\Phi' := \psi \Rightarrow \Phi$.

As augmented games can always be solved via their "implication-form" $(G, \psi \Rightarrow \Phi)$, it is not immediately obvious that augmentations can fundamentally change the given reactive synthesis problem, in particular their complexity class. However, game graphs resulting from CPS abstractions tend to be very large while specifications are rather simple, i.e., typically induced by 'reach-while-avoid' problems. Given for example a reachability game augmented with strong transition fairness assumptions, results in a Rabin game (with one Rabin pair per fair transition) when re-written in implication form, hence moving from a PTIME problem to solve the original (non-augmented) game to an NP-complete problem for its augmented version. Given a very large graph with augmentations on almost every vertex, this can very quickly lead to computational intractability. On the other hand, augmentations are typically very structured and local (as e.g., strong transition fairness). Using these features, it was recently shown by Banerjee et al. [6] that strong transition fairness assumptions can actually be handled 'for free' in synthesis – i.e., solving reachability games augmented with strong transition fairness assumptions can be solved in PTIME.

Motivated by this result, this paper investigates which other classes of augmentations can be "handled for free" in *reachability and parity games*, as summarized in Table 1. In particular, we show that parity games (resp. reachability games) augmented with live edges (strong transition fairness), co-live edges (strong transition *co-fairness*), singleton-source live groups or persistent live groups can be solved in quasi-polynomial time (resp. polynomial time), which largely extends the known class of efficiently solvable augmented games. In addi-

Table 1. Summary of complexity classes for solving the augmented games discussed in this paper.

Augmentation Type	Reachability Games	Parity Games
no augmentations	PTIME (Lemma 1)	QP (Lemma 2)
live edges (Definition 2)	PTIME (Theorem 1)	QP (Theorem 1)
co-live edges (Definition 3)	PTIME (Corollary 2)	QP (Corollary 2)
live groups (Definition 4)	NP-complete (Theorem 3)	NP-complete (Theorem 3)
singleton-source live groups (Sect. 6.2)	PTIME (Theorem 4)	QP (Theorem 4)
persistent live groups (Definition 5)	PTIME (Corollary 3)	QP (Theorem 6)

tion, we also prove NP-completeness of a slightly more general, but most frequent class of assumptions, called live groups (group transition fairness) assumptions. Unfortunately, this result shows that solving augmented games resulting from the product of different components augmented with strong transition fairness is an NP-hard problem, even for simple reachability objectives.

We note that most of the progress assumptions we consider in this paper solely restrict the moves of the environment player, and are hence non-falsifiable by the system player. This implies that, in most cases, the system player cannot vacuously win the implication-version $(G, \psi \Rightarrow \Phi)$ of an augmented game (G, Φ, ψ) by enforcing $\neg \psi$ (instead of Φ). This distinguishes our results from the large body of work investigating how winning strategies should be synthesized in augmented games such that (general LTL) assumptions are "handled correctly", as e.g. in [8–10,15,16,23,29]. In addition, we are also not concerned with the problem of *computing* assumptions that render non-realizable synthesis problems realizable, as e.g. in [2,14,30]. In this paper, assumptions are assumed to be *given* and resulting from an abstraction process of existing component dynamics.

2 Preliminaries

Notation. Given two real numbers $a, b \in \mathbb{R}$ with $a < b$, we use $[a; b]$ to denote the set $\{n \in \mathbb{N} \mid a \leq n \leq b\}$ of all integers between a and b. For any given set $[a; b]$, we write $i \in_{\text{even}} [a; b]$ and $i \in_{\text{odd}} [a; b]$ as shorthand for $i \in [a; b] \cap \{0, 2, 4, \ldots\}$ and $i \in [a; b] \cap \{1, 3, 5, \ldots\}$ respectively. Given two sets A and B, a relation $R \subseteq A \times B$, and an element $a \in A$, we write $R(a)$ to denote the set $\{b \in B \mid (a, b) \in R\}$.

Languages. Let Σ be a finite alphabet. Σ^* and Σ^ω denote, respectively, the set of finite and infinite words over Σ, and Σ^∞ is equal to $\Sigma^* \cup \Sigma^\omega$. Given two words $u \in \Sigma^*$ and $v \in \Sigma^\infty$, their concatenation is written as the word uv.

Game Graphs. A *game graph* is a tuple $G = (V = V_0 \uplus V_1, E)$ where (V, E) is a finite directed graph with *vertices* V partitioned into two sets and *edges* $E \subseteq V \times V$. For vertices $u, v \in V$, we write (u, v) or $u \to v$ to denote an edge from u to v. For a set $E' \subseteq E$ of edges, we write $\text{src}(E') = \{u \mid (u, v) \in E'\}$ to denote the sources of the edges in E'. Furthermore, for each $i \in \{0, 1\}$, we write E_i to denote the set of edges originating from V_i. A *play* ρ originating at a vertex v_0 is a finite or infinite sequence of vertices where consecutive ones are connected by edges, denoted by $v_0 v_1 \ldots \in V^\infty$ or $v_0 \to v_1 \to \cdots \in V^\infty$. We sometimes fix a designated initial vertex v_0 and disregard the subgraph not reachable from v_0.

Specifications/Objectives. Given a game graph G, we consider specifications/objectives specified using a formula Φ in *linear temporal logic* (LTL) over the vertex set V, that is, we consider LTL formulas whose atomic propositions are sets of vertices V. In this case the set of desired infinite plays is given by the semantics of Φ which is an ω-regular language $\mathcal{L}(\Phi) \subseteq V^\omega$. Every game graph with an arbitrary ω-regular set of desired infinite plays can be reduced to a game graph (possibly with a different set of vertices) with an LTL specification, as above[1]. The standard definitions of ω-regular languages and LTL are omitted for brevity and can be found in standard textbooks [5]. To simplify notation we use $e = (u, v)$ in LTL formulas as syntactic sugar for $u \wedge \bigcirc v$, with $\bigcirc$ as the LTL *next* operator. We further use a set of edges $E' = \{e_i\}_{i \in [0;k]}$ as an atomic proposition to denote $\bigvee_{i \in [0;k]} e_i$.

Games and Strategies. A *two-player (turn-based) game* is a pair $\mathcal{G} = (G, \Phi)$ where G is a game graph and Φ is a *specification* over G. A *strategy* of Player i, $i \in \{0, 1\}$, is a function $\pi_i \colon V^* V_i \to V$ such that for every $\rho v \in V^* V_i$ holds that $\pi_i(\rho v) \in E(v)$. Furthermore, a strategy π_i is *memoryless/positional* if $\pi_i(\rho v) = \pi(v)$ for every $\rho v \in V^* V_i$. We write such memoryless strategies as functions of the form $\pi_i \colon V_i \to V$. Given a strategy π_i, we say that the play $\rho = v_0 v_1 \ldots$ is *compliant* with π_i if $v_{k-1} \in V_i$ implies $v_k = \pi_i(v_0 \ldots v_{k-1})$ for all k. We refer to a play compliant with π_i and a play compliant with both π_0 and π_1 as a π_i-*play* and a $\pi_0\pi_1$-*play*, respectively. We collect all plays originating in a set S and compliant with π_i, (and compliant with both π_0 and π_1) in the sets $\mathcal{L}(S, \pi_i)$ (and $\mathcal{L}(S, \pi_0\pi_1)$, respectively). When $S = V$, we drop the mention of the set in the previous notation, and when S is singleton $\{v\}$, we simply write $\mathcal{L}(v, \pi_i)$ (and $\mathcal{L}(v, \pi_0\pi_1)$, respectively).

Winning. Given a game $\mathcal{G} = (G, \Phi)$, a play ρ in $\mathcal{G}$ is *winning for* Player 0, if $\rho \in \mathcal{L}(\Phi)$, and it is winning for Player 1, otherwise. A strategy π_i for Player i is *winning from a vertex* $v \in V$ if all plays compliant with π_i and originating from v are winning for Player i. We say that a vertex $v \in V$ is *winning for* Player i, if there exists a winning strategy π_i from v. We collect all winning vertices of Player i in the Player i *winning region* $Win_i \subseteq V$. A (uniform) *winning strategy* for Player i is a strategy that is winning for Player i from all vertices in Win_i. We always interpret winning w.r.t. Player 0 if not stated otherwise. Furthermore, in games with a designated initial vertex v_0, when we discuss the winner of the game we refer to the winner of v_0.

Reachability Games. A *reachability game* is a game $\mathcal{G} = (G, \Phi)$ with reachability objective $\Phi = \Diamond T$ for some set T of vertices. A play is winning for Player 0 in such a game if it visits any vertex in T. It is well-known that reachability games can be solved in linear time in the size of the game graph.

Lemma 1 ([40]). *The problem of solving reachability games lies in* PTIME.

Parity Games. A *parity game* is a game $\mathcal{G} = (G, \Phi)$ with parity objective $\Phi = Parity(\mathbb{P})$, s.t. $Parity(\mathbb{P}) := \bigwedge_{i \in_{\mathrm{odd}} [0;d]} \left(\Box \Diamond P_i \implies \bigvee_{j \in_{\mathrm{even}} [i+1;d]} \Box \Diamond P_j \right)$, with $P_i = \{v \in V \mid \mathbb{P}(v) = i\}$ for some priority function $\mathbb{P} \colon V \to [0;d]$ that assigns each vertex a priority. A play is winning (for Player 0) in such a game if the maximum of priorities seen infinitely often is even. There are several quasi-polynomial algorithms to solve such parity games giving us the following result.

---

[1] This is because every ω-regular specification can be reduced to a parity specification [3], which can be written as an LTL specification over V as given later.

Lemma 2 ([11,24,34]). *The problem of solving parity games lies in* QP.

Rabin Games. A *Rabin game* is a game $\mathcal{G} = (G, \Phi)$ with Rabin objective $\Phi = Rabin(\Omega)$ defined by a set of *Rabin pairs* $\Omega = \{(F_i, R_i) \subseteq V \times V \mid i \in [1; k]\}$ s.t. $Rabin(\Omega) := \bigvee_{i \in [1;k]} (\Box \Diamond F_i \wedge \neg \Box \Diamond R_i)$. A play is winning for Player 0 in such a game if there exists an $i \in [1; k]$ such that the plays visits the set F_i infinitely often and R_i only finitely often. The dual of such an objective, i.e. $\neg Rabin(\Omega)$, is known as a *Streett* objective. It is well-known that solving Rabin games for Player 0 is an NP-complete problem whereas solving it for Player 1, which corresponds to solving Streett games for Player 0, is co-NP-complete.

Lemma 3 ([18,19,40]). *The problem of solving Rabin games for Player* 0 *is* NP-*complete and for Player* 1 *it is* co-NP-*complete.*

Computing Attractors and Winning Regions. For a set T of vertices, $\mathsf{pre}(T) \subseteq V$ is the set of vertices from which there is an edge to T. Furthermore, the attractor function $\mathrm{ATTR}^i(G, T)$ solves the (non-augmented) reachability game $(G, \Diamond T)$ for Player i, i.e., it returns the attractor set, i.e., winning region, $A := \mathsf{attr}^i(G, T) \subseteq V$ and a memoryless attractor strategy, i.e., winning strategy, π_A of Player i. Intuitively, A collects all vertices from which Player i has a strategy (i.e., π_A) to force every play starting in A to visit T in a finite number of steps. Moreover, the function $\mathrm{SOLVE}(G, \Phi)$ returns the winning region and a winning strategy (that is memoryless if possible) in a game (G, Φ). Both the functions ATTR and SOLVE (for a parity objective) solve classical synthesis problems with standard algorithms (see, e.g. [3]).

3 Problem Statement

In this paper we consider the problem of efficiently solving reachability and parity games (G, Φ) which are augmented with assumptions, as formalized next.

Definition 1. *An* augmented game $\mathfrak{G} = (G, \Phi, \psi)$ *is a game* (G, Φ) *augmented with an* assumption ψ *given as an LTL formula over the vertex set V of G. $\mathfrak{G}$ is interpreted as the (non-augmented) game* $\mathcal{G} = (G, \psi \Rightarrow \Phi)$, *i.e., Player i wins in $\mathfrak{G}$ if s/he wins in $\mathcal{G}$. An augmented game with reachability (resp. parity) objective Φ is called an* augmented reachability (resp. parity) game.

Problem 1. Given a class of assumptions, does the problem of solving the respective augmented reachability (resp. parity) game lie in PTIME (resp. QP)?

It is easy to see that the answer to Problem 1 is negative for arbitrary combinations of Φ and ψ. A simple example is a reachability game $\mathfrak{G}$ augmented with a Streett assumption ψ. Solving such games is an NP-complete problem, as it is equivalent to solving Rabin games, due to the duality of Rabin and Streett objectives (Lemma 3), while solving (non-augmented) reachability games lies in PTIME (Lemma 1). We therefore restrict our attention to particular classes of *progress* assumptions, explain their relevance in the context of CPS design and establish the worst-case time complexity of the resulting augmented games. Our results are summarized in Table 1. On a higher level, the goal of this paper is to pave the way towards a more comprehensive understanding of assumption classes that allow for efficient solution algorithms of augmented games.

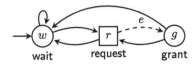

Fig. 2. A reachability game with Player 1 (squares) vertices, Player 0 (circles) vertices, and specification $\Phi = \Diamond g$ augmented with (dashed) live edge $E^\ell = \{e\}$.

4 Strong Transition Fairness Assumptions

Fairness assumptions have proven to be useful to prevent reactive synthesis problems to become unrealizable for 'uninteresting reasons'. For example, synthesizing a mutual exclusion protocol might fail because a process that entered the critical section might decide to never leave it, allowing no other process which requested access to enter. This can be circumvented by a fairness constraint which asserts that every process will eventually leave the critical section again (which must be ensured by its local implementation). This reasoning analogously holds for CPS, where a triggered landing maneuver will lead to successfully landing (assuming the low-level controller being correct), or attempting to grasp an object will eventually succeed (assuming the robot to be programmed well).

Depending on the game graph used to model the overall synthesis problem, the outlined 'fair progress' can be captured by a local fairness notion called *strong transition fairness* [5,20,35], as used for resource management [13], path following [17,27,33] or robot manipulators [1]. This assumption class is given by a set of *live edges* and requires that whenever the source vertex of a live edge is visited infinitely often along a play, the edge itself is traversed infinitely often along the play as well (see Fig. 1 for different illustrations). Whenever strong transition fairness is used as an *assumption*, it is solely restricting the environment player in the resulting game, as formalized next.

Definition 2. *Given a game graph $G=(V,E)$, strong transition fairness assumptions $\psi_{\mathrm{LIVE}}(E^\ell)$ are represented by a set $E^\ell \subseteq E_1$ of live edges and captured by the LTL formula*

$$\psi_{\mathrm{LIVE}}(E^\ell) := \bigwedge_{e=(u,v)\in E^\ell} (\Box \Diamond u \Rightarrow \Box \Diamond e). \tag{1}$$

We call $\mathfrak{G} = (G, \Phi, \psi_{\mathrm{LIVE}}(E^\ell))$ a game augmented with live edges E^ℓ.

Example 1. A simple example of a game augmented with live edges is shown in Fig. 2. Intuitively, in this game graph, from every Player 0 vertex, she can choose to either wait (by going to vertex w) or request (by going to vertex r), and from Player 1 vertex r, he can either grant the request (by going to vertex g) or make Player 0 wait (by going to vertex w). Furthermore, starting from vertex w, Player 0's objective, i.e., $\Phi = \Diamond g$, is to finally get her request granted by visiting vertex g. It is easy to see that without any assumption, Player 0 does not have a winning strategy from w, as Player 1 can always choose to make Player 0 wait. Now consider the assumption $\psi_{\mathrm{LIVE}}(E^\ell)$ for live edges $E^\ell = \{e\}$, which says that if Player 0 requests infinitely often, then Player 1 grants infinitely often. Under this assumption, now, Player 0 can win by requesting again and again until Player 1 grants. Hence, Player 0 has a winning strategy in this augmented game with assumption $\psi_{\mathrm{LIVE}}(E^\ell)$.

It turns out, that for this assumption class the answer to Problem 1 is positive.

Theorem 1. *Reachability (resp. parity) games augmented with live edges can be solved in* PTIME *(resp.* QP*).*

Theorem 1 is actually a corollary of known results, due to the following observations. First, it was recently shown by Banerjee et al. [6] that both reachability and parity games augmented with live edges can be solved using a symbolic fixed-point algorithm. Their algorithm encodes winning regions of such games as the solutions of fixpoint equations, and shows that one can solve the equations obtained with almost the same computational (worst case) complexity as the standard fixed-point algorithm for the corresponding *non-augmented* games.

The winning region of parity games augmented with live edges can be encoded further as the solution of *nested fixpoint equations*, where the nesting depth of the obtained equations depend only on the number of priorities [6, Sec.3.4]. Following the quasi-polynomial algorithms for parity games, several results show that these techniques can be extended to solve arbitrary nested fixpoint equations in quasi-polynomial time [4,21,22]. As a corollary of these results, the existence of quasi-polynomial time algorithms for such augmented parity games follows. Furthermore, as reachability games augmented with live edges can be encoded by a nested fixpoint equation whose nesting depth is fixed, they can be solved in polynomial time.

As a byproduct of this result, qualitative winning in stochastic parity games can also be decided in similar time complexity by reducing them to augmented games with live edges (see [6, Sec.5] for the formal reduction). Stochastic two-player games (also known as $2\frac{1}{2}$-player games) generalize two-player graph games with an additional category of "random" vertices: whenever the game reaches a random vertex, a random process picks one of the outgoing edges (uniformly at random, w.l.o.g.). The qualitative winning problem asks whether a vertex of the game graph is almost surely (with probability 1) winning for Player 0.

Corollary 1. *The qualitative winning problem in stochastic parity games can be solved in* QP.

5 Strong Transition Co-fairness Assumptions

Even though fairness assumptions have arguably received much more attention in the reactive synthesis community, their dual – *co-fairness assumptions* – also provide a simple and local abstraction of (non)-progress behaviour. An examples are capacity or energy restrictions, which prevent a robot manipulator to place infinitely many pieces into a buffer before it is emptied. Formally, strong transition *co-fairness* requires a particular set of *co-live edges* to be taken only finitely often in every play, as formalized next.

Definition 3. *Given a game graph* $G = (V, E)$, *strong transition co-fairness assumptions are represented by a set* $E^c \subseteq E_1$ *of co-live edges and captured by the LTL formula*

$$\psi_{\text{COLIVE}}(E^c) := \bigwedge_{e \in E^c} \neg \Box \Diamond e. \tag{2}$$

We call $\mathfrak{G} = (G, \Phi, \psi_{\text{COLIVE}}(E^c))$ *a game augmented with co-live edges* E^c.

The concept of co-live edges along with their induced assumption on the environment player was recently introduced by Anand et al. [2]. In their work, the problem of computing adequately permissive assumptions which render a given non-realizable (non-augmented) synthesis problem realizable, was studied. In contrast, we are interested in the problem where such assumptions are *given* (as a result of the modelling process) and need to be utilized in synthesis.

Algorithm 1 provides an algorithm to solve parity games augmented with co-live edges. The idea of the algorithm is to first solve the parity game w.r.t. the game graph obtained by removing the co-live edges (line 2) to get the current winning regions (Win'_0, Win'_1). As Player 1 can use the co-live edges finitely many times to reach his winning region, we compute the Player 1 attractor set A for Player 1's current winning region Win'_1 (line 3), which gives us a subset of his complete winning region. If $A \subseteq V$ is a non-trivial (i.e., non-empty and strict subset), then we re-solve the game on the game restricted to its complement $B = V \setminus A$ (line 5). This is formalized in the following theorem, the proof of which can be found in the extended version of this paper [37], in App. A.

Theorem 2. *Given an augmented parity game* $\mathfrak{G} = (G, \Phi, \psi_{\text{COLIVE}}(E^c))$ *with game graph* $G = (V, E)$ *and co-live edges* E^c, *the algorithm* SOLVECOLIVE(G, Φ, E^c) *returns the winning region and a memoryless winning strategy in game* $\mathfrak{G}$.

As in each iteration, SOLVECOLIVE (Algorithm 1) restricts the game graph to a smaller vertex set, the algorithm terminates within $|V|$ iterations. As reachability games can be reduced to simple parity games [3], and since non-augmented parity (resp. reachability) games can be solved in quasi-polynomial (resp. polynomial) time (Lemma 2 and 1), Algorithm 1 can solve parity (resp. reachability) games augmented with co-live edges in quasi-polynomial (resp. polynomial) time.

Algorithm 1. SOLVECOLIVE(G, Φ, E^c)

Require: Augmented parity game $\mathfrak{G} = (G = (V, E), \Phi, \psi_{\text{COLIVE}}(E^c))$
Ensure: Winning region and winning strategy in the augmented game $\mathfrak{G}$
1: $G' \leftarrow (V, E \setminus E^c)$
2: $Win'_0, \pi'_0 \leftarrow \text{SOLVE}(G', \Phi); \; Win'_1 \leftarrow V \setminus Win'_0$
3: $A \leftarrow \text{attr}^1(G, Win'_1); \; B \leftarrow V \setminus A$
4: **if** $B = \emptyset$ or $B = V$ **then return** B, π'_0
5: **else return** SOLVECOLIVE($\mathcal{G}|_B$)

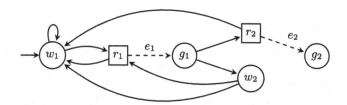

Fig. 3. Augmented reachability game $(G_2, \Phi_2, \psi_{\text{GLIVE}}(\mathcal{H}))$ with Player 1 vertices (squares), Player 0 vertices (circles), specification $\Phi_2 = \Diamond g_2$ and live group $\mathcal{H} = \{\{e_1, e_2\}\}$. This game is equivalent to the augmented game $(G, \Phi_1, \psi_{\text{LIVE}}(E^\ell))$, with G, $E^\ell = \{e\}$ as depicted in Fig. 2 and $\Phi_1 := \Diamond (g \wedge \bigcirc \bigcirc g)$.

Corollary 2. *Reachability (resp. parity) games augmented with strong transition co-fairness assumptions can be solved in* PTIME *(resp.* QP*).*

6 Group Transition Fairness Assumptions

In the previous sections, we saw that augmenting reachability or parity games with live or co-live edges allows for solution algorithms with the same worst-case time complexity as for the respective *non-augmented* games. However, in many scenarios, live and co-live edges are not expressive enough to capture the intended progress assumption, as illustrated in the following example.

Example 2. Consider again the game graph G as shown in Fig. 2 with a new specification $\Phi_1 = \Diamond(g \wedge \bigcirc \bigcirc g)$. Intuitively, Player 0 wants to ensure that at some point she gets two grants consecutively. However, under the live edge assumption $\psi_{\text{LIVE}}(E^\ell)$ with $E^\ell = \{e\}$, Player 0 does not have a winning strategy. This live edge only ensures infinitely many grants, but does not ensure consecutive grants. As the specification is not a reachability/parity objective, the standard way to solve the game (G, Φ_1) is to translate Φ_1 into a parity game $\mathcal{G}'$ and then take the product of $\mathcal{G}'$ with G. In the given example, the result turns out to be the reachability game (G_2, Φ_2) with $\Phi_2 = \Diamond g_2$, as depicted Fig. 3. In this product game graph, live edge e of G has been split into two copies, i.e., edge e_1 and e_2. If we consider the live edge assumption $\psi_{\text{LIVE}}(E_2^\ell)$ with $E_2^\ell = \{e_1, e_2\}$, then Player 0 can satisfy Φ_2 by requesting every time – this will force Player 1 to take edge e_1 infinitely often and then finally take edge e_2. However, Player 0 should actually not have a winning strategy as she does not have one in game (G, Φ_1). This tells us that the live edge e in G does not translate to live edges $\{e_1, e_2\}$ in the product game. In particular, in order to satisfy the fairness assumption inherited from the original game, it is sufficient for Player 1 to take only one of the edges in $\{e_1, e_2\}$ infinitely often if their sources are visited infinitely often. This disjunction over live edges can be expressed by a *live group* $H = \{e_1, e_2\}$, as formalized next. With this, the augmented game $(G_2, \Phi_2, \psi_{\text{GLIVE}}(\{H\}))$ becomes equivalent to $(G, \Phi_1, \psi_{\text{LIVE}}(E^\ell))$, and is therefore again unrealizable.

6.1   Live Group Assumptions

Live group assumptions, a generalization of live edge assumptions, are defined by a finite set $\mathcal{H}$ of edge groups H in a game. The assumptions require that for each live group $H \in \mathcal{H}$, if at least one source vertex in H is visited infinitely often along a play, at least one of the edges in H is traversed infinitely often as well.

Definition 4. *Given a game graph $G = (V, E)$, the assumptions represented by a set $\mathcal{H}$ of live groups H are captured by the LTL formula*

$$\psi_{\text{GLIVE}}(\mathcal{H}) := \bigwedge_{H \in \mathcal{H}} (\Box \Diamond \, \text{src}(H) \Rightarrow \Box \Diamond \, H). \tag{3}$$

We call $\mathfrak{G} = (G, \Phi, \psi_{\text{GLIVE}}(\mathcal{H}))$ a game augmented with live groups $\mathcal{H}$.

Unfortunately, it turns out that for such games the answer to Problem 1 is negative.

Theorem 3. *Solving reachability and parity games augmented with live groups is NP-complete.*

In order to prove this theorem, we will introduce Lemma 4 and Lemma 5. Since reachability specifications give one of the easiest infinite games, this result can be perceived as a negative result for all meaningful games augmented with live group assumptions. However, the NP-completeness result also holding for parity implies that going from reachability to parity does not add up to the complexity.

Lemma 4. *Parity games augmented with live groups can be solved in NP.*

Proof sketch [Full proof in [37], App. B]. The intuition is that we can encode both the parity conditions and the live group conditions as Rabin pairs. For an augmented parity game $\mathfrak{G} = (G = (V, E), \Phi = Parity(\mathbb{P}), \psi_{\text{GLIVE}}(\mathcal{H}))$, an equivalent Rabin game is $\mathcal{G}' = ((V', E'), \Phi' = Rabin(\Omega_1 \cup \Omega_2))$ with $V' = V \uplus E$ and $E' = \{(u, e), (e, v) \mid (u, v) \in E\}$ where the Rabin pairs in $\Omega_1 = \{(\text{src}(H), H) \mid H \in \mathcal{H}\}$ represent the live group conditions and the ones in $\Omega_2 = \{(P_{2i}, \cup_{j \ni 2i} P_j) \mid 0 \leq 2i \leq d\}$ where $P_i = \{q \in Q \mid \mathbb{P}(q) = i\}$ represent the parity conditions.   □

Lemma 5. *Solving reachability games augmented with live groups is NP-hard.*

Proof sketch [Full proof in [37], App. B]. We will give a polynomial-time reduction from the 3-SAT problem to reachability games augmented with live groups. Towards this goal, we consider the 3-SAT instance $\varphi = C_1 \wedge C_2 \wedge \cdots \wedge C_k$ where $C_i = (y_{1i} \vee y_{2i} \vee y_{3i})$ for $i \in [1;k]$ and each y_{1i}, y_{2i}, y_{3i} is a literal from the set $\{x_1, x_2, \ldots, x_m, \neg x_1, \neg x_2, \ldots, \neg x_m\}$. We construct the augmented reachability game $\mathfrak{G}^\varphi = (G = (V, E), \Phi = \Diamond \odot, \psi_{\mathrm{GLIVE}}(\mathcal{H}))$ with vertex set

$$V = \{v_0\} \cup \{C_i \mid i \in [1;k]\} \cup \{y, y' \mid y \in \{x_i, \neg x_i\} \text{ for } i \in [1;m]\} \cup \{\odot\}$$

where Player 0 vertices consists of $\{C_i \mid i \in [1;m]\}$ and $\odot$; edge set [2]

$$E = \{(v_0, C_i) \mid i \in [1;k]\} \cup \{(y, \odot), (y, y') \mid y \in \{x_i, \neg x_i\} \text{ for } i \in [1;m]\} \cup$$
$$\{(C_j, y) \mid j \in [1;m], y \in \{x_i, \neg x_i\} \text{ for } i \in [1;m] \text{ and } y \text{ is in clause } C_j\} \cup \{(\odot, \odot)\};$$

and live groups $\mathcal{H} = \{H_i^1, H_i^2 \mid i \in [1;m]\}$ where $H_i^1 := \{(x_i, \odot), (\neg x_i', v_0)\}$ and $H_i^2 := \{(\neg x_i, \odot), (x_i', v_0)\}$. The game $\mathfrak{G}^\varphi$ for the 3-SAT formula $\varphi = C_1 \wedge C_2 \wedge C_3$ with $C_1 = (x_1 \vee x_2 \vee \neg x_3)$, $C_2 = (\neg x_1 \vee x_2 \vee \neg x_3)$, $C_3 = (\neg x_1 \vee \neg x_2 \vee x_3)$ is given in Fig. 4 for illustration.

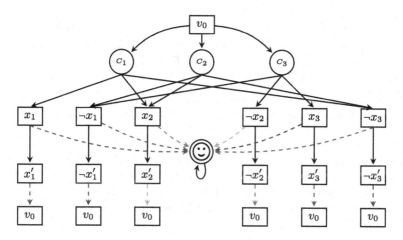

Fig. 4. Game $\mathfrak{G}^\varphi$. Each live group is denoted by dashed edges of a different color, e.g. H_1^1 is blue. φ has a satisfying assignment $L = \{x_1, x_2, x_3\}$ and thus, the positional strategy with $\pi_0(C_1) = x_1, \pi_0(C_2) = x_2$ and $\pi_0(C_3) = x_3$ is winning. (Color figure online)

Game $\mathfrak{G}^\varphi$ starts from the Player 1 vertex v_0. v_0 has an outgoing edge to each Player 0 vertex C_i, each representing the respective clause in the 3-SAT formula. By taking the edge (v_0, C_i), Player 1 challenges Player 0 to satisfy clause C_i. C_i

---

[2] The edges $x_i \to x_i'$ mainly serve illustrative purposes, and the live outgoing edge of x_i' can actually be attributed directly to x_i. Further, distributing live edges to separate vertices underscores the result's validity for live edges with disjoint sources.

has an outgoing edge to every literal in the clause. By taking edge (C_i, y), Player 0 decides through which literal it will satisfy C_i. Then from each Player 1 vertex y, there exist two edges: one to ☺, and the other to y'. From vertex y' there is only one outgoing edge, i.e. to v_0. The live groups assert the following condition: Whenever a literal or its negation (say, y or y') is visited infinitely often in the game, Player 1 must take one of the edges $(y, ☺)$ or $(\neg y', v_0)$ infinitely often. This condition forces Player 1 to either for each literal the game visits infinitely often, also visit the negation of the literal infinitely often; or, to go to ☺.

If φ has a satisfying assignment, then the game is won by the positional strategy π_0 that sends each C_i to a literal satisfied in C_i (see Fig. 4). If φ is unsatisfiable, then every positional Player 0 strategy (it is sufficient to consider positional strategies for Player 0 as $\mathfrak{G}^\varphi$ can be viewed as a Rabin game – see the proof of Lemma 4, and Rabin games are half positional) there exist C_i, C_j, y with $\pi_0(C_i) = y$ and $\pi_0(C_j) = \neg y$, or else, π_0 gives a satisfying assignment for φ. For each such π_0, Player 1 has a winning strategy π_1: Whenever v_0 is visited it alternates between $\pi_1(v_0) = C_i$ and $\pi_1(v_0) = C_j$ and never takes an edge to ☺. A $\pi_0\pi_1$-play neither violates any assumptions nor visits ☺. Thus, Player 0 has no winning strategy in $\mathfrak{G}^\varphi$ and therefore, Player 1 wins $\mathfrak{G}^\varphi$.   □

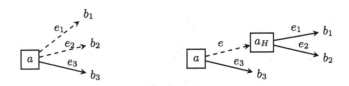

Fig. 5. Part of a game augmented with live group $H = \{e_1, e_2\}$ with $\mathsf{src}(e_1) = \mathsf{src}(e_2) = a$ (left), part of an equivalent game augmented with live edge e (right).

Since reachability games are a special case of parity games, Lemma 4 implies reachability games augmented with live groups lies in NP and Lemma 5 implies parity games augmented with live groups is NP-hard. Hence, combining these results with the ones from Lemma 4 and Lemma 5 proves Theorem 3 (see [37] for the complete proof).

6.2   Singleton-Source Live Groups

The proof for NP-completeness in Sect. 6.1 shows this hardness result already for live groups where each group only contains two edges (indicated by different colors in Fig. 4). Hence, one of the simplest possible generalization of live edges to live groups makes the problem already harder to solve. However, these edges had two different source vertices. The next theorem shows that this distinction in source vertices is actually necessary for NP-hardness. In particular, it shows that *singleton-source live groups*, i.e., live groups $\mathcal{H}$ s.t. $|\mathsf{src}(H)| = 1$ for all $H \in \mathcal{H}$, can be reduced to games annotated with live edges, and hence can be solved efficiently.

The corresponding construction is illustrated with an example in Fig. 5. Its proof can be found in [37].

Theorem 4. *Reachability (resp. parity) games augmented with singleton-source live groups can be solved in* PTIME *(resp. QP).*

Intuitively, a live group H with a singleton source models a disjunctive form of transition fairness, while the classical definition of live edge assumptions (as in Definition 2) amounts to a conjunctive form of transition fairness. The construction in Fig. 5 shows that both are equally expressive. It is therefore not surprising that the complexity result from Theorem 4 generalizes to combinations of conjunctions and disjunctions of live edges from the same source in CNF (Conjunctive Normal Form). This is formalized and proven in [37], App. C.

6.3   Live Groups in Product Games

As motivated in the introduction as well as in Example 2, the use of progress assumptions, especially in the context of CPS control, typically stems from a modelling step where such annotations help to capture progress ensured by the underlying physical system in an abstract manner. In this context, the high-level synthesis game is typically obtained by, first, constructing a game graph G annotated with assumption ψ (as an abstraction of the underlying dynamics), and second, taking the product of the annotated graph (G, ψ) with a parity game $\mathcal{G}'$ obtained from the translation of the specification Φ, which is typically given as an arbitrary LTL formula over the vertex set V of G. This process was examplified in Example 2, illustrating the need for live group assumptions.

Further, even if the specification Φ is, for example, a simple reachability objective, a product construction might still be needed prior to synthesis, if the controlled system consists of multiple interacting components, some of which annotated with *live edge* assumptions. Then the product of all component models, and hence the final reachability game, is annotated with *live group* assumption.

While we have seen in the previous subsection that live group assumptions render the respective solution problem NP-complete (even for reachability games), one could hope that the progress assumptions stemming from the product construction described above, is a more tractable subclass of live group assumptions. This would be a reasonable anticipation as the liveness assumptions arising from the product construction can easily be represented as live group assumptions, and these assumptions appear to carry more structure, as they inherit the edges in each live group from a single live edge in one automaton/game during the product construction. Unfortunately, in what is to come we show that this is not the case.

Before we move on to Theorem 5 stating the main result of this section, we introduce some concepts needed to state the theorem. In particular, we introduce alternations and labels for games, which are the de-facto standard in synthesis games stemming from CPS design problems, to easily formalize the above discussed product construction in the usual way.

Alternating Games. A game graph $G = (V, E)$ is called *alternating* if for each $(u, v) \in E$ either $u \in V_0$ and $v \in V_1$ or vice versa. A game (G, Φ) or an augmented game (G, Φ, ψ) is called alternating if G is alternating.[3]

Labeled Games. Let $G = (V, E)$ be a game graph, then the tuple $(G, \kappa : E \to \Sigma)$ is called *the game graph G labeled with* κ, where the function κ assigns an action a to each edge from the finite alphabet Σ. Similarly, the tuples $(\mathcal{G} = (G, \Phi), \kappa)$ and $(\mathfrak{G} = (G, \Phi, \psi), \kappa)$ are called *the games $\mathcal{G}$ and $\mathfrak{G}$ labeled with* κ. For brevity, we denote these tuples by $G_\kappa, \mathcal{G}_\kappa$ and $\mathfrak{G}_\kappa$, respectively. In what is to come, we will also look at labeled game graphs augmented with assumptions $(G = (V, E), \psi, \kappa)$, and denote them by $G_{\psi,\kappa}$. Intuitively, in a labeled game as given above, Player i moves to a vertex v from a vertex $u \in V_i$ with action a (denoted by $u \xrightarrow{a} v$) if and only if there exists an edge $(u, v) \in E$ with $\kappa(u, v) = a$.

Product Games. Let $\mathcal{G}_{\kappa^1} = ((V^1, E^1), \Phi, \kappa^1 : E^1 \to \Sigma)$ be an alternating labeled game with initial vertex $v_0^1 \in V_1^1$ and $G_{\psi_{\mathrm{LIVE}}(E^\ell),\kappa^2} = ((V^2, E^2), \psi_{\mathrm{LIVE}}(E^\ell), \kappa^2 : E^2 \to \Sigma)$ be an alternating labeled augmented game graph with initial vertex $v_0^2 \in V_1^2$. The product $\mathcal{G}_\kappa \times G_{\psi_{\mathrm{LIVE}}(E^\ell),\kappa^2}$ is defined in the usual way to be the alternating augmented game $\mathcal{P} = (((V_\mathcal{P}, E_\mathcal{P}), \Phi_\mathcal{P}), \psi_{\mathrm{GLIVE}}(\mathcal{H}_\mathcal{P}))$ s.t.

- $V_\mathcal{P}$ is the cartesian product $V_0^1 \times V_0^2 \cup V_1^1 \times V_1^2$,
- the initial vertex of $\mathcal{P}$ is (v_0^1, v_0^2),
- a vertex $(v, v') \in V_\mathcal{P}$ belongs to Player 0 if and only if $(v, v') \in V_0^1 \times V_0^2$,
- an edge $(v, v') \to (w, w')$ is in $E_\mathcal{P}$ iff there exists an action $a \in \Sigma$ for which $v \xrightarrow{a} w \in E^1$ and $v' \xrightarrow{a} w' \in E^2$,
- for each $(v', w') \in E^\ell \subseteq E^2$, there exists a live group $H_{(v',w')} \in \mathcal{H}_\mathcal{P}$ such that $H_{(v',w')} = \{(v, v') \to (w, w') \in E_\mathcal{P} \mid v, w \in V^1\}$, and
- a play $(v_1, v_1')(v_2, v_2') \ldots$ satisfies the specification $\Phi_\mathcal{P}$ if and only if the play $v_1 v_2 \ldots$ satisfies Φ.

For a play $\rho = (v_1, v_1')(v_2, v_2') \ldots$, we denote the projection of ρ to V^1 by $\mathrm{proj}^1(\rho) = v_1 v_2 \ldots$ and the projection to V^2 by $\mathrm{proj}^2(\rho) = v_1' v_2' \ldots$.

With this, we are now ready to use the formalized product construction and state the main result of this subsection.

Theorem 5. *Any alternating augmented game $\mathfrak{G} = ((V, E), \Phi, \psi_{\mathrm{GLIVE}}(\mathcal{H}))$ where $\mathcal{H} = \{H_i \mid i \in [1; m]\}$ with initial vertex $v_{\mathrm{init}} \in V_1$ can be obtained as a product of a labeled non-augmented game $\mathcal{G}_{\kappa^1}$ of same size, and a labeled game graph $G_{\psi_{\mathrm{LIVE}}(E^\ell),\kappa^2}$ of size $m + 1$ augmented with live edges E^ℓ.*

Proof sketch [Full proof in [37], App. D]. We construct a non-augmented game and the labeled augmented game graph from a given $\mathfrak{G} = ((V, E), \Phi, \psi_{\mathrm{GLIVE}}(\mathcal{H}))$ and show that their product is equivalent to the original game. Towards this

---

[3] We note that all games can be converted to an equivalent alternating game by at most doubling the size of the vertex and edge sets of the game graph.

goal we define $\Sigma := \{a\} \cup \{h_i | i \in [0; m]\}$ and $\mathcal{G}_{\kappa^1} := (((V, E), \Phi), \kappa^1 : E \to \Sigma)$ s.t. $v_0^1 := v_{\text{init}}$,

$$\kappa^1 : E \to \Sigma \text{ where } \kappa^1(u, v) = \begin{cases} h_i, & \text{if } (u, v) \in H_i \text{ for some } i \in [1; m], \\ a, & \text{otherwise,} \end{cases}$$

where a is a fixed letter different from each h_i. Further, define
$G_{\psi_{\text{LIVE}}(E^\ell), \kappa^2} := ((V^2, E^2), \psi_{\text{LIVE}}(E^\ell), \kappa^2 : E^2 \to \Sigma)$ s.t. $v_0^2 = u_0 \in V_1^2$ and

- $V^2 = \{u_0\} \cup \{u_*\} \cup \{x_1, \dots, x_m\}$, where only u_0 is a Player 1 vertex,
- $E^2 = \{(u_0, x_i), (x_i, u_0) \mid i \in [1; m]\} \cup \{(u_0, u_*), (u_*, u_0)\}$
  where $E^\ell = \{(u_0, x_i) \mid i \in [1; m]\}$,
- $\kappa^2(u_0, x_i) = h_i$ and all other edges in E^2 are labeled with the letter a

(see Fig. 6 for an illustration). In order to prove the equivalence of $\mathfrak{G}$ and $\mathcal{P}$, it is enough to prove the equivalence of plays, i.e. a play $\rho = (v_0, v_0')(v_1, v_1') \dots$ is winning in $\mathcal{P}$ iff $\text{proj}^1(\rho)$, as a play on the identical game graph (V, E) of $\mathcal{G}_{\kappa^1}$ and $\mathfrak{G}$, is winning in $\mathfrak{G}$. Let ρ be winning. Then $\text{proj}^1(\rho)$ satisfies Φ or ρ violates a live group assumption $H_{(u_0, x_i)}$. This implies that there exists a $w \to v \in H_i$ with $(w, u_0) \to (v, x_i) \in H_{(u_0, x_i)}$ that is enabled infinitely often in $\text{proj}^1(\rho)$, but no edges in H_i are taken infinitely often. Therefore, $\text{proj}^1(\rho)$ violates the group liveness assumptions in $\mathfrak{G}$, and thus, it is winning in $\mathfrak{G}$.

Now let ρ be winning for Player 1. Then $\text{proj}^1(\rho)$ violates Φ and ρ satisfies all live group assumptions $H_{(u_0, x_i)}$. This translates to $\text{proj}^1(\rho)$ similar to before: for each $i \in [1; m]$, if an edge $w \to v \in H_i$ is visited infinitely often, then an edge in H_i is taken infinitely often. Thus, $\text{proj}^1(\rho)$ is winning for Player 1 in $\mathfrak{G}$. □

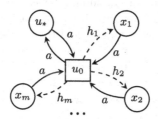

Fig. 6. $G_{\psi_{\text{LIVE}}(E^\ell), \kappa^2}$: Labeled alternating game graph (V^2, E^2) augmented with live edges (shown by dashed lines) labeled $h_1, \dots, h_m$.

With Theorem 3 and Theorem 5 in place, it follows that games resulting from the product of multiple graphs – some of which annotated with live edges – renders the resulting augmented synthesis problem NP-complete.

6.4 A Remark on Co-live Groups

As we considered live groups in the last sections, it would be natural to also consider co-live groups, i.e., groups of edges where the assumption is to ensure that at least one of them is co-live. However, unlike in the case of live edge assumptions, taking the product of a game with co-live edges and another non-augmented game graph does actually result in a game augmented with co-live *edges* not co-live *groups*. This is due to the fact that *every* edge in the product game which originates from a co-live edge in the original game, is only allowed to be taken finitely often. Furthermore, given a live group H, its dual, $\neg\,\square\,\lozenge H$ can actually be expressed by a set of co-live edges, since

$$\neg\,\square\,\lozenge H = \neg\,\square\,\lozenge \bigvee_{e \in H} e = \neg \bigvee_{e \in H} \square\,\lozenge\, e = \bigwedge_{e \in H} \neg\,\square\,\lozenge e.$$

Hence, co-live groups do not seem relevant in the considered context.

7 Persistent Live Group Assumptions

Given the practical relevance of product constructions in CPS synthesis problems, as outlined in Sect. 6.3, the negative result of Theorem 3 and Theorem 5 is rather discouraging, as the more tractable subclass in Sect. 6.2 seems rather restrictive.

In contrast, this section discusses a different version of live groups, called *persistent live groups*, which originally appeared in the work of Ozay et al. [33, 38] in the context of abstraction-based control design (ABCD) and which (i) have very nice computational properties while (ii) being closed under product constructions. Before formally defining this assumption class and formalizing its properties, we give an example[4] from ABCD to motivate their relevance.

Example 3. Consider a robot that should be controlled to react to changes in its desired goal location. For simplicity, suppose there are only two (non-intersecting) target locations T_1 and T_2 and there exist (already designed) low-level controllers C_1 and C_2 which ensure that, whenever a target is persistently activated, the robot will reach the respective target region. The high-level synthesis problem is to design a strategy to trigger the low level controllers C_i in response to the currently activated target (signaled by proposition A_i set to

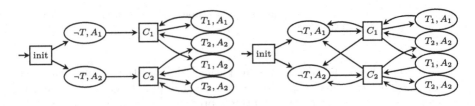

Fig. 7. Two abstract game graphs to model the robot motion control problem from Example 3: without persistent live groups (left) and with persistent live groups $(\mathsf{S}_i, \mathsf{C}_i, \mathsf{T}_i) = (V, E \cap (V \times \{C_i\}), \{(T_i, \cdot)\})$ for each $i \in [1; 2]$ (right).

---

[4] See [32] for a more in-depth version of this example.

'true' by the environment), such that the target region T_i is actually reached, if the respective target is persistently activated. While the correct strategy is obvious in this case – choose C_i iff A_i is true – constructing a correct abstract game graph that returns this strategy and allows to conclude that T_i is actually reached under this strategy, is surprisingly cumbersome.

Two possible abstract game graphs are depicted in Fig. 7. Here, Player 1 chooses the activated target (i.e., which (single) A_i is 'true') and in which region of the state space the robot currently is (either in one of the target regions T_i or in the non-target region $\neg T$). Further, Player 0 chooses which controller C_i to activate. In favor of readability, we only depict the Player 0 edges corresponding to the strategy of choosing C_i iff A_i is activated.

In Fig. 7 (left) the controller only allows the environment to activate a new target after it reached a target location, while Fig. 7 (right) models the more realistic scenario that the environment might activate a different target at any point, which then allows Player 0 to activate the other controller, even if the robot is still in $\neg T$. However, with the additional edges in Fig. 7 (right), Player 1 can force the play to loop between state $(\neg T, A_i)$ and C_i, thereby preventing the play from ever reaching T_i even if A_i is persistently active. In the physical system, however, we know that the robot will reach T_i if C_i is persistently used. Intuitively, this assumption can be expressed by a *persistent live group* which models that in the source region $S_i = V$ (complete vertex set in this case), if Player 0 persistently chooses edges from $C_i = E \cap (V \times \{C_i\})$ (representing the choice of controller C_i), then the play will eventually reach the target region $T_i = \{(T_i, \cdot)\}$. This augmentation allows to synthesize the correct strategy from the abstract augmented game.

Definition 5. *Given a game graph $G = (V, E)$, a persistent live group is a tuple (S, C, T) consisting of sets $S, T \subseteq V$ and $C \subseteq E_0$ such that $T \subseteq S$. The assumption represented by such a persistent live group is expressed by the LTL formula*

$$\psi_{\mathrm{PERS}}(S, C, T) := \Box\big(\Box\,(S \land \psi_{\mathrm{CONT}}(C)) \Rightarrow \Diamond T\big), \tag{4}$$

where $\psi_{\mathrm{CONT}}(C) := \bigwedge_{(u,v) \in C} u \Rightarrow \bigcirc v$. Furthermore, the assumptions represented by a set Λ of persistent live groups is denoted by $\psi_{\mathrm{PERS}}(\Lambda) := \bigwedge_{(S,C,T) \in \Lambda} \psi_{\mathrm{PERS}}(S, C, T)$. Moreover, we write games augmented with set Λ of persistent live groups to refer to the augmented games $\mathfrak{G} = (G, \Phi, \psi_{\mathrm{PERS}}(\Lambda))$.

Intuitively, $\psi_{\mathrm{CONT}}(C)$ ensures that edges in C are chosen when possible, as this is only possible for Player 0 vertices in S. Furthermore, (4) ensures that if Player 0 satisfies the safety constraints as in left side of the implication, i.e., persistently choosing the edges in C from the source vertices S, will eventually make progress and reach a vertex in T. Moreover, once it visits T, Player 0 can choose to not satisfy the safety constraint anymore unless she wants to visit T again (see [32] for more details).

In particular, such persistent live groups are helpful to Player 0 only if she satisfies the safety constraints described by it. Intuitively, this safety part of the live group makes games augmented with such assumptions easier to solve – Player 0 needs to stick to a particular live group by persistently choosing the corresponding edges until it reaches the corresponding targets. In (normal) live groups, Player 0 can visit the sources of multiple live groups infinitely often, and hence potentially activated multiple ones at the same time. Furthermore, unlike live groups, Player 0 can utilize persistent live groups in series – one after another – as they are safety-type constraint that are not conditioned on visiting vertices infinitely often. Using this observation, an algorithm to solve a restricted version of augmented games with persistent live groups was already provided in [33]. In a recent work, Nayak et al. [32] provide another algorithm for the general case. Their algorithm works in polynomial time for reachability objectives but is exponential for parity objectives. In this section, we will show that using their algorithm for reachability objectives, one can also obtain a quasi-polynomial algorithm for parity objectives.

Before going into the details of the algorithm, let us show a quick remark about persistent live groups in product games. By considering the same scenario as in Sect. 6.3 but starting with a game graph G annotated with a *persistent live group* (instead of live edges as in Sect. 6.3), the product of this annotated graph with any other graphs or games (also only potentially augmented with persistent live groups), still results in a product game augmented by *persistent live groups*. This is due to the fact, that the definition of this assumption class already accounts for groups of vertices, which is retained under product constructions.

Remark 1. Persistent live groups assumptions are closed under product (in the sense of Sect. 6.3).

7.1 Augmented Reachability Games

We first consider the augmented reachability game $\mathfrak{G} = (G, \Diamond T, \psi_{\text{PERS}}(\Lambda))$ with persistent live group assumptions. Following [32], the recursive algorithm that solves such an augmented reachability game $\mathfrak{G}$ is given in Algorithm 2. The main idea of the algorithm is to first compute the set of vertices A from which Player 0 can reach T even without the help of any persistent live group assumptions (line 2) along with the corresponding strategy σ for Player 0 (line 3). Afterwards, the algorithm computes the set of states B from which Player 0 has a strategy (i.e. σ_B) to reach A with the help of a persistent live group (lines 5–6). If this set B enlarges the winning state set A (line 7), we use recursion to solve another such augmented reachability game with target $T := A \cup B$ (line 11).

Within Algorithm 2, we use the following notation. Given a game graph $G = (V, E)$ and a persistent live group $(\mathtt{S}, \mathtt{C}, \mathtt{T})$, we write $G|_{\mathtt{C}}$ to denote the restricted game graph (V, E') such that $E' \subseteq E$ and for every edge $e = (u, v) \in E'$, either $e \in \mathtt{C}$ or there is no edge in $\mathtt{C}$ starting from u. Furthermore, we use the function $\text{SOLVE}(G, \Phi)$ to solve the game (G, Φ) with $\Phi = \Diamond A \vee \Box S$ for some $A, S \subseteq V$, which can be done by reducing it to safety game (see [37], App. E - Remark 2). With this, we can state the result for augmented reachability game given in [32].

Algorithm 2. ATTRPERS(G, T, Λ)

Require: An augmented reachability game $\mathfrak{G} = (G, \Diamond T, \psi_{\text{PERS}}(\Lambda))$
Ensure: Winning region and memoryless winning strategy in the augmented game $\mathfrak{G}$
 1: Initialize a random Player 0 strategy π
 2: $A, \pi_A \leftarrow \text{ATTR}^0(G, T)$
 3: $\pi(v) \leftarrow \pi_A(v)$ for every $v \in A \setminus T$
 4: **for** $(\text{S}, \text{C}, \text{T}) \in \Lambda$ **do**
 5:     **if** $(\text{S} \setminus A) \cap \text{pre}(A) \neq \emptyset$ **then**
 6:         $B, \pi_B \leftarrow \text{SOLVE}(G|_{\text{C}}, \Phi_B)$ with $\Phi_B = \Diamond A \vee \Box(\text{S} \setminus \text{T})$
 7:         **if** $B \not\subseteq A$ **then**
 8:             $\pi(v) \leftarrow \pi_B(v)$ for every $v \in B \setminus A$
 9:             $C, \pi_C \leftarrow \text{ATTRPERS}(G, A \cup B, \Lambda)$
10:             $\pi(v) \leftarrow \pi_C(v)$ for every $v \in C \setminus (A \cup B)$
11:             **return** (C, π)
12: **return** A, π

Proposition 1. *[32, Theorem 2] Given an augmented game $\mathfrak{G} = (G, \Phi, \psi_{\text{PERS}}(\Lambda))$ with game graph $G = (V, E)$, specification $\Phi = \Diamond T$, and persistent live groups Λ, the algorithm* ATTRPERS(G, T, Λ) *returns the winning region and a memoryless winning strategy in game $\mathfrak{G}$. The algorithm terminates in $\mathcal{O}(|\Lambda| \cdot |V| \cdot |E|)$ time.*

The proof of Proposition 1 with our notation can be found in [37], App. E. Now, Problem 1 for such augmented reachability games can be answered as follows.

Corollary 3. *Reachability games augmented with persistent live group assumptions can be solved in* PTIME.

7.2 Augmented Parity Games

To solve parity games augmented with persistent live groups, we use the fact that most of the algorithms to solve (non-augmented) parity games are based on the attractor functions $\text{ATTR}^0(G, T)$ and $\text{attr}^0(G, T)$ (as defined in Sect. 2). Furthermore, the only difference between the attractor function $\text{ATTR}^0(G, T)$ and our new function ATTRPERS(G, T, Λ) from Algorithm 2 is the utilization of augmented persistent live groups to solve reachability games. Hence, in many of these algorithms for parity games, one can simply replace every use of $\text{ATTR}^0(G, T)$ with ATTRPERS(G, T, Λ) to obtain an algorithm to solve parity games augmented with persistent live groups. In particular, the authors in [32], obtained an exponential algorithm by using ATTRPERS(G, T, Λ) in Zielonka's algorithm [41]. In this section, we show that applying the same technique to the quasi-polynimial algorithm given by Lehtinen et al. [24] and Parys [34] gives a quasi-polynomial algorithm for such augmented games.

As the algorithm given by Lehtinen et al. [24,34] recursively solves the parity games restricted to a smaller set of vertices, we need to define how the persistent live groups are transformed when we restrict the game graph in such a way. In addition, we also need to ensure that the transformed persistent live groups still express the same assumption w.r.t. the restricted game graph. Intuitively, one easy way to do this is by restricting all three sets S, C, T of a persistent live group to the set of vertices and edges in the restricted game graph. However, if there is an edge $e = (v, w)$ in C, then the constraint $\psi_{\text{CONT}}(C)$ enforces that edge e is taken from vertex v. If the restricted game graph contains v but not the edge e, then to satisfy the constraint $\psi_{\text{CONT}}(C)$, we need to ensure that vertex v is not visited. Hence, we need to removed such vertices from S in the transformed persistent live groups. This is formalized below.

Definition 6. *Given a game graph $G = (V, E)$ augmented with a set Λ of persistent live groups, and a set $U \subseteq V$, we define the set of persistent live groups restricted to U as $\Lambda|_U = \{(S|_U, C|_U, T|_U) \mid (S, C, T) \in \Lambda\}$ s.t.*

$$T|_U = T \cap U, \quad C|_U = \{(u, v) \in E \mid u, v \in U\}, \quad S|_U = (S \cap U) \setminus (\text{src}(C) \setminus \text{src}(C')).$$

Furthermore, the restriction applies as usual to game graph and parity objectives, i.e., $G|_U = (U, E \cap U \times U)$ and parity objective $Parity(\mathbb{P}|_U)$ is defined such that $\mathbb{P}|_U(u) = \mathbb{P}(u)$ for all $u \in U$.

One can show that $\Lambda|_U$ indeed captures the same assumption as Λ w.r.t. the game restricted to U as formalized below. For the proof, see [37], App. F.

Lemma 6. *Given an augmented parity game $\mathfrak{G} = (G, \Phi, \psi_{\text{PERS}}(\Lambda))$ and a set $U \subseteq V$, let ρ be a play in $G|_U$. Then, ρ is winning in augmented game $\mathfrak{G}|_U = (G|_U, \Phi|_U, \psi_{\text{PERS}}(\Lambda|_U))$ if and only if it is winning in augmented game $\mathfrak{G}$.*

With this well-defined restrictions, one can replace every use of $\text{ATTR}^0(G, T)$ with $\text{ATTRPERS}(G, T, \Lambda)$ in the algorithm given by Lehtinen et al. [24,34] to obtain a quasi-polynomial result for parity games augmented with persistent live groups.

Although the proof of correctness is almost identical to the proof given by Parys in [34], the adapted algorithm and its proof is provided explicitly in [37].

Theorem 6. *Parity games augmented with persistent live group assumptions can be solved in QP.*

References

1. Aminof, B., Giacomo, G.D., Rubin, S.: Stochastic fairness and language-theoretic fairness in planning in nondeterministic domains. In: Beck, J.C., Buffet, O., Hoffmann, J., Karpas, E., Sohrabi, S. (eds.) Proceedings of the Thirtieth International Conference on Automated Planning and Scheduling, Nancy, France, 26–30 October 2020, pp. 20–28. AAAI Press (2020)
2. Anand, A., Mallik, K., Nayak, S.P., Schmuck, A.: Computing adequately permissive assumptions for synthesis. In: Sankaranarayanan, S., Sharygina, N. (eds.) TACAS 2023. LNCS, vol. 13994, pp. 211–228. Springer, Cham (2023). https://doi.org/10.1007/978-3-031-30820-8_15

3. Apt, K.R., Grädel, E. (eds.): Lectures in Game Theory for Computer Scientists. Cambridge University Press, Cambridge (2011)
4. Arnold, A., Niwiński, D., Parys, P.: A quasi-polynomial black-box algorithm for fixed point evaluation. In: Baier, C., Goubault-Larrecq, J. (eds.) 29th EACSL Annual Conference on Computer Science Logic, CSL 2021, 25–28 January 2021, Ljubljana, Slovenia (Virtual Conference). LIPIcs, vol. 183, pp. 9:1–9:23. Schloss Dagstuhl - Leibniz-Zentrum für Informatik (2021)
5. Baier, C., Katoen, J.: Principles of Model Checking. MIT Press, Cambridge (2008)
6. Banerjee, T., Majumdar, R., Mallik, K., Schmuck, A., Soudjani, S.: Fast symbolic algorithms for omega-regular games under strong transition fairness. TheoretiCS 2 (2023)
7. Belta, C., Sadraddini, S.: Formal methods for control synthesis: an optimization perspective. Annu. Rev. Control Robot. Auton. Syst. 2, 115–140 (2019)
8. Bloem, R., et al.: Synthesizing robust systems. Acta Informatika 51(3–4), 193–220 (2014)
9. Bloem, R., Ehlers, R., Jacobs, S., Könighofer, R.: How to handle assumptions in synthesis. In: SYNT 2014, Vienna, Austria, pp. 34–50 (2014)
10. Bloem, R., Ehlers, R., Könighofer, R.: Cooperative reactive synthesis. In: ATVA 2015, Shanghai, China, pp. 394–410 (2015)
11. Calude, C.S., Jain, S., Khoussainov, B., Li, W., Stephan, F.: Deciding parity games in quasipolynomial time. In: Hatami, H., McKenzie, P., King, V. (eds.) Proceedings of the 49th Annual ACM SIGACT Symposium on Theory of Computing, STOC 2017, Montreal, QC, Canada, 19–23 June 2017, pp. 252–263. ACM (2017). https://doi.org/10.1145/3055399.3055409
12. Belta, C., Yordanov, B., Gol, E.: Formal Methods for Discrete-Time Dynamical Systems. Studies in Systems, Decision and Control, vol. 15. Springer, Cham (2017). https://doi.org/10.1007/978-3-319-50763-7
13. Chatterjee, K., de Alfaro, L., Faella, M., Majumdar, R., Raman, V.: Code aware resource management. Formal Methods Syst. Des. 42(2), 146–174 (2013)
14. Chatterjee, K., Henzinger, T., Jobstmann, B.: Environment assumptions for synthesis. In: CONCUR, pp. 147–161 (2008)
15. Chatterjee, K., Horn, F., Löding, C.: Obliging games. In: Gastin, P., Laroussinie, F. (eds.) CONCUR 2010. LNCS, vol. 6269, pp. 284–296. Springer, Heidelberg (2010). https://doi.org/10.1007/978-3-642-15375-4_20
16. D'Ippolito, N., Braberman, V., Piterman, N., Uchitel, S.: Synthesis of live behavior models. In: 18th International Symposium on Foundations of Software Engineering, pp. 77–86. ACM (2010)
17. D'Ippolito, N., Rodríguez, N., Sardiña, S.: Fully observable non-deterministic planning as assumption-based reactive synthesis. J. Artif. Intell. Res. 61, 593–621 (2018)
18. Emerson, E.A., Jutla, C.S.: Tree automata, mu-calculus and determinacy (extended abstract). In: 32nd Annual Symposium on Foundations of Computer Science, San Juan, Puerto Rico, 1–4 October 1991, pp. 368–377. IEEE Computer Society (1991). https://doi.org/10.1109/SFCS.1991.185392
19. Emerson, E.A., Jutla, C.S.: The complexity of tree automata and logics of programs. SIAM J. Comput. 29(1), 132–158 (1999). https://doi.org/10.1137/S0097539793304741
20. Francez, N.: Fairness. Springer, Heidelberg (1986)
21. Hausmann, D., Schröder, L.: Quasipolynomial computation of nested fixpoints. In: TACAS 2021. LNCS, vol. 12651, pp. 38–56. Springer, Cham (2021). https://doi.org/10.1007/978-3-030-72016-2_3

22. Jurdzinski, M., Morvan, R., Thejaswini, K.S.: Universal algorithms for parity games and nested fixpoints. In: Raskin, J., Chatterjee, K., Doyen, L., Majumdar, R. (eds.) Principles of Systems Design. LNCS, vol. 13660, pp. 252–271. Springer, Cham (2022). https://doi.org/10.1007/978-3-031-22337-2_12

23. Klein, U., Pnueli, A.: Revisiting synthesis of GR(1) specifications. In: Barner, S., Harris, I., Kroening, D., Raz, O. (eds.) HVC 2010. LNCS, vol. 6504, pp. 161–181. Springer, Heidelberg (2011). https://doi.org/10.1007/978-3-642-19583-9_16

24. Lehtinen, K., Parys, P., Schewe, S., Wojtczak, D.: A recursive approach to solving parity games in quasipolynomial time. Log. Methods Comput. Sci. **18**(1) (2022). https://doi.org/10.46298/lmcs-18(1:8)2022

25. Lindemann, L., Pappas, G.J., Dimarogonas, D.V.: Reactive and risk-aware control for signal temporal logic. IEEE Trans. Autom. Control **67**(10), 5262–5277 (2022). https://doi.org/10.1109/TAC.2021.3120681

26. Liu, J.: Closing the gap between discrete abstractions and continuous control: completeness via robustness and controllability. In: Dima, C., Shirmohammadi, M. (eds.) FORMATS 2021. LNCS, vol. 12860, pp. 67–83. Springer, Cham (2021). https://doi.org/10.1007/978-3-030-85037-1_5

27. Majumdar, R., Mallik, K., Schmuck, A., Soudjani, S.: Symbolic control for stochastic systems via parity games. CoRR abs/2101.00834 (2021)

28. Majumdar, R., Mallik, K., Schmuck, A., Soudjani, S.: Symbolic qualitative control for stochastic systems via finite parity games. In: Jungers, R.M., Ozay, N., Abate, A. (eds.) 7th IFAC Conference on Analysis and Design of Hybrid Systems, ADHS 2021, Brussels, Belgium, 7–9 July 2021. IFAC-PapersOnLine, vol. 54, pp. 127–132. Elsevier (2021). https://doi.org/10.1016/j.ifacol.2021.08.486

29. Majumdar, R., Piterman, N., Schmuck, A.-K.: Environmentally-friendly GR(1) synthesis. In: Vojnar, T., Zhang, L. (eds.) TACAS 2019. LNCS, vol. 11428, pp. 229–246. Springer, Cham (2019). https://doi.org/10.1007/978-3-030-17465-1_13

30. Maoz, S., Ringert, J.O., Shalom, R.: Symbolic repairs for GR(1) specifications. In: ICSE (2019)

31. Mohajerani, S., Malik, R., Wintenberg, A., Lafortune, S., Ozay, N.: Divergent stutter bisimulation abstraction for controller synthesis with linear temporal logic specifications. Autom. **130**, 109723 (2021). https://doi.org/10.1016/j.automatica.2021.109723

32. Nayak, S., Egidio, L., Rossa, M.D., Schmuck, A.K., Jungers, R.: Context-triggered abstraction-based control design. IEEE Open J. Control Syst. 1–21 (2023). https://doi.org/10.1109/OJCSYS.2023.3305835

33. Nilsson, P., Ozay, N., Liu, J.: Augmented finite transition systems as abstractions for control synthesis. Discret. Event Dyn. Syst. **27**(2), 301–340 (2017). https://doi.org/10.1007/s10626-017-0243-z

34. Parys, P.: Parity games: Zielonka's algorithm in quasi-polynomial time. In: Rossmanith, P., Heggernes, P., Katoen, J.P. (eds.) 44th International Symposium on Mathematical Foundations of Computer Science (MFCS 2019). Leibniz International Proceedings in Informatics (LIPIcs), vol. 138, pp. 10:1–10:13. Schloss Dagstuhl-Leibniz-Zentrum fuer Informatik, Dagstuhl, Germany (2019). https://doi.org/10.4230/LIPIcs.MFCS.2019.10

35. Queille, J., Sifakis, J.: Fairness and related properties in transition systems - a temporal logic to deal with fairness. Acta Informatica **19**, 195–220 (1983)

36. Sanfelice, R.G.: Hybrid Feedback Control. Princeton University Press, Princeton (2020)

37. Schmuck, A.K., Thejaswini, K.S., Sağlam, I., Nayak, S.P.: Solving two-player games under progress assumptions (extended version). arXiv:2310.12767 (2023)

38. Sun, F., Ozay, N., Wolff, E.M., Liu, J., Murray, R.M.: Efficient control synthesis for augmented finite transition systems with an application to switching protocols. In: 2014 American Control Conference, pp. 3273–3280 (2014). https://doi.org/10.1109/ACC.2014.6859428

39. Tabuada, P.: Verification and Control of Hybrid Systems: A Symbolic Approach. Springer, New York (2009). https://doi.org/10.1007/978-1-4419-0224-5d

40. Thomas, W.: Languages, Automata, and Logic. In: Rozenberg, G., Salomaa, A. (eds.) Handbook of Formal Languages, pp. 389–455. Springer, Heidelberg (1997). https://doi.org/10.1007/978-3-642-59126-6_7

41. Zielonka, W.: Infinite games on finitely coloured graphs with applications to automata on infinite trees. Theor. Comput. Sci. **200**(1–2), 135–183 (1998). https://doi.org/10.1016/S0304-3975(98)00009-7

SAT, SMT, and Automated Reasoning

Interpolation and Quantifiers
in Ortholattices

Simon Guilloud$^{(\boxtimes)}$ ⓘ, Sankalp Gambhir ⓘ, and Viktor Kunčak ⓘ

EPFL, School of Computer and Communication Sciences, Lausanne, Switzerland
{simon.guilloud,sankalp.gambhir,Viktor.Kuncak}@epfl.ch

Abstract. We study quantifiers and interpolation properties in *ortho-logic*, a non-distributive weakening of classical logic that is sound for formula validity with respect to classical logic, yet has a quadratic-time decision procedure. We present a sequent-based proof system for quantified orthologic, which we prove sound and complete for the class of all complete ortholattices. We show that orthologic does not admit quantifier elimination in general. Despite that, we show that interpolants always exist in orthologic. We give an algorithm to compute interpolants efficiently. We expect our result to be useful to quickly establish unreachability as a component of verification algorithms.

1 Introduction

Interpolation-based techniques are important in hardware and software model checking [17,19,25,30,31,31,32,34,39]. The interpolation theorem for classical propositional logic states that, for two formulas A and B such that $A \implies B$ is valid, there exists a formula I, with free variables among only those common to both A and B, such that both $A \implies I$ and $I \implies B$ are valid. All known algorithms for propositional logic have worst-case exponential size of proofs they construct, which is not surprising given that the validity problem is coNP-hard. Interpolation algorithms efficiently construct interpolants from such exponentially-sized proofs [38], which makes the overall process exponential. It is therefore interesting to explore whether there are logical systems for which proof search and interpolation are polynomial-time in the size of the input formulas.

Orthologic is a relaxation of classical logic, corresponding to the algebraic structure of ortholattices, where the law of distributivity does not necessarily hold, but where the weaker absorption law (V9) does (Table 1). In contrast to classical and intuitionistic logic, where the problem of deciding the validity of a formula is, respectively, coNP-complete and PSPACE-complete, there is a *quadratic-time* algorithm to decide validity in orthologic [14,16]. Orthologic was first studied as a candidate for quantum logic, where distributivity fails [1,2]. Due to its advantageous computational properties, orthologic has recently been suggested as a tool to reason about proofs and programs in formal verification in a way that is sound, efficient and predictable [14,15], even if incomplete.

As a step towards enabling the use of state-of-the-art model checking techniques backed by orthologic, this paper studies interpolation, as well as properties

R. Dimitrova et al. (Eds.): VMCAI 2024, LNCS 14499, pp. 235–257, 2024.
https://doi.org/10.1007/978-3-031-50524-9_11

of quantifiers in orthologic. The quantifier elimination property would immediately lead to the existence of interpolants. We show, however, that quantified orthologic does *not* admit quantifier elimination. To do so, we define semantics of quantified orthologic (QOL) using complete ortholattices. Furthermore, we present a natural sequent calculus proof system for QOL that we show to be sound and complete with respect to this semantics. We then consider the question of interpolation. We show that a refutation-based notion of interpolation fails. However, a natural notion of interpolants based on the lattice ordering of formulas yields interpolants in orthologic. Namely, if $A \leq B$ is provable, then there exists an interpolant I such that $A \leq I$ and $I \leq B$, where $\leq$ corresponds to implication. Moreover, these interpolants can be computed efficiently from a proof of $A \leq B$. We expect that this notion of interpolation can be used in future verification algorithms.

Table 1. Axioms of orthologic, a generalization of classical logic corresponding to the algebraic variety of ortholattices. As lattices, ortholattices admit a partial order $\leq_{OL}$ defined by $a \leq_{OL} b$ iff $a \wedge b = a$ or, equivalently, $a \vee b = b$.

V1:	$x \vee y = y \vee x$		V1':	$x \wedge y = y \wedge x$
V2:	$x \vee (y \vee z) = (x \vee y) \vee z$		V2':	$x \wedge (y \wedge z) = (x \wedge y) \wedge z$
V3:	$x \vee x = x$		V3':	$x \wedge x = x$
V4:	$x \vee 1 = 1$		V4':	$x \wedge 0 = 0$
V5:	$x \vee 0 = x$		V5':	$x \wedge 1 = x$
V6:	$\neg\neg x = x$			
V7:	$x \vee \neg x = 1$		V7':	$x \wedge \neg x = 0$
V8:	$\neg(x \vee y) = \neg x \wedge \neg y$		V8':	$\neg(x \wedge y) = \neg x \vee \neg y$
V9:	$x \vee (x \wedge y) = x$		V9':	$x \wedge (x \vee y) = x$

In some cases, ortholattices may be not only a relaxation of propositional logic but a direct intended interpretation of formulas. Indeed, lattices already play a crucial role in abstract interpretation [9,37] and have been adopted by the Flix programming language [29]. Furthermore, De Morgan bi-semilattices and lattices (generalizations of ortholattices where law V6 of Table 1 does not necessarily hold) have been used to model multivalued logics with undefined states [4,6]. Lattice automata [26] map final automaton states to elements of a finite distributive De Morgan lattice, which admits a notion of complement, but, in contrast to ortholattices, need not satisfy V7 or V7' of Table 1 (the chain $0 \leq 1/2 \leq 1$ is a finite distributive de Morgan lattice but not an ortholattice).

Proof-theoretic properties of propositional orthologic are presented in [16], but without discussion of interpolation and without the treatment of quantifiers as lattice operators. These topics are the subject of the present paper.

Contributions. We make the following contributions:

1. We define quantified ortholattice, in the spirit of QBF, presenting its semantics in terms of validity in all complete ortholattices. We present a proof system for quantified ortholattice, which extends an existing polynomial-time proof system for quantifier-free ortholattice [16, 40] with rules for quantifier introduction and elimination. We show soundness and completeness of our proof system.
2. We show that quantified ortholattice does not admit quantifier elimination. Consequently, quantifiers increase the class of definable relationships between ortholattice elements. This also makes the existence of interpolants a more subtle question than in classical propositional logic, where quantifier elimination alone guarantees that quantifier-free interpolants exist.
3. We consider a refutation-based interpolation property, which reduces to the usual one in classical logic. We show that ortholattice does *not* satisfy this variant of interpolation.
4. We consider another notion of interpolation, one which is natural in any lattice-based logic. In the language of ortholattices (Table 2), given two formulas A and B such that $A \leq B$, an interpolant I is a formula such that $A \leq I$, $I \leq B$, and $\mathsf{FV}(I) \subseteq \mathsf{FV}(A) \cap \mathsf{FV}(B)$. While it is known [36] that ortholattice admits such interpolants, we show using the sequent calculus proof system for OL that (a generalization of) such interpolants can always be computed efficiently. Specifically, we present an algorithm to compute interpolants from proofs of sequents in time linear in the size of the proof (where finding proofs in ortholattice is worst-case quadratic time in the size of the inequality).

The final result yields an end-to-end polynomial-time algorithm that first finds a proof and then computes an interpolant I, where validity of both the input $A \implies B$ and the result $A \implies I, I \implies B$ is with respect to OL axioms.

Table 2. Axiomatization of ortholattices in the signature $(S, \leq, \wedge, \vee, 0, 1, \neg)$ as partially ordered sets. & denotes conjunction between atomic formulas of axioms, to differentiate it from the term-level lattice operation $\wedge$.

P1:	$x \leq x$		
P2:	$x \leq y \;\&\; y \leq z \implies x \leq y$		
P3:	$0 \leq x$	P3':	$x \leq 1$
P4:	$x \wedge y \leq x$	P5':	$x \leq x \vee y$
P5:	$x \wedge y \leq y$	P6':	$y \leq x \vee y$
P6:	$x \leq y \;\&\; x \leq z \implies x \leq y \wedge z$	P6':	$x \leq z \;\&\; y \leq z \implies x \vee y \leq z$
P7:	$x \leq \neg\neg x$	P7':	$\neg\neg x \leq x$
P8:	$x \leq y \implies \neg y \leq \neg x$		
P9:	$x \wedge \neg x \leq 0$	P9':	$1 \leq x \vee \neg x$

Preliminaries. We follow the definitions and notation of [16]. An ortholattice is an algebraic variety with language $(\wedge, \vee, \neg, 0, 1)$ and axioms in Table 1. As lattices, ortholattices can be described as a partially ordered set whose order relation, noted $\leq_{OL}$, is defined by $a \leq_{OL} b$ iff $a \wedge b = a$ or equivalently $a \vee b = b$ [22,35]. In both Boolean and Heyting algebras, this order relation corresponds to the usual logical implication. The (equivalent) axiomatization of ortholattices as a poset can be found in Table 2. We denote by $\mathcal{T}_{OL}$ the set of terms built as trees with nodes labelled by either by $(\wedge, \vee, \neg, 0, 1)$ or by symbols in a fixed, countably infinite set of variable symbols $Var = \{x, y, z, ...\}$. This corresponds precisely to the set of propositional formulas. Note that since $\wedge$ and $\vee$ are commutative, children of a node are described for simplicity as an unordered set. In particular, $x \wedge y$ and $y \wedge x$ denote the same term. Since 0 can always be represented as $x \wedge \neg x$, we sometimes omit it from case analysis for brevity, and similarly for 1.

2   Quantified Ortholologic: Syntax, Semantics, and a Complete Proof System

We consider the extension of propositional ortholologic to quantified ortholologic, noted QOL, the analogue of QBF [8] for classical logic, or of System F [13] for intuitionistic logic. To do so, we extend the proof system of [16] by adding axiomatization of an existential quantifier $(\bigvee)$ and a universal quantifier $(\bigwedge)$. The deduction rules of QOL are in Fig. 1. It is folklore that the sequent calculus LK [12] with arbitrarily many formulas on both sides corresponds to classical logic, while, if we restrict the right sides of sequents to only contain at most one formula, we obtain intuitionistic logic [42, section 7.1]. Ortholologic exhibits a different natural restriction: sequents can only contain at most two formulas in both sides of the sequent combined. For this reason, it is convenient to represent sequents by decorating the formulas with superscript ^{L} or ^{R}, depending on whether they appear on the left or right side.

Definition 1 (From [16]). If ϕ is a formula, we call ϕ^L and ϕ^R annotated formulas. A *sequent* is a set of at most two annotated formulas. We use uppercase Greek letters (e.g. Γ and Δ) to represent sets that are either empty or contain exactly one annotated formula ($|\Gamma| \leq 1, |\Delta| \leq 1$).

Given formulas ϕ and ψ, we thus write ϕ^L, ϕ^R for a sequent often denoted $\phi \vdash \psi$.

Our use of quantifiers in this paper (quantified ortholologic) is different from considering the first-order theory of ortholattices. In particular, the semantic of an existential quantifier $(\bigwedge x. t)$ in QOL corresponds to the least upper bound of a (possibly infinite) family of *lattice* elements given by values of term t. In contrast, when considering a classical first-order theory of ortholattices, we would build an atomic formula such as $t_1 \leq t_2$, obtaining a definite truth or falsehood in the metatheory, and only then apply quantifiers to build formulas such as $\exists x.(t_1 \leq t_2)$. Such difference also exists in the case of Boolean algebras [24].

2.1   Complete Ortholattices

To model quantified Ortholologic, we restrict ortholattices to complete ones.

$$\frac{}{\phi^L, \phi^R} \text{ Hyp}$$

$$\frac{\Gamma, \psi^R \qquad \psi^L, \Delta}{\Gamma, \Delta} \text{ Cut}$$

$$\frac{\Gamma}{\Gamma, \Delta} \text{ Weaken}$$

$$\frac{\Gamma, \phi^L}{\Gamma, (\phi \wedge \psi)^L} \text{ LeftAnd} \qquad \frac{\Gamma, \phi^R \qquad \Gamma, \psi^R}{\Gamma, (\phi \wedge \psi)^R} \text{ RightAnd}$$

$$\frac{\Gamma, \phi^L \qquad \Gamma, \psi^L}{\Gamma, (\phi \vee \psi)^L} \text{ LeftOr} \qquad \frac{\Gamma, \phi^R}{\Gamma, (\phi \vee \psi)^R} \text{ RightOr}$$

$$\frac{\Gamma, \phi^R}{\Gamma, (\neg\phi)^L} \text{ LeftNot} \qquad \frac{\Gamma, \phi^L}{\Gamma, (\neg\phi)^R} \text{ RightNot}$$

(a) Deduction rules of propositional Orthologic.

$$\frac{\Gamma, \phi[x := \gamma]^L}{\Gamma, (\bigwedge x.\phi)^L} \text{ LeftForall} \qquad \frac{\Gamma, \phi[x := x']^R}{\Gamma, (\bigwedge x.\phi)^R} \text{ RightForall} \atop (x' \text{ not free in } \Gamma)$$

$$\frac{\Gamma, \phi[x := x']^L}{\Gamma, (\bigvee x.\phi)^L} \text{ LeftExists} \atop (x' \text{ not free in } \Gamma) \qquad \frac{\Gamma, \phi[x := \gamma]^R}{\Gamma, (\bigvee x.\phi)^R} \text{ RightExists}$$

(b) Deduction rules of Quantified Orthologic.

Fig. 1. Deduction rules of Orthologic. Each holds for arbitrary Γ, Δ, ϕ, ψ. Sets Γ and Δ are either empty or contain a single annotated formula.

Definition 2 (Complete Ortholattice). An ortholattice $\mathcal{O} = (O, \sqsubseteq, \sqcup, \sqcap, -)$ is *complete* if and only if for any possibly infinite set of elements $X \subseteq O$, there exist two elements noted $\bigsqcup X$ and $\bigsqcap X$ which are respectively the *lowest upper bound* and *greatest lower bound* of elements of X, with respect to $\sqsubseteq$:

$$\forall x \in X. \ x \sqsubseteq \bigsqcup X \text{ and } \bigsqcap X \sqsubseteq x \, ,$$

and, for all $y \in O$:

$$(\forall x \in X. x \sqsubseteq y) \implies (\bigsqcup X \sqsubseteq y)$$

$$(\forall x \in X. y \sqsubseteq x) \implies (y \sqsubseteq \bigsqcap X)$$

This definition coincides with the usual definition in complete lattices. Note that, in particular, all finite ortholattices are complete with bounds computed by iterating the binary operators $\sqcup$ and $\sqcap$.

Definition 3. $\mathcal{T}_{QOL}$ denotes the set of quantified orthologic formulas, i.e. $\mathcal{T}_{OL} \subset \mathcal{T}_{QOL}$ and for any $x \in Var$ and $\phi \in \mathcal{T}_{QOL}$,

$$\bigwedge x.\phi \in \mathcal{T}_{QOL} \quad \text{and} \quad \bigvee x.\phi \in \mathcal{T}_{QOL}.$$

Note that $\mathcal{T}_{QOL} = \mathcal{T}_{QBF}$, the set of quantified Boolean formulas. For two formulas $\phi, \psi \in \mathcal{T}_{QOL}$ and a variable x, let $\phi[x := \psi]$ denotes the usual capture-avoiding substitution of x by ψ inside ϕ.

We assume a representation of quantified formulas where alpha-equivalent terms are equal, so that capture-avoiding substitution is well-defined. It is easy to check that any construction in this paper (and in particular, provability) is consistent across alpha-equivalent formulas.

Definition 4 (Models and Interpretation). A *model* for QOL is a complete ortholattice $\mathcal{O} = (O, \sqsubseteq, \sqcup, \sqcap, -)$ and an assignment $\sigma : Var \to O$. The interpretation of a formula ϕ with respect to an assignment σ is defined recursively as usual:

$$\begin{aligned}
[\![x]\!]_\sigma &:= \sigma(x) \\
[\![\phi \wedge \psi]\!]_\sigma &:= [\![\phi]\!]_\sigma \sqcap [\![\psi]\!]_\sigma \\
[\![\phi \vee \psi]\!]_\sigma &:= [\![\phi]\!]_\sigma \sqcup [\![\psi]\!]_\sigma \\
[\![\neg\phi]\!]_\sigma &:= -[\![\phi]\!]_\sigma \\
[\![\bigvee x.\phi]\!]_\sigma &:= \bigsqcup\{[\![\phi]\!]_{\sigma[x:=e]} \mid e \in O\} \\
[\![\bigwedge x.\phi]\!]_\sigma &:= \bigsqcap\{[\![\phi]\!]_{\sigma[x:=e]} \mid e \in O\}
\end{aligned}$$

where $\sigma[x := e]$ denotes the assignment σ with its value at x changed to e and all other values unchanged.

The interpretation of a sequent is defined in the following way, as in [16]:

$$\begin{aligned}
[\![\phi^L, \psi^R]\!]_\sigma &:= [\![\phi]\!]_\sigma \sqsubseteq [\![\psi]\!]_\sigma \\
[\![\phi^L, \psi^L]\!]_\sigma &:= [\![\phi]\!]_\sigma \sqsubseteq -[\![\psi]\!]_\sigma \\
[\![\phi^R, \psi^R]\!]_\sigma &:= -[\![\phi]\!]_\sigma \sqsubseteq [\![\psi]\!]_\sigma \\
[\![\phi^L]\!]_\sigma &:= [\![\phi]\!]_\sigma \sqsubseteq 0_\mathcal{O} \\
[\![\phi^R]\!]_\sigma &:= 1_\mathcal{O} \sqsubseteq [\![\phi]\!]_\sigma \\
[\![\emptyset]\!]_\sigma &:= 1_\mathcal{O} \sqsubseteq 0_\mathcal{O}
\end{aligned}$$

Definition 5 (Entailment). If the sequent Γ, Δ is provable, we write $\vdash \Gamma, \Delta$. If for every complete ortholattice $\mathcal{O}$ and assignment $\sigma : Var \to O$, $[\![\Gamma, \Delta]\!]_\sigma$ is true, we write $\models \Gamma, \Delta$.

By slight abuse of notation, we sometimes write, e.g., $\phi \vdash \psi$ in place of $\vdash \phi^L, \psi^R$ to help readability.

Definition 6. Given formulas ϕ and ψ, let $\phi \dashv\vdash \psi$ denote the fact that both $\phi \vdash \psi$ and $\psi \vdash \phi$ are provable.

We show soundness and completeness of QOL with respect to the class of all complete ortholattices. Soundness is easy and direct, completeness less so.

2.2   Soundness

Lemma 1 (Soundness). *For every sequent S, if $\vdash S$ then $\vDash S$.*

Proof. We simply verify that every deduction rule of Fig. 1 preserves truth of interpretation in any model. We show the case of LeftAnd as an example, as well as LeftForall and LeftExists. Other cases are easy or analogous.

Fix an arbitrary ortholattice $\mathcal{O} = (O, \sqsubseteq, \sqcup, \sqcap, -)$.

LeftAnd: For any assignment $\sigma : Var \to O$, the interpretation of the conclusion of a LeftAnd rule is

$$\llbracket \Gamma, (\phi \wedge \psi)^L \rrbracket_\sigma$$

Γ can be empty, a left formula or a right formula. If it is empty then

$$\llbracket \Gamma, (\phi \wedge \psi)^L \rrbracket_\sigma \iff \llbracket 0^R, (\phi \wedge \psi)^L \rrbracket_\sigma$$

If $\Gamma = \gamma^L$, then we have

$$\llbracket \Gamma, (\phi \wedge \psi)^L \rrbracket_\sigma \iff \llbracket (\neg\gamma)^R, (\phi \wedge \psi)^L \rrbracket_\sigma .$$

So without loss of generality we can assume $\Gamma = \gamma^R$ to be a right formula. Now,

$$\llbracket \gamma^R, (\phi \wedge \psi)^L \rrbracket_\sigma \iff$$
$$\llbracket \phi \wedge \psi \rrbracket_\sigma \sqsubseteq \llbracket \gamma \rrbracket_\sigma \iff$$
$$\llbracket \phi \rrbracket_\sigma \sqcap \llbracket \psi \rrbracket_\sigma \sqsubseteq \llbracket \gamma \rrbracket_\sigma$$

But using the premise of the LeftAnd rule and the induction hypothesis, we know $\llbracket \gamma^R, \phi^L \rrbracket_\sigma$ holds true. Hence,

$$\llbracket \gamma^R, \phi^L \rrbracket_\sigma \iff$$
$$\llbracket \phi \rrbracket_\sigma \sqsubseteq \llbracket \gamma \rrbracket_\sigma \implies$$
$$\llbracket \phi \rrbracket_\sigma \sqcap \llbracket \psi \rrbracket_\sigma \sqsubseteq \llbracket \gamma \rrbracket_\sigma$$

where the implication holds by the laws of ortholattices (Table 2).

LeftForall: For any assignment $\sigma : Var \to O$,

$$\llbracket \Gamma, (\textstyle\bigwedge x.\phi)^L \rrbracket_\sigma$$

where we can again assume $\Gamma = \gamma^R$ without loss of generality. Then

$$\llbracket \gamma^R, (\textstyle\bigwedge x.\phi)^L \rrbracket_\sigma =$$
$$\llbracket \textstyle\bigwedge x.\phi \rrbracket_\sigma \sqsubseteq \llbracket \gamma \rrbracket_\sigma =$$
$$\textstyle\bigsqcap \{\llbracket \phi \rrbracket_{\sigma[x:=e]} \mid e \in O\} \sqsubseteq \llbracket \gamma \rrbracket_\sigma$$

Again, by hypothesis, there exists a formula ψ such that $\llbracket \gamma^R, \phi[x := \psi]^L \rrbracket_\sigma$ holds true. Finally,

$$\llbracket \gamma^R, \phi[x := \psi]^L \rrbracket_\sigma \iff$$
$$\llbracket \phi[x := \psi] \rrbracket_\sigma \sqsubseteq \llbracket \gamma \rrbracket_\sigma \iff$$
$$\llbracket \phi \rrbracket_{\sigma[x := \llbracket \psi \rrbracket_\sigma]} \sqsubseteq \llbracket \gamma \rrbracket_\sigma \implies$$
$$\bigsqcap \{ \llbracket \phi \rrbracket_{\sigma[x := e]} \mid e \in O \} \sqsubseteq \llbracket \gamma \rrbracket_\sigma$$

where the last implication holds by definition of $\bigsqcap$.
LeftExists: For any assignment $\sigma : Var \to O$,

$$\llbracket \Gamma, (\bigvee x.\phi)^L \rrbracket_\sigma$$

where we assume one last time without loss of generality that $\Gamma = \gamma^R$. Then

$$\llbracket \gamma^R, (\bigvee x.\phi)^L \rrbracket_\sigma \iff$$
$$\llbracket \bigvee x.\phi \rrbracket_\sigma \sqsubseteq \llbracket \gamma \rrbracket_\sigma \iff$$
$$\bigsqcup \{ \llbracket \phi \rrbracket_{\sigma[x := e]} \mid e \in O \} \sqsubseteq \llbracket \gamma \rrbracket_\sigma$$

By hypothesis, $\llbracket \gamma^R, \phi^L \rrbracket_\tau$ holds for any assignment τ, and in particular for any τ of the form $\sigma[x := e]$. Since x does not appear in γ, $\llbracket \gamma \rrbracket_{\sigma[x := e]} = \llbracket \gamma \rrbracket_\sigma$. Hence, for any $e \in O$, each of the following line holds true:

$$\llbracket \gamma^R, \phi^L \rrbracket_{\sigma[x := e]} \iff$$
$$\llbracket \phi \rrbracket_{\sigma[x := e]} \sqsubseteq \llbracket \gamma \rrbracket_{\sigma[x := e]} \iff$$
$$\llbracket \phi \rrbracket_{\sigma[x := e]} \sqsubseteq \llbracket \gamma \rrbracket_\sigma$$

By the least upper bound property of $\bigsqcup$, we obtain as desired the truth of:

$$\bigsqcup \{ \llbracket \phi \rrbracket_{\sigma[x := e]} \mid e \in O \} \sqsubseteq \llbracket \gamma \rrbracket_\sigma$$

2.3   Completeness

In classical propositional logic, we can show completeness with respect to the $\{0, 1\}$ Boolean algebra, which is straightforward. In orthologic, however, we do not have completeness with respect to a simple finite structure; we will need to build an infinite complete ortholattices. The construction is distinct but not entirely unlike that of models for predicate orthologic [1,36]. In particular, Mac-Neille completion [28] is used to transform the initial incomplete model into a complete one.

Lemma 2 (Completeness). *For any sequent S, $\vDash S$ implies $\vdash S$.*

Proof. We prove the contraposition: if the sequent S is not provable, then there exists a complete ortholattice $\mathcal{O} = (O, \sqsubseteq, \sqcup, \sqcap, -)$ and an assignment $\sigma : Var \to O$ such that $[\![S]\!]_\sigma$ does not hold. We construct O from the set of syntactic terms of complete ortholattices themselves, similarly to a free algebra (but with quantifiers). Formally, let O be $\mathcal{T}_{QOL}/\dashv\vdash$, i.e. the quotient set of $\mathcal{T}_{QOL}$ by the relation $\dashv\vdash$.

It is immediate that the function symbols $\wedge, \vee, \neg$ and relation symbol $\vdash$ of $\mathcal{T}_{QOL}$ are consistent over the equivalence classes of O, allowing us to extend them to O:

$$[\phi]_{\dashv\vdash} \sqcap [\psi]_{\dashv\vdash} := [\phi \wedge \psi]_{\dashv\vdash}$$
$$[\phi]_{\dashv\vdash} \sqcup [\psi]_{\dashv\vdash} := [\phi \vee \psi]_{\dashv\vdash}$$
$$-[\phi]_{\dashv\vdash} := [\neg\phi]_{\dashv\vdash}$$
$$[\phi]_{\dashv\vdash} \sqsubseteq [\psi]_{\dashv\vdash} := (\phi \vdash \psi) \text{ is provable}$$

It is also immediate that $\mathcal{O} = (O, \sqsubseteq, \sqcup, \sqcap, -)$ satisfies all the laws of ortholattices of Table 1. However, to interpret a quantified formula into $\mathcal{O}$, we would need $\mathcal{O}$ to be complete. It might not be complete, but it is "complete enough" to define all upper bounds of interest, as the following lemma shows.

Lemma 3. *For any $\sigma : Var \to O$, let $\sigma' : Var \to \mathcal{T}_{QOL}$ be such that for any x $[\sigma'(x)]_{\dashv\vdash} = \sigma(x)$. Let $\phi[\sigma']$ denote the simultaneous capture-avoiding substitution of variables in the formula ϕ with the assignments in σ'.*
Then, for any $\phi \in \mathcal{T}_{QOL}$, $[\![\phi]\!]_\sigma$ exists and $[\![\phi]\!]_\sigma = [\phi[\sigma']]_{\dashv\vdash}$.

Proof. First note that $[\phi[\sigma']]_{\dashv\vdash}$ is well-defined: it does not depend on the specific choice of assignment we make for σ'. Then, the proof works by structural induction on ϕ. If it is a variable x,

$$[\![x]\!]_\sigma = \sigma(x) = [\sigma'(x)]_{\dashv\vdash} = [x[\sigma']]_{\dashv\vdash}$$

by definition. Then, if $\phi = \phi_1 \wedge \phi_2$,

$$[\![\phi_1 \wedge \phi_2]\!]_\sigma = [\![\phi_1]\!]_\sigma \sqcap [\![\phi_2]\!]_\sigma = [\phi_1[\sigma']]_{\dashv\vdash} \sqcap [\phi_2[\sigma']]_{\dashv\vdash} = [\phi_1[\sigma'] \wedge \phi_2[\sigma']]_{\dashv\vdash}$$

where the first equality is the definition of $[\![\cdot]\!]$, the second equality the induction hypothesis and the third equality is the definition of $\sqcap$ in $\mathcal{O}$. $\vee$ and $\neg$ are similar. Consider now the interpretation of a formula $[\![\bigvee x.\phi]\!]_\sigma$. Since alpha-equivalence holds in our proof system and in the definition of the least upper bound, we assume to ease notation that x is fresh with respect to σ, i.e., that we don't need to signal explicitly capture-avoiding substitution. By definition of $[\![\cdot]\!]_\sigma$, we should have:

$$\left[\!\!\left[\bigvee x.\phi\right]\!\!\right]_\sigma = \bigsqcup \{[\![\phi]\!]_{\sigma[x:=e]} \mid e \in O\}$$

Does the right-hand side always exist in $\mathcal{O}$? We claim that it does, and that it is equal to $[(\bigvee x.\phi)[\sigma']]_{\dashv\vdash}$. Mainly, we need to show that it satisfies the two properties of the least upper bound. First, the *upper bound* property:

$$\forall a \in \{[\![\phi]\!]_{\sigma[x:=e]} \mid e \in O\}, \quad a \sqsubseteq \left[(\bigvee x.\phi)[\sigma']\right]_{\dashv\vdash}$$

Which is equivalent to $\forall e \in O$,

$$[\![\phi]\!]_{\sigma[x:=e]} \sqsubseteq \left[(\bigvee x.\phi)[\sigma'] \right]_{\dashv\vdash} \qquad\qquad \Longleftrightarrow \qquad (1)$$

$$\left[\phi[\sigma'_{[x:=e]}] \right]_{\dashv\vdash} \sqsubseteq \left[(\bigvee x.\phi)[\sigma'] \right]_{\dashv\vdash} \qquad\qquad \Longleftrightarrow \qquad (2)$$

$$\phi[\sigma'_{[x:=e]}] \vdash (\bigvee x.\phi)[\sigma'] \text{ is provable} \qquad\qquad \Longleftrightarrow \qquad (3)$$

$$\phi[\sigma'_{[x:=e]}] \vdash \bigvee x.\phi[\sigma'_{[x:=x]}] \text{ is provable} \qquad\qquad (4)$$

where (1) is the desired least upper bound property, (2) is equivalent by induction hypothesis and definition of σ', (3) by definition of $\sqsubseteq$ in $\mathcal{O}$ and (4) by definition of substitution. The last statement is indeed provable:

$$\frac{\dfrac{}{\phi[\sigma_{[x:=e]}]^L, \phi[\sigma_{[x:=e]}]^R} \text{Hyp}}{\phi[\sigma_{[x:=e]}]^L, (\bigvee x.\phi[\sigma'_{[x:=x]}])^R} \text{RightExists}$$

Secondly, we need to show the *least* upper bound property:

$$\forall a \in O. (\forall e \in O. [\![\phi]\!]_{\sigma[x:=e]} \sqsubseteq a) \implies \left(\left[(\bigvee x.\phi)[\sigma'] \right]_{\dashv\vdash} \sqsubseteq a \right)$$

which is equivalent to

$$\forall \psi \in \mathcal{T}_{\mathcal{QOL}}. (\forall e \in O. [\![\phi]\!]_{\sigma[x:=e]} \sqsubseteq [\psi]_{\dashv\vdash}) \implies \left(\left[(\bigvee x.\phi)[\sigma'] \right]_{\dashv\vdash} \sqsubseteq [\psi]_{\dashv\vdash} \right)$$

Fix an arbitrary ψ and assume $\forall e \in O. [\![\phi]\!]_{\sigma[x:=e]} \sqsubseteq [\psi]_{\dashv\vdash}$. Consider a variable x_2 which does not appear in ψ. Then, we have in particular,

$$[\![\phi]\!]_{\sigma[x:=x_2]} \sqsubseteq [\psi]_{\dashv\vdash}.$$

Then,

$$[\![\phi]\!]_{\sigma[x:=x_2]} \sqsubseteq [\psi]_{\dashv\vdash} \qquad\qquad \Longleftrightarrow$$

$$\left[\phi[\sigma'_{[x:=x_2]}] \right]_{\dashv\vdash} \sqsubseteq [\psi]_{\dashv\vdash} \qquad\qquad \Longleftrightarrow$$

$$\phi[\sigma'_{[x:=x_2]}] \vdash \psi \text{ is provable} \qquad\qquad \Longleftrightarrow$$

Then using a proof of the last line, we can construct:

$$\frac{\phi[\sigma'_{[x:=x_2]}]^L, \psi^R}{(\bigvee x.\phi[\sigma'_{[x:=x]}])^L, \psi^R} \text{LeftExists}$$

We finally obtain our second property as desired:

$$\left[(\bigvee x.\phi)[\sigma'] \right]_{\dashv\vdash} \sqsubseteq [\psi]_{\dashv\vdash}$$

To conclude the proof of Lemma 3, the case with $\bigwedge$ instead of $\bigvee$ is symmetrical.

Hence, our interpretation in $\mathcal{O}$ is guaranteed to be well-defined. However, $\mathcal{O}$ is not guaranteed to be complete for arbitrary sets of elements, which our definition of a model requires. To obtain a complete ortholattice, we will apply MacNeille completion to $\mathcal{O}$.

Definition 7 (MacNeille Completion, [28]). Given a lattice L, there exists a smallest complete lattice L' containing L as a sublattice with an embedding $i : L \to L'$ preserving the least upper bounds and greatest lower bounds of arbitrary (possibly infinite) subsets of L. This is the MacNeille completion of L.

Hence, there exists a complete lattice $\mathcal{O}'$ containing $\mathcal{O}$ as a sublattice and preserving the existing least upper bounds and greatest lower bound. But we also need $\mathcal{O}'$ to be an ortholattice, containing $\mathcal{O}$ as a subortholattice. Fortunately, this is true thanks to a theorem of Bruns.

Lemma 4 (Theorem 4.2 of [5]). *For every ortholattice $\mathcal{O}$, its MacNeille completion $\mathcal{O}'$ admits an orthocomplementation which extends the orthocomplementation of $\mathcal{O}$.*

Corollary 1. *There exists an injective ortholattice homomorphism $i : \mathcal{O} \to \mathcal{O}'$ such that*

$$\forall a, b \in O. a \leq_{\mathcal{O}} b \iff i(a) \leq_{\mathcal{O}'} i(b)$$

and for any $X \subset O$ such that $\bigsqcup X$ (resp. $\bigsqcap X$) exists, and

$$i\left(\bigsqcup X\right) = \bigsqcup(\{i(x) \mid x \in X\})$$
$$i\left(\bigsqcap X\right) = \bigsqcap(\{i(x) \mid x \in X\}).$$

We can now finish our completeness proof. Define $\sigma : Var \to O$ by $\sigma(x) = [x]_{\dashv\vdash}$. Then by Lemma 3, $[\![\phi]\!]_\sigma = [\phi]_{\dashv\vdash}$. Let γ, δ be the two formulas such that $[\![S]\!]_\sigma = ([\![\gamma]\!]_\sigma \sqsubseteq [\![\delta]\!]_\sigma)$, according to Definition 4. Remember that the sequent S is not provable by assumption, i.e. $[\![\gamma]\!]_\sigma \not\sqsubseteq [\![\delta]\!]_\sigma$, and hence:

$$[\gamma]_{\dashv\vdash} \not\leq_{\mathcal{O}} [\delta]_{\dashv\vdash}$$

from which we deduce

$$i([\gamma]_{\dashv\vdash}) \not\leq_{\mathcal{O}'} i([\delta]_{\dashv\vdash})$$

in the ortholattice $\mathcal{O}'$. We now define $\tau : Var \to O$ such that $\tau(x) = i(\sigma(x))$, implying (by induction and Corollary 1) that for any ϕ,

$$i([\phi]_{\dashv\vdash}) = [\![\phi]\!]_\tau$$

and therefore, in $\mathcal{O}'$:

$$[\![\gamma]\!]_\tau \not\sqsubseteq [\![\delta]\!]_\tau$$

We have hence built a model with the complete ortholattice $\mathcal{O}'$ and the assignment τ in which $[\![\gamma]\!]_\tau \not\sqsubseteq [\![\delta]\!]_\tau$, so $[\![S]\!]_\tau$ does not hold, as desired.

Theorem 1. *QOL is sound and complete for complete ortholattices, i.e. for any sequent S:*

$$\vdash S \iff \models S$$

3 No Quantifier Elimination for Orthologic

Definition 8 (Quantifier Elimination). A quantified propositional logic admits *quantifier elimination* if for any term Q there exists a quantifier-free term E such that $Q \dashv\vdash E$.

Example 1. QBF, the theory of quantified classical propositional logic, admits quantifier elimination that replaces the quantified proposition $\exists x.F$ with the proposition $F[x := 0] \vee F[x := 1]$. This quantifier elimination approach is sound over Boolean algebras in general, thanks to the distributivity law.

Example 2. The theory of quantified intuitionistic propositional logic does not admit quantifier elimination. Whereas provability in quantifier-free intuitionistic propositional logic corresponds closely to inhabitation in simply typed lambda calculus and is PSPACE-complete [43], the quantified theory corresponds to System F, and is undecidable [11].

Note that quantifier elimination provides a solution to the interpolant problem for QBF (and QOL, if it were to admit quantifier elimination). Indeed, consider a provable sequent $A_{(x,y)} \vdash B_{(y,z)}$, and x, y, z the free variables in A and B. We ask for an interpolant such that

$$A_{(x,y)} \vdash I_y \text{ and } I_y \vdash B_{(y,z)}$$

By quantifier elimination, there exists a quantifier-free formula I_y equivalent to $\bigwedge z.B_{(y,z)}$. This I_y satisfies the interpolant condition.

$$\frac{A_{(x,y)} \vdash B_{(y,z)}}{A_{(x,y)} \vdash (\bigwedge z.B_{(y,z)})} \text{ RightForall} \qquad \frac{\dfrac{}{B_{(y,z)} \vdash B_{(y,z)}} \text{ Hyp}}{(\bigwedge z.B_{(y,z)}) \vdash B_{(y,z)}} \text{ LeftForall}$$

However, the next theorem will show that QOL does not admit quantifier elimination in general, even though it still admits interpolation.

Theorem 1. *QOL does not admit quantifier elimination. In particular, there exists no quantifier free formula E such that*

$$E \dashv\vdash \bigvee x.\big(\neg x \wedge (y \vee x)\big)$$

Proof. For the sake of contradiction, suppose such an E exists. Let $y, w_1, ..., w_n$ be the free variables appearing in E. Since $\bigvee x.\neg x \wedge (y \vee x)$ is constant with respect to $w_1, ..., w_n$, E must be as well, and hence we can assume E only uses y as a variable. Moreover, the laws of OL in Table 1 imply that any quantifier free formula whose only variable is y is equivalent to one of $0, 1, y$ or $\neg y$. This can easily be shown by induction on the structure of the formula:

$$0 \wedge 0 \; = 0 \qquad 0 \wedge 1 \quad = 1$$
$$0 \wedge y \; = 0 \qquad 0 \wedge \neg y = 0$$
$$1 \wedge 1 \; = 1 \qquad 1 \wedge y \quad = y$$
$$1 \wedge \neg y = \neg y \qquad y \wedge y \quad = y$$
$$y \wedge \neg y = 0 \qquad \neg y \wedge \neg y = \neg y$$
$$\neg 0 \quad = 1 \qquad \neg 1 \quad = 0$$
$$\neg\neg y \quad = y$$

and similarly for disjunction.

Now, consider the ortholattices M_2 and M_4 in Fig. 2:

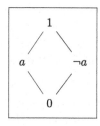 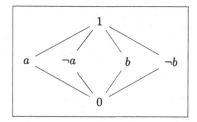

Fig. 2. The ortholattices M_2 and M_4. M_2 is distributive, but M_4 is not.

We use soundness of orthologic over ortholattices (Lemma 1) so that the formula E (if it exists) needs to be equal to $\bigvee x.\neg x \wedge (y \vee x)$ in all models. Since the model is finite, it is straightforward to compute in the ortholattice M_2 with the assignment $y := a$ that:

$$\left[\!\!\left[\bigvee x.\neg x \wedge (y \vee x) \right]\!\!\right]_{M_2, y:=a} = a$$

And hence the only compatible formula for E is the atom y.

However, in M_4:

$$\left[\!\!\left[\bigvee x.\neg x \wedge (y \vee x) \right]\!\!\right]_{M_4, y:=a} = 1$$

Hence, any expression for E among $0, 1, y, \neg y$ will fail to satisfy at least one of the two examples, and we conclude that there is no quantifier free formula E that is equivalent to $\bigvee x.\neg x \wedge (y \vee x)$. $\qquad \square$

On one hand, this result shows that we can use quantifiers to define new operators, such as

$$\lceil y \rceil \;\equiv\; \bigvee x.(\neg x \wedge (y \vee x))$$

while Theorem 1 shows that $\lceil y \rceil$ is not expressible without quantifiers. On the other hand, this result implies that we cannot use quantifier elimination to compute quantifier-free interpolants; such interpolants require a different approach.

4   Failure of a Refutation-Based Interpolation

We now consider a notion of interpolation based on orthologic with axioms. Using axioms makes the assumptions stronger and is a closer approximation of classical propositional logic.

Definition 9 (Refutation-Based Interpolation).   Given an inconsistent pair of sequents A and B, i.e. there exists a proof of contradiction (the empty sequent) assuming them, a sequent I is said to be a refutation-based interpolant of (A, B) if

- I can be deduced from A alone,
- I and B are inconsistent, and
- $\mathsf{FV}(I) \subseteq \mathsf{FV}(A) \cap \mathsf{FV}(B)$.

We show, by counterexample, that such an interpolant does not exist in general in orthologic.

Theorem 2. *Given any inconsistent pair (A, B) of sequents, a refutation-based interpolant for it does not necessarily exist in orthologic. In particular, a refutation-based interpolant does not exist for the choice $A = (z \vee \neg y) \wedge (\neg z \vee \neg y)^R$ and $B = (x \wedge y) \vee (\neg x \wedge y)^R$.*

Proof. For the counterexample, let A be the sequent

$$(z \vee \neg y) \wedge (\neg z \vee \neg y)^R$$

and let B be

$$(x \wedge y) \vee (\neg x \wedge y)^R .$$

We show the proof of inconsistency of A and B in orthologic in Fig. 3. For readability and space constraints, the proof is split into four parts, with Fig. 3a showing a proof of y^R from B, Fig. 3b and Fig. 3c showing proofs of z^R and $\neg z^R$ respectively from y^R and A, and Fig. 3d finally deriving the empty sequent from z^R and $\neg z^R$.

Given the inconsistent pair of A and B, however, we find that the only common variable between A and B is y. Sequents built only from the variable y are equivalent to one of $0^R, 1^R, y^R, (\neg y)^R$, none of which is a consequence of A while also being inconsistent with B.

Hence, the inconsistent sequent pair A, B as chosen does not admit a refutation-based interpolant according to Definition 9.

5   Interpolation for Orthologic Formulas

An arguably more natural definition of interpolation for a lattice-based logic (such as orthologic) using the $\leq$ relation is the following:

$$\cfrac{(x \wedge y) \vee (\neg x \wedge y)^R \qquad \cfrac{\cfrac{\cfrac{\cfrac{}{y^L, y^R}\ \text{Hyp}}{x \wedge y^L, y^R}\ \text{LeftAnd} \qquad \cfrac{\cfrac{\cfrac{}{y^L, y^R}\ \text{Hyp}}{\neg x \wedge y^L, y^R}\ \text{LeftAnd}}{}\ \text{LeftOr}}{(x \wedge y) \vee (\neg x \wedge y)^L, y^R}}{}}{y^R}\ \text{Cut}$$

(a) Proof of y^R from B.

$$\cfrac{(\neg z \vee \neg y) \wedge (z \vee \neg y)^R \qquad \cfrac{\cfrac{\cfrac{}{z^L, z^R}\ \text{Hyp} \qquad \cfrac{\cfrac{\cfrac{y^R}{\neg y^L}\ \text{LeftNot}}{\neg y^L, z^R}\ \text{Weaken}}{}}{z \vee \neg y^L, z^R}\ \text{LeftOr}}{(\neg z \vee \neg y) \wedge (z \vee \neg y)^L, z^R}\ \text{LeftAnd}}{z^R}\ \text{Cut}$$

(b) Proof of z^R from y^R and A.

$$\cfrac{(\neg z \vee \neg y) \wedge (z \vee \neg y)^R \qquad \cfrac{\cfrac{\cfrac{}{\neg z^L, \neg z^R}\ \text{Hyp} \qquad \cfrac{\cfrac{\cfrac{y^R}{\neg y^L}\ \text{LeftNot}}{\neg y^L, \neg z^R}\ \text{Weaken}}{}}{\neg z \vee \neg y^L, \neg z^R}\ \text{LeftOr}}{(\neg z \vee \neg y) \wedge (z \vee \neg y)^L, \neg z^R}\ \text{LeftAnd}}{\neg z^R}\ \text{Cut}$$

(c) Proof of $\neg z^R$ from y^R and A.

$$\cfrac{\neg z^R \qquad \cfrac{\cfrac{z^R}{\neg z^L}\ \text{LeftNot}}{}}{\emptyset}\ \text{Cut}$$

(d) Proof of $\emptyset$ from z^R and $\neg z^R$.

Fig. 3. Proof of inconsistency of $A = (\neg z \vee \neg y) \wedge (z \vee \neg y)$ and $B = (x \wedge y) \vee (\neg x \wedge y)$.

Definition 1 (Implicational interpolation). *Given two propositional quantifier-free formulas A and B such that $A \leq B$, an interpolant is a formula I such that $FV(I) \subseteq FV(A) \cap FV(B)$ and $A \leq I \leq B$.*

For classical logic, this definition is equivalent to the one of Sect. 4, since $A \leq_{CL} B$ if and only if the empty sequent is provable from $(A, \neg B)$. These definitions are however not equivalent in intuitionistic logic and orthologic.

We now prove that the theory of ortholattices admits this form of interpolation by showing a procedure constructing the interpolant inductively from a proof of the sequent A^L, B^R. For induction, we prove a slightly more general statement:

Theorem 3 (Interpolant for orthologic sequents). *There exists an algorithm (interpolate in Fig. 4), which, given a proof of a sequent Γ, Δ, computes a formula I called an* interpolant *for the ordered pair (Γ, Δ) such that $FV(I) \subseteq FV(\Gamma) \cap FV(\Delta)$ and the sequents Γ, I^R and I^L, Δ are provable. The algorithm has runtime linear in the size of the given proof of Γ, Δ.*

Note that interpolants for (Γ, Δ) and for (Δ, Γ) are distinct. In fact, in orthologic, they are negations of each other. Note also that if the sequent Γ, Δ is provable, then by [16] there is a proof of it of at most quadratic size. Hence, the interpolation algorithm runs in the worst-case in time quadratic in sizes of Γ and Δ.

We have made an executable Scala implementation of the interpolation algorithm alongside orthologic proof search (as described in [16]) open-source on GitHub[1].

Proof. We show correctness of the algorithm in Fig. 4 with inputs Γ, Δ, P where P is a proof of the sequent $S = \Gamma, \Delta$. By the cut-elimination theorem of orthologic [16,40], we assume that P is cut-free. We show that the result of interpolate(Γ, Δ, P) is an interpolant for (Γ, Δ).

We first deal with the particular case where either Γ or Δ is empty or when $\Gamma = \Delta$, as it will simplify the rest of the proof to assume that they are both non-empty and distinct.

- Suppose $(\Gamma, \Delta) = (\Pi, \emptyset)$. Then 0 is an interpolant, as both $\Pi, 0^R$ and $0^L, \emptyset$ are provable.
- Suppose $(\Gamma, \Delta) = (\emptyset, \Pi)$. Then 1 is an interpolant, as both $\emptyset, 1^R$ and $1^L, \Pi$ are provable
- Suppose $(\Gamma, \Delta) = (\Pi, \Pi)$. Then any formula ψ (in particular 0 and 1) is an interpolant as both Π, ψ^R and ψ^L, Π are provable by weakening.

In all other cases, the algorithm works recursively on the proof tree of P, starting from the concluding (root) step. At every step, the algorithm reduces the construction of the interpolant of S to those of its premises. By induction, assume that for a given proof P, the algorithm is correct for all proofs of smaller size (and in particular for the premises of P) and consider every proof step from Fig. 1 with which P can be concluded:

- Hyp: suppose the concluding step is

$$\frac{}{\phi^L, \phi^R} \text{ Hyp },$$

We must have $(\Gamma, \Delta) = (\phi^L, \phi^R)$, or $(\Gamma, \Delta) = (\phi^R, \phi^L)$. Assuming the former, consider the interpolant $I = \phi$. We then trivially have proofs of $(\Gamma, I^R) = (\phi^R, \phi^L)$ and $(I^L, \Delta^R) = (\phi^L, \phi^R)$:

$$\frac{}{\phi^L, \phi^R} \text{ Hyp }, \quad \text{and} \quad \frac{}{\phi^L, \phi^R} \text{ Hyp },$$

---

[1] https://github.com/sankalpgambhir/ol-interpolation.

```
1   def interpolate(
2       Γ: Option[AnnotatedFormula],
3       Δ: Option[AnnotatedFormula], // the input sequent Γ, Δ
4       p: ProofStep // proof of validity of the input sequent
5   ): Formula =
6   (Γ, Δ) match
7       case (Some(Π), None) ⇒ 0
8       case (None, Some(Π)) ⇒ 1
9       case (Some(Π), Some(Π)) ⇒ 0 // or 1
10      case _ ⇒
11          p match
12              case Hypothesis(ϕ) ⇒
13                  Γ match
14                      case Some(`ϕ`^L) ⇒ ϕ
15                      case Some(`ϕ`^R) ⇒ ¬ϕ
16              case Weaken(Σ, p') ⇒
17                  Γ match
18                      case `Σ` ⇒ interpolate(None, Δ, p')
19                      case _ ⇒ interpolate(Γ, None, p')
20              case LeftAnd(ϕ, ψ, p') ⇒
21                  Γ match
22                      case Some(`ϕ` ∧ `ψ`^L) ⇒ interpolate(ϕ^L, Δ, p')
23                      case _ ⇒ interpolate(Γ, ϕ^L, p')
24              case RightAnd(ϕ, ψ, p_1, p_2) ⇒
25                  Γ match
26                      case Some(`ϕ` ∧ `ψ`^R) ⇒
27                          interpolate(ϕ^R, Δ, p_1) ∨ interpolate(ψ^R, Δ, p_2)
28                      case _ ⇒
29                          interpolate(Δ, ϕ^R, p_1) ∧ interpolate(Δ, ψ^R, p_2)
30              case LeftOr(ϕ, ψ, p_1, p_2) ⇒
31                  Γ match
32                      case Some(`ϕ` ∨ `ψ`^L) ⇒
33                          interpolate(ϕ^L, Δ, p_1) ∨ interpolate(ψ^L, Δ, p_2)
34                      case _ ⇒
35                          interpolate(Δ, ϕ^L, p_1) ∧ interpolate(Δ, ψ^L, p_2)
36              case RightOr(ϕ, ψ, p') ⇒
37                  Γ match
38                      case Some(`ϕ` ∨ `ψ`^R) ⇒ interpolate(ϕ^R, Δ, p')
39                      case _ ⇒ interpolate(Γ, ϕ^R, p_1)
40              case LeftNot(ϕ, p') ⇒
41                  Γ match
42                      case Some(¬`ϕ`^L) ⇒ interpolate(ϕ^R, Δ, p')
43                      case _ ⇒ interpolate(Γ, ϕ^R, p')
44              case RightNot(ϕ, p') ⇒
45                  Γ match
46                      case Some(¬`ϕ`^R) ⇒ interpolate(ϕ^L, Δ, p')
47                      case _ ⇒ interpolate(Γ, ϕ^L, p')
```

Fig. 4. The algorithm interpolate to produce an interpolant for any valid sequent, given a partition as an ordered pair and a proof. `ϕ` in a pattern match is Scala syntax to indicate that ϕ is an existing variable to be tested for equality, and not a fresh variable free to be assigned.

The latter case is symmetrical, with $I = \neg\phi$.

– Weaken: suppose the final inference is

$$\frac{\Pi}{\Pi, \Sigma} \text{ Weaken } .$$

As before, we must have $(\Gamma, \Delta) = (\Pi, \Sigma)$ or $(\Gamma, \Delta) = (\Sigma, \Pi)$. In the former case, consider the interpolant C for $(\Pi, \emptyset)$, the premise (in fact $C = 0$ or $C = 1$). By the hypothesis, the sequents

$$\Pi, C^R \text{ , and } \quad C^L, \emptyset$$

are provable. Taking $I = C$, and applying Weaken on the second sequent, we obtain proofs of Γ, I^R and I^L, Δ:

$$\Pi, C^R \quad \text{ and } \quad \frac{C^L}{C^L, \Delta} \text{ Weaken}$$

The case $\Gamma = \Delta$ is analogous, with $I = \neg C$.

– LeftAnd: suppose the final inference is

$$\frac{\Pi, \phi^L}{\Pi, \phi \wedge \psi^L} \text{ LeftAnd } .$$

We must have $(\Gamma, \Delta) = ((\phi \wedge \psi)^L, \Pi)$ or swapped. In the former case, by the induction hypothesis, consider an interpolant C for (ϕ^L, Π), such that the sequents

$$\phi^L, C^R \qquad C^L, \Pi$$

are provable. For $I = C$ as interpolant, we have proofs of Γ, I^R and I^L, Δ:

$$\frac{\phi^L, C^R}{\phi \wedge \psi^L, C^R} \text{ LeftAnd } \quad \text{ and } \quad C^L, \Pi \quad .$$

Since $\mathsf{FV}(\phi) \subseteq \mathsf{FV}(\phi \wedge \psi)$, $I = C$ is an interpolant for the conclusion as required. The case where $(\Gamma, \Delta) = (\Pi, (\phi \wedge \psi)^L)$ is analogous.

– RightAnd: suppose the final inference is

$$\frac{\Pi, \phi^R \qquad \Pi, \psi^R}{\Pi, (\phi \wedge \psi)^R} \text{ RightAnd } .$$

We have $(\Gamma, \Delta) = (\Pi, (\phi \wedge \psi)^R)$, or the other way round. Assume the former. Applying the induction hypothesis twice, we obtain an interpolant for each of the premises, C_ϕ and C_ψ, such that the sequents

$$\Pi, C_\phi^R \qquad C_\phi^L, \phi^R$$
$$\Pi, C_\psi^R \qquad C_\psi^L, \psi^R$$

are valid. Take $I = C_\phi \wedge C_\psi$ as interpolant. Indeed, its free variables are contained in $\mathsf{FV}(\Pi) \cap (\mathsf{FV}(\phi) \cup \mathsf{FV}(\psi)) = \mathsf{FV}(\Gamma) \cap \mathsf{FV}(\phi \wedge \psi)$.
We then need proofs for Γ, I^R and I^L, Δ:

$$\frac{\Pi, C_\phi^R \qquad \Pi, C_\psi^R}{\Pi, (C_\phi \wedge C_\psi)^R} \text{ RightAnd}$$

and

$$\frac{\dfrac{C_\phi^L, \phi^R}{C_\phi \wedge C_\psi^L, \phi^R} \text{ LeftAnd} \qquad \dfrac{C_\psi^L, \psi^R}{C_\phi \wedge C_\psi^L, \psi^R} \text{ LeftAnd}}{(C_\phi \wedge C_\psi)^L, (\phi \wedge \psi)^R} \text{ RightAnd.}$$

showing that I is an interpolant for the pair (Γ, Δ).
Now in the other case $(\Gamma, \Delta) = ((\phi \wedge \psi)^R, \Pi)$, the induction hypothesis gives us the following interpolants:

$$\begin{array}{cc} \phi^R, D_\phi^R & D_\phi^L, \Pi \\ \psi^R, D_\psi^R & D_\psi^L, \Pi \end{array}$$

We can then take $I = (D_\phi \vee D_\psi)$ to obtain proofs of Γ, I^R and I^L, Δ:

$$\frac{\dfrac{\phi^R, D_\phi^R}{\phi^R, (D_\phi \vee D_\psi)^R} \text{ RightOr} \qquad \dfrac{\psi^R, D_\psi^R}{\psi^R, (D_\phi \vee D_\psi)^R} \text{ RightOr}}{(\phi \wedge \psi)^R, (D_\phi \vee D_\psi)^R} \text{ RightAnd.}$$

and

$$\frac{D_\phi^L, \Pi \qquad D_\psi^L, \Pi}{(D_\phi \vee D_\psi)^L, \Pi} \text{ LeftOr}$$

Note that we can show by induction that $D_\phi \vee D_\psi = \neg(C_\phi \wedge C_\psi)$.
– LeftNot: suppose the final inference is

$$\frac{\Pi, \phi^R}{\Pi, \neg\phi^L} \text{ LeftNot .}$$

We have $(\Gamma, \Delta) = (\Pi, (\neg\phi)^L)$, or the other way round. Assume the former. We apply the induction hypothesis as before to obtain an interpolant C for (Π, ϕ^R) such that

$$\Pi, C^R \qquad\qquad C^L, \phi^R$$

are valid. $I = C$ suffices as an interpolant for the concluding sequent, with the proofs of Γ, I^R and I^L, Δ

$$\Pi, C^R \quad \text{and} \quad \frac{C^L, \phi^R}{C^L, \neg\phi^L} \text{ LeftNot} \ .$$

The proofs for the remaining proof rules, LeftOr, RightOr, and RightNot, are analogous to the cases listed above.

Corollary 2 (Interpolation for Ortholattices). *Ortholattices admit interpolation, i.e., for any pair of formulas A, B in an ortholattice with $A \le B$, there exists a formula I such that $A \le I$ and $I \le B$, with $FV(I) \subseteq FV(A) \cap FV(B)$.*

6 Further Related Work

The best known interpolation result is Craig's interpolation theorem for first order logic [10], of which interpolation for classical propositional logic is a special case. Interpolation for predicate intuitionistic logic was first shown by [41]. The propositional case was further studied by [27].

Interpolation can be leveraged, among other applications, to solve constrained Horn clauses [30], for model checking [3,31] or for invariant generation [23,33]. Interpolants are computed by many existing solvers and provers such as Eldarica [19], Vampire [18] and Wolverine [25].

A sequent calculus proof system for orthologic was first described in [40], with cut elimination. [7] studied implication symbols in orthologic. [35] showed that orthologic with axioms is decidable. [36] showed that Orthologic admits the super amalgamation property, which implies that it admits interpolants. They also show that a predicate logic extension to orthologic admits interpolants, although the proof of both theorems are non-constructive and contain no discussion of algorithms or space and time complexity. Recently, orthologic was used in practice in a proof assistant [15], for modelling of epistemic logic [20,21] and for normalizing formulas in software verification [14].

7 Conclusion

We showed that quantified orthologic, with a sequent-based proof system, is sound and complete with respect to all complete ortholattices. A soundness and completeness theorem typically allows demonstrating further provability results by using semantic arguments. We then showed that orthologic does not admit, in general, quantifier elimination. If such a procedure existed, it would have also allowed computing strongest and weakest interpolants. We instead presented an efficient algorithm, computing orthologic interpolants for two formulas, given a proof that one formula implies the other. Computing interpolants is a key part of some algorithms in model checking and program verification. Since orthologic has efficient algorithms to decide validity and compute proofs, which are necessary to compute interpolants, we expect that our present results will allow further development of orthologic-based tools and efficient algorithms for model checking and program verification.

References

1. Bell, J.L.: Orthologic, Forcing, and The Manifestation of Attributes. In: Chong, C.T., Wicks, M.J. (eds.) Studies in Logic and the Foundations of Mathematics. Studies in Logic and the Foundations of Mathematics, vol. 111, pp. 13–36. Elsevier, Singapore (1983). https://doi.org/10.1016/S0049-237X(08)70953-4
2. Birkhoff, G., Von Neumann, J.: The logic of quantum mechanics. Ann. Math. **37**(4), 823–843 (1936). https://doi.org/10.2307/1968621
3. Bradley, A.R.: SAT-based model checking without unrolling. In: Jhala, R., Schmidt, D. (eds.) VMCAI 2011. LNCS, vol. 6538, pp. 70–87. Springer, Heidelberg (2011). https://doi.org/10.1007/978-3-642-18275-4_7
4. Bruns, G., Godefroid, P.: Model checking with multi-valued logics. In: Díaz, J., Karhumäki, J., Lepistö, A., Sannella, D. (eds.) ICALP 2004. LNCS, vol. 3142, pp. 281–293. Springer, Heidelberg (2004). https://doi.org/10.1007/978-3-540-27836-8_26
5. Bruns, G.: Free Ortholattices. Can. J. Math. **28**(5), 977–985 (1976). https://doi.org/10.4153/CJM-1976-095-6
6. Brzozowski, J.: De Morgan bisemilattices. In: Proceedings 30th IEEE International Symposium on Multiple-Valued Logic (ISMVL 2000), pp. 173–178 (2000). https://doi.org/10.1109/ISMVL.2000.848616
7. Chajda, I., Halaš, R.: An implication in orthologic. Int. J. Theor. Phys. **44**(7), 735–744 (2005). https://doi.org/10.1007/s10773-005-7051-1
8. Cook, S., Morioka, T.: Quantified propositional calculus and a second-order theory for NC1. Arch. Math. Logic **44**(6), 711–749 (2005). https://doi.org/10.1007/s00153-005-0282-2
9. Cousot, P., Cousot, R.: Abstract interpretation: a unified lattice model for static analysis of programs by construction or approximation of fixpoints. In: Proceedings of the 4th ACM SIGACT-SIGPLAN Symposium on Principles of Programming Languages, pp. 238–252. POPL '77, Association for Computing Machinery, New York, NY, USA (1977). https://doi.org/10.1145/512950.512973
10. Craig, W.: Three uses of the Herbrand-Gentzen theorem in relating model theory and proof theory. J. Symb. Log. **22**(3), 269–285 (1957). https://doi.org/10.2307/2963594
11. Dudenhefner, A., Rehof, J.: A Simpler Undecidability Proof for System F Inhabitation. In: TYPES, p. 11. Schloss Dagstuhl - Leibniz-Zentrum fuer Informatik GmbH, Wadern/Saarbruecken, Germany (2019). https://doi.org/10.4230/LIPICS.TYPES.2018.2
12. Gentzen, G.: Untersuchungen über das logische Schließen I. Math. Z. **39**, 176–210 (1935)
13. Girard, J.Y., Taylor, P., Lafont, Y.: Proofs and Types. Cambridge University Press, New York, USA (1989)
14. Guilloud, S., Bucev, M., Milovančević, D., Kunčak, V.: Formula Normalizations in Verification. In: Enea, C., Lal, A. (eds.) Computer Aided Verification, pp. 398–422. Springer Nature Switzerland, Cham (2023). https://doi.org/10.1007/978-3-031-37709-9_19
15. Guilloud, S., Gambhir, S., Kunčak, V.: LISA - A modern proof system. In: Naumowicz, A., Thiemann, R. (eds.) 14th International Conference on Interactive Theorem Proving (ITP 2023). Leibniz International Proceedings in Informatics (LIPIcs), vol. 268, pp. 17:1–17:19. Schloss Dagstuhl - Leibniz-Zentrum für Informatik, Dagstuhl, Germany (2023). https://doi.org/10.4230/LIPIcs.ITP.2023.17, https://drops.dagstuhl.de/opus/volltexte/2023/18392

16. Guilloud, S., Kunčak, V.: Orthologic with axioms. Proc. ACM Program. Lang. 8(POPL) (2024)

17. Henzinger, T.A., Jhala, R., Majumdar, R., McMillan, K.L.: Abstractions from proofs. In: Jones, N.D., Leroy, X. (eds.) Proceedings of the 31st ACM SIGPLAN-SIGACT Symposium on Principles of Programming Languages, POPL 2004, Venice, Italy, January 14–16, 2004, pp. 232–244. ACM (2004). https://doi.org/10.1145/964001.964021

18. Hoder, K., Kovács, L., Voronkov, A.: Interpolation and Symbol Elimination in Vampire. In: Giesl, J., Hähnle, R. (eds.) Automated Reasoning, 5th International Joint Conference, IJCAR 2010, Edinburgh, UK, July 16–19, 2010. Proceedings. Lecture Notes in Computer Science, vol. 6173, pp. 188–195. Springer (2010). https://doi.org/10.1007/978-3-642-14203-1_16

19. Hojjat, H., Rummer, P.: The ELDARICA Horn Solver. In: 2018 Formal Methods in Computer Aided Design (FMCAD), pp. 1–7 (2018). https://doi.org/10.23919/FMCAD.2018.8603013

20. Holliday, W.H.: A fundamental non-classical logic. Logics 1(1), 36–79 (2023). https://doi.org/10.3390/logics1010004

21. Holliday, W.H., Mandelkern, M.: The Orthologic of Epistemic Modals (2022). https://doi.org/10.48550/ARXIV.2203.02872

22. Kalmbach, G.: Orthomodular Lattices. Academic Press Inc, London; New York (1983)

23. Kovács, L., Voronkov, A.: Finding Loop Invariants for Programs over Arrays Using a Theorem Prover. In: Chechik, M., Wirsing, M. (eds.) Fundamental Approaches to Software Engineering, pp. 470–485. Lecture Notes in Computer Science, Springer, Berlin, Heidelberg (2009). https://doi.org/10.1007/978-3-642-00593-0_33

24. Kozen, D.: Complexity of Boolean algebras. Theor. Comput. Sci. 10, 221–247 (1980). https://doi.org/10.1016/0304-3975(80)90048-1

25. Kroening, D., Weissenbacher, G.: Interpolation-based software verification with WOLVERINE. In: Gopalakrishnan, G., Qadeer, S. (eds.) CAV 2011. LNCS, vol. 6806, pp. 573–578. Springer, Heidelberg (2011). https://doi.org/10.1007/978-3-642-22110-1_45

26. Kupferman, O., Lustig, Y.: Lattice automata. In: Cook, B., Podelski, A. (eds.) VMCAI 2007. LNCS, vol. 4349, pp. 199–213. Springer, Heidelberg (2007). https://doi.org/10.1007/978-3-540-69738-1_14

27. de Lavalette, G.R.R.: Interpolation in fragments of intuitionistic propositional logic. J. Symbolic Logic 54(4), 1419–1430 (1989). https://doi.org/10.2307/2274823

28. MacNeille, H.M.: Partially ordered sets. Trans. Am. Math. Soc. 42(3), 416–460 (1937). https://doi.org/10.1090/S0002-9947-1937-1501929-X

29. Madsen, M., Yee, M.H., Lhoták, O.: From Datalog to flix: a declarative language for fixed points on lattices. In: Proceedings of the 37th ACM SIGPLAN Conference on Programming Language Design and Implementation, pp. 194–208 (2016). https://doi.org/10.1145/2908080.2908096

30. McMillan, K., Rybalchenko, A.: Solving Constrained Horn Clauses using Interpolation. Tech. rep, Microsoft Research (2013)

31. McMillan, K.L.: Interpolation and SAT-based model checking. In: Hunt, W.A., Somenzi, F. (eds.) CAV 2003. LNCS, vol. 2725, pp. 1–13. Springer, Heidelberg (2003). https://doi.org/10.1007/978-3-540-45069-6_1

32. McMillan, K.L.: Interpolants and symbolic model checking. In: Cook, B., Podelski, A. (eds.) VMCAI 2007. LNCS, vol. 4349, pp. 89–90. Springer, Heidelberg (2007). https://doi.org/10.1007/978-3-540-69738-1_6

33. McMillan, K.L.: Quantified invariant generation using an interpolating saturation prover. In: Ramakrishnan, C.R., Rehof, J. (eds.) TACAS 2008. LNCS, vol. 4963, pp. 413–427. Springer, Heidelberg (2008). https://doi.org/10.1007/978-3-540-78800-3_31
34. McMillan, K.L.: Interpolation and model checking. In: Handbook of Model Checking, pp. 421–446. Springer, Cham (2018). https://doi.org/10.1007/978-3-319-10575-8_14
35. Meinander, A.: A solution of the uniform word problem for ortholattices. Math. Struct. Comput. Sci. **20**(4), 625–638 (2010). https://doi.org/10.1017/S0960129510000125
36. Miyazaki, Y.: The super-amalgamation property of the variety of ortholattices. Reports Math. Log. **33**, 45–63 (1999)
37. Nielson, F., Nielson, H.R., Hankin, C.: Principles of Program Analysis. Springer, Berlin Heidelberg (1999). https://doi.org/10.1007/978-3-662-03811-6
38. Pudlák, P.: The lengths of proofs. In: Studies in Logic and the Foundations of Mathematics, vol. 137, pp. 547–637. Elsevier (1998). https://doi.org/10.1016/S0049-237X(98)80023-2
39. Rümmer, P., Hojjat, H., Kuncak, V.: Disjunctive interpolants for horn-clause verification. In: Computer Aided Verification (CAV) (2013)
40. Schulte Mönting, J.: Cut elimination and word problems for varieties of lattices. Algebra Univers. **12**(1), 290–321 (1981). https://doi.org/10.1007/BF02483891
41. Schütte, K.: Der Interpolationssatz der intuitionistischen Prädikatenlogik. Math. Ann. **148**(3), 192–200 (1962). https://doi.org/10.1007/BF01470747
42. Sørensen, M., Urzyczyn, P.: Lectures on the curry-howard isomorphism. Stud. Logic Found. Math. **149** (2010). https://doi.org/10.1016/S0049-237X(06)80005-4
43. Urzyczyn, P.: Inhabitation in typed lambda-calculi (a syntactic approach). In: de Groote, P., Roger Hindley, J. (eds.) TLCA 1997. LNCS, vol. 1210, pp. 373–389. Springer, Heidelberg (1997). https://doi.org/10.1007/3-540-62688-3_47

Function Synthesis for Maximizing Model Counting

Thomas Vigouroux[ID], Marius Bozga[✉][ID], Cristian Ene[ID],
and Laurent Mounier[ID]

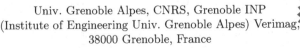

Univ. Grenoble Alpes, CNRS, Grenoble INP
(Institute of Engineering Univ. Grenoble Alpes) Verimag,
38000 Grenoble, France
{thomas.vigouroux,marius.bozga,cristian.ene,
laurent.mounier}@imag.fr

Abstract. Given a boolean formula $\phi(X, Y, Z)$, the *Max#SAT* problem [10,29] asks for finding a partial model on the set of variables X, maximizing its number of projected models over the set of variables Y. We investigate a strict generalization of *Max#SAT* allowing dependencies for variables in X, effectively turning it into a synthesis problem. We show that this new problem, called *DQMax#SAT*, subsumes both the *DQBF* [23] and *DSSAT* [19] problems. We provide a general resolution method, based on a reduction to *Max#SAT*, together with two improvements for dealing with its inherent complexity. We further discuss a concrete application of *DQMax#SAT* for symbolic synthesis of adaptive attackers in the field of program security. Finally, we report preliminary results obtained on the resolution of benchmark problems using a prototype *DQMax#SAT* solver implementation.

Keywords: Function synthesis · Model counting · *Max#SAT* ·
DQBF · *DSSAT* · Adaptive attackers

1 Introduction

A major concern in software security are active adversaries, i.e., adversaries that can *interact* with a target program by feeding inputs. Moreover, these adversaries can often make observations about the program execution through side-channels and/or legal outputs. In this paper, we consider *adaptive* adversaries, i.e., adversaries that choose their inputs by taking advantage of previous observations.

In order to get an upper bound of the insecurity of a given program with respect to this class of adversaries, a possible approach is to synthesize the *best* adaptive attack strategy. This can be modelled as finding a function A (corresponding to the adversarial strategy) satisfying some logical formula Φ (capturing some combination of attack objectives). Actually, this corresponds to a classical functional synthesis problem.

This work was supported by the French national projects TAVA (ANR-20-CE25-0009) and SECUREVAL (ANR-22-PECY-05).

Informally, in our case, given a Boolean relation Φ between output variables (observables) and input variables (attacker provided), our goal is to synthesize each input variable as a function on preceeding outputs satisfying Φ. In the literature, this synthesis problem is captured by the so-called Quantified Boolean Formulae (QBF) satisfiability problem [11,12] and its generalization, the Dependency Quantified Boolean Formulae ($DQBF$) satisfiability problem [23].

These existing qualitative frameworks are not sufficient in a security context: we are not only interested by adversaries able to succeed *in all cases*, but rather for adversaries succeeding with "a good probability". The Stochastic SAT ($SSAT$) problem [21] was therefore proposed and replaces the classical universal (resp. existential) quantifiers by *counting* (resp. *maximizing*) quantifiers. This corresponds to finding the optimal inputs, depending on preceeding outputs, that maximize the number of models of Φ, hence the succeeding probability of the attack. More recently, the Dependency Stochastic SAT ($DSSAT$) problem [19] has been proposed as a strict generalization of the $SSAT$ problem by allowing explicit dependencies for maximizing variables, in a similar way the $DQBF$ problem generalizes the QBF problem.

Nonetheless, an additional complication is hindering the use of quantitative stochastic frameworks in our security context. In general, the output variables in a program may hold expressions computed from one or more secret variables. Consequently, they rarely translate as counting variables in a stochastic formula. Most likely, the above-mentioned secret variables translate into counting variables whereas the observable variables need to be projected out when counting the models. Yet, the output variables are mandatory to express the knowledge available and the dependencies for synthesizing the attacker's optimal inputs.

As an example, we are interested in solving counting problems of the form:

$$\max^{\{z_1\}} x_1.\ \max^{\{z_2\}} x_2.\ Яy_1.\ Яy_2.\ \exists z_1.\ \exists z_2.$$

$$(x_1 \Rightarrow y_2) \wedge (y_1 \Rightarrow x_2) \wedge (y_1 \vee z_2 \Leftrightarrow y_2 \wedge z_1)$$

which involve three types of quantified variables and which are interpreted as follows: synthesize for x_1 (resp. x_2) a boolean expression e_1 (resp. e_2), depending only on z_1 (resp. z_2), such that the formula obtained after replacing x_i by e_i has a maximal number of *partial* models over the counting variables y_1, y_2.

Notice that this problem generalizes in a non-trivial way three well-known existing problems: (i) it generalizes the $Max\#SAT$ problem [10,29] by allowing the maximizing variables to depend *symbolically* on other variables; (ii) it lifts the $DQBF$ problem [23] to a quantitative problem, we do not want to check if there exist expressions e_i working for all y_1, y_2, but to find expressions e_i maximizing the number of models on y_1, y_2; (iii) it extends the $DSSAT$ problem [19] with the additional category of *existential* variables, which can occur in the dependencies of maximizing variables, but which are projected for model counting.

Our contributions are the following:

- We introduce formally the $DQMax\#SAT$ problem as a new problem that arises naturally in the field of software security, and we show that it subsumes the $Max\#SAT$, $DQBF$ and $DSSAT$ problems.

- We develop a general resolution method based on a reduction to *Max#SAT* and further propose two improvements in order to deal with its inherent complexity: (i) an incremental method, that enables anytime resolution; (ii) a local method, allowing to split the initial problem into independent smaller sub-problems, enabling parallel resolution.
- We provide applications of *DQMax#SAT* to software security: we show that *quantitative robustness* [3] and *programs as information leakage-channels* [24, 26] can be systematically cast as instances of the *DQMax#SAT* problem.
- We provide a first working prototype solver for the *DQMax#SAT* problem and we apply it to the examples considered in this paper.

The paper is organized as follows. Section 2 introduces formally the *DQMax#SAT* problem and its relation with the *Max#SAT*, *DQBF* and *DSSAT* problems. Sections 3 to 5 present the three different approaches we propose for solving *DQMax#SAT*. Section 6 shows concrete applications of *DQMax#SAT* in software security, that is, for the synthesis of adaptive attackers. Finally, Sect. 7 provides preliminary experimental results obtained with our prototype *DQMax#SAT* solver. Section 8 discusses some references to related work and Sect. 9 concludes and proposes some extensions to address in the future.

2   Problem Statement

2.1   Preliminaries

Given a set V of Boolean variables, we denote by $\mathcal{F}\langle V \rangle$ (resp. $\mathcal{M}\langle V \rangle$) the set of Boolean formulae (resp. complete monomials) over V. A model of a boolean formula $\phi \in \mathcal{F}\langle V \rangle$ is an assignement $\alpha_V : V \to \mathbb{B}$ of variables to Boolean values such that ϕ evaluates to $\top$ (that is, *true*) on α_V, it is denoted by $\alpha_V \models \phi$. A formula is satisfiable if it has at least one model α_V. A formula is valid (i.e., tautology) if any assignement α_V is a model.

Given a formula $\phi \in \mathcal{F}\langle V \rangle$ we denote by $|\phi|_V$ the number of its models, formally $|\phi|_V \overset{def}{=} |\{\alpha_V : V \to \mathbb{B} \mid \alpha_V \models \phi\}|$. For a partitioning $V = V_1 \uplus V_2$ we denote by $|\exists V_2.\ \phi|_{V_1}$ the number of its V_1-projected models, formally $|\exists V_2.\ \phi|_{V_1} \overset{def}{=} |\{\alpha_{V_1} : V_1 \to \mathbb{B} \mid \exists \alpha_{V_2} : V_2 \to \mathbb{B}.\ \alpha_{V_1} \uplus \alpha_{V_2} \models \phi\}|$. Note that $|\exists V_2.\ \phi|_{V_1} \leq |\phi|_V$ in general, with equality only in some restricted situations (e.g. when V_1 is an independent support of the formula [6]).

Let V, V', V'' be arbitrary sets of Boolean variables. Given a Boolean formula $\phi \in \mathcal{F}\langle V \rangle$ and a substitution $\sigma : V' \to \mathcal{F}\langle V'' \rangle$ we denote by $\phi[\sigma]$ the Boolean formula in $\mathcal{F}\langle (V \setminus V') \cup V'' \rangle$ obtained by replacing in ϕ all occurrences of variables v' from V' by the associated formula $\sigma(v')$.

2.2   Problem Formulation

Definition 1 (*DQMax#SAT* Problem). *Let $X = \{x_1, ..., x_n\}$, Y, Z be pairwise disjoint finite sets of Boolean variables, called respectively* maximizing, counting *and* existential *variables. The DQMax#SAT problem is specified* *as:*

$$\max^{H_1} x_1. \ ... \ \max^{H_n} x_n. \ \text{Я}Y. \ \exists Z. \ \Phi(X,Y,Z) \tag{1}$$

where $H_1, ..., H_n \subseteq Y \cup Z$ and $\Phi \in \mathcal{F}\langle X \cup Y \cup Z \rangle$ are respectively the dependencies *of maximizing variables and the* objective *formula.*

A solution to the problem is a substitution $\sigma_X^* : X \to \mathcal{F}\langle Y \cup Z \rangle$ associating formulae on counting and existential variables to maximizing variables such that (i) $\sigma_X^*(x_i) \in \mathcal{F}\langle H_i \rangle$, for all $i \in [1,n]$ and (ii) $|\exists Z. \ \Phi[\sigma_X^*]|_Y$ is maximal. That means, the chosen substitution conforms to dependencies on maximizing variables and guarantees the objective holds for the largest number of models projected on the counting variables.

Example 1. Consider the problem:

$$\max^{\{z_1,z_2\}} x_1. \ \text{Я}y_1. \ \text{Я}y_2. \ \exists z_1. \ \exists z_2. \ (x_1 \Leftrightarrow y_1) \wedge (z_1 \Leftrightarrow y_1 \vee y_2) \wedge (z_2 \Leftrightarrow y_1 \wedge y_2)$$

Let Φ denote the objective formula. In this case, $\mathcal{F}\langle\{z_1,z_2\}\rangle = \{\top, \bot, z_1, \overline{z_1}, z_2, \overline{z_2}, z_1 \vee z_2, \overline{z_1} \vee z_2, z_1 \vee \overline{z_2}, \overline{z_1} \vee \overline{z_2}, z_1 \wedge z_2, \overline{z_1} \wedge z_2, z_1 \wedge \overline{z_2}, \overline{z_1} \wedge \overline{z_2}, z_1 \Leftrightarrow z_2, \overline{z_1} \Leftrightarrow \overline{z_2}\}$, and one shall consider every possible substitution. One can compute for instance $\Phi[x_1 \mapsto \overline{z_1} \wedge \overline{z_2}] \equiv ((\overline{z_1} \wedge \overline{z_2}) \Leftrightarrow y_1) \wedge (z_1 \Leftrightarrow y_1 \vee y_2) \wedge (z_2 \Leftrightarrow y_1 \wedge y_2)$ which only has one model $(\{y_1 \mapsto \bot, y_2 \mapsto \top, z_1 \mapsto \top, z_2 \mapsto \bot\})$ and henceforth $|\exists z_1. \ \exists z_2. \ \Phi[x_1 \mapsto \overline{z_1} \wedge \overline{z_2}]|_{\{y_1,y_2\}} = 1$. Overall, for this problem there exists four possible maximizing substitutions σ^* respectively $x_1 \mapsto z_1, x_1 \mapsto z_2, x_1 \mapsto z_1 \vee z_2,$ $x_1 \mapsto z_1 \wedge z_2$ such that for all of them $|\exists z_1. \ \exists z_2. \ \Phi[\sigma^*]|_{\{y_1,y_2\}} = 3$.

Example 2. Let us consider the following problem:

$$\max^{\{z_1\}} x_1. \ \max^{\{z_2\}} x_2. \ \text{Я}y_1. \ \text{Я}y_2. \ \exists z_1. \ \exists z_2.$$
$$(x_1 \Rightarrow y_2) \wedge (y_1 \Rightarrow x_2) \wedge (y_1 \vee z_2 \Leftrightarrow y_2 \wedge z_1)$$

Let Φ denote the objective formula. An optimal solution is $x_1 \mapsto \bot, x_2 \mapsto \overline{z_2}$ and observe that $|\exists z_1. \ \exists z_2. \ \Phi[x_1 \mapsto \bot, x_2 \mapsto \overline{z_2}]|_{\{y_1,y_2\}} = 3$. Moreover, one can notice that there do not exist expressions $e_1 \in \mathcal{F}\langle\{z_1\}\rangle$ (respectively $e_2 \in \mathcal{F}\langle\{z_2\}\rangle$), such that $\exists z_1. \ \exists z_2. \ \Phi[x_1 \mapsto e_1, x_2 \mapsto e_2]$ admits the model $y_1 \mapsto \top, y_2 \mapsto \bot$.

2.3 Hardness of *DQMax#SAT*

We briefly discuss now the relationship between the *DQMax#SAT* problem and the *Max#SAT*, *DQBF* and *DSSAT* problems. It turns out that *DQMax#SAT* is at least as hard as all of them, as illustrated by the following reductions.

DQMax#SAT is At Least as Hard as Max#SAT: Let $X = \{x_1, ..., x_n\}$, Y, Z be pairwise disjoint finite sets of Boolean variables, called *maximizing*, *counting* and *existential* variables. The *Max#SAT* problem [10] specified as

$$\max x_1. \ ... \max x_n. \ \text{Я}Y. \ \exists Z. \ \Phi(X,Y,Z) \tag{2}$$

asks for finding an assignement $\alpha_X^* : X \to \mathbb{B}$ of maximizing variables to Boolean values such that $|\exists Z. \ \Phi[\alpha_X^*]|_Y$ is maximal. It is immediate to see that the

Max#SAT problem is the particular case of the *DQMax#SAT* problem where there are no dependencies, that is, $H_1 = H_2 = ... = H_n = \emptyset$.

DQMax#SAT is At Least as Hard as DQBF: Let $X = \{x_1, ..., x_n\}$, Y be disjoint finite sets of Boolean variables and let $H_1, ..., H_n \subseteq Y$. The *DQBF* problem [23] asks, given a *DQBF* formula:

$$\forall Y. \exists^{H_1} x_1. \ ... \ \exists^{H_n} x_n. \ \Phi(X, Y) \tag{3}$$

to synthesize a substitution $\sigma_X^* : X \to \mathcal{F}\langle Y \rangle$ whenever one exists such that (i) $\sigma_X^*(x_i) \in \mathcal{F}\langle H_i \rangle$, for all $i \in [1, n]$ and (ii) $\Phi[\sigma_X^*]$ is valid. The *DQBF* problem is reduced to the *DQMax#SAT* problem:

$$\max{}^{H_1} x_1. \ ... \ \max{}^{H_n} x_n. \ \text{Я}Y. \ \Phi(X, Y) \tag{4}$$

By solving (4) one can solve the initial *DQBF* problem (3). Indeed, let $\sigma_X^* : X \to \mathcal{F}\langle Y \rangle$ be a solution for (4). Then, the *DQBF* problem admits a solution if and only if $|\Phi[\sigma_X^*]|_Y = 2^{|Y|}$. Moreover, σ_X^* is a solution for the problem (3) because (i) σ_X^* satisfies dependencies and (ii) $\Phi[\sigma_X^*]$ is valid as it belongs to $\mathcal{F}\langle Y \rangle$ and has $2^{|Y|}$ models. Note that through this reduction of *DQBF* to *DQMax#SAT*, the maximizing quantifiers in *DQMax#SAT* can be viewed as Henkin quantifiers [14] in *DQBF* with a quantitative flavor.

DQMax#SAT is At Least as Hard as DSSAT: Let $X = \{x_1, ..., x_n\}$, $Y = \{y_1, ..., y_m\}$ be disjoint finite sets of variables. A *DSSAT* formula is of the form:

$$\max{}^{H_1} x_1. \ ... \ \max{}^{H_n} x_n. \ \text{Я}^{p_1} y_1. \ ... \ \text{Я}^{p_m} y_m. \ \Phi(X, Y) \tag{5}$$

where $p_1, ..., p_m \in [0, 1]$ are respectively the probabilities of variables $y_1, ..., y_m$ to be assigned $\top$ and $H_1, ..., H_n \subseteq Y$ are respectively the dependency sets of variables $x_1, ..., x_n$. Given a *DSSAT* formula (5), the probability of an assignement $\alpha_Y : Y \to \mathbb{B}$ is defined as

$$\mathbb{P}[\alpha_Y] \overset{def}{=} \prod_{i=1}^{m} \begin{cases} p_i & \text{if } \alpha_Y(y_i) = \top \\ 1 - p_i & \text{if } \alpha_Y(y_i) = \bot \end{cases}$$

This definition is lifted to formula $\Psi \in \mathcal{F}\langle Y \rangle$ by summing up the probabilities of its models, that is, $\mathbb{P}[\Psi] \overset{def}{=} \sum_{\alpha_Y \models \Psi} \mathbb{P}[\alpha_Y]$.

The *DSSAT* problem [19] asks, for a given formula (5), to synthesize a substitution $\sigma_X^* : X \to \mathcal{F}\langle Y \rangle$ such that (i) $\sigma_X^*(x_i) \in \mathcal{F}\langle H_i \rangle$, for all $i \in [1, n]$ and (ii) $\mathbb{P}[\Phi[\sigma_X^*]]$ is maximal. If $p_1 = ... = p_m = \frac{1}{2}$ then for any substitution $\sigma_X : X \to \mathcal{F}\langle Y \rangle$ it holds $\mathbb{P}[\Phi[\sigma_X]] = \frac{|\Phi[\sigma_X]|_Y}{2^m}$. In this case, it is immediate to see that solving (5) as a *DQMax#SAT* problem (i.e., by ignoring probabilities) would solve the original *DSSAT* problem. Otherwise, in the general case, one can use existing techniques such as [4] to transform arbitrary *DSSAT* problems (5) into equivalent ones where all probabilities are $\frac{1}{2}$ and solve them as above.

Note that while the reduction above from *DSSAT* to *DQMax#SAT* seems to indicate the two problems are rather similar, a reverse reduction from

$DQMax\#SAT$ to $DSSAT$ seems not possible in general. That is, recall that $DQMax\#SAT$ allows for a third category of *existential* variables Z which can occur in the dependencies sets H_i and which are not used for counting but are projected out. Yet, such problems arise naturally in our application domain as illustrated later in Sect. 6. If no such existential variables exists or if they do not occur in the dependencies sets then one can apriori project them from the objective Φ and syntactically reduce $DQMax\#SAT$ to $DSSAT$ (i.e., adding $\frac{1}{2}$ probabilities on counting variables). However, projecting existential variables in a brute-force way may lead to an exponential blow-up of the objective formula Φ, an issue already explaining the hardness of projected model counting vs model counting [2,18]. Otherwise, in case of dependencies on existential variables, it is an open question if any direct reduction exists as these variables do not fit into the two categories of variables (counting, maximizing) occurring in $DSSAT$ formula.

From the complexity point of view, the decision version of $DQMax\#SAT$ can be shown to be NEXPTIME-complete and hence it lies in the same complexity class as $DQBF$ [22] and $DSSAT$ [19].

Theorem 1. *The decision version of DQMax#SAT is NEXPTIME-complete.*

Proof. Given a $DQMax\#SAT$ problem, notice that the solution corresponding to the optimal substitution for the *maximizing variables* can be guessed and constructed as a truth table in nondeterministic exponential time. Given the guessed substitution, the counting of the number of models projected on the *counting variables* and comparison against the threshold can be performed in exponential time, too. Overall, the whole procedure is done in nondeterministic exponential time, and hence $DQMax\#SAT$ belongs to the NEXPTIME complexity class.

Second, to see why $DQMax\#SAT$ is NEXPTIME-hard, one can check that the above reduction of the NEXPTIME-complete problem $DQBF$ to $DQMax\#SAT$ can be done in polynomial time wrt the size of the initial problem. □

The following proposition provides an upper bound on the number of models corresponding to the solution of (1) computable using projected model counting.

Proposition 1. *For any substitution $\sigma_X : X \to \mathcal{F}\langle Y \cup Z \rangle$ it holds*

$$|\exists Z. \ \Phi[\sigma_X]|_Y \le |\exists X. \ \exists Z. \ \Phi|_Y.$$

3   Global Method

We show in this section that the $DQMax\#SAT$ problem can be directly reduced to a $Max\#SAT$ problem with an exponentially larger number of maximizing variables and exponentially bigger objective formula.

First, recall that any boolean formula $\varphi \in \mathcal{F}\langle H \rangle$ can be written as a finite disjunction of a subset M_φ of complete monomials from $\mathcal{M}\langle H \rangle$, that is, such that the following equivalences hold:

$$\varphi \iff \vee_{m \in M_\varphi} m \iff \vee_{m \in \mathcal{M}\langle H \rangle} (\llbracket m \in M_\varphi \rrbracket \wedge m)$$

Therefore, any formula $\varphi \in \mathcal{F}\langle H \rangle$ is uniquely *encoded* by the set of boolean values $[\![m \in M_\varphi]\!]$ denoting the membership of each complete monomial m to M_φ. We use this idea to encode the substitution of a maximizing variable x_i by some formula $\varphi_i \in \mathcal{F}\langle H_i \rangle$ by using a set of boolean variables $(x'_{i,m})_{m \in \mathcal{M}\langle H_i \rangle}$ denoting respectively $[\![m \in M_{\varphi_i}]\!]$ for all $m \in \mathcal{M}\langle H_i \rangle$. We now define the following *Max#SAT* problem:

$$(\max x'_{1,m} \cdot)_{m \in \mathcal{M}\langle H_1 \rangle} \ ... (\max x'_{n,m} \cdot)_{m \in \mathcal{M}\langle H_n \rangle} \ \text{Я}Y. \ \exists Z. \ \exists X.$$

$$\Phi(X,Y,Z) \wedge \bigwedge_{i \in [1,n]} \left(x_i \Leftrightarrow \vee_{m \in \mathcal{M}\langle H_i \rangle}(x'_{i,m} \wedge m) \right) \quad (6)$$

The next theorem establishes the relation between the two problems.

Theorem 2. $\sigma_X^* = \{x_i \mapsto \varphi_i^*\}_{i \in [1,n]}$ *is a solution to the problem DQMax#SAT (1) if and only if* $\alpha_{X'}^* = \{x'_{i,m} \mapsto [\![m \in M_{\varphi_i^*}]\!]\}_{i \in [1,n], m \in \mathcal{M}\langle H_i \rangle}$ *is a solution to Max#SAT problem (6).*

Proof. Let us denote

$$\Phi'(X',X,Y,Z) \overset{def}{=} \Phi(X,Y,Z) \wedge \bigwedge_{i \in [1,n]} \left(x_i \Leftrightarrow \vee_{m \in \mathcal{M}\langle H_i \rangle}(x'_{i,m} \wedge m) \right)$$

Actually, for any $\Phi \in \mathcal{F}\langle X \cup Y \cup Z \rangle$ for any $\varphi_1 \in \mathcal{F}\langle H_1 \rangle, ..., \varphi_n \in \mathcal{F}\langle H_n \rangle$ the following equivalence is valid:

$$\Phi(X,Y,Z)[\{x_i \mapsto \varphi_i\}_{i \in [1,n]}] \Leftrightarrow$$
$$(\exists X. \ \Phi'(X',X,Y,Z)) \left[\{x'_{i,m} \mapsto [\![m \in M_{\varphi_i}]\!]\}_{i \in [1,n], m \in \mathcal{M}\langle H_i \rangle} \right]$$

Consequently, finding the substitution σ_X which maximize the number of Y-models of the left-hand side formula (that is, of $\exists Z. \ \Phi(X,Y,Z)$) is actually the same as finding the valuation $\alpha_{X'}$ which maximizes the number of Y-models of the right-hand side formula (that is, $\exists Z. \ \exists X. \ \Phi'(X',X,Y,Z)$). □

Example 3. Example 1 is reduced to the following:

$$\max x'_{1,z_1 z_2} \cdot \ \max x'_{1,z_1 \overline{z_2}} \cdot \ \max x'_{1,\overline{z_1} z_2} \cdot \ \max x'_{1,\overline{z_1 z_2}} \cdot \ \text{Я}y_1. \ \text{Я}y_2. \ \exists z_1. \ \exists z_2. \ \exists x_1.$$
$$(x_1 \Leftrightarrow y_1) \wedge (z_1 \Leftrightarrow y_1 \vee y_2) \wedge (z_2 \Leftrightarrow y_1 \wedge y_2) \wedge (x_1 \Leftrightarrow ((x'_{1,z_1 z_2} \wedge z_1 \wedge z_2) \vee$$
$$(x'_{1,z_1 \overline{z_2}} \wedge z_1 \wedge \overline{z_2}) \vee (x'_{1,\overline{z_1} z_2} \wedge \overline{z_1} \wedge z_2) \vee (x'_{1,\overline{z_1 z_2}} \wedge \overline{z_1} \wedge \overline{z_2})))$$

One possible answer is $x'_{1,z_1 z_2} \mapsto \top, x'_{1,z_1 \overline{z_2}} \mapsto \top, x'_{1,\overline{z_1} z_2} \mapsto \bot, x'_{1,\overline{z_1 z_2}} \mapsto \bot$. This yields the solution $\sigma_X(x_1) = (z_1 \wedge z_2) \vee (z_1 \wedge \overline{z_2}) = z_1$ which is one of the optimal solutions as explained in Example 1.

4   Incremental Method

In this section we propose a first improvement with respect to the reduction in the previous section. It allows to control the blow-up of the objective formula in the reduced $Max\#SAT$ problem through an incremental process. Moreover, it allows in practice to find good approximate solutions early.

The incremental method consists in solving a sequence of related $Max\#SAT$ problems, each one obtained from the original $DQMax\#SAT$ problem and a reduced set of dependencies $H_1' \subseteq H_1, \ldots, H_n' \subseteq H_n$. Actually, if the sets of dependencies $H_1', \ldots, H_n'$ are chosen such that to augment progressively from $\emptyset, \ldots, \emptyset$ to $H_1, \ldots, H_n$ by increasing only one of H_i' at every step then (i) it is possible to build every such $Max\#SAT$ problem from the previous one by a simple syntactic transformation and (ii) most importantly, it is possible to steer the search for its solution knowing the solution of the previous one.

The incremental method relies therefore on an oracle procedure `max#sat` for solving $Max\#SAT$ problems. We assume this procedure takes as inputs the sets X, Y, Z of maximizing, counting and existential variables, an objective formula $\Phi \in \mathcal{F}\langle X \cup Y \cup Z \rangle$, an initial assignment $\alpha_0 : X \to \mathbb{B}$ and a filter formula $\Psi \in \mathcal{F}\langle X \rangle$. The last two parameters are essentially used to restrict the search for maximizing solutions and must satisfy:

- $\Psi[\alpha_0] = \top$, that is, the initial assignment α_0 is a model of Ψ and
- forall $\alpha : X \to \mathbb{B}$ if $\alpha \nvDash \Psi$ then $|\exists Z.\ \Phi[\alpha]|_Y \leq |\exists Z.\ \Phi[\alpha_0]|_Y$, that is, any assignment α outside the filter Ψ is at most as good as the assignement α_0.

Actually, whenever the conditions hold, the oracle can safely restrict the search for the optimal assignements within the models of Ψ. The oracle produces as output the optimal assignement $\alpha^* : X \to \mathbb{B}$ solving the $Max\#SAT$ problem.

The incremental algorithm proposed in Algorithm 1 proceeds as follows:

- at lines 1–5 it prepares the arguments for the first call of the $Max\#SAT$ oracle, that is, for solving the problem where $H_1' = H_2' = \ldots = H_n' = \emptyset$,
- at line 7 it calls to the $Max\#SAT$ oracle,
- at lines 9–10 it chooses an index i_0 of some dependency set $H_i' \neq H_i$ and a variable $u \in H_{i_0} \setminus H_{i_0}'$ to be considered in addition for the next step, note that these steps open the door to variable selection heuristics (see [17]),
- at lines 11–19 it prepares the argument for the next call of the $Max\#SAT$ oracle, that is, it updates the set of maximizing variables X', it refines the objective formula Φ', it defines the new initial assignement α_0' and the new filter Ψ' using the solution of the previous problem,
- at lines 6,20,22 it controls the main iteration, that is, keep going as long as sets H_i' are different from H_i,
- at line 23 it builds the expected solution, that is, convert the Boolean solution α'^* of the final $Max\#SAT$ problem where $H_i' = H_i$ for all $i \in [1, n]$ to the corresponding substitution σ_X^*.

Finally, note that the application of substitution at line 15 can be done such that to preserve the CNF form of Φ'. That is, the application proceeds clause by clause by using the following equivalences, for every formula ψ and substitution

```
input  : X = {x₁, ..., xₙ}, Y, Z, H₁, ..., Hₙ, Φ
output: σ*ₓ
```

$\textbf{input}\ : X = \{x_1, ..., x_n\}, Y, Z, H_1, ..., H_n, \Phi$
$\textbf{output}: \sigma_X^*$

1 $H_i' \leftarrow \emptyset$ for all $i \in [1, n]$
2 $X' \leftarrow \{x_{i,\top}'\}_{i \in [1,n]}$
3 $\Phi' \leftarrow \Phi \wedge \bigwedge_{i \in [1,n]}(x_i \Leftrightarrow x_{i,\top}')$
4 $\alpha_0' \leftarrow \{x_{i,\top}' \mapsto \bot\}_{i \in [1,n]}$
5 $\Psi' \leftarrow \top$
6 **repeat**
7     $\alpha'^* \leftarrow \texttt{max\#sat}(X', Y, Z \cup X, \Phi', \alpha_0', \Psi')$
8   **if** $H_i' \neq H_i$ *for some* $i \in [1, n]$ **then**
9        $i_0 \leftarrow \texttt{choose}(\{i \in [1, n] \mid H_i' \neq H_i\})$
10       $u \leftarrow \texttt{choose}(H_{i_0} \setminus H_{i_0}')$
11       $\alpha_0' \leftarrow \alpha'^*$
12       $\Psi' \leftarrow \bot$
13   **foreach** $m \in \mathcal{M}\langle H_{i_0}'\rangle$ **do**
14          $X' \leftarrow (X' \setminus \{x_{i_0,m}'\}) \cup \{x_{i_0,mu}^t, x_{i_0,m\bar{u}}'\}$
15          $\Phi' \leftarrow \Phi'[x_{i_0,m}' \mapsto (x_{i_0,mu}' \wedge u) \vee (x_{i_0,m\bar{u}}' \wedge \bar{u})]$
16          $\alpha_0' \leftarrow (\alpha_0' \setminus \{x_{i_0,m}' \mapsto \_\}) \cup \{x_{i_0,mu}', x_{i0,m\bar{u}}' \mapsto \alpha_0'(x_{i_0,m}')\}$
17          $\Psi' \leftarrow \Psi' \vee (x_{i_0,mu}' \not\Leftrightarrow x_{i_0,m\bar{u}}')$
18   **end**
19       $\Psi' \leftarrow \Psi' \vee \bigwedge_{x \in X'}(x \Leftrightarrow \alpha_0'(x))$
20       $H_{i_0}' \leftarrow H_{i_0}' \cup \{u\}$
21   **end**
22 **until** $H_i' = H_i$ for all $i \in [1, n]$
23 $\sigma_X^* \leftarrow \{x_i \mapsto \vee_{m \in \mathcal{M}\langle H_i\rangle}(\alpha'^*(x_{i,m}') \wedge m)\}_{i \in [1,n]}$

Algorithm 1: Incremental Algorithm

$$\sigma_{i_0,m,u} \overset{def}{=} \{x_{i_0,m}' \mapsto (x_{i_0,mu}' \wedge u) \vee (x_{i_0,m\bar{u}}' \wedge \bar{u})\}:$$

$$(\psi \vee x_{i_0,m}')[\sigma_{i_0,m,u}] \Leftrightarrow (\psi \vee x_{i_0,mu}' \vee x_{i_0,m\bar{u}}') \wedge (\psi \vee x_{i_0,mu}' \vee \bar{u}) \wedge (\psi \vee x_{i_0,m\bar{u}}' \vee u)$$

$$(\psi \vee \overline{x_{i_0,m}'})[\sigma_{i_0,m,u}] \Leftrightarrow (\psi \vee \overline{x_{i_0,mu}'} \vee \bar{u}) \wedge (\psi \vee \overline{x_{i_0,m\bar{u}}'} \vee u)$$

Theorem 3. *Algorithm 1 is correct for solving the DQMax#SAT problem (1).*

Proof. The algorithm terminates after $1 + \sum_{i \in [1,n]} |H_i|$ oracle calls. Moreover, every oracle call solves correctly the *Max#SAT* problem corresponding to *DQMax#SAT* problem

$$\max^{H_1'} x_1. \ ... \ \max^{H_n'} x_n. \ \text{Я} Y. \ \exists Z. \ \Phi(X, Y, Z)$$

This is an invariance property provable by induction. It holds by construction of X', Φ', α_0', Ψ' at the initial step. Then, it is preserved from one oracle call to the next one i.e., X' and Φ' are changed such that to reflect the addition of the variable u of the set H_{i_0}'. The new initial assignement α_0' is obtained (i) by replicating the optimal value $\alpha'^*(x'_{i_0,m})$ to the newly introduced $x'_{i_0,mu}, x'_{i_0,m\bar{u}}$ variables derived from $x'_{i_0,m}$ variable (line 16) and (ii) by keeping the optimal value $\alpha'^*(x'_{i,m})$ for other variables (line 11). As such, for the new problem, the assignement α'_0 has exactly the same number of Y-projected models as the optimal assignement α'^* had on the previous problem. The filter Ψ' is built such that to contain this new initial assignment α'_0 (line 19) as well as any other assignement that satisfies $x'_{i_0,mu} \not\Leftrightarrow x'_{i_0,m\bar{u}}$ for some monomial m (lines 12, 17). This construction guarantees that, any assignment which does not satisfy the filter Ψ' reduces precisely to an assignment of the previous problem, other than the optimal one α'^*, and henceforth at most as good as α'_0 regarding the number of Y-projected models. Therefore, it is a sound filter and can be used to restrict the search for the new problem. The final oracle call corresponds to solving the complete $Max\#SAT$ problem (6) and it will therefore allow to derive a correct solution to the initial $DQMax\#SAT$ problem (1). $\qquad\square$

Example 4. Let reconsider Example 1. The incremental algorithm will perform 3 calls to the $Max\#SAT$ oracle. The first call corresponds to the problem

$$\max x'_{1,\top}.\ \text{Я}y_1.\ \text{Я}y_2.\ \exists z_1.\ \exists z_2.\ \exists x_1.$$

$$(x_1 \Leftrightarrow y_1) \wedge (z_1 \Leftrightarrow y_1 \vee y_2) \wedge (z_2 \Leftrightarrow y_1 \wedge y_2) \wedge (x_1 \Leftrightarrow x'_{1,\top})$$

A solution found by the oracle is e.g., $x'_{1,\top} \mapsto \bot$ which has 2 projected models. If z_1 is added to H'_1, the second call corresponds to the refined $Max\#SAT$ problem:

$$\max x'_{1,z_1}.\ \max x'_{1,\bar{z}_1}\ \text{Я}y_1.\ \text{Я}y_2.\ \exists z_1.\ \exists z_2.\ \exists x_1.$$

$$(x_1 \Leftrightarrow y_1) \wedge (z_1 \Leftrightarrow y_1 \vee y_2) \wedge (z_2 \Leftrightarrow y_1 \wedge y_2) \wedge (x_1 \Leftrightarrow x'_{1,z_1} \wedge z_1 \vee x'_{1,\bar{z}_1} \wedge \bar{z}_1)$$

A alignfound by the oracle is e.g., $x'_{1,z_1} \mapsto \top, x'_{1,\bar{z}} \mapsto \bot$ which has 3 projected models. Finally, z_2 is added to H'_1 therefore the third call corresponds to the complete $Max\#SAT$ problem as presented in Example 3. The solution found by the oracle is the same as in Example 3.

A first benefit of Algorithm 1 is the fact that it opens the door to any-time approaches to solve the $DQMax\#SAT$ problem. Indeed, the distance between the current and the optimal solution (that is, the relative ratio between the corresponding number of Y-projected models) can be estimated using the upper bound provided by Prop. 1. Hence, one could stop the search at any given iteration as soon as some threshold is reached and construct the returned value σ_X similarly as in Algorithm 23 of Algorithm 1. In this case the returned σ_X would be defined as $\sigma_X = \{x_i \mapsto \vee_{m \in \mathcal{M}\langle H'_i \rangle}(\alpha'^*(x'_{i,m}) \wedge m)\}_{i \in [1,n]}$ (note here that the monomials are selected from H'_i instead of H_i).

Another benefit of the incremental approach is that it is applicable without any assumptions on the underlying $Max\#SAT$ solver. Indeed, one can use Ψ' in Algorithm 1 by solving the $Max\#SAT$ problem corresponding to $\Phi' \wedge \Psi'$, and return the found solution. Even though the α'_0 parameter requires an adaptation of the $Max\#SAT$ solver in order to ease the search of a solution, one could still benefit from the incremental resolution of $DQMax\#SAT$. Notice that a special handling of the Ψ' parameter by the solver would avoid complexifying the formula passed to the $Max\#SAT$ solver and still steer the search properly.

5   Local Method

The local resolution method allows to compute the solution of an initial $DQMax\#SAT$ problem by combining the solutions of two strictly smaller and independent $DQMax\#SAT$ sub-problems derived syntactically from the initial one. The local method applies only if either 1) some counting or existential variable u is occurring in all dependency sets; or 2) if there is some maximizing variable having an empty dependency set. That is, in contrast to the global and incremental methods, the local method is applicable only in specific situations.

Given a $DQMax\#SAT$ problem of form (1) and a variable v, let $\Phi_v \overset{def}{=} \Phi[v \mapsto \top]$, $\Phi_{\bar{v}} \overset{def}{=} \Phi[v \mapsto \bot]$ be the two cofactors on variable v of the objective Φ.

5.1   Reducing Common Dependencies

Let us consider now a variable u which occurs in all dependency sets H_i and let us consider the following u-reduced $DQMax\#SAT$ problems:

$$\max^{H_1 \setminus \{u\}} x_1. \ldots \max^{H_n \setminus \{u\}} x_n. \; \text{Я}\, Y \setminus \{u\}. \; \exists Z \setminus \{u\}. \; \Phi_u \qquad (7)$$

$$\max^{H_1 \setminus \{u\}} x_1. \ldots \max^{H_n \setminus \{u\}} x_n. \; \text{Я}\, Y \setminus \{u\}. \; \exists Z \setminus \{u\}. \; \Phi_{\bar{u}} \qquad (8)$$

Let $\sigma^*_{X,u}$, $\sigma^*_{X,\bar{u}}$ denote respectively the solutions to the problems above.

Theorem 4. *If either*

(i) $u \in Y$ or
(ii) $u \in Z$ and u is functionally dependent on counting variables Y within the objective Φ (that is, for any valuation $\alpha_Y : Y \to \mathbb{B}$, at most one of $\Phi[\alpha_Y][u \mapsto \top]$ and $\Phi[\alpha_Y][u \mapsto \bot]$ is satisfiable).

then σ^*_X defined as

$$\sigma^*_X(x_i) \overset{def}{=} \left(u \wedge \sigma^*_{X,u}(x_i) \right) \vee \left(\bar{u} \wedge \sigma^*_{X,\bar{u}}(x_i) \right) \text{ for all } i \in [1,n]$$

is a solution to the $DQMax\#SAT$ problem (1).

Proof. First, any formula $\varphi_i \in \mathcal{F}\langle H_i \rangle$ can be equivalently written as $u \wedge \varphi_{i,u} \vee \bar{u} \wedge \varphi_{i,\bar{u}}$ where $\varphi_{i,u} \overset{def}{=} \varphi_i[u \mapsto \top] \in \mathcal{F}\langle H_i \setminus \{u\} \rangle$ and $\varphi_{i,\bar{u}} \overset{def}{=} \varphi_i[u \mapsto \bot] \in \mathcal{F}\langle H_i \setminus \{u\} \rangle$. Second, we can prove the equivalence:

$$\Phi[x_i \mapsto \varphi_i] \Leftrightarrow \left((u \wedge \Phi_u) \vee (\bar{u} \wedge \Phi_{\bar{u}}) \right) [x_i \mapsto u \wedge \varphi_{i,u} \vee \bar{u} \wedge \varphi_{i,\bar{u}}]$$
$$\Leftrightarrow (u \wedge \Phi_u[x_i \mapsto \varphi_{i,u}]) \vee (\bar{u} \wedge \Phi_{\bar{u}}[x_i \mapsto \varphi_{i,\bar{u}}])$$

by considering the decomposition of Φ_u, $\Phi_{\bar{u}}$ according to the variable x_i.

The equivalence above can then be generalized to a complete substitution $\sigma_X = \{x_i \mapsto \varphi_i\}_{i \in [1,n]}$ of maximizing variables. Let us denote respectively $\sigma_{X,u} \overset{def}{=} \{x_i \mapsto \varphi_{i,u}\}_{i \in [1,n]}$, $\sigma_{X,\bar{u}} \overset{def}{=} \{x_i \mapsto \varphi_{i,\bar{u}}\}_{i \in [1,n]}$. Therefore, one obtains

$$
\begin{aligned}
\Phi[\sigma_X] &\Leftrightarrow (u \wedge \Phi_u \vee \bar{u} \wedge \Phi_{\bar{u}})[x_i \mapsto \varphi_i]_{i \in [1,n]} \\
&\Leftrightarrow \left(u \wedge \Phi_u[x_i \mapsto \varphi_{i,u}]_{i \in [1,n]}\right) \vee \left(\bar{u} \wedge \Phi_{\bar{u}}[x_i \mapsto \varphi_{i,\bar{u}}]_{i \in [1,n]}\right) \\
&\Leftrightarrow (u \wedge \Phi_u[\sigma_{X,u}]) \vee (\bar{u} \wedge \Phi_{\bar{u}}[\sigma_{X,\bar{u}}])
\end{aligned}
$$

Third, the later equivalence provides a way to compute the number of Y-models of the formula $\exists Z.\ \Phi[\sigma_Z]$ as follows:

$$
\begin{aligned}
|\exists Z.\ \Phi[\sigma_X]|_Y &= |\exists Z.\ (u \wedge \Phi_u[\sigma_{X,u}] \vee \bar{u} \wedge \Phi_{\bar{u}}[\sigma_{X,\bar{u}}])|_Y \\
&= |\exists Z.\ (u \wedge \Phi_u[\sigma_{X,u}]) \vee \exists Z.\ (\bar{u} \wedge \Phi_{\bar{u}}[\sigma_{X,\bar{u}}])|_Y \\
&= |\exists Z.\ (u \wedge \Phi_u[\sigma_{X,u}])|_Y + |\exists Z.\ (\bar{u} \wedge \Phi_{\bar{u}}[\sigma_{X,\bar{u}}])|_Y \\
&= |\exists Z \setminus \{u\}.\ \Phi_u[\sigma_{X,u}]|_{Y \setminus \{u\}} + |\exists Z \setminus \{u\}.\ \Phi_{\bar{u}}[\sigma_{X,\bar{u}}]|_{Y \setminus \{u\}}
\end{aligned}
$$

Note that the third equality holds only because $u \in Y$ or $u \in Z$ and functionally dependent on counting variables Y. Actually, in these situations, the sets of Y-projected models of respectively, $u \wedge \Phi_u[\sigma_{X,u}]$ and $\bar{u} \wedge \Phi_{\bar{u}}[\sigma_{X,\bar{u}}]$ are disjoint. Finally, the last equality provides the justification of the theorem, that is, finding σ_X which maximizes the left hand side reduces to finding $\sigma_{X,u}$, $\sigma_{X,\bar{u}}$ which maximizes independently the two terms of right hand side, and these actually are the solutions of the two u-reduced problems (7) and (8).   □

Example 5. Let us reconsider Example 1. It is an immediate observation that existential variables z_1, z_2 are functionally dependent on counting variables y_1, y_2 according to the objective. Therefore the local method is applicable and henceforth since $H_1 = \{z_1, z_2\}$ one reduces the initial problem to four smaller problems, one for each valuation of z_1, z_2, as follows:

$$z_1, z_2 \mapsto \top, \top: \quad \max^{\emptyset} x_1.\ я y_1.\ я y_2\ .(x_1 \Leftrightarrow y_1) \wedge (\top \Leftrightarrow y_1 \vee y_2) \wedge (\top \Leftrightarrow y_1 \wedge y_2)$$

$$z_1, z_2 \mapsto \top, \bot: \quad \max^{\emptyset} x_1.\ я y_1.\ я y_2\ .(x_1 \Leftrightarrow y_1) \wedge (\top \Leftrightarrow y_1 \vee y_2) \wedge (\bot \Leftrightarrow y_1 \wedge y_2)$$

$$z_1, z_2 \mapsto \bot, \top: \quad \max^{\emptyset} x_1.\ я y_1.\ я y_2\ .(x_1 \Leftrightarrow y_1) \wedge (\bot \Leftrightarrow y_1 \vee y_2) \wedge (\top \Leftrightarrow y_1 \wedge y_2)$$

$$z_1, z_2 \mapsto \bot, \bot: \quad \max^{\emptyset} x_1.\ я y_1.\ я y_2\ .(x_1 \Leftrightarrow y_1) \wedge (\bot \Leftrightarrow y_1 \vee y_2) \wedge (\bot \Leftrightarrow y_1 \wedge y_2)$$

The four problems are solved independently and have solutions e.g., respectively $x_1 \mapsto c_1 \in \{\top\}$, $x_1 \mapsto c_2 \in \{\top, \bot\}$, $x_1 \mapsto c_3 \in \{\top, \bot\}$, $x_1 \mapsto c_4 \in \{\bot\}$. By recombining these solutions according to Theorem 4 one obtains several solutions to the original *DQMax#SAT* problem of the form:

$$x_1 \mapsto (z_1 \wedge z_2 \wedge c_1) \vee (z_1 \wedge \bar{z}_2 \wedge c_2) \vee (\bar{z}_1 \wedge z_2 \wedge c_3) \vee (\bar{z}_1 \wedge \bar{z}_2 \wedge c_4)$$

They correspond to solutions already presented in Example 3, that is:

$$x_1 \mapsto (z_1 \wedge z_2 \wedge \top) \vee (z_1 \wedge \overline{z_2} \wedge \bot) \vee (\overline{z_1} \wedge z_2 \wedge \bot) \vee (\overline{z_1} \wedge \overline{z_2} \wedge \bot) \quad (\equiv z_1 \wedge z_2)$$

$$x_1 \mapsto (z_1 \wedge z_2 \wedge \top) \vee (z_1 \wedge \overline{z_2} \wedge \bot) \vee (\overline{z_1} \wedge z_2 \wedge \top) \vee (\overline{z_1} \wedge \overline{z_2} \wedge \bot) \qquad (\equiv z_2)$$

$$x_1 \mapsto (z_1 \wedge z_2 \wedge \top) \vee (z_1 \wedge \overline{z_2} \wedge \top) \vee (\overline{z_1} \wedge z_2 \wedge \bot) \vee (\overline{z_1} \wedge \overline{z_2} \wedge \bot) \qquad (\equiv z_1)$$

$$x_1 \mapsto (z_1 \wedge z_2 \wedge \top) \vee (z_1 \wedge \overline{z_2} \wedge \top) \vee (\overline{z_1} \wedge z_2 \wedge \top) \vee (\overline{z_1} \wedge \overline{z_2} \wedge \bot) \quad (\equiv z_1 \vee z_2)$$

Finally, note that the local resolution method has potential for parallelization. It is possible to eliminate not only one but all common variables in the dependency sets as long as they fulfill the required property. This leads to several strictly smaller sub-problems that can be solved in parallel. The situation has been already illustrated in the previous example, where by the elimination of z_1 and z_2 one obtains 4 smaller sub-problems.

5.2   Solving Variables with No Dependencies

Let us consider now a maximizing variable which has an empty dependency set. Without lack of generality, assume x_1 has an empty dependency set, i.e. $H_1 = \emptyset$. Thus, the only possible values that can be assigned to x_1 are $\top$ or $\bot$. Let us consider the following x_1-reduced $DQMax\#SAT$ problems:

$$\max^{H_2} x_2. \ \ldots \ \max^{H_n} x_n. \ \text{Я}\, Y. \ \exists Z. \ \Phi_{x_1}$$
$$\max^{H_2} x_2. \ \ldots \ \max^{H_n} x_n. \ \text{Я}\, Y. \ \exists Z. \ \Phi_{\overline{x_1}}$$

and let $\sigma^*_{X,x_1}, \sigma^*_{X,\overline{x_1}}$ denote respectively the solutions to the problems above. The following proposition is easy to prove, and provides the solution of the original problem based on the solutions of the two smaller sub-problems.

Proposition 2. *The substitution σ^*_X defined as*

$$\sigma^*_X \stackrel{def}{=} \begin{cases} \sigma^*_{X,x_1} \uplus \{x_1 \mapsto \top\} & \text{if } |\exists Z. \ \Phi_{x_1}[\sigma^*_{X,x_1}]|_Y \geq |\exists Z. \ \Phi_{\overline{x_1}}[\sigma^*_{X,\overline{x_1}}]|_Y \\ \sigma^*_{X,\overline{x_1}} \uplus \{x_1 \mapsto \bot\} & \text{otherwise} \end{cases}$$

is a solution to the DQMax#SAT problem (1).

6   Application to Software Security

In this section, we give a concrete application of $DQMax\#SAT$ in the context of *software security*. More precisely, we show that finding an optimal strategy for an adaptive attacker trying to break the security of some program can be naturally encoded as specific instances of the $DQMax\#SAT$ problem.

In our setting, we allow the attacker to interact multiple times with the target program. Moreover, we assume that the adversary is able to make *observations*, either from the legal outputs or using some side-channel leaks. Adaptive attackers [9,24,25] are a special form of active attackers considered in security that are able to select their inputs based on former observations, such that they maximize their chances to reach their goals (i.e., break some security properties).

First we present in more details this attacker model we consider, and then we focus on two representative attack objectives the attacker aims to maximize:

- either the probability of reaching a specific point in the target program, while satisfying some objective function (Sect. 6.2),
- or the amount of information it can get about some fixed secret used by the program (Sect. 6.3).

At the end of the section, we show that the improvements presented in the previous sections apply in both cases.

6.1   Our Model of Security in Presence of an Adaptive Adversary

The general setting we consider is the one of *active* attackers, able to provide *inputs* to the program they target. Such attacks are then said *adaptive* when the attacker is able to deploy a more succesfull attack strategy, which continuously relies on some knowledge gained from previous interactions with the target program. Moreover, we consider the more powerful attacker model where the adversary is assumed to know the code of the target program.

Note that such an attacker model is involved in most recent concrete attack scenarios, where launching an exploit or disclosing some sensitive data requires to chain several (interactive) attack steps in order to defeat some protections and/or to gain some intermediate privileges on the target platform. Obviously, from the defender side, quantitative measures about the "controllability" of such attacks is of paramount importance for exploit analysis or vulnerability triage.

When formalizing the process of *adaptatively attacking* a given program, one splits the program's variables between those *controlled* and those *uncontrolled* by the attacker. Among the *uncontrolled* variables one further distinguishes those *observable* and those *non-observable*, the former ones being available to the attacker for producing its (next) inputs. The *objective* of the attacker is a formula, depending on the values of program variables, and determining whether the attacker has successfully conducted the attack.

For the sake of simplicity – in our examples – we restrict ourselves to non-looping sequential programs operating on variables with bounded domains (such as finite integers, Boolean's, etc.). We furthermore consider the programs are written in SSA form, assuming that each variable is assigned before it is used. These hypothesis fit well in the context of a code analysis technique like *symbolic execution* [15], extensively used in software security.

Finally, we also rely on explicit (user-given) annotations by predefined functions (or macros) to identify the different classes of program variables and the attacker's objective. In the following code excerpts, we assume that:

- The `random` function produces an uncontrolled non-observable value; it allows for instance to simulate the generation of both long term keys and nonces in a program using cryptographic primitives.

```
1  y₁ ← random()
2  y₂ ← random()
3  z₁ ← output(y₁ + y₂)
4  x₁ ← input()
5  if y₁ ≤ x₁ then
6  |   win(x₁ ≤ y₂)
7  end
```

Program 2: A first program example

- The input function feeds the program with an attacker-controlled value.
- The output function simulates an observation made by the adversary and denotes a value obtained through the evaluation of some expression of program variables.

6.2 Security as a Rechability Property

We show in this section how to encode *quantitative reachability* defined in [3] as an instance of the *DQMax#SAT* problem. In *quantitative reachability*, the goal of an adversary is to reach some target location in some program such that some *objective property* get satisfied. In order to model this target location of the program that the attacker wants to reach, we extend our simple programming language with a distinguished win function. The win function can take a predicate as argument (the objective property) and this predicate is omitted whenever it is the True predicate. In practice such a predicate may encode some extra conditions required to trigger and exploit some vulnerability at the given program location (e.g., overflowing a buffer with a given payload).

Example 6. In Program 2 one can see an example of annotated program. y_1 and y_2 are uncontrollable non-observable variables. z_1 is an observable variable holding the sum $y_1 + y_2$. x_1 is a variable controlled by the attacker. The *attacker's objective* corresponds to the *path predicate* $y_1 \leq x_1$ denoting the condition to reach the win function call and the *argument predicate* $x_1 \leq y_2$ denoting the *objective property*. Let us observe that a successful attack exists: taking $x_1 \leftarrow \frac{z_1}{2}$ maximizes the probability of reaching the objective (whenever $y_1 \leq y_2$).

When formalizing adaptive attackers, the *temporality* of interactions (that is, the order of inputs and outputs) is important, as the attacker can only synthesize an input value from the output values that were observed *before* it is asked to provide that input. To track the temporal dependencies in our formalization, for every controlled variable x_i one considers the set H_i of observable variables effectively known at the time of defining x_i, that is, representing the accumulation of attacker's knowledge throughout the interactions with the program.

We propose hereafter a systematic way to express the problem of synthesis of an optimal attack (that is, with the highest probability of the objective property to get satisfied), as a *DQMax#SAT* instance. Let Y be the set of uncontrolled variables being assigned to `random()` in the program, which in this section is assumed to uniformly sample values in their domain. Similarly, let Z be the set of other expressions (possibly involving variables in Y). For a variable $z \in Z$ let moreover e_z be the unique expression assigned to it in the program, either through an assignment of the form $z \leftarrow e_z$ or $z \leftarrow$ `output`(e_z). Let $X = \{x_1, ..., x_n\}$ be the set of controlled variables with their temporal dependencies respectively subsets $H_1, \dots, H_n \subseteq Z$ of uncontrollable variables. Finally, let Ψ be the attacker objective, that is, the conjunction of the argument of the `win` function and the path predicate leading to the `win` function call. Consider the next most likely generalized $DQMax\#SAT$ problem:

$$\max^{H_1} x_1. \ \dots \ \max^{H_n} x_n. \ \text{Я}Y. \ \exists Z. \ \Psi \wedge \bigwedge_{z \in Z} (z = e_z) \tag{9}$$

Example 7. Consider the annotated problem from Program 2. The encoding of the optimal attack leads to the generalized $DQMax\#SAT$ problem:

$$\max^{\{z_1\}} x_1. \ \text{Я}y_1. \ \text{Я}y_2. \ \exists z_1. \ (y_1 \leq x_1 \wedge x_1 \leq y_2) \wedge (z_1 = y_1 + y_2)$$

Note that in contrast to the $DQMax\#SAT$ problem (1), the variables are not restricted to Booleans (but to some finite domains) and the expressions are not restricted to Boolean terms (but involve additional operators available in the specific domain theories e.g., $=$, $\geq$, $+$, $-$, etc.). Nevertheless, as long as both variables and additional operators can be respectively, represented by and interpreted as operations on bitvectors, one can use *bitblasting* and transform the generalized problem into a full-fledged $DQMax\#SAT$ problem and then solve it by the techniques introduced earlier in the paper.

Finally, note also that in the $DQMax\#SAT$ problems constructed as above, the maximizing variables are dependent by definition on existential variables only. Therefore, as earlier discussed in Sect. 2, these problems cannot be actually reduced to similar $DSSAT$ problems. However, they compactly encode the quantitative reachability properties subject to input/output dependencies.

6.3 Security as a Lack of Leakage Property

In this section, we extend earlier work on adaptive attackers from [25] by effectively synthesizing the *strategy* the attacker needs to deploy in order to maximize its knowledge about some secret value used by the program. Moreover, we show that in our case, we are able to keep symbolic the trace corresponding to the attack strategy, while in [24], the attacker strategy is a concretized tree, which explicitly states, for each concrete program output, what should be the next input provided by the adversary. Following ideas proposed in [24], symbolic execution can be used to generate constraints characterizing partitions on the secrets values, where each partition corresponds to the set of secrets leading to the same *sequences* of side-channel observations.

274   T. Vigouroux et al.

```
1 z ← random() //the secret
2 x ← input()
3 if x ≥ z then
4 |   ... some computation taking 10
  |   seconds
5 else
6 |   ... some computation taking 20
  |   seconds
7 end
```

Program 3: A leaking program

```
1 z ← random() ;
2 x₁ ← input();
3 y₁ ← output(x₁ ≥ z);
4 x₂ ← input();
5 y₂ ← output(x₂ ≥ z);
6 x₃ ← input();
7 y₃ ← output(x₃ ≥ z);
```

Program 4: An iterated leaking program

Example 8. Let us consider the excerpt Program 3 taken from [24]. This program is not *constant-time*, namely it executes a branching instruction whose condition depends on the secret z. Hence an adversary able to learn the branch taken during the execution, either by measuring the time or doing some cache-based attack, will get some information about the secret z. A goal of an adversary interacting several times with the program could be to maximize the amount of information leaked about the secret value z. When the program is seen as a channel leaking information, the channel capacity theorem [26] states that the information leaked by a program is upper-bounded by the number of different observable outputs of the program (and the maximum is achieved whenever the secret is the unique randomness used by the program). In our case, it means that an optimal adaptive adversary interacting k-times with the program should maximize the number of different observable outputs. Hence, for example, if as in [24], we fix $k = 3$ and if we assume that the secret z is uniformly sampled in the domain $1 \leq z \leq 6$, then the optimal strategy corresponds to maximize the number of different observable outputs y of the Program 4, which corresponds to the following generalized *DQMax#SAT* instance:

$$\max^\emptyset x_1. \max^{\{y_1\}} x_2. \max^{\{y_1,y_2\}} x_3. Яy_1. Яy_2. Яy_3. \exists z .$$
$$(y_1 = (x_1 \geq z)) \wedge (y_2 = (x_2 \geq z)) \wedge (y_3 = (x_3 \geq z)) \wedge (1 \leq z \leq 6)$$

Our prototype provided the following solution: $x_1 = 100$, $x_2 = y_1 10$, $x_3 = y_1 y_2 1$, that basically says: the attacker should first input 4 to the program, then the input corresponding to the integer whose binary encoding is y_1 concatenated with 10, and the last input x_3 is the input corresponding to the integer whose binary encoding is the concatenation of y_2, y_1 and 1. In [24] the authors obtain an equivalent attack encoded as a tree-like strategy of concrete values.

We now show a systematic way to express the problem of the synthesis of an optimal attack expressed as the maximal channel capacity of a program seen as an information leakage channel, as a *DQMax#SAT* instance. Contrary to the previous section, the roles of Y and Z are now switched: Y is a set of variables encoding the observables output by the program; Z is the set of variables uniformly sampled by random() or assigned to other expressions in the program.

For a variable $y \in Y$, let e_y be the unique expression assigned to it in the program through an assignment of the form $y \leftarrow \text{output}(e_y)$. For a variable $z \in Z$, let moreover e_z be the unique expression assigned to it in the program through an assignment of the form $z \leftarrow e_z$ or the constraint encoding the domain used to sample values in $z \leftarrow \text{random}()$. Let $X = \{x_1, ..., x_n\}$ be the set of controlled variables with their temporal dependencies respectively subsets $H_1, \ldots, H_n \subseteq Y$. Consider now the following most likely generalized $DQMax\#SAT$ problem:

$$\max^{H_1} x_1. \; ... \; \max^{H_n} x_n. \; \text{Я} Y. \; \exists Z. \; \bigwedge_{y \in Y} (y = e_y) \wedge \bigwedge_{z \in Z} (z = e_z)$$

Finally, in contrast to reachability properties, in the $DQMax\#SAT$ problems obtained as above for evaluating leakage properties, the maximizing variables are by definition dependent on counting variables only. Consequently, for these problems, the existential variables can be apriori eliminated so that to obtain an equivalent $DSSAT^1$ problem as discussed in Sect. 2.

6.4   Some Remarks About the Applications to Security

Let us notice some properties of the attacker synthesis's $DQMax\#SAT$ problems. If controlled variables $x_1, ..., x_n$ are input in this order within the program, then necessarily $H_1 \subseteq ... \subseteq H_n$. That is, the knowledge of the attacker only increases as long as newer observable values became available to it. Moreover, since we assumed that variables are used only after they were initialized, the sets H_i contain observable variables that are dependent only on the counting variables Y. Hence we can apply iteratively the following steps from the local resolution method described in Sect. 5:

- While $H_1 \neq \emptyset$, apply the local resolution method described in Sect. 5.1 iteratively until H_1 becomes empty. For example, it is the case of Example 6 where z_1 is dependent only on counting variables y_1 and y_2.
- When H_1 becomes $\emptyset$, apply the local resolution method described in Sect. 5.2 in order to eliminate the first maximizing variable.

7   Implementation and Experiments

In our implementation [27] of Algorithm 1, we leave generic the choice of the underlying $Max\#SAT$ solver. For concrete experiments, we used both the approximate solver BaxMC^2 [29] and the exact solver D4Max [1].

In the implementation of Algorithm 1 in our tool, the filter Ψ' is handled as discussed at the end of Sect. 4: the formula effectively solved is $\Phi' \wedge \neg\Psi'$, allowing to use any $Max\#SAT$ solver without any prior modification. Remark that none of BaxMC and D4Max originally supported exploiting the α_0 parameter of

---

[1] Actually these problems can even be reduced to $SSAT$ instances.
[2] Thanks to specific parametrization and the oracles [5] used internally by BaxMC, it can be considered an exact solver on the small instances of interest in this section.

Algorithm 1 out of the box. While D4MAX is used of the shelf, we modified BAXMC to actually support this parameter for the purpose of the experiment.

We use the various examples used in this paper as benchmark instances for the implemented tool. Examples 1 and 2 are used as they are. We consider Examples 7 8 from Sect. 6 and perform the following steps to convert them into $DQMax\#SAT$ instances: (i) bitblast the formula representing the security problem into a $DQMax\#SAT$ instance over boolean variables; (ii) solve the later formula; (iii) propagate the synthesized function back into a function over bit-vectors for easier visual inspection of the result. We also add the following security related problems (which respectively correspond to Program 5 from [28] and a relaxed version of Example 8 in Sect. 6) into our benchmark set:

Example 9. $\max^{\emptyset} x_1.\ \max^{\{z_1\}} x_2.\ \max^{\{z_1, z_2\}} x_3.\ \Pi y_1.\ \exists z_1.\ \exists z_2.$

$$(x_3 = y_1) \wedge (z_1 = (x_1 \geq y_1) \wedge z_2 = (x_2 \geq y))$$

Example 10. $\max^{\emptyset} x_1.\ \max^{\{y_1\}} x_2.\ \max^{\{y_1, y_2\}} x_3.\ \Pi y_1.\ \Pi y_2.\Pi y_3.\ \exists z\ .$

$$(y_1 = (x_1 \geq z)) \wedge (y_2 = (x_2 \geq z)) \wedge (y_3 = (x_3 \geq z))$$

When bitblasting is needed for a given benchmark, the number of bits used for bitblasting is indicated in parentheses. After the bitblasting operation, the problems can be considered medium sized.

Table 1. Summary of the performances of the tool. $|\Phi|$ denotes the number of clauses. The last two columns indicate the running time using the specific $Max\#SAT$ oracle.

Benchmark name	$|X|$	$|Y|$	$|Z|$	$|\Phi|$	Max Models	Time (BAXMC)	Time (D4MAX)
Example 1	1	2	2	7	3	32 ms	121 ms
Example 2	2	2	2	7	3	25 ms	134 ms
Example 7 (3 bits)	3	6	97	329	6	378 ms	79.88 s
Example 7 (4 bits)	4	8	108	385	28	638.63 s	>30 min
Example 8 (3 bits)	9	3	150	487	6	18.78 s	74.58 s
Example 9 (3 bits)	9	3	93	289	4	74.00 s	18.62 s
Example 10 (3 bits)	9	3	114	355	8	9.16 s	93.48 s

As you can see in Table 1, the implemented tool can effectively solve all the examples presented in this paper. The synthesized answers (i.e. the monomials selected in Algorithm 1, Line 23) returned by both oracles are the same.

For security examples, one key part of the process is the translation of the synthesized answer (over boolean variables) back to the original problem (over bit-vectors). In order to do that, one can simply concatenate the generated sub-functions for each bit of the bit-vector into a complete formula, but that would lack explainability because the thus-generated function would be a concatenation of potentially big sums of monomials. In order to ease visual inspection, we run a generic simplification step [13] for all the synthesized sub-function, before con-catenation. This simplification allows us to directly derive the answers explicited in Examples 7 and 8 instead of their equivalent formulated as sums of monomials, and better explain the results returned by the tool.

Unfortunately, we could not compare our algorithm against the state-of-the-art $DSSAT$ solver DSSATPRE [19] on the set of example described in this paper because (i) as discussed in Sect. 2.3, some $DQMax\#SAT$ instances cannot be converted into $DSSAT$ instances, (ii) for the only $DQMax\#SAT$ instance (Example 10) that can be converted into a $DSSAT$ instance, we were not able to get an answer using DSSATPRE.

8 Related Work

As shown in Sect. 2, $DQMax\#SAT$ subsumes the $DSSAT$ and $DQBF$ problems. This relation indicates a similarity of the three problems, and thus some related works can be extracted from here.

Comparing the performances of existing $DQBF$ algorithms with the proposed algorithms for $DQMax\#SAT$ is not yet realistic since they address different objectives. However, one can search for potential improvements for solving $DQMax\#SAT$ by considering the existing enhancements proposed in [16] to improve the resolution of $DQBF$. For example, dependency schemes [31] are a way to change the dependency sets in $DQBF$ without changing the *truth value* compared to the original formula. Thus, adaptations of these dependency schemes could be applied to our problem as well and potentially lead to a significant decrease of the size of the resulting $Max\#SAT$ problems.

The $DSSAT$ problem is currently receiving an increased attention by the research community. A sound and complete resolution procedure has been proposed in [20], but it is not yet implemented. The only available $DSSAT$ solver is DSSATPRE [8], and it relies on preprocessing to get rid of dependencies and to produce equivalent $SSAT$ problems. These problems are then accurately solved by existing $SSAT$ solvers [7,30], some of them being also able to compute the optimal assignments for maximizing variables. In contrast, our tool for solving $DQMax\#SAT$ relies on existing $Max\#SAT$ solvers, always synthesizes the assignments for maximizing variables and provide support for approximate solving. Moreover, due to the presence of existential variables, note that $DQMax\#SAT$ and $DSSAT$ are fundamentally different problems. Existential variables are already pinpointing the difference between the two pure counting problems $\#SAT$ and $\#\exists SAT$ [2,18]. In cases where maximizing variables depend on existential variables no trivial reduction from $DQMax\#SAT$ to $DSSAT$ seems to exists.

From the security point of view, the closest works to our proposal are the ones decribed in [24,25]. As the authors in these papers, we are able to effectively synthesize the optimal adaptive strategy the attacker needs to deploy in order to maximize its knowledge about some secret value used by the program. In addition, we show that in our case, we are able to keep symbolic the trace corresponding to the attack strategy, while in [24], the attacker strategy is a concretized tree which explicitly states, for each concrete program output, what should be the next input provided by the adversary.

9   Conclusions

We exposed a new problem called *DQMax#SAT* that subsumes both *DQBF* and *DSSAT*. We presented three different resolution methods based on reductions to *Max#SAT* and showed the effectiveness of one of them, the incremental method, by implementing a prototype solver. An application of *DQMax#SAT* lies in the context of software security, allowing to assess the robustness of a program by synthesizing the optimal adversarial strategy of an adaptive attacker.

Our work can be expanded in several directions. First, we would like to enhance our prototype with strategies for dependency expansion in the incremental algorithm. Second, we plan to integrate the local resolution method in our prototype. Third, we shall apply these techniques on more realistic security related examples, and possibly getting further improvement directions.

Data-Availability Statement

The prototype tool and all the benchmark data from Sect. 7 are openly available on Zenodo [27].

References

1. Audemard, G., Lagniez, J., Miceli, M.: A new exact solver for (weighted) max#SAT. In: SAT. LIPIcs, vol. 236, pp. 28:1–28:20. Schloss Dagstuhl - Leibniz-Zentrum für Informatik (2022)
2. Aziz, R.A., Chu, G., Muise, C., Stuckey, P.: #∃SAT: projected model counting. In: Heule, M., Weaver, S. (eds.) SAT 2015. LNCS, vol. 9340, pp. 121–137. Springer, Cham (2015). https://doi.org/10.1007/978-3-319-24318-4_10
3. Bardin, S., Girol, G.: A Quantitative Flavour of Robust Reachability. arXiv preprint arXiv:2212.05244 (2022)
4. Chakraborty, S., Fried, D., Meel, K.S., Vardi, M.Y.: From weighted to unweighted model counting. In: IJCAI, pp. 689–695. AAAI Press (2015)
5. Chakraborty, S., Meel, K.S., Vardi, M.Y.: A Scalable Approximate Model Counter. arXiv preprint arXiv:1306.5726 (2013)
6. Chakraborty, S., Meel, K.S., Vardi, M.Y.: Balancing Scalability and Uniformity in SAT Witness Generator. In: DAC, pp. 60:1–60:6. ACM (2014)
7. Chen, P., Huang, Y., Jiang, J.R.: A sharp leap from quantified boolean formula to stochastic boolean satisfiability solving. In: AAAI, pp. 3697–3706. AAAI Press (2021)
8. Cheng, C., Jiang, J.R.: Lifting (D)QBF preprocessing and solving techniques to (D)SSAT. In: AAAI, pp. 3906–3914. AAAI Press (2023)
9. Dullien, T.: Weird machines, exploitability, and provable unexploitability. IEEE Trans. Emerg. Top. Comput. 8(2), 391–403 (2020)
10. Fremont, D.J., Rabe, M.N., Seshia, S.A.: Maximum model counting. In: AAAI, pp. 3885–3892. AAAI Press (2017)
11. Garey, M.R., Johnson, D.S.: Computers and Intractability: A Guide to the Theory of NP-Completeness. W.H. Freeman (1979)

12. Garey, M.R., Johnson, D.S., So, H.C.: An application of graph coloring to printed circuit testing (working paper). In: FOCS, pp. 178–183 (1975)
13. Gario, M., Micheli, A.: PySMT: a solver-agnostic library for fast prototyping of SMT-based algorithms. In: SMT Workshop, vol. 2015 (2015)
14. Henkin, L., Karp, C.R.: Some remarks on infinitely long formulas. J. Symb. Log. **30**(1), 96–97 (1965). https://doi.org/10.2307/2270594
15. King, J.C.: Symbolic execution and program testing. Commun. ACM **19**(7), 385–394 (1976)
16. Kovásznai, G.: What is the state-of-the-art in DQBF solving. In: MaCS-16. Joint Conference on Mathematics and Computer Science (2016)
17. Kullmann, O.: Fundamentals of branching heuristics. In: Handbook of Satisfiability, Frontiers in Artificial Intelligence and Applications, vol. 336, pp. 351–390. IOS Press (2021)
18. Lagniez, J., Marquis, P.: A recursive algorithm for projected model counting. In: AAAI, pp. 1536–1543. AAAI Press (2019)
19. Lee, N., Jiang, J.R.: Dependency stochastic boolean satisfiability: a logical formalism for NEXPTIME decision problems with uncertainty. In: AAAI, pp. 3877–3885. AAAI Press (2021)
20. Luo, Y., Cheng, C., Jiang, J.R.: A resolution proof system for dependency stochastic boolean satisfiability. J. Autom. Reason. **67**(3), 26 (2023)
21. Papadimitriou, C.H.: Games against nature. J. Comput. Syst. Sci. **31**(2), 288–301 (1985)
22. Peterson, G., Reif, J., Azhar, S.: Lower bounds for multiplayer noncooperative games of incomplete information. Comput. Math. Appl. **41**(7–8), 957–992 (2001)
23. Peterson, G.L., Reif, J.H.: Multiple-person alternation. In: FOCS, pp. 348–363. IEEE Computer Society (1979)
24. Phan, Q., Bang, L., Pasareanu, C.S., Malacaria, P., Bultan, T.: Synthesis of adaptive side-channel attacks. In: IACR Cryptol. ePrint Arch, p. 401 (2017)
25. Saha, S., Eiers, W., Kadron, I.B., Bang, L., Bultan, T.: Incremental adaptive attack synthesis. arXiv preprint arXiv:1905.05322 (2019)
26. Smith, G.: On the foundations of quantitative information flow. In: de Alfaro, L. (ed.) FoSSaCS 2009. LNCS, vol. 5504, pp. 288–302. Springer, Heidelberg (2009). https://doi.org/10.1007/978-3-642-00596-1_21
27. Vigouroux, T., Bozga, M., Ene, C., Mounier, L.: DQMaxMC Solver (2023). https://doi.org/10.5281/zenodo.8405351
28. Vigouroux, T., Bozga, M., Ene, C., Mounier, L.: Function synthesis for maximizing model counting. arXiv preprint arXiv:2305.10003 (2023)
29. Vigouroux, T., Ene, C., Monniaux, D., Mounier, L., Potet, M.: BaxMC: a CEGAR approach to Max#SAT. In: FMCAD, pp. 170–178. IEEE (2022)
30. Wang, H., Tu, K., Jiang, J.R., Scholl, C.: Quantifier elimination in stochastic boolean satisfiability. In: SAT. LIPIcs, vol. 236, pp. 23:1–23:17. Schloss Dagstuhl - Leibniz-Zentrum für Informatik (2022)
31. Wimmer, R., Scholl, C., Wimmer, K., Becker, B.: Dependency schemes for DQBF. In: Creignou, N., Le Berre, D. (eds.) SAT 2016. LNCS, vol. 9710, pp. 473–489. Springer, Cham (2016). https://doi.org/10.1007/978-3-319-40970-2_29

Boosting Constrained Horn Solving
by Unsat Core Learning

Parosh Aziz Abdulla[1]([⊠]), Chencheng Liang[1], and Philipp Rümmer[1,2]

[1] Uppsala University, Uppsala, Sweden
{parosh.abdulla,chencheng.liang}@it.uu.se, philipp.ruemmer@ur.de
[2] University of Regensburg, Regensburg, Germany

Abstract. The Relational Hyper-Graph Neural Network (R-HyGNN) was introduced in [1] to learn domain-specific knowledge from program verification problems encoded in Constrained Horn Clauses (CHCs). It exhibits high accuracy in predicting the occurrence of CHCs in counterexamples. In this research, we present an R-HyGNN-based framework called MUSHyperNet. The goal is to predict the Minimal Unsatisfiable Subsets (MUSes) (i.e., unsat core) of a set of CHCs to guide an abstract symbolic model checking algorithm. In MUSHyperNet, we can predict the MUSes once and use them in different instances of the abstract symbolic model checking algorithm. We demonstrate the efficacy of MUSHyperNet using two instances of the abstract symbolic model-checking algorithm: Counter-Example Guided Abstraction Refinement (CEGAR) and symbolic model-checking-based (SymEx) algorithms. Our framework enhances performance on a uniform selection of benchmarks across all categories from CHC-COMP, solving more problems (6.1% increase for SymEx, 4.1% for CEGAR) and reducing average solving time (13.3% for SymEx, 7.1% for CEGAR).

Keywords: Automatic program verification · Constrained Horn clauses · Graph Neural Networks

1 Introduction

Constrained Horn Clauses (CHCs) [2] are logical formulas that can describe program behaviors and specifications. Encoding program verification problems in CHCs and solving them (checking CHCs' satisfiability) has been an active research area for a number of years [3,4]. If the encoded CHCs are satisfiable, the corresponding program verification problem is safe; if not, it is unsafe.

Solving a set of CHCs means that we either find an interpretation for the predicate (relation) symbols and variables that satisfies all the clauses, or prove that no such interpretation exists. Various techniques, such as Counterexample

---

Author names in alphabetical order. The theory presented in the paper was developed by Abdulla, Liang, and Rümmer, the implementation is by Liang and Rümmer, evaluation was done by Liang.

R. Dimitrova et al. (Eds.): VMCAI 2024, LNCS 14499, pp. 280–302, 2024.
https://doi.org/10.1007/978-3-031-50524-9_13

Guided Abstraction Refinement (CEGAR) [5] and IC3 [6], have been utilized for this purpose. However, due to the undecidability of solving CHCs, we need carefully designed or tuned heuristics for specific instances.

In this paper, we consider Minimal Unsatisfiable Subsets (MUSes) [7] of sets of CHCs to support the solving process. Given an unsatisfiable set of CHCs, each MUS is a subset that is again unsatisfiable, but removing any CHC from an MUS makes it satisfiable. Understanding MUSes of a set of CHCs can guide solvers to focus on error-prone clauses [8, Section 3.1]: for an unsatisfiable set of CHCs, evaluating the satisfiability of MUSes first can quickly identify unsatisfiable CHCs, eliminating the need for a comprehensive check. In a satisfiable set of CHCs, examining MUSes first can provide a better starting point for refining potentially problematic constraints. This guidance can help the solver converge towards a solution more efficiently. Manually designed heuristics to find MUSes involves summarizing and generalizing the features of a set of CHCs from examples, which can be replaced by data-driven methods.

Various studies that apply deep learning to formal methods for verification have been published in the recent past, e.g., [9–12]. Primarily, deep learning serves as a feature extractor, which can automatically summarize and generalize program features from examples, alleviating the need for manual crafting and tuning of various heuristics. In addition, the idea of representing logic formulas as graphs and using Graph Neural Networks (GNNs) [13] to guide solvers has been employed in many successful studies [14–17].

However, we are not aware of any study that applies GNNs to guide CHC-based symbolic model checking techniques. The main contribution of this paper is to train a GNN model to predict MUSes of a set of CHCs. We train a GNN using the CHC-R-HyGNN framework [1] to predict values between 0 and 1, representing the probabilities of CHCs being elements of MUSes. For example, we assume that a set $C = \{c_1, c_2, c_3\}$ of CHCs has one MUS $\{c_1, c_2\}$. The predicted probabilities for c_1, c_2, and c_3 being in the MUS could be 0.9, 0.8, and 0.1, respectively. We propose several strategies that use the predicted probability to guide an abstract symbolic model checking algorithm. The strategies can be instantiated for different algorithms on CHCs, such as CEGAR- and symbolic execution-based satisfiability checkers.

Figure 1 depicts an overview of our framework MUSHyperNet. Firstly, we encode the program verification problem into a set of CHCs. Secondly, we use the graph encoder in the CHC-R-HyGNN framework [1] to convert the set of CHCs into a graph format. Then, we train a GNN model named Relational Hypergraph Neural Network (R-HyGNN) to predict the probability of each CHC occurring in MUSes. Finally, we employ the predicted probabilities to guide the abstract symbolic model checking algorithm by determining the sequence for processing each CHC in the set of CHCs.

We utilize the same benchmarks as in [1], comprising 17 130 Linear Integer Arithmetic (LIA) problems from the CHC-COMP 2021 repository [18] for training and evaluating our approach. Further details about the benchmark can be found in [19, Table 1]. The problems for evaluation are uniformly selected

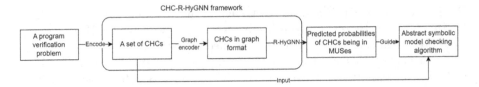

Fig. 1. The CHC-R-HyGNN framework [1] (represented by the round box) comprises a CHC graph encoder and a GNN known as the Relational Hypergraph Neural Network (R-HyGNN), which is capable of handling hypergraphs. In our previous work, we introduced the CHC-R-HyGNN framework that employs various proxy tasks to generalize valuable information for constructing heuristics for CHC-encoded program verification problems.

from this benchmark and can be found in a public repository [20]. The experimental results show an improvement of up to 4.1% and 6.1% in the number of solved problems for the CEGAR and SymEx algorithms, respectively. Additionally, the average solving time demonstrates enhancements of 7.1% and 13.3% for the CEGAR and SymEx algorithms, respectively. In other words, MUSHyperNet can increase the number of solved problems and decrease the solving time for problems similar to those in the CHC-COMP benchmark. To the best of our knowledge, this is the first time unsat core learning has been used successfully in the context of CHCs.

In summary, our contributions are as follows:

- We develop a GNN-based framework named MUSHyperNet, which trains a GNN to predict the MUSes of CHCs and utilizes the predicted probabilities to guide an abstract symbolic model checking algorithm.
- We explore GNN models trained on different datasets and methods for applying predicted MUSes to guide two instances of the model checking algorithm.
- We evaluate MUSHyperNet on 383 linear and 488 non-linear LIA problems, uniformly sampled from the CHC-COMP benchmark [19]. The improvements in the number of solved problems and average solving time are up to 6.1% and 13.3% for the SymEx, and 4.1% and 7.1% for the CEGAR.

2 Preliminaries

We first introduce required notation for multi-sorted first-order logic, and define Constrained Horn Clauses (CHCs) and the encoding of a program verification problem as CHCs. Finally, we explain basic concepts of Graph Neural Network (GNN) and introduce Relational Hyper Graph Neural Network (R-HyGNN).

2.1 Notations

We assume familiarity with standard multi-sorted first-order logic (e.g., see [21]). A first-order language $\mathcal{L}$ is defined by a signature $\Sigma = (\mathcal{S}, \mathcal{R}, \mathcal{F}, \mathcal{X})$, where $\mathcal{S}$ is a non-empty set of sorts; $\mathcal{R}$ is a set of fixed-arity predicate (relation) symbols, each of which is associated with a list of argument sorts; $\mathcal{F}$ is a set of fixed-arity function symbols, each of which is associated with a list of argument sorts and a result sort; and $\mathcal{X} = \bigcup_{s \in \mathcal{S}} \mathcal{X}_s$ is a set of sorted variables, where $\mathcal{X}_s$ are the variables of sort s. A *term* t is a variable from $\mathcal{X}$, or an n-ary function symbol $f \in \mathcal{F}$ applied to terms $t_1, \ldots, t_n$ of the right sorts. An *atomic formula* (*atom*) of $\mathcal{L}$ is of the form $p(t_1, \ldots, t_n)$, where $p \in \mathcal{R}$ is an n-ary predicate symbol and $t_1, \ldots, t_n$ are terms of the right sorts. A *formula* is a Boolean literal *true*, *false*, an atom, or obtained by applying logical connectives $\neg$, $\wedge$, $\vee$, $\rightarrow$ and quantifiers $\forall$, $\exists$ to formulas. We write implications both left-to-right ($\varphi \rightarrow \psi$) and right-to-left ($\psi \leftarrow \varphi$). A formula is *closed* if all variables occurring in the formula are bound by quantifiers.

A *multi-sorted structure* $\mathcal{M} = (\mathcal{U}, I)$ for $\mathcal{L}$ consists of a set $\mathcal{U} = \bigcup_{s \in \mathcal{S}} \mathcal{U}_s$, being the union of non-empty domains $\mathcal{U}_s$ of each sort $s \in \mathcal{S}$, and an interpretation I such that $I(s) = \mathcal{U}_s$ for every sort $s \in \mathcal{S}$; for each n-ary predicate symbol $p \in \mathcal{R}$ with argument sorts $s_1, \ldots, s_n$, $I(p) \subseteq U_{s_1} \times \cdots \times U_{s_n}$; and for each n-ary function symbol $f \in \mathcal{F}$ with argument sorts $s_1, \ldots, s_n$ and result sort s, $I(f) \in U_{s_1} \times \cdots \times U_{s_n} \rightarrow U_s$. A *variable assignment* β for the structure $\mathcal{M} = (\mathcal{U}, I)$ is a function $\mathcal{X} \rightarrow \mathcal{U}$ that maps each variable $x \in \mathcal{X}_s$ to an element of the corresponding domain $\mathcal{U}_s$. Given $\mathcal{L}$, a structure $\mathcal{M} = (\mathcal{U}, I)$, and a variable assignment β, the *evaluation* of a term or formula is performed by the function $val_{\mathcal{M},\beta}$, defined by $val_{\mathcal{M},\beta}(x) = \beta(x)$ for a variable $x \in \mathcal{X}$; $val_{\mathcal{M},\beta}(f(t_1, \ldots, t_n)) = I(f)[val_{\mathcal{M},\beta}(t_1), \ldots, val_{\mathcal{M},\beta}(t_n)]$ for a function $f \in \mathcal{F}$; and $val_{\mathcal{M},\beta}(p(t_1, \ldots, t_n)) = true$ iff $(val_{\mathcal{M},\beta}(t_1), \ldots, val_{\mathcal{M},\beta}(t_n)) \in I(p)$. The evaluation of compound formulas is defined as is common. When $\mathcal{M}$ is clear from the context, we also write val_β instead of $val_{\mathcal{M},\beta}$.

We say that a formula φ is *satisfied* in $\mathcal{M}, \beta$ if $val_{\mathcal{M},\beta}(\varphi) = true$, and that it is *satisfiable* (SAT) if it is satisfied by some $\mathcal{M}, \beta$. We say a set Γ of formulae *entails* a formula φ, denoted $\Gamma \models \varphi$, if φ is satisfied whenever all formulas in Γ are satisfied.

Example 1. We assume a language $\mathcal{L}$ consisting of a sort s, two constants a and b, a unary function symbol f with argument and result of sort s; a binary predicate symbol p with two arguments of sort s. A structure $\mathcal{M} = (U_s, I)$ can be defined by $U_s = \mathbb{Z}$, $I(a) = 1$, $I(b) = 2$, $I(f)[x] = x + 2$, and $I(p)[x, y] = x \leq y$. The formula $\varphi = p(a, b) \rightarrow p(x, f(a))$ is satisfied in $\mathcal{M}$ by a variable assignment $\beta(x) = 1$ since $val_\beta(\varphi) = I(p)[val_\beta(a), val_\beta(b)] \rightarrow I(p)[val_\beta(x), I(f)[val_\beta(x)]] = \neg(1 \leq 2) \vee 1 \leq (1 + 2)$ is true. The formula φ is satisfiable since $val_\beta(\varphi) = true$ for $\mathcal{M}$ and a variable assignment $\beta(x) = 1$.

2.2   Constraint Horn Clauses

To introduce the notion of constrained Horn clauses, we assume a fixed base signature $\Sigma = (\mathcal{S}, \mathcal{R}, \mathcal{F}, \mathcal{X})$, as well as a unique structure $\mathcal{M}$ over this signature, forming the background theory. In this paper, we mainly consider the background theory of Linear Integer Arithmetic (LIA), following the SMT-LIB standard [22]. We further assume a set $\mathcal{R}_C$ of additional relation symbols that is disjoint from $\mathcal{R}$, which will be used to formulate the head and body of clauses.

Definition 1 (Constrained Horn clause). *Given signature Σ and the set $\mathcal{R}_C$, a Constrained Horn Clause (CHC) is a closed formula in the form*

$$\forall \bar{x}.\ H \leftarrow p_1(\bar{t}_1) \wedge \cdots \wedge p_n(\bar{t}_n) \wedge \varphi, \tag{1}$$

where $\bar{x}$ is a vector of variables; H is either false or an atom $p(t_1, \ldots, t_n)$ with $p \in \mathcal{R}_C$; the relation symbols $p_1, \ldots, p_n$ are elements of $\mathcal{R}_C$; and $\bar{t}_1, \ldots, \bar{t}_n$ and φ are vectors of terms and a formula over Σ, respectively. We call H the head *and $p_1(\bar{t}_1) \wedge \cdots \wedge p_n(\bar{t}_n) \wedge \varphi$ the* body *of the clause, respectively. We call the formula φ in the body the* constraint *of the clause.*

For convenience, in many places we leave out the quantifiers $\forall \bar{x}$ when writing clauses. A CHC without atoms in its body (the case $n = 0$) is called a *fact*. If the body of a CHC contains zero or one atom, the CHC is called *linear*. Otherwise, it is called *non-linear*.

Solving CHCs boils down to searching interpretations of the relation symbols $\mathcal{R}_C$ that satisfy the CHCs, assuming that all background symbols from Σ are interpreted by the fixed structure $\mathcal{M}$:

Definition 2 (Satisfiability of a CHC). *A CHC h is* satisfiable *if there is a structure $\mathcal{M}_C = (\mathcal{U}, I_C)$ for the extended signature $\Sigma_C = (\mathcal{S}, \mathcal{R} \cup \mathcal{R}_C, \mathcal{F}, \mathcal{X})$ such that (i) I_C coincides with I on Σ, and (ii) $\mathcal{M}_C$ satisfies h. A set $\mathcal{C}$ of CHCs is* satisfiable *if there is an extended structure $\mathcal{M}_C$ simultaneously satisfying all clauses in $\mathcal{C}$.*

Encoding Program Verification Problems using CHCs. A program verification problem involves checking whether a program adheres to its specified behavior. One approach to verification is to transform the problem into determining the satisfiability of a set of CHCs. This can be done, for instance, by encoding the partial correctness of a procedural imperative program into a negated Existential positive Least Fixed-point Logic (E+LFP) formula [4] using the weakest precondition calculus. Generally, encodings are designed such that the set of CHCs is satisfiable if and only if a program is safe. Various encoding schemes for different programming languages have been introduced in the literature, e.g., [4].

2.3   Graph Neural Networks

A Graph Neural Network (GNN) [13] is a type of neural network that consists of Multi-Layer Perceptrons (MLPs) [23]. GNNs operate on graph-structured data with nodes and edges, making them suitable for logic formulas which can naturally be represented as graphs. A GNN can take a set of typed nodes and edges as input and output a set of feature representations (vectors of real numbers) associated with the properties of the nodes. We refer to them as *node representations* in the rest of the sections.

The Message-Passing based GNN (MP-GNN) [24] is a type of GNN model. It utilizes an iterative message-passing algorithm in which each node in the graph aggregates messages from its neighboring nodes to update its own node representation. This mechanism assists in identifying the inner connections within substructures, such as terms and atoms, in graph represented logic formulae.

Formally, let $G = (V, E)$ be a graph, where V is the set of nodes and E is the set of edges. Let x_v be the initial node representation (a vector of random real numbers) for node v in the graph, and let N_v be the set of neighbors of node v. An MP-GNN consists of a series of T message-passing steps. At each step t, every node v in the graph updates its node representation as follows:

$$h_v^t = \phi_t(\rho_t(\{h_u^{t-1} \mid u \in N_v\}), h_v^{t-1}), \tag{2}$$

where $h_v^t \in \mathbb{R}^n$ is the updated node representation for node v after t iterations. The initial node representation, h_v^0, is usually derived from the node type and given by x_v. The node representation of u in the previous iteration $t - 1$ is h_u^{t-1}, and node u is a neighbor of node v. $\rho_t : (\mathbb{R}^n)^{|N_v|} \to \mathbb{R}^n$ is a aggregation function with trainable parameters (e.g., a MLP followed by sum, min, or max) that aggregates the node representations of v's neighboring nodes at the t-th iteration. $\phi_t : \mathbb{R}^n \to \mathbb{R}^n$ is a function with trainable parameters (e.g., a MLP) that takes the aggregated node representation from ρ_t and the node representation of v in previous iteration as input, and outputs the updated node representation of v at the t-th iteration. MP-GNN assumes a node can capture local structural information from t-hop's neighbors by updating the node representation using aggregated representations of the neighbor nodes.

The final output of the MP-GNN could be the set of updated node representations for all nodes in the graph after T iterations. These node representations can be used for a variety of downstream tasks, such as node classification or graph classification.

Relational Hyper-Graph Neural Network (R-HyGNN) [1] is an extension of one MP-GNN called Relational Graph Convolutional Networks (R-GCN) [25], and it is specifically designed to handle labeled hypergraphs.

A labeled (typed) hypergraph is a hypergraph where each vertex (node) and hyperedge is assigned a type from a predefined set of types. Formally, a labeled hypergraph LHG is defined as a tuple $LHG = (V, E, \lambda_V, L_V, L_E)$, where V is a set of elements called vertices (or nodes), L_E is a set of pair consisting of a label (type) r and the number of nodes k under the label r, $E \subseteq V^* \times L_E$ is a set of hyperedges in which each hyperedge consists of a non-empty subsets of V and a pair $(r, k) \in L_E$. Here, $\lambda_V : V \to L_V$ is a labeling function that assigns a type from the set L_V to each vertex in V. The L_V is a set of possible types (labels) for the vertices V.

The node representation updating rule of R-HyGNN for one node v at timestep t is

$$h_v^t = \text{ReLU}(\sum_{\substack{(w_1,\ldots,w_k,(r,k)) \\ \in E}} \sum_{\substack{i \in \{1,\ldots,k\}, \\ w_i = v}} W_{r,i}^t \cdot \|(h_{w_1}^{t-1}, \ldots, h_{w_{i-1}}^{t-1}, h_{w_{i+1}}^{t-1}, \ldots, h_{w_k}^{t-1})), \tag{3}$$

where the pair $(r, k) \in L_E$ is the edge type (relation) and the number of node for a edge $(w_1, \ldots, w_k, (r, k)) \in E$, $W_{r,i}^t$ is a matrix of learnable parameters in time step t for node $v = w_i$ in the edge with type r. There are $|L_E| \times \sum_{(r,k) \in L_E}(k) \times t$ matrices of learnable parameters in total. Here, $\|(h_{w_1}^{t-1}, \ldots, h_{w_{i-1}}^{t-1}, h_{w_{i+1}}^{t-1}, \ldots, h_{w_k}^{t-1})$ means concatenate v' neighbour node representations in time step $t - 1$. The initial node representation h_v^0 is derived from the node types L_V.

Intuitively, to update the representation of a node, R-HyGNN first concatenates the neighbor representations of node v for each edge from the previous time step $t-1$. It then multiplies the concatenated neighbor representations by the corresponding matrix of trained parameters (i.e., $W_{r,i}^t$) to derive a local representation of v. Next, it aggregates the local node representations (e.g., by addition). In other words, $\sum_{\substack{(w_1, \ldots, w_k, (r,k)) \\ \in E}} \sum_{\substack{i \in \{1, \ldots, k\}, \\ w_i = v}} W_{r,i}^t \cdot \|(h_{w_1}^{t-1}, \ldots, h_{w_{i-1}}^{t-1}, h_{w_{i+1}}^{t-1}, \ldots, h_{w_k}^{t-1})$ can be abstract to $\rho_t(\|(h_{w_1}^{t-1}, \ldots, h_{w_{i-1}}^{t-1}, h_{w_{i+1}}^{t-1}, \ldots, h_{w_k}^{t-1}))$ where ρ_t is an aggregation function with trainable parameters in $W_{r,i}^t$. Finally, it applies a ReLU function [26] as the update function ϕ_t. This function takes the aggregated local node representations as input and produces the final node representation h_v^t. This updating process for a single node recurs t times.

3 Abstract Symbolic Model Checking for CHCs

The goal of our work is to utilize GNNs to obtain improved state space exploration methods for CHCs. To this end, in this section we introduce an abstract formulation of CHC state space exploration, covering both the classical CEGAR approach and exploration in the style of symbolic execution.

3.1 Satisfiability Checking for CHCs

Our Algorithm 1 checks the satisfiability of a given a set $\mathcal{C}$ of CHCs by constructing an abstract reachability hyper-graph (ARG). An ARG is an over-approximation of the facts $p(\bar{a})$ that are logically entailed by $\mathcal{C}$; by demonstrating that the atom *false* does not follow from $\mathcal{C}$, it can be shown that $\mathcal{C}$ is satisfiable. Since each node of an ARG can represent a whole set of facts, a finite ARG can be a representation even of infinite models of a set $\mathcal{C}$.

We first give an abstract definition of ARGs that does not mandate any particular symbolic representation of sets of $p(\bar{a})$. We later introduce two instances of this abstract framework.

Like in Sect. 2.2, we denote the set of relation symbols used in CHCs by $\mathcal{R}_C$. For a k-ary relation symbol $p \in \mathcal{R}_C$ with argument sorts $s_1, \ldots, s_k$, we write $\mathcal{R}_p = \mathcal{P}(U_{s_1} \times \cdots \times U_{s_k})$ for the set of possible relations represented by p.

Definition 3. *An abstract reachability graph for a set $\mathcal{C}$ of CHCs is a hyper-graph (V, E), where*

- *the set V of nodes is a set of pairs (p, R), with p being a relation symbol and $R \in \mathcal{R}_p$;*

- $E \subseteq V^* \times C \times V$ is a set of hyper-edges labelled with clauses. For every edge $(v_1, \ldots, v_n, h, v_0) \in E$, it is the case that:
  - the head of a clause h is not false, i.e., h is of the form $\forall \bar{x}.\ p_0(\bar{t}_0) \leftarrow p_1(\bar{t}_1) \wedge \cdots \wedge p_n(\bar{t}_n) \wedge \varphi$;
  - nodes $v_0, \ldots, v_n$ correspond to the head and the body of h, i.e., for $i \in \{0, \ldots, n\}$ it is the case that $v_i = (p_i, R_i)$ for some $R_i \in \mathcal{R}_{p_i}$;
  - R_0 over-approximates the facts implied by the clause h:

$$R_0 \supseteq \left\{ val_\beta(\bar{t}_0) \ \middle| \ \begin{array}{l} val_\beta(\varphi) = true \text{ and } val_\beta(\bar{t}_i) \in R_i \text{ for } i = \{1, \ldots, n\} \\ \text{for some variable assignment } \beta \end{array} \right\}. \tag{4}$$

The algorithm starts with an empty ARG, and then adds nodes and edges to it until the ARG is *complete*, which intuitively means that all possible non-trivial edges are present in the graph. To define the notion of a complete ARG, we first need to characterize what it means for a clause to correspond to a feasible edge of the ARG:

Definition 4. *A clause $h \in C$ is* feasible *for nodes $v_1, \ldots, v_n \in V$ of an ARG (V, E) if*

- *h is of the form $\forall \bar{x}.\ H \leftarrow p_1(\bar{t}_1) \wedge \cdots \wedge p_n(\bar{t}_n) \wedge \varphi$;*
- *nodes $v_1, \ldots, v_n$ correspond to the body of h, i.e., for $i \in \{1, \ldots, n\}$ it is the case that $v_i = (p_i, R_i)$ for some $R_i \in \mathcal{R}_{p_i}$;*
- *the constraints imposed by φ and $v_1, \ldots, v_n$ are not contradictory, i.e., there is a variable assignment β such that $val_\beta(\varphi) = true$ and $val_\beta(\bar{t}_i) \in R_i$ for $i = \{1, \ldots, n\}$.*

Definition 5. *An ARG (V, E) is* complete *for a set C of CHCs if*

- *for every CHC $h \in C$ that has a head different from false, and that is feasible for $v_1, \ldots, v_n \in V$, there is some edge $\langle (v_1, \ldots, v_n), h, v_0 \rangle \in E$;*
- *there is no CHC $h \in C$ with head false that is feasible for any $v_1, \ldots, v_n \in V$.*

We can finally observe that complete ARGs correspond to models of the clause set C.

Lemma 1. *A set C of CHCs has a complete ARG iff C is satisfiable.*

Algorithm 1 describes how ARGs can be constructed for a given clause set C. The algorithm starts with an empty ARG (V, E), and maintains a queue $Q \subseteq C \times V^*$ of feasible edges to be added to the graph next. The queue is initialized with the clauses with empty body, representing the initial states of a program. Once the queue runs empty, the constructed ARG is complete and the set C has been shown to be satisfiable.

In each iteration, in lines 5–6 an element $(h, \bar{v})$ is picked and removed from the queue Q. If the head of h is *false* (line 7), the edge to be added might be part of a witness of unsatisfiability of C. In this case, it has to be checked whether the nodes $\bar{v}$ are over-approximate (line 8); this can happen when the

Input: A set $\mathcal{C}$ of CHCs
Output: Satisfiability of $\mathcal{C}$
Initialise: $V := \emptyset, E := \emptyset, Q := \{(h, ()) \mid h \in \mathcal{C}, h = \forall \bar{x}.\ H \leftarrow \varphi\}$

```
 1  while true do
 2  │   if Q is empty then
 3  │   │   return satisfiable
 4  │   else
 5  │   │   Pick (h, v̄) ∈ Q to be considered next (guided by GNNs)
 6  │   │   Q := Q \ {(h, v̄)}
 7  │   │   if the head in h is false then
 8  │   │   │   if derivation of false is genuine then
 9  │   │   │   │   return unsatisfiable
10  │   │   │   else
11  │   │   │   │   Refine over-approximations
12  │   │   │   │   Delete all affected nodes in (V, E)
13  │   │   │   │   Regenerate elements in Q
14  │   │   │   end
15  │   │   else
16  │   │   │   Assume h = ∀x̄. p₀(t̄₀) ← p₁(t̄₁) ∧ ··· ∧ pₙ(t̄ₙ) ∧ φ
17  │   │   │   Compute new node u = (p₀, R₀) for (h, v̄)
18  │   │   │   if u ∉ V then
19  │   │   │   │   V := V ∪ {u}
20  │   │   │   │   Q := Q ∪ {(d, (w₁,...,wₘ)) | d ∈ C is feasible for w₁,...,wₘ, u ∈ {w₁,...,wₘ} ⊆ V}
21  │   │   │   end
22  │   │   │   E := E ∪ {(v̄, h, u)}
23  │   │   end
24  │   end
25  end
```

Algorithm 1: Abstract symbolic model checking algorithm for checking satisfiability of CHCs

containment in (4) is sometimes strict in the constructed ARG, and occurs in particular when instantiating the abstract algorithm as CEGAR (see Sect. 3.2). The over-approximation can then be refined in lines 11–13.

If the head of h is not *false*, a further edge is added to the graph by computing in line 17 some set R_0 satisfying (4). If the resulting node u is new in the graph, the queue Q is updated by adding possible outgoing edges for u (line 19–20).

In this paper, we apply the GNN-based guidance in line 5. Specifically, we presume that the GNN can predict the probability of h being in MUSes for each $q = (h, \bar{v}) \in Q$. We then combine this probability with certain features of q, such as the number of iterations $q \in Q$ has been waiting, to calculate a priority of q. When selecting an element q from Q, we consult this priority. We explain more details in Sect. 4.

3.2   CEGAR- And Symbolic Execution-Style Exploration

We now discuss two concrete instantiations of Algorithm 1. The first one, in this paper called SymEx, resembles the symbolic execution [27] of a program, and represents the relation R in an ARG node (p, R), for a k-ary relation symbol p, as a formula over free variables $z_1, \ldots, z_k$.

To obtain SymEx, in line 17 of Algorithm 1 the relation R_0 is derived from the nodes $\bar{v}$ by simple symbolic manipulation. Assuming that $v_i = (p_i, R_i)$ for $i \in \{1, \ldots, n\}$, and the relations R_i are all represented as formulas, we can define:

$$R_0[z_1, \ldots, z_k] = \exists \bar{x}. \ \bar{z} = \bar{t}_0 \land R_1[\bar{t}_1] \land \cdots \land R_n[\bar{t}_n] \land \varphi$$

where the notation $R_i[\bar{t}_i]$ expresses that the terms $\bar{t}_i$ are substituted for the free variables $\bar{z}$. In our implementation on top of the CHC solver Eldarica [28], the formula $R_0[z_1, \ldots, z_k]$ is afterwards simplified by eliminating the quantifiers, albeit only in a best-effort way by running the built-in formula simplifier of Eldarica. In SymEx, since no over-approximation is applied, the test in line 8 always succeeds, and lines 11–13 are never executed.

Our second instantiation, called CEGAR, is designed following counter example-guided abstraction refinement [5,29] with Cartesian predicate abstraction [30]. In this version of the algorithm, we assume that a finite pre-defined set Π_p of predicates is available for every relation symbol p. If p is k-ary, then the elements of Π_p are formulas over free variables $z_1, \ldots, z_k$. The relation R in a node (p, R) is now represented as a subset of Π_p.

In line 17, the set R_0 is computed by determining the elements of Π_{p_0} that are entailed by the body of clause h:

$$R_0 = \{\phi \in \Pi_{p_0} \mid \bar{z} = \bar{t}_0 \land \bigwedge R_1[\bar{t}_1] \land \cdots \land \bigwedge R_n[\bar{t}_n] \land \varphi \models \phi\}$$

The notation $\bigwedge R_i[\bar{t}_i]$ denotes the conjunction of the elements of R_i, with $\bar{t}_i$ substituted for the free variables $\bar{z}$.

Over-approximation in CEGAR stems from the fact that a chosen set of predicates Π_p will oftentimes not be able to exactly represent a relation; the constructed ARG might then include facts $p(\bar{a})$ that are not actually entailed by $\mathcal{C}$. In line 8, the algorithm therefore has to verify that discovered derivations of *false* are genuine. This is done by collecting the clauses that were used to derive the nodes $\bar{v}$ in the ARG and constructing a counterexample tree. If the counterexample turns out to be *spurious*, further predicates are added to the sets Π_p, for instance using tree interpolation [31], in lines 11–13.

4   Guiding CHC Solvers Using MUSes

In this section, we begin by defining the notion of Minimal Unsatisfiable Sets (MUSes), then we detail the process of collecting three types of training labels using the MUSes. Following that, we explain the various strategies of employing the predicted probability of a CHC being in MUSes to guide Algorithm 1.

4.1   Minimal Unsatisfiable Sets

Throughout the section, assume that $\mathcal{C}$ is an unsatisfiable set of CHCs.

Definition 6 (Minimal Unsatisfiable Set). *A subset $U \subseteq \mathcal{C}$ is a* Minimal Unsatisfiable Set *(MUS) if U is unsatisfiable and for all CHCs $h \in U$ it is the case that $U \backslash \{h\}$ is satisfiable.*

Intuitively, MUSes of a set of CHCs encoding a program correspond to minimal counterexamples (i.e., a subset of program statements) witnessing the incorrectness of the program. MUSes are therefore good candidates for guiding CHC solvers towards the critical clauses, and we aim at predicting MUSes using GNNs.

The number of MUSes can, however, be exponential in the number of CHCs in $\mathcal{C}$. We therefore consider the *union, intersection*, and a particular *single* MUS for $\mathcal{C}$. Denoting the set of all MUSes of $\mathcal{C}$ by MUS$(\mathcal{C})$, those are:

$$\mathcal{C}_{\mathrm{MUSes}}^{\mathrm{union}} = \bigcup \mathrm{MUS}(\mathcal{C}), \qquad \mathcal{C}_{\mathrm{MUSes}}^{\mathrm{intersection}} = \bigcap \mathrm{MUS}(\mathcal{C}),$$

$$\mathcal{C}_{\mathrm{MUSes}}^{\mathrm{single}} = \operatorname*{argmax}_{U \in \mathrm{MUS}(\mathcal{C})} numAtom(U),$$

where $numAtom(U)$ is the total number of atoms of the CHCs in U, and $\mathcal{C}_{\mathrm{MUSes}}^{\mathrm{single}}$ is some MUS that maximizes the total number of atoms. The three clause sets can be computed using the OptiRica extension of the Eldarica Horn solver [32].

Intuitively, $\mathcal{C}_{\mathrm{MUSes}}^{\mathrm{union}}$ includes all information about possible MUSes and encourages the algorithm to go through all possible error-prone areas. In contrast, $\mathcal{C}_{\mathrm{MUSes}}^{\mathrm{intersection}}$ only takes the intersection of all MUSes which can guide the algorithm to only focus on the most suspicious areas. $\mathcal{C}_{\mathrm{MUSes}}^{\mathrm{single}}$ is one of MUSes and corresponds to a long path in the ARG, given that a high number of atoms is associated with a large number of nodes. We believe a long path contains intricate information, which is challenging for human to parse, but easier for a deep-learning-based model to find.

We form three types of Boolean clause labelings by using $\mathcal{C}_{\mathrm{MUSes}}^{\mathrm{union}}$, $\mathcal{C}_{\mathrm{MUSes}}^{\mathrm{intersection}}$, and $\mathcal{C}_{\mathrm{MUSes}}^{\mathrm{single}}$, respectively. For instance, for $\mathcal{C}_{\mathrm{MUSes}}^{\mathrm{union}}$, we obtain the labels

$$l^{\mathrm{union}}(c) = \begin{cases} 1 & \textit{if } h \in \mathcal{C}_{\mathrm{MUSes}}^{\mathrm{union}} \\ 0 & \textit{if } h \in \mathcal{C} \setminus \mathcal{C}_{\mathrm{MUSes}}^{\mathrm{union}} \end{cases}$$

4.2   MUS-Based Guidance for CHC Solvers

Eldarica Heuristics The implementations of CEGAR and SymEx in the Horn solver Eldarica [28] by default use fixed, hand-written selection heuristics in line 5 of Algorithm 1. Such heuristics are defined by a ranking function $r : Q \to \mathbb{Z}$ that maps every element in the queue to an integer; in line 5, the element of Q is picked that minimizes r. The standard implementation of r used for CEGAR is defined by

$$EldCEGARRank(q) = numPredicate(q) + birthTime(q) + falseClause(q) \,,$$

where $numPredicate((h, \bar{v})) = \sum_{(p,R) \in \bar{v}} |R|$ is the total number of predicates occurring in the considered nodes of the ARG; $birthTime(q)$ is the iteration (as an integer) in which q was added to Q;[1] and $falseClause((h, \bar{v}))$ is 0 if the head of h is not *false*, and some big integer otherwise. The rationale behind the ranking function is that nodes with few predicates tend to subsume nodes with many predicates, every clause should be picked eventually, and clauses with head *false* will trigger either termination of the algorithm or abstraction refinement.

In SymEx, for a node (p, R), we define $numConstraint(R)$ to be the number of conjuncts of R. For instance, for $R = (x > 1 \land y < 0)$ we would get $numConstraint(R) = 2$. The default Eldarica ranking function for SymEx is defined by

$$EldSymExRank((h, \bar{v})) = \sum_{(p,R) \in \bar{v}} numConstraint(R) \,.$$

Similar to $numPredicate$, the intuition behind $EldSymExRank$ is that nodes with larger formulas (more restrictions) tend to be subsumed by nodes with smaller formulas (fewer restrictions).

MUS-Guided Heuristics We now introduce several new ranking functions defined with the help of MUSes. For this, suppose that $\mathcal{C}$ is the set of CHCs that a Horn solver is applied on. Under the assumption that $\mathcal{C}$ is unsatisfiable, we obtain three labeling functions l^{union}, $l^{intersection}$, l^{single} that are able to point out clauses to be prioritized in Algorithm 1.

In practice, of course, the status of $\mathcal{C}$ will initially be unknown. We therefore use three GNNs to *predict* labels $l^{union}(h)$, $l^{intersection}(h)$, $l^{single}(h)$, respectively, given just the set $\mathcal{C}$ and some clause $h \in \mathcal{C}$ as input. We interpret the prediction of a GNN as a *probability* $P(h)$ of a clause h to be in the union, intersection, or the single MUS set, and use those probabilities to define ranking functions. Table 1 lists several candidate ranking functions in terms of this membership probability P, where P stands for one of P^{union}, P^{single}, or $P^{intersection}$.

We consider two ways to convert probabilities to integers that can be used in the ranking function. In $rank_P(q)$, the elements $q = (h, \bar{v})$ of the queue Q are first sorted in descending order of $P(h)$; the number $rank_P(q)$ is the position of q in this sequence. This means that $rank_P(q)$ ranges from 0 to $|Q| - 1$, and elements with large probability $P(h)$ will be assigned small rank.

In $norm_P(q)$, we assume that $min_P = \min\{P(h) \mid (h, \bar{v}) \in Q\}$ and $max_P = \max\{P(h) \mid (h, \bar{v}) \in Q\}$ denote the minimum and maximum probability, respectively, among elements of Q. The normalized value $norm_P(q) \in [0, 1]$ is defined by

$$norm_P((h, \bar{v})) = \frac{P(h) - min_P}{max_P - min_P}$$

In the ranking functions, we multiple $norm_P(q)$ with a negative coefficient *coef* and round the result to the nearest integer.

---

[1] Strictly speaking, this information cannot be computed from q, it is in practice stored as an additional field of the queue elements.

Table 1. Ranking function for queue elements $q \in Q$ in Algorithm 1, where $P \in \{P^{\text{union}}, P^{\text{single}}, P^{\text{intersection}}\}$.

Algorithm	Name	Ranking function
CEGAR	Fixed	$EldCEGARRank(q)$
	Random	$RandomRank$
	Score	$coef \cdot norm_P(q)$
	Rank	$rank_P(q)$
	R-Plus	$rank_P(q) + EldCEGARRank(q)$
	S-Plus	$coef \cdot norm_P(q) + EldCEGARRank(q)$
	R-Minus	$rank_P(q) - EldCEGARRank(q)$
	S-Minus	$coef \cdot norm_P(q) - EldCEGARRank(q)$
SymEx	Fixed	$EldSymExRank(q)$
	Random	$RandomRank$
	Score	$coef \cdot norm_P(q)$
	Rank	$rank_P(q)$
	R-Plus	$rank_P(q) + EldSymExRank(q) + birthTime(q)$
	S-Plus	$coef \cdot norm_P(q) + EldSymExRank(q) + birthTime(q)$
	R-Minus	$rank_P(q) - EldSymExRank(q) - birthTime(q)$
	S-Minus	$coef \cdot norm_P(q) - EldSymExRank(q) - birthTime(q)$
	Two-queue	R-Minus, 80% *probability* Random, 20% *probability*

The *RandomRank* function ensures that each CHC has an equal opportunity to be selected in each iteration. The Two-queue case denotes a setup with two queues, Q_1 and Q_2, each used with a certain probability in line 5 of Algorithm 1. We list just one such combination, which alternates between queues with the R-Minus and Random functions: there's an 80% chance we use the R-Minus queue and a 20% chance we use the Random queue.

5   Design of Model

As shown in Fig. 1, within the CHC-R-HyGNN framework, we first encode a set of CHCs into a graph format, and then we build a GNN model consisting of an encoder, GNN layers, and a decoder.

5.1   Encode CHCs to Graph Representation

We apply the two (hyper-)graph representations of CHCs defined in [1]. We will briefly describe their main features here.

The first graph representation is called a *constraint graph* (CG). This graph encapsulates syntactic information by using nodes to represent each symbol and connecting them with binary edges. Each CHC and each predicate symbol is represented by a unique node. Terms and formulas are represented using their abstract syntax trees (AST). CHC nodes are connected to the constituent atoms and constraints by binary edges. Within one CHC, common sub-expressions are represented by the same nodes. Different hyper-graph node and edges types are used to distinguish the various encoded operators (see Sect. 2.3).

The second graph representation is the *control- and data-flow hypergraph* (CDHG). This graph is designed to capture both control- and data-flow within CHCs using hyperedges, and therefore captures the semantics of CHCs more directly than GCs. Similar to the CG, in the CDHG, each CHC is represented by a unique node, and the atoms are rendered in the same way as in CG. Unlike the CG, the CDHG uses *control-flow hyperedges* (CFHEs) to describe the control-flow from the body to the head in each CHC, guarded by the constraint of the CHC. Furthermore, the CDHG uses *data-flow hyperedges* (DFHEs) to represent data-flow from the terms in the body to the terms in the head. These data-flows are also guarded by the constraint.

5.2 Model Structure

The R-HyGNN model consists of three sub-components: (i) encoder, (ii) R-HyGNN [1] (a GNN), and (iii) decoder:

$$\text{(i) } \mathcal{H}_0 = \text{encoder}(V, \lambda_V, L_V), \qquad \text{(ii) } \mathcal{H}_t = \text{R-HyGNN}(\mathcal{H}_0, E, L_E),$$

$$\text{(iii) } \hat{\mathcal{L}} = \text{decoder}(\mathcal{H}_t) .$$

The encoder in (i) first maps each node in V to an integer according to the node's type determined by λ_V and L_V. Then, it passes the encoded integers to a single-layer neural network (embedding layer) to compute initial node representations $\mathcal{H}_0$. The R-HyGNN in (ii) is a GNN with its node representation updating rule defined in (3). It takes the initial node representations ($\mathcal{H}_0$), edges (E), and edge types (L_E) as input and outputs the updated node representations $\mathcal{H}_t$. The decoder in (iii) first identifies the node that denote the CHC instead of the variables, atoms, or other elements in the CHC, then we collect the representations of these CHC nodes $\mathcal{H}_t^{\text{CHCs}}$ from all node representations $\mathcal{H}_t$. Finally, we pass $\mathcal{H}_t^{\text{CHCs}}$ to a set of fully connected neural networks to compute the probability of each CHC being in the MUSes, and $\hat{\mathcal{L}}$ is a set of probabilities.

The parameters of neural networks in (i), (ii), and (iii) are optimized together by minimizing the binary cross-entropy loss [33] between $\hat{\mathcal{L}}$ and the true labels $\mathcal{L}$, i.e.,

$$loss = -\frac{1}{N} \sum_{i=1}^{N} \mathcal{L}_i log(\hat{\mathcal{L}}_i) + (1 - \mathcal{L}_i)log(1 - \hat{\mathcal{L}}_i). \tag{5}$$

6 Evaluation

We first describe how the benchmarks are split for training and evaluation, then list some important parameters. Finally, we show and explain the experimental results. This work can be reproduced by following the instructions in [34].

Table 2. Distribution of the number of problems for both training and evaluation. T/O, N/A, and Evail. denote timeout, not available, and evaluation, respectively.

Linear LIA problems					Non-linear LIA problems				
8705					8425				
Benchmarks for training			Holdout set		Benchmarks for training			Holdout set	
7834 (90%)			871 (10%)		7579 (90%)			846 (10%)	
UNSAT	SAT	T/O	Eval.	N/A	UNSAT	SAT	T/O	Eval.	N/A
1585	4004	2245	383	488	3315	4010	254	488	358
Train	Valid	N/A			Train	Valid	N/A		
782	87	716			1617	180	1518		

6.1 Benchmark

The training and evaluation data are specified in Table 2, and available in [20]. There are 8705 linear and 8425 non-linear LIA problems, taken from CHC-COMP 2021 [19]. We first uniformly reserved 10% of the benchmarks as hold-out set for the final evaluation. We ran the CEGAR in Eldarica using a 3-hour timeout to solve the remaining 90% of benchmarks, leading to three groups of benchmarks: SAT, UNSAT, and timeout. For UNSAT problems, we also computed the MUS sets $C_{\text{MUSes}}^{\text{union}}$, $C_{\text{MUSes}}^{\text{intersection}}$, and $C_{\text{MUSes}}^{\text{single}}$. Some problems in UNSAT were eliminated in this process (N/A, for both training and evaluation) because the problems were trivial (already solved by the Eldarica preprocessor), the process of extracting MUSes timed out (3 h), or a timeout occurred when encoding CHCs as graphs. The remaining problems are divided into training (90%) and validation (10%) datasets.

6.2 Parameters

We select the hyper-parameters for the GNN empirically, and according to the experimental results from [1]. We set the vector size of the initial node representation and the number of neurons in all intermediate neural network layers to 32; we also set the number of message-passing steps to 8 (i.e., applying (3) 8 times). The constant *coef* in Table 1 is -1000. Other parameters and the instructions to reproduce these results can be found in [20].

6.3 Experimental Results

In our experiment, we measure the number of solved problems and the average solving time for the holdout evaluation set. This included 383 and 488 problems in the linear and non-linear LIA datasets, respectively. The timeout for evaluating each problem is 1200 s. The additional overhead for reading and applying the GNN predicted results in each iteration is included in the solving time. The numerical results are shown in Tables 3 and 4. We also visualize some numerical results by scatter plots in Fig. 2.

Table 3. Overview of the best ranking function and improvement in number of solved problems compared to the Eldarica. A ranking function marked with * (e.g., S-Plus*) denotes that there are multiple ranking functions with the same performance.

Benchmark Algorithm	MUS data set (best count)	Best ranking function (improvement in %)						
		Number of Solved Problems			Average Time			
		Total	SAT	UNSAT	All	Common	SAT	UNSAT
Linear CEGAR	Union (0)	R-Plus (1.4%)	R-Plus (2.4%)	R-Minus (1.0%)	R-Plus (1.3%)	S-Plus (19.1%)	S-Minus (46.5%)	Rank (31.1%)
	Single (3)	Rank (3.6%)	**R-Plus (4.0%)**	Rank (8.2%)	R-Plus (1.9%)	S-Plus (26.6%)	**R-Minus (57.9%)**	**Rank (36.3%)**
	Intersection (4)	**R-Plus (4.1%)**	S-Plus (0.8%)	**R-Plus (9.3%)**	**R-Plus (3.1%)**	**S-Plus (27.6%)**	R-Minus (45.0%)	S-Plus (0.0%)
Linear SymEx	Union (4)	**Two-Q (1.0%)**	**S-Plus* (0.0%)**	**Random (2.0%)**	Two-Q (0.9%)	R-Minus (12.7%)	**R-Minus (30.2%)**	S-Plus (26.5%)
	Single (3)	S-Minus* (0.5%)	**S-Plus* (0.0%)**	**Random (2.0%)**	Random (0.8%)	**S-Plus (12.9%)**	Random (28.4%)	S-Plus (17.6%)
	Intersection (5)	**S-Plus* (1.0%)**	**S-Plus* (0.0%)**	**S-Plus* (2.0%)**	**S-Plus (1.3%)**	Score (9.5%)	Random (28.4%)	**R-Plus (35.8%)**
Non-Linear CEGAR	Union (7)	**S-Plus (0.5%)**	**S-Plus (0.8%)**	**S-Plus* (0.0%)**	**S-Plus** B	**R-Minus (20.8%)**	**Rank (53.5%)**	**S-Plus (19.4%)**
	Single (1)	R-Plus (0.2%)	R-Plus (0.4%)	**R-Plus* (0.0%)**	R-Plus (6.6%)	S-Plus (18.4%)	R-Minus (52.8%)	R-Minus (14.2%)
	Intersection (1)	R-Plus* (0.0%)	S-Plus (0.5%)	**S-Plus* (0.0%)**	R-Plus (5.9%)	R-Plus (20.3%)	Rank (45.8%)	S-Plus (16.8%)
Non-Linear SymEx	Union (6)	**Two-Q (6.1%)**	**S-Minus* (1.6%)**	Random (12.3%)	**Two-Q (13.3%)**	**R-Minus (7.3%)**	**Score (5.1%)**	**R-Plus (19.9%)**
	Single (3)	**Two-Q (6.1%)**	**Score (1.6%)**	**Two-Q (12.9%)**	Two-Q (12.4%)	Rank (-2.2%)	R-Minus (0.2%)	Two-Q (11.2%)
	Intersection (3)	**Two-Q (6.1%)**	**S-Plus (1.6%)**	**Two-Q (12.9%)**	Two-Q (12.7%)	S-Minus (0.6%)	Two-Q (1.7%)	S-Plus (5.4%)

In Tables 3 and 4, under the Number of Solved Problems column, the Total, and SAT, UNSAT columns denote the number of totals solved, solved SAT, and solved UNSAT problems, respectively. Under the Average Time column, the All column denotes the average solving time for all problems, including those that timed out; the Common column means the average solving time for problems that were commonly solved using one of the ranking functions in Table 1, and the default Eldarica ranking function; the SAT and UNSAT columns are the average solving times for SAT and UNSAT problems, respectively. In certain cells, the percentage in brackets represents the improvement compared to the corresponding default ranking function. The bold text highlights the best performance in a Benchmark Algorithm block for each measurement.

Table 3 displays the best ranking function and its improvement over the default Eldarica ranking function in different measurements for various combinations of benchmarks (linear and non-linear LIA), algorithms (CEGAR and SymEx), and MUS datasets (union, single, and intersection of MUSes). In the MUS dataset column, the numbers in brackets represent the count of bold text cells in the row, indicating the number of best performances achieved by that type of MUS dataset. For instance, in the last row (i.e., the Intersection row of the Non-Linear SymEx block), the number in the bracket is counted from the bold text highlighted cells in columns Total, SAT, and UNSAT under the Number of Solved Problems. This suggests that using the intersection of the MUSes

Table 4. Evaluation on holdout problems using union dataset. The time unit is second.

Benchmark Algorithm	Ranking Function	Number of Solved Problems (improvement %)			Average Time(improvement %)			
		Total	SAT	UNSAT	All	Common	SAT	UNSAT
Linear CEGAR	Default	222	125	97	519.38	25.77	38.97	8.77
	Random	221 (−0.5%)	124 (−0.8%)	97 (0.0%)	523.58 (−0.8%)	27.49 (−29.5%)	37.05 (4.9%)	15.85 (−80.7%)
	R-Plus	**225** (**1.4%**)	**128** (**2.4%**)	97 (0.0%)	**512.41** (**1.3%**)	21.65 (16.0%)	42.89 (−10.1%)	11.99 (−36.7%)
	R-Minus	220 (−0.9%)	122 (−2.4%)	**98** (**1.0%**)	526.08 (−1.3%)	18.02 (−24.4%)	30.93 (20.6%)	21.60 (−146.3%)
	S-Plus	222 (0.0%)	125 (0.0%)	97 (0.0%)	517.43 (0.4%)	**20.92** (**19.1%**)	34.13 (12.4%)	**7.32** (**16.5%**)
	S-Minus	219 (−1.4%)	122 (−2.4%)	97 (0.0%)	522.97 (−0.7%)	12.56 (2.4%)	**20.86** (**46.5%**)	9.81 (−11.9%)
	Portfolio	229 (3.2%)	130 (4.0%)	99 (2.1%)	503.16 (3.1%)	18.28 (29.1%)	45.67 (−17.2%)	19.94 (−127.4%)
Linear SymEx	Default	200	101	99	590.68	33.16	55.42	10.44
	Random	201 (0.5%)	100 (−1.0%)	101 (2.0%)	586.12 (0.8%)	30.08 (−8.5%)	39.69 (28.4%)	20.95 (−100.7%)
	R-Plus	192 (−4.0%)	101 (0.0%)	91 (−8.1%)	617.60 (−4.6%)	38.59 (−10.9%)	52.87 (4.6%)	21.99 (−110.6%)
	R-Minus	200 (0.0%)	100 (−1.0%)	100 (1.0%)	586.24 (0.8%)	**24.67** (**12.7%**)	**38.69** (**30.2%**)	10.60 (−1.5%)
	S-Plus	198 (−1.0%)	101 (0.0%)	97 (−2.0%)	595.02 (−0.7%)	30.22 (11.6%)	50.97 (8.0%)	**7.67** (**26.5%**)
	S-Minus	201 (0.5%)	101 (0.0%)	100 (1.0%)	586.35 (0.7%)	30.64 (7.8%)	50.57 (8.8%)	10.65 (−2.0%)
	Two-queue	**202** (**1.0%**)	101 (0.0%)	101 (2.0%)	**585.58** (**0.9%**)	35.11 (−5.9%)	49.94 (9.9%)	20.14 (−92.9%)
	Portfolio	206 (3%)	101 (0.0%)	105 (6.1%)	569.1 (3.7%)	25.79 (22.2%)	44.58 (19.6%)	10.16 (2.6%)
Non Linear CEGAR	Default	432	250	182	131.12	42.05	43.34	40.28
	Random	425 (−1.6%)	243 (−2.8%)	**182** (**0.0%**)	143.42 (−9.4%)	34.27 (−11.1%)	34.84 (19.6%)	38.75 (3.8%)
	R-Plus	432 (0.0%)	250 (0.0%)	**182** (**0.0%**)	122.29 (6.7%)	31.74 (17.8%)	28.59 (34.0%)	37.82 (6.1%)
	R-Minus	417 (−3.5%)	240 (−4.0%)	177 (−2.7%)	154.07 (−17.5%)	**26.20** (**20.8%**)	21.46 (50.5%)	**32.51** (**19.3%**)
	S-Plus	**434** (**0.5%**)	**252** (**0.8%**)	82 (0.0%)	**121.75** (**7.1%**)	34.64 (13.1%)	35.97 (17.0%)	39.10 (2.9%)
	S-Minus	421 (−2.5%)	242 (−3.2%)	179 (−1.6%)	149.02 (−13.7%)	31.76 (−2.0%)	26.33 (39.2%)	38.95 (3.3%)
	Portfolio	435 (0.7%)	253 (1.2%)	182 (0.0%)	113.49 (13.4%)	28.24 (29.1%)	30.57 (29.5%)	31.75 (21.2%)
Non Linear SymEx	Default	342	187	155	343.82	28.39	29.05	27.59
	Random	362 (5.8%)	188 (0.5%)	**174** (**12.3%**)	301.90 (12.2%)	32.67 (−15.1%)	**36.24** (**−24.8%**)	41.83 (−51.6%)
	R-Plus	339 (−0.9%)	**190** (**1.6%**)	149 (−3.9%)	357.18 (−3.9%)	27.88 (0.3%)	47.71 (−64.2%)	**22.10** (**19.9%**)
	R-Minus	361 (5.6%)	189 (1.1%)	172 (11.0%)	299.86 (12.8%)	**26.35** (**7.3%**)	37.68 (−29.7%)	27.98 (−1.4%)
	S-Plus	340 (−0.6%)	189 (1.1%)	151 (−2.6%)	352.84 (−2.6%)	29.04 (−0.3%)	41.41 (−42.5%)	24.54 (11.1%)
	S-Minus	362 (5.8%)	**190** (**1.6%**)	172 (11.0%)	303.65 (11.7%)	28.62 (−0.4%)	44.11 (−51.8%)	37.95 (−37.5%)
	Two-queue	**363** (**6.1%**)	189 (1.1%)	**174** (**12.3%**)	**297.93** (**13.3%**)	30.15 (−6.2%)	41.14 (−41.6%)	32.51 (−17.8%)
	Portfolio	366 (7.0%)	191 (2.1%)	175 (12.9%)	288.85 (16.0%)	22.29 (21.4%)	42.42 (−46.0%)	26.75 (3.0%)

dataset achieves the best performance when the evaluation set is non-linear and the algorithm is SymEx. Across the entire table, there are 17, 10, and 13 bold text counts for the union, single, and intersection MUS datasets, respectively. This indicates that the union is the most effective MUS dataset for better performance across different benchmarks and algorithms. Consequently, we provide further numerical details for the union MUS dataset in Table 4. Evaluation results for both the single and intersection MUS datasets can be found in [20].

Table 4 illustrates the evaluation results using MUS-guided ranking functions (see Table 1), compared to the default and random ranking functions. In terms of the total number of solved problems, the improvement for the Linear dataset is at most 1.4%, achieved by the CEGAR algorithm with the R-Plus ranking function. Meanwhile, for the Non-linear dataset, the improvement is 6.1%, achieved by the SymEx algorithm with the two-queue ranking function. This is consistent with the average solving time for all benchmarks. In each Benchmark Algorithm block, we also show the results obtained by a virtual portfolio that selects the best ranking function for each benchmark.

The biggest improvement in Average Time for SAT and UNSAT problems are 50.5% and 26.5%, achieved by the CEGAR with R-Minus and the SymEx with S-Plus ranking function, respectively. When combined with the corresponding numbers of total solved problems (i.e., −3.5% and −1.0%), it suggests that these ranking functions either solve the problems quickly or not at all. The Average Times in the Common column often differ significantly from the times in the All column. This suggests that the number of newly solved problems has a greater impact on the improvement in the average solving time for all problems than the commonly solved problems.

Figure 2 shows the solving time scatter plots for the problems from the best configurations in each Benchmark Algorithm block in Table 4. Notably, a majority of the dots lie below the diagonal lines in each scatter plot, indicating the solving time is improved by the MUS-guided ranking function for more than half of the problems. This is consistent with the numerical results. The plots also show that the MUS-guided ranking functions achieve speedups in CEGAR in particular for long-running problems that are SAT, while MUS-guidance in SymEx makes it possible to solve a significant number of UNSAT problems on which the default configuration times out.

In summary, for both algorithms in different datasets, there is always at least one of the MUS-guided ranking functions that achieves the best result in terms of all aspects of measurements. Using the predicted probabilities alone (i.e., using the Rank and Score ranking function) performs weaker than other MUS-guided ranking functions that combine the predicted probability and the default heuristics. Currently, the MUS-guided ranking functions in Table 1 are designed by simply varying the relation symbols "+" and "−" between different elements (e.g., ranking functions S-Plus and S-Minus for both CEGAR and SymEx) or by setting the restart point randomly (i.e., ranking function Two-queue in SymEx). We believe MUSHyperNet has more potential if the ranking functions are designed carefully or learned from some good tasks.

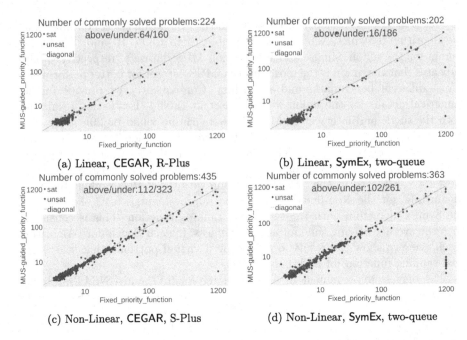

(a) Linear, CEGAR, R-Plus

(b) Linear, SymEx, two-queue

(c) Non-Linear, CEGAR, S-Plus

(d) Non-Linear, SymEx, two-queue

Fig. 2. All benchmark average solving time scatter plots for best ranking functions in different dataset and algorithms. "above/under" means the number of dots above and under the diagonal line.

7   Related Work

Machine learning techniques have been adapted in various ways to assist in formal verification. For example, the study in [35] employs Support Vector Machines (SVM) [36] and Gaussian processes [37] to select heuristics for theorem proving. Similarly, [38] introduces the use of a Recurrent Neural Network Based Language Model (RNNLM) to derive finite-state automaton-based specifications from execution traces. In the domain of selecting algorithms for program verification, [39] apply the Transformer architecture [40], while [41] uses kernel-based methods [42]. With the thriving of deep learning techniques, an increasing number of works are utilizing GNNs to learn the features from programs and logic formulae. This trend is attributed to the inherent structure of these languages, which can be naturally represented as graphs and subsequently learned by GNNs. For instance, studies like [14–16,43], and [44] use GNNs [24,45,46] to learn features from graph-represented logic formulas and programs, aiding in tasks such as theorem proving, SAT solving, and loop invariant reasoning.

One closed idea is NeuroSAT [16,17], which trains a GNN to predict the probability of variables appearing in unsat cores. This prediction can guide the variable branching decisions for Conflict-Driven Clause Learning (CDCL) [47]-based SAT solvers. In a similar vein, our study trains a GNN to predict the probability of a CHC appearing in MUSes. This aids in determining the processing sequence of CHCs in Algorithm 1 used for solving a set of CHCs.

8   Conclusion and Future Works

In this study, we train a GNN model to predict the probability of each CHC in a set of CHCs being in the MUSes. We then utilize these predicted probabilities to guide the abstract symbolic model-checking algorithm in selecting a CHC during each iteration. Extensive experiments demonstrate improvements in both the number of solved problems and average solving time when using the MUS-guided ranking functions, compared to the default ranking function. This was observed in two instances of the abstract symbolic model checking algorithm: CEGAR and SymEx. We believe that this approach can be extended to other algorithms, as many could benefit from understanding more about the MUSes of a set of CHCs.

There are several ways to further enhance the performance of MUSHyperNet. One of our future work is to integrate the work of manually designing the ranking functions in Table 1 to the learning process. Regarding the GNN model, we believe that incorporating an attention mechanism could bolster its performance, subsequently refining the quality of the predicted probabilities. Another avenue to explore involves integrating the GNN with the solver in a more interactive manner. Instead of predicting something at once and then using them in each iteration, we could query the GNN model to predict something in real-time based on the current context during each iteration.

Acknowledgement. We thank Zafer Esen for providing assistance in using the symbolic execution engine of Eldarica, and Marc Brockschmidt for various discussions on this work. The computations and data handling were enabled by resources provided by the Swedish National Infrastructure for Computing (SNIC) at UPPMAX and C3SE. The research was partially funded by the Swedish Research Council through grant agreement no. 2018–05973, by a Microsoft Research PhD grant, the Swedish Foundation for Strategic Research (SSF) under the project WebSec (Ref. RIT17-0011), and the Wallenberg project UPDATE.

References

1. Liang, C., Rümmer, P., Brockschmidt, M.: Exploring representation of Horn clauses using GNNs (extended technique report). CoRR, abs/2206.06986 (2022)
2. Horn, A.: On sentences which are true of direct unions of algebras. J. Symbol. Logic 16(1), 14–21 (1951)
3. De Angelis, E., Fioravanti, F., Gallagher, J.P., Hermenegildo, M.V., Pettorossi, A., Proietti, M.: Analysis and transformation of constrained Horn clauses for program verification. CoRR, abs/2108.00739 (2021)
4. Bjørner, N., Gurfinkel, A., McMillan, K., Rybalchenko, A.: Horn clause solvers for program verification. In: Beklemishev, L.D., Blass, A., Dershowitz, N., Finkbeiner, B., Schulte, W. (eds.) Fields of Logic and Computation II. LNCS, vol. 9300, pp. 24–51. Springer, Cham (2015). https://doi.org/10.1007/978-3-319-23534-9_2

5. Clarke, E.M.: SAT-based counterexample guided abstraction refinement in model checking. In: Baader, F. (ed.) CADE 2003. LNCS (LNAI), vol. 2741, pp. 1–1. Springer, Heidelberg (2003). https://doi.org/10.1007/978-3-540-45085-6_1

6. Bradley, A.R.: SAT-based model checking without unrolling. In: Jhala, R., Schmidt, D. (eds.) VMCAI 2011. LNCS, vol. 6538, pp. 70–87. Springer, Heidelberg (2011). https://doi.org/10.1007/978-3-642-18275-4_7

7. Liffiton, M.H., Sakallah, K.A.: Algorithms for computing minimal unsatisfiable subsets of constraints. J. Autom. Reason. **40**(1), 1–33 (2008)

8. Gurfinkel, A.: Program verification with constrained horn clauses (invited paper). In: Shoham, S., Vizel, Y. (eds.) Computer Aided Verification. CAV 2022. LNCS, vol. 13371, pp. 19–29. Springer, Cham (2022). https://doi.org/10.1007/978-3-031-13185-1_2

9. Chvalovský, K., Jakubův, J., Suda, M., Urban, J.: ENIGMA-NG: efficient neural and gradient-boosted inference guidance for E. In: Fontaine, P. (ed.) CADE 2019. LNCS (LNAI), vol. 11716, pp. 197–215. Springer, Cham (2019). https://doi.org/10.1007/978-3-030-29436-6_12

10. Jakubův, J., Urban, J.: ENIGMA: efficient learning-based inference guiding machine. In: Geuvers, H., England, M., Hasan, O., Rabe, F., Teschke, O. (eds.) CICM 2017. LNCS (LNAI), vol. 10383, pp. 292–302. Springer, Cham (2017). https://doi.org/10.1007/978-3-319-62075-6_20

11. Jakubův, J., Chvalovský, K., Olšák, M., Piotrowski, B., Suda, M., Urban, J.: ENIGMA anonymous: symbol-independent inference guiding machine (system description). In: Peltier, N., Sofronie-Stokkermans, V. (eds.) IJCAR 2020. LNCS (LNAI), vol. 12167, pp. 448–463. Springer, Cham (2020). https://doi.org/10.1007/978-3-030-51054-1_29

12. Suda, M.: Vampire with a brain is a good ITP hammer. In: Konev, B., Reger, G. (eds.) FroCoS 2021. LNCS (LNAI), vol. 12941, pp. 192–209. Springer, Cham (2021). https://doi.org/10.1007/978-3-030-86205-3_11

13. Battaglia, P.W., et al.: Relational inductive biases, deep learning, and graph networks. CoRR, abs/1806.01261 (2018)

14. Wang, M., Tang, Y., Wang, J., Deng, J.: Premise selection for theorem proving by deep graph embedding. In: Guyon, I., et al. (eds.), Advances in Neural Information Processing Systems, vol. 30, pp. 2786–2796. Curran Associates Inc. (2017)

15. Paliwal, A., Loos, S.M. Rabe, M.N., Bansal, K., Szegedy, C.: Graph representations for higher-order logic and theorem proving. CoRR, abs/1905.10006 (2019)

16. Selsam, D., Bjørner, N.: Neurocore: guiding high-performance SAT solvers with unsat-core predictions. CoRR, abs/1903.04671 (2019)

17. Selsam, D., Lamm, M., Bünz, B., Liang, P., de Moura, L., Dill, D.L.: Learning a SAT solver from single-bit supervision. In 7th International Conference on Learning Representations, ICLR 2019, New Orleans, LA, USA, 6–9 May 2019. OpenReview.net (2019)

18. CHC-COMP benchmarks. https://chc-comp.github.io/. Accessed 07 Sept 2023

19. Fedyukovich, G., Rümmer, P.: Competition report: CHC-COMP-21. In: Hojjat, H., Kafle, B., (eds.), Proceedings 8th Workshop on Horn Clauses for Verification and Synthesis, HCVS@ETAPS 2021, Virtual, 28th March 2021, vol. 344. EPTCS, pp. 91–108 (2021)

20. Repository for the training and evaluation dataset in this work. https://github.com/ChenchengLiang/Horn-graph-dataset. Accessed 07 Sept 2023

21. Harrison, J.: Handbook of Practical Logic and Automated Reasoning. Cambridge University Press, Cambridge (2009)

22. Barrett, C., Fontaine, P., Tinelli, C.: The SMT-LIB Standard: Version 2.6. Technical report, Department of Computer Science, The University of Iowa (2017). www.SMT-LIB.org
23. Goodfellow, I.J., Bengio, Y., Courville, A.: Deep Learning. MIT Press, Cambridge (2016). http://www.deeplearningbook.org
24. Gilmer, J., Schoenholz, S.S., Riley, P.F., Vinyals, O., Dahl, G.E.: Neural message passing for quantum chemistry. CoRR, abs/1704.01212 (2017)
25. Schlichtkrull, M., Kipf, T.N., Bloem, P., van den Berg, R., Titov, I., Welling, M.: Modeling relational data with graph convolutional networks. In: Gangemi, A., et al. (eds.) ESWC 2018. LNCS, vol. 10843, pp. 593–607. Springer, Cham (2018). https://doi.org/10.1007/978-3-319-93417-4_38
26. Agarap, A.F.: Deep Learning using Rectified Linear Units (ReLU). arXiv e-prints: arXiv:1803.08375, March 2018
27. King, J.C.: Symbolic execution and program testing. Commun. ACM 19(7), 385–394 (1976)
28. Hojjat, H., Rümmer, P.: The ELDARICA Horn solver. In: 2018 Formal Methods in Computer Aided Design (FMCAD), pp. 1–7 (2018)
29. Graf, S., Saidi, H.: Construction of abstract state graphs with PVS. In: Grumberg, O. (ed.) CAV 1997. LNCS, vol. 1254, pp. 72–83. Springer, Heidelberg (1997). https://doi.org/10.1007/3-540-63166-6_10
30. Ball, T., Podelski, A., Rajamani, S.K.: Boolean and Cartesian abstraction for model checking C programs. Int. J. Softw. Tools Technol. Transf. 5(1), 49–58 (2003). https://doi.org/10.1007/s10009-002-0095-0
31. Rümmer, P., Hojjat, H., Kuncak, V.: Disjunctive interpolants for horn-clause verification. In: Sharygina, N., Veith, H. (eds.) CAV 2013. LNCS, vol. 8044, pp. 347–363. Springer, Heidelberg (2013). https://doi.org/10.1007/978-3-642-39799-8_24
32. Hojjat, H., Rümmer, P.: OptiRica: towards an efficient optimizing Horn solver. In: Hamilton, G.W., Kahsai, T., Proietti, M., (eds.), Proceedings 9th Workshop on Horn Clauses for Verification and Synthesis and 10th International Workshop on Verification and Program Transformation, HCVS/VPT@ETAPS 2022, and 10th International Workshop on Verification and Program TransformationMunich, Germany, 3rd April 2022, vol. 373. EPTCS, pp. 35–43 (2022)
33. David R. Cox. The regression analysis of binary sequences. J. R. Statist. Soc. Ser. B (Methodol.) 20(2), 215–242 (1958)
34. Code repository for reproduce this work. https://github.com/ChenchengLiang/Relational-Hypergraph-Neural-Network-PyG. Accessed 07 Sept 2023
35. Bridge, J., Holden, S., Paulson, L.: Machine learning for first-order theorem proving. J. Autom. Reason. 53, 08 (2014)
36. Cortes, C., Vapnik, V.: Support-vector networks. Mach. Learn. 20(3), 273–297 (1995)
37. Rasmussen, C.E., Williams, C.K.I.: Gaussian Processes for Machine Learning (Adaptive Computation and Machine Learning). The MIT Press, Cambridge (2005)
38. Le, T.-D.B., Lo, D.: Deep specification mining. In: Proceedings of the 27th ACM SIGSOFT International Symposium on Software Testing and Analysis, ISSTA 2018, pp. 106–117. Association for Computing Machinery, New York, NY, USA (2018)
39. Richter, C., Wehrheim, H.: Attend and represent: a novel view on algorithm selection for software verification. In: 2020 35th IEEE/ACM International Conference on Automated Software Engineering (ASE), pp. 1016–1028 (2020)

40. Vaswani, A., et al.: Attention is all you need. In: Guyon, I., et al. (eds.), Advances in Neural Information Processing Systems, vol. 30. Curran Associates Inc. (2017)
41. Richter, C., Hüllermeier, E., Jakobs, M.-C., Wehrheim, H.: Algorithm selection for software validation based on graph kernels. Autom. Softw. Eng. **27**(1), 153–186 (2020)
42. Scholkopf, B., Smola, A.J.: Learning with Kernels: Support Vector Machines, Regularization, Optimization, and Beyond. MIT Press, Cambridge (2001)
43. Selsam, D., Lamm, M., Bünz, B., Liang, P., de Moura, L., Dill, D.L.: Learning a SAT solver from single-bit supervision. CoRR, abs/1802.03685 (2018)
44. Si, X., Dai, H., Raghothaman, M., Naik, M., Song, L.: Learning loop invariants for program verification. In: Bengio, S., Wallach, H., Larochelle, H., Grauman, K., Cesa-Bianchi, N., Garnett, R., (eds.), Advances in Neural Information Processing Systems, vol. 31. Curran Associates Inc. (2018)
45. Scarselli, F., Gori, M., Tsoi, A.C., Hagenbuchner, M., Monfardini, G.: The graph neural network model. IEEE Trans. Neural Netw. **20**(1), 61–80 (2009)
46. Li, Y., Tarlow, D., Brockschmidt, M., Zemel, R.S.: Gated graph sequence neural networks. In: Bengio, Y., LeCun, Y., (eds.), 4th International Conference on Learning Representations, ICLR 2016, San Juan, Puerto Rico, 2–4 May 2016, Conference Track Proceedings (2016)
47. Marques-Silva, J.P., Sakallah, K.A.: GRASP: a search algorithm for propositional satisfiability. IEEE Trans. Comput. **48**(5), 506–521 (1999)

On the Verification of the Correctness of a Subgraph Construction Algorithm

Lucas Böltz[✉], Viorica Sofronie-Stokkermans, and Hannes Frey

University of Koblenz, Koblenz, Germany
{boeltz,sofronie,frey}@uni-koblenz.de

Abstract. We automatically verify the crucial steps in the original proof of correctness of an algorithm which, given a geometric graph satisfying certain additional properties removes edges in a systematic way for producing a connected graph in which edges do not (geometrically) intersect. The challenge in this case is representing and reasoning about geometric properties of graphs in the Euclidean plane, about their vertices and edges, and about connectivity. For modelling the geometric aspects, we use an axiomatization of plane geometry; for representing the graph structure we use additional predicates; for representing certain classes of paths in geometric graphs we use linked lists.

1 Introduction

We present an approach for automatically verifying the main steps in the correctness proof of an algorithm (the CP-algorithm) which, given a geometric graph satisfying two properties (the redundancy property and the coexistence property) removes edges in a systematic way for producing a connected graph in which edges do not (geometrically) intersect. The challenge in this case is representing and reasoning about properties of *geometric graphs* with vertices in the Euclidean plane. For modelling the geometric aspects, we use an axiomatization of plane geometry related to Hilbert's and Tarski's axiomatization [23,39]. The vertices and edges are modeled using unary resp. binary predicates; the properties of geometric graphs we consider refer both to geometric conditions and to conditions on the vertices and edges. In addition, we use linked lists to represent certain paths in such geometric graphs. We show that this axiomatization can be used for describing the classes of geometric graphs we are interested in, discuss the reasoning tasks occurring in the verification of the graph algorithm, and the way we automatically proved the crucial steps in the original proof of correctness of the algorithm, using the prover Z3 [5,6]. The main contributions of this paper can be described as follows:

- We present an axiomatization of a part of Euclidean geometry, referring to bounded domains of the Euclidean plane, which is necessary to describe graph drawings in the Euclidean plane. We show how this axiomatization can be used to prove theorems in plane geometry (Axiom of Pasch, transitivity of the inside predicate, symmetry of the intersection predicate).

R. Dimitrova et al. (Eds.): VMCAI 2024, LNCS 14499, pp. 303–325, 2024.
https://doi.org/10.1007/978-3-031-50524-9_14

- We extend this axiomatization with additional predicate symbols for edges and vertices, and formulate properties of graphs also containing geometric conditions in this extended language. We show how this axiomatization can be used to prove simple properties (e.g. the clique property inside a triangle) in the class of geometric graphs we consider.
- We formalize constructions used in the correctness proof of a planarization algorithm for geometric graphs, and use Z3 to perform the corresponding proof tasks and to obtain counterexamples if the hypotheses are weakened.
- We extend the axiomatization with a specification of list structures for expressing lists of nodes along the convex hull of certain specified sets of points. We use this extended axiomatization for proving the connectedness of the graphs obtained as a result of applying the algorithm we analyze.

The main challenge in this paper was to find a suitable axiomatization of notions from geometric graph theory which would allow us to automatically verify the main steps in an existing correctness proof of an algorithm for geometric graphs.

Related work: Existing approaches to the study of graphs and graph algorithms are often based on monadic second-order logic or higher-order logic.

In [13–16] (these are only a few of the numerous publications of Courcelle on this topic) monadic second-order logic is used; many of the results are for graphs with bounded tree-width or bounded clique-width. The classes of graphs we study do not necessarily have such properties. For the verification of graph algorithms, higher-order theorem provers like Isabelle/HOL and Coq were used (cf. e.g. [1]; the possibility of checking graph theoretical properties in Coq is discussed in [18]). An overview on checking graph transformation systems is given in [22]. We are not aware of research that combines reasoning about graphs and reasoning about geometric properties of graphs. In our work we focus on the study of geometric graphs and identify a situation in which the use of higher-order logic and higher-order logic systems can be avoided. We show that for the type of graphs we consider (geometric graphs with vertices in the Euclidean plane) and for the verification problems we consider we can avoid second-order logic also for handling connectedness, a property known to not be first-order definable: in our context it will be sufficient to use certain, uniquely determined paths along the convex hull of determined sets of points, which we represent using linked data structures.

When modelling and reasoning about geometric graphs we need to consider in addition to the encoding of the topological aspects of graphs (vertices and edges) also the geometric aspects, and combine reasoning about the graphs themselves with reasoning about geometric properties. For reasoning in geometry, one can use approaches using coordinates and ultimately analytic geometry or axiomatic approaches. Among the approaches to automated reasoning in *analytic geometry*, we mention methods based on Gröbner bases (cf. e.g. [17]) or approaches based on quantifier elimination in real closed fields (cf. e.g. [19,38]). Systems of *axioms for plane geometry* have been proposed by Tarski [39] or Hilbert [23]; such systems were used for theorem proving in geometry for instance by Beeson and Wos in [3]. We decided to use an axiomatic approach. Compared to classical axiom systems for plane geometry such as those proposed by Tarski [39] and Hilbert [23] we only use axioms of order, but did not include axioms for incidence, congruence, parallelity and continuity because our goal is to describe the part of plane Euclidean geometry necessary to describe geometrically bounded graph drawings in the Euclidean plane. Due to technical considerations (which we explain in Sect. 3), the axioms we propose are slightly different from the axioms of Hilbert [23], but all necessary conditions to prove correctness of the algorithm are satisfied. An advantage of our axiomatic approach to plane geometry is that the axiom system is simple and only information about the relative positions of nodes is required, we do not need to consider concrete coordinates, lengths and angles between the nodes. So this axiom system is stable for small changes, where the relative positions of the nodes and therefore the predicates do not change. The simple axiom system makes it easier for a prover to decide whether a condition for a specific combination of the predicates is satisfiable or not. We perform proofs by contradiction; if unsatisfiability cannot be proved we use the possibilities of generating models with Z3 - we used this, for instance, to show that the axioms we consider can not be relaxed further.

Structure of the Paper: In Sect. 2 we define geometric graphs and describe the CP-algorithm. In Sect. 3 we describe the representation of geometric graphs we use and define the axiom system we use to model geometric properties, properties of geometric graphs and properties used for proving correctness of the CP-algorithm. In Sect. 4 we discuss the fragments in which these axioms are, and describe the automated reasoning tools we use and the steps of verifying the correctness proof of the CP-algorithm. Section 5 contains conclusions and plans for future work. For further details we refer to the extended version of the paper [10].

2 The CP-Algorithm

We start by presenting the algorithm we will verify. Before doing so, we introduce the main notions related to geometric graphs needed in what follows.

Geometric graphs. A *geometric graph* is a graph $G = (V, E)$ in which vertices are seen as points (usually in the Euclidean plane) and an edge uv is represented as *straight line segment* between the vertices u and v. We consider undirected graphs without self-loops, i.e. graphs $G = (V, E)$ in which $uv \in E$ iff $vu \in E$, and $uu \notin E$ for all $u, v \in V$.

Plane Graphs. A *plane graph* is a geometric graph on a plane without intersecting edges (where edges are seen as line segments). Plane graphs have applications in the context of wireless networks. Representing a wireless network as a plane graph is a sufficient condition to ensure the correctness of routing approaches such as face routing [21]. Other applications are distributed data storage [26], mobile object tracking [12] and clustering algorithms [31]. These methods have in common that they are implemented distributed across the graph vertices and that they rely only on the local view of the vertices. Locality means that each vertex has only information about vertices and edges of the graph that can be reached via a path of a fixed, limited path length. All of these methods achieve a network-wide objective by vertex-local decisions, such as, e.g., delivery of messages to the destination, retrieving data from the distributed storage, seamless tracking of passing by objects, or unique assignment of nodes into clusters.

Given a global view of an arbitrary geometric graph on a plane, constructing a plane subgraph alone is not difficult, however, ensuring connectedness at the same time can be a conflicting goal. When only a local view on the graph is available, even edge intersections are not always detectable.

Considering wireless networks, however, the graph topology is not arbitrary. Due to limited transmission range, a vertex can only be connected to spatially close by vertices. For this reason, graph models for wireless network graphs usually have this spatial locality inherent, including unit disk graphs being the simplest such model, quasi unit disk graph as an extension thereof [2,27], or log-normal shadowing graphs, a more practical model [4,7].

Local solutions for wireless networks can then be studied in view of such model constraints. In particular, regarding the question in how far a network graph can be planarized by just local information, the specific models *redundancy property* and *coexistence property* had been introduced and studied in the context of wireless network graphs [7,20,28,29,33,35]. These two models also fundamentally take the above-mentioned locality into account.

A question that was not completely clarified until a few years ago is in how far graphs with redundancy and coexistence property can be planarized and if this can be done with local rules. The first question was answered to the positive in [9]. Moreover, in [8] a local algorithm was described which as proved there manually guarantees planarity and connectivity provided that the underlying graph satisfies both redundancy and coexistence property.

In this work we take the manual proof from there as a blue print and verify each of the proof steps. This requires an axiomatic formalization of the redundancy and coexistence property, as well as a formalization of the considered algorithm.

Definition 1 (Redundancy property). *We say that a geometric graph $G = (V, E)$ satisfies the **redundancy property** if for every pair of intersecting edges uv and wx of G one of the four vertices u, v, w, x is connected to the other three (see Fig. 1a). The edges uw, ux, vw, vx are called the redundancy edges for the intersection of uv and wx.*

Definition 2 (Triangle in a graph; interior of a triangle). *Let $G = (V, E)$ be a geometric graph and let $u, v, w \in V$ such that the edges uv, vw, wu exist. We say that u, v, w build a triangle. The interior of the triangle is the set $Cl(u, v, w) \backslash \{u, v, w\}$, where $Cl(u, v, w)$ is the closed convex set in the Euclidean plane delimited by the points associated with the vertices u, v, w. For three collinear vertices, where w is located between u and v, the interior of the triangle $\triangle uvw$ is the open segment between u and w, while the interior of the triangle $\triangle vuw$ is the open segment between w and v.*

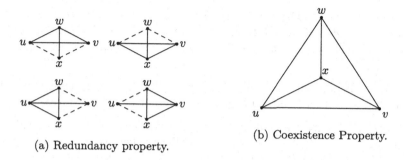

(a) Redundancy property.

(b) Coexistence Property.

Fig. 1. Redundandcy and Coexistence property [9].

Definition 3 (Coexistence property). *Let* $G = (V, E)$ *be a geometric graph. We say that* G *satisfies the* **coexistence property** *if for every triangle formed by three edges* uv, vw, wu *of* G *every vertex* x *located in the interior of the triangle is connected to all three border vertices of the triangle, i.e. the edges* ux, vx, wx *have to exist (see Fig. 1b).*

Definition 4 (RCG). *A geometric graph satisfying the redundancy and coexistence property is called redundancy-coexistence graph (RCG).*

2.1 The CP-Algorithm; Idea of the Correctness Proof

The *connected planarization (CP)* algorithm from [8] we verify in this work is sketched by the following pseudo code. We assume the given graph to be a finite, connected RCG.

Algorithm 1 Global CP-algorithm (**Input:** $G = (V, E)$; **Output:** $G' = (V, F)$)

$F \leftarrow \emptyset$ (* initialize the result *)
$W \leftarrow E$ (* set up the working set *)
while $W \neq \emptyset$ **do**
    choose $uv \in W$
    remove uv from W
    if for all $w_i x_i \in E$ intersecting uv : $((uw_i \in E, ux_i \in E)$ or $(vw_i \in E, vx_i \in E))$ **then**
        add uv to F
        remove all intersecting edges $w_i x_i$ from W
    end if
end while
$G' \leftarrow (V, F)$

In the initial state, the CP-algorithm has a working set W containing all edges E and an empty result set F. In every iteration step of the algorithm an edge uv of W is chosen and removed from W, and it is decided whether uv is added to F or not. This decision depends on the CP-condition.

Definition 5 (The CP-condition). *The CP-condition is satisfied in a graph $G = (V, E)$ for an edge $uv \in E$ if for all intersecting edges $w_i x_i \in E$ the two redundancy edges uw_i and ux_i or the two redundancy edges vw_i and vx_i exist. (see Fig. 2a).*

If the CP-condition holds, uv is added to F and all intersecting edges are removed from W. On the other hand, an edge uv can be not in F for two reasons:

(a) An edge wx intersecting with uv was added to F before uv was considered.
(b) The edge uv does not satisfy the CP-condition: there is an intersecting edge wx such that $(uw \notin E$ and $vw \notin E)$ or $(ux \notin E$ and $vx \notin E)$.

Definition 6 (Deleting edges). *If there is an edge wx intersecting with uv with the properties of (a) or (b), then wx deletes uv.*

One can prove (cf. [8]) that if an edge uv is not in F because it does not satisfy the CP-condition, then there also exists an edge intersecting with uv that is in F. One can therefore assume that every edge that is not in F was not added to F because of condition (a). In order to make the terminology even clearer, we will in this context consider the order of the vertices w, x when wx deletes uv.

Definition 7 (Deleting edges (oriented)). *Assume that wx and uv intersect. If wx deletes uv then $wu, wv \in E$, and if xw deletes uv then $xu, xv \in E$.*

The order in which the algorithm chooses the edges from the working set W determines which edges are added to F and which not.

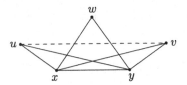 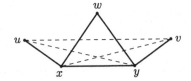

(a) The edge uv does not satisfy the CP-condition; all other edges satisfy it.

(b) A possible output graph after the execution of the CP-algorithm.

Fig. 2. Illustration of the CP-condition and the CP-algorithm.

Termination. Since the graph G is finite and in every iteration of the while loop at least one edge is removed from the finite set W of edges, it is easy to see that the CP-algorithm terminates.

Idea of the Correctness Proof. For the proof of the correctness of the algorithm we need to show that if the input graph $G = (V, E)$ is connected and satisfies the redundancy and the coexistence property, then the graph $G' = (V, F)$ constructed by the CP-algorithm is connected and plane.

Planarity is easy to prove: if an edge uv is added to F, then all intersecting edges are removed from the working set W. Therefore at any time W only contains

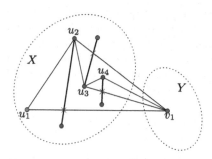

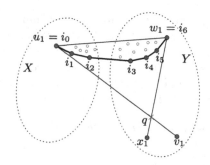

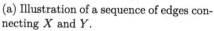

(a) Illustration of a sequence of edges connecting X and Y.

(b) Illustration of the path on the convex hull of the triangle $\Delta u_1 q w_1$ connecting X and Y.

Fig. 3. Connecting paths from X to Y.

edges which are not intersecting with edges so far added to F. This means that only edges can be added to F, which are not intersecting with edges that are already in F, which leads to a plane graph as output.

Connectivity. The proof is given by contradiction. Assume that after the termination of the algorithm, the resulting graph $G' = (V, F)$ has $k \geq 2$ components which are not connected. Let X be one of the components and Y be the induced subgraph of $G' \setminus X$, which contains the other $k - 1$ components. One can prove that a path with edges in F connecting a vertex in X and a vertex in Y exists, which is a contradiction. For proving this, the following lemmas are needed. The correctness of the lemmas is verified in Sect. 4.4 (Lemma 1), Sect. 4.5 (Lemma 2) and Sect. 4.6 (Lemma 3).

Lemma 1. *For every edge uv that is not in F there exists an edge wx in F that intersects with uv.*

Lemma 2. *There exists an edge uv in E with $u \in X$ and $v \in Y$, such that all edges in F intersecting with uv are located in Y (Example: the edge $u_4 v_1$ in Fig. 3a which is resulting from edge sequence construction $u_1 v_1$, $u_2 v_1$, ..., $u_4 v_1$ described in Sect. 4.5).*

Lemma 3. *For an edge uv and the intersecting edge $wx \in F$ with the redundancy edges wu and wv, and intersection point q closest to u among all intersecting edges in F, the finest path P from u to w on the convex hull of the triangle Δuqw contains only edges in F (Example: the path $i_0 i_1 \ldots i_6$ from u_1 to w_1 in Fig. 3b, constructed for the edge $u_1 v_1$ and the intersecting edge $w_1 x_1 \in F$).*

Combining the results of the Lemmas 1–3 leads to the connectivity of G' with the following argumentation: Since the original graph $G = (V, E)$ before applying the CP-algorithm was connected, there has to exist an edge $u_1 v_1$ in E but not in F with $u_1 \in X$ and $v_1 \in Y$. Lemma 1 ensures that $u_1 v_1$ is intersected by an edge $w_1 x_1$ in F and by Lemma 2 we can choose $u_1 v_1$ such that all intersecting edges in F are located in Y. Lemma 3 ensures now – for the intersecting edge $w_1x_1 \in F$ (located in Y) with redundancy edges w_1u_1 and w_1v_1 and intersection point q closest to u_1 – the existence of a path P from u_1 to w_1 with all edges in F (see Fig. 3b). This path is connecting u_1 and w_1 and therefore connects X and Y in G'. This is a contradiction to the assumption that G' is not connected. Therefore the following theorem holds.

Theorem 1. *The edges of F form a connected plane spanning subgraph of G after the termination of the CP-algorithm.*

A local CP-algorithm. In [8] a local version of the CP-algorithm is proposed. In the local variant, each vertex u takes its own decision to keep or remove an edge based on the information about the vertices connected to u in at most 2 steps. The only problem can occur if for two intersecting edges, uv and wx, two nodes might take their decision of adding these edges to F at the same time. Such conflicts can be solved by defining a priority order for the edges. Since in the global version of the CP-algorithm described in Algorithm 1 the order in which the edges are chosen from the working set W is arbitrary, the correctness proof of Theorem 1 can be adapted for the local version without problems.

3   Modeling Geometric Graphs

In order to automatically verify the main steps of the correctness proof in the CP-algorithm, we have to find an efficient way of representing geometric graphs and their properties, in particular the redundancy and coexistence property. For expressing the redundancy property we need to be able to express the fact that two edges geometrically intersect when regarded as segments. For expressing the coexistence property we need to be able to describe the interior of a triangle according to Definition 2.

Expressing such conditions in analytic geometry using coordinates and distances may lead to a high computational effort. One reason is that for a line passing through two points A and B first an equation of the form $l(x,y) = ax + by + c = 0$ has to be computed and then for other points C, D with coordinates (x_C, y_C) resp. (x_D, y_D), the values $l(x_C, y_C)$ and $l(x_D, y_D)$ have to be computed to decide if C and D are on the same side of the line AB or not. An important problem in this context is to correctly cover the limit cases such as: "Is a point located on the line passing through two other points?" or "Is a point located on the border of a triangle formed by three other points?", and to obtain the desired conditions for describing the interior of a triangle according to Definition 2. In order to minimize the computational effort, alternative approaches like the axiomatization of Hilbert [23] and Tarski [39] can be considered. The advantage of these approaches is that for each point instead of its concrete position only the relative position to other points is considered. This might be enough to distinguish if two lines intersect or if a point is located inside the triangle formed by three other points.

Fig. 4. w_1, w_2 and w_3 are left of uv, while w_3, w_4 and w_5 are left of vu.

Fig. 5. Illustration of the axioms **A1**, **A2**, **A3** and **A4**.

We show that this axiomatic approach is sufficient for automatically checking the correctness proof of the CP-algorithm and its distributed version. We start with axioms describing geometric aspects, then add axioms describing graph properties, and finally notions and constructions needed for describing the CP-algorithm and for checking the main steps in its correctness proof.

3.1 Geometric Axioms

The geometric axioms are structured in a hierarchical way, starting with an axiomatization of the properties of a predicate **left** of arity 3. Using this predicate additional notions are axiomatized (**intersection, inside**).

Axioms for left. The intended meaning for $\mathsf{left}(u, v, w)$ is "w is on the left side of the (oriented) line through u and v or on the ray starting in u and passing through v, but $u \neq w$", see Fig. 4. This includes the case that $v = w$ (if $u \neq v$). Note that $\mathsf{left}(u, v, w)$ is false for $u = v$ and $u = w$. If $u \neq v$ then for every point w at least one of of the predicates $\mathsf{left}(u, v, w)$ or $\mathsf{left}(v, u, w)$ is true. If both $\mathsf{left}(u, v, w)$ and $\mathsf{left}(v, u, w)$ are true then w is located strictly between u and v.

We proposed a set of axioms for the predicate **left**. When analyzing these axioms we noticed that there is a minimal choice of 6 axioms **A1**–**A6** from which the other axioms can be derived.[1] These axioms are given below (the variables u, v, w, x, y, z are universally quantified), see also Fig. 5.

A1 : $(\mathsf{left}(u, v, w) \wedge \mathsf{left}(v, u, w) \rightarrow \neg\mathsf{left}(w, u, v))$
A2 : $(u \neq w \wedge v \neq w \wedge \neg\mathsf{left}(w, u, v) \wedge \neg\mathsf{left}(w, v, u) \rightarrow \mathsf{left}(u, v, w))$
A3 : $(\mathsf{left}(u, v, w) \wedge \mathsf{left}(v, u, w) \wedge \mathsf{left}(u, x, w) \rightarrow \mathsf{left}(u, x, v))$
A4 : $(\mathsf{left}(u, v, w) \wedge \mathsf{left}(v, u, w) \wedge \mathsf{left}(u, x, v) \rightarrow \mathsf{left}(u, x, w))$
A5 : $(\mathsf{left}(u, v, w) \wedge \mathsf{left}(v, w, u) \wedge \mathsf{left}(w, u, v)$
 $\rightarrow \mathsf{left}(u, v, x) \vee \mathsf{left}(v, w, x) \vee \mathsf{left}(w, u, x))$
A6 : $(\mathsf{left}(u, v, z) \wedge \mathsf{left}(v, w, z) \wedge \mathsf{left}(w, u, z) \wedge$
 $\mathsf{left}(x, y, u) \wedge \mathsf{left}(x, y, v) \wedge \mathsf{left}(x, y, w) \rightarrow \mathsf{left}(x, y, z))$

Axiom $A5$ can be derived from the other axioms if there exist at least 5 different points.

Based on the **left** predicate, the axiomatization of two additional predicates, **intersection** and **inside**, can be given.

---

[1] The other axioms, which are derived from **A1**, **A2**, **A3**, **A4**, **A5**, **A6** are listed in the extended version of this paper [10, Appendix 6.1].

Axioms for intersection: The intended meaning for intersection(u, v, w, x) for the segments uv and wx is *"the segments uv and wx intersect"*. This is the case iff one of points w or x is located left of uv and one left of vu and one of points u or v is located left of wx and one left of xw.

With this definition two segments intersect if they have a point in common, which is not an end vertex of both segments. Moreover, a segment can not intersect with itself, therefore intersection(u, v, u, v) and intersection(u, v, v, u) are false. The axioms for the intersection predicate (names starting with **I**) are given below (the variables u, v, w, x are universally quantified).[2]

I1-2 : intersection$(u, v, w, x) \rightarrow (u \neq w \lor v \neq x) \land (u \neq x \lor v \neq w)$

I3 : intersection$(u, v, w, x) \rightarrow$ left$(u, v, w) \lor$ left(u, v, x)

I4 : intersection$(u, v, w, x) \rightarrow$ left$(v, u, w) \lor$ left(v, u, x)

I5 : intersection$(u, v, w, x) \rightarrow$ left$(w, x, u) \lor$ left(w, x, v)

I6 : intersection$(u, v, w, x) \rightarrow$ left$(x, w, u) \lor$ left(x, w, v)

I11-12 : left$(u, v, w) \land$ left$(v, u, x) \land$ left$(w, x, u) \land$ left$(x, w, v) \rightarrow$
intersection$(u, v, w, x) \lor (u = x \land v = w)$

I13 : left$(u, v, w) \land$ left$(v, u, x) \land$ left$(w, x, v) \land$ left$(x, w, u) \rightarrow$
intersection(u, v, w, x)

I14 : left$(u, v, x) \land$ left$(v, u, w) \land$ left$(w, x, u) \land$ left$(x, w, v) \rightarrow$
intersection(u, v, w, x)

I15-16 : left$(u, v, x) \land$ left$(v, u, w) \land$ left$(w, x, v) \land$ left$(x, w, u) \rightarrow$
intersection$(u, v, w, x) \lor (u = w \land v = x)$

Axioms for Inside: The intended meaning for inside(u, v, w, x) is *"the point x is located inside or on the boundary the triangle Δuvw, but x is not equal to u, v or w"*. We can therefore define inside by the following axioms (the variables u, v, w, x are universally quantified):

T1-6 : (inside$(u, v, w, x) \leftrightarrow$ left$(u, v, x) \land$ left$(v, w, x) \land$ left(w, u, x))

The triangles we consider are oriented, so the order in which the vertices are written is important: In the (oriented) triangle Δuvw, left(u, v, w), left(v, w, u) and left(w, u, v) are all true. A point x is located inside the triangle Δuvw iff left(u, v, x), left(v, w, x) and left(w, u, x) are all true; x is located inside the triangle Δuwv iff left(u, w, x), left(w, v, x) and left(v, u, x) are all true. The triangle Δuvw is equal to the triangles Δvwu and Δwuv. We could prove:

- The inside predicate is transitive, i.e. if y is located inside the triangle Δuvx and x is located inside the triangle Δuvw, then y is also located inside the triangle Δuvw (**T21-T23**).
- If x and y are located inside a triangle Δuvw, where $x \neq y$, then y is also located in at least one of the triangles Δuvx, Δvwx or Δwux (**T24**).

In what follows we will refer to the union of the axioms for **left**, **intersection** and **inside** by AxGeom. A list of all the axioms we considered can be found in [10, Appendix 6.1-6.3].

---

[2] Since the clause form of some of the axioms contain more than one formula, and for doing the tests we used axioms in clause form, in such cases a range of numbers is used for the axioms.

Link to Other Axiom Systems for Plane Geometry. Tarski's [39] and Hilbert's [23] axiom systems for plane geometry use both a predicate "between". Tarski's betweenness notion is non-strict: a point c is between a and b iff c is on the closed segment $[ab]$, while Hilbert's notion is strict: c is between a and b iff c is on the open segment (ab). The relation "c is between a and b" can be expressed in our axiomatization by "$\mathsf{left}(a, b, c) \land \mathsf{left}(b, a, c)$" and is strict, so Hilbert's axiomatization is closer to ours than Tarski's. Since we only consider line segments the axioms of Hilbert's group 1,3 and 4 (cf. [23]) are not important for us; the related axioms can be found in group 2 (cf. [23]). Hilbert's axioms 2.1, 2.3 and 2.4 can be derived from our axioms (cf. also the remarks in Sect. 4.3 on Pasch's axiom). Since we only consider a finite set of points, Hilbert's axiom 2.2 (a $\forall\exists$-sentence) was not needed for the correctness proof, so no equivalent was included. Pasch's axiom could also be added in our axiom set and replace the axioms **A4** and **A6**, because they can be derived by the remaining axioms of AxGeom and Pasch's axiom. But we decided to use the axioms in the proposed way, because the axioms for left can be introduced independently first and afterwards the axioms for intersection and inside can be formulated based on the axioms for left.

3.2   Axiomatizing Graph Properties

In the next step conditions for vertices and edges are defined. We use a unary predicate V for representing the vertices[3] and a binary predicate E for representing the edges. Axioms for the edge predicate (denoted by **E**) are given below (the variables u, v are universally quantified):

E1: $\neg\mathsf{E}(u, u)$ (self-loops are not allowed)
E2: $\mathsf{E}(u, v) \to \mathsf{E}(v, u)$ (E is symmetric)
E3: $\mathsf{E}(u, v) \to \mathsf{V}(u) \land \mathsf{V}(v)$ (edges exist only between vertices)

From the definitions of intersection, inside and edges the conditions for the two properties redundancy and coexistence are defined. The redundancy property expresses the fact that if two edges uv and wx intersect, then one of the vertices u, v, w or x is connected to the other 3 ones. This is expressed with the following axiom (the variables u, v, w, x are universally quantified):

R1: $\mathsf{E}(u, v) \land \mathsf{E}(w, x) \land \mathsf{intersection}(u, v, w, x) \to \mathsf{E}(u, w) \lor \mathsf{E}(v, x)$

The coexistence property expresses the fact that if $\mathsf{inside}(u, v, w, x)$ is true and the edges uv, vw and wu exist, then also the edges ux, vx and wx exist. By the rotation of the triangle it is sufficient that only one of the edges has to exist. Axiom **C1** (the variables u, v, w, x are universally quantified) expresses this:

C1: $\mathsf{inside}(u, v, w, x) \land \mathsf{E}(u, v) \land \mathsf{E}(v, w) \land \mathsf{E}(w, u) \land \mathsf{V}(x) \to \mathsf{E}(u, x)$

---

[3] The predicate for the vertices is used here to distinguish between the axioms for geometry, which hold for arbitrary points, and properties for edges between vertices in a graph.

3.3 Axioms for Notions Used in the CP-Algorithm

For graphs satisfying the redundancy and coexistence property the CP algorithm constructs a connected intersection-free subgraph $G' = (V, F)$. The properties of F are described by a set of axioms **F1–F6** (in these formulae the variables u, v, w, x are universally quantified). Axioms **F1–F3** express the fact that the edges in F are a subset of E and F is symmetric and intersection-free. Axiom **F4** expresses the fact that no vertex can be located on an edge in F.

F1: $F(u, v) \rightarrow F(v, u)$
F2: $F(u, v) \rightarrow E(u, v)$
F3: $F(u, v) \wedge F(w, x) \rightarrow \neg\text{intersection}(u, v, w, x)$
In addition, no vertex can be located on an edge in F:
F4: $\text{left}(u, v, w) \wedge \text{left}(v, u, w) \wedge V(w) \rightarrow \neg F(u, v)$

Furthermore, the edges in F have to satisfy the CP-condition, i.e. for each edge wx intersecting with an edge uv in F, both edges uw and ux or both edges vw and vx have to exist. This can be expressed (after removing the implications already entailed by **R1** and **F2**) by the following axioms:

F5: $F(u, v) \wedge E(w, x) \wedge \text{intersection}(u, v, w, x) \rightarrow E(u, w) \vee E(v, w)$ (CP-condition)
F6: $F(u, v) \wedge E(w, x) \wedge \text{intersection}(u, v, w, x) \rightarrow E(u, x) \vee E(v, x)$ (CP-condition)
(only one of the conditions **F5** or **F6** is needed due to symmetry)

3.4 Axioms for Notions Used in the Correctness Proof

For the proof of correctness a further predicate **deleting** of arity 4 is defined, and notions such as path along the convex hull of points in a certain triangle are formalized. We describe the axiomatizations we used for this below.

Axioms for deleting: The intended meaning for $\text{deleting}(u, v, w, x)$ is *"uv and wx are intersecting edges and the edge wx prohibits that uv is in F"*.

Note that if $\text{deleting}(u, v, w, x)$ is true then the two redundancy edges wu and wv exist, while if $\text{deleting}(u, v, x, w)$ is true then the two redundancy edges xu and xv exist. The edge wx can prohibit that the edge uv is in F if wx is in F or if only the redundancy edges wu and wv, but neither xu nor xv exist for the intersection of uv and wx. Thus, $\text{deleting}(u, v, w, x)$ should hold iff $E(u, v) \wedge E(w, x) \wedge \text{intersection}(u, v, w, x) \wedge E(u, w) \wedge E(v, w) \wedge ((\neg E(u, x) \wedge \neg E(v, x)) \vee F(w, x))$ holds. The axioms for the deleting predicate are denoted by **D** and represent the condition above (the variables u, v, w, x are universally quantified).

D1-3: $\text{deleting}(u, v, w, x) \rightarrow E(u, v) \wedge E(w, x) \wedge \text{intersection}(u, v, w, x)$
D4-5: $\text{deleting}(u, v, w, x) \rightarrow E(u, w) \wedge E(v, w)$
D11: $E(u, x) \wedge \text{deleting}(u, v, w, x) \rightarrow F(w, x)$
D12: $E(v, x) \wedge \text{deleting}(u, v, w, x) \rightarrow F(w, x)$
D13: $E(u, v) \wedge E(w, x) \wedge \text{intersection}(u, v, w, x)$
    $\rightarrow \text{deleting}(u, v, w, x) \vee E(u, x) \vee E(v, x)$

D14: $E(u,v) \wedge F(w,x) \wedge \mathsf{intersection}(u,v,w,x) \wedge E(u,w) \wedge E(v,w)$
$\qquad \rightarrow \mathsf{deleting}(u,v,w,x)$

In what follows, we will refer to the axioms describing properties of E and F, and of the deleting predicate (axioms **E1–E3**, **F1–F6**, and **D1–D14**) by AxGraphs.

Path Along the Convex Hull of a Set of Points. Consider the triangle $\Delta u_1 q w_1$, where q is the intersection of edges $u_1 v_1$ and $w_1 x_1$. For representing a path from u_1 to w_1 along the convex hull of all points located inside of this triangle we use a list structure, with a function symbol next, where for every point i, $\mathsf{next}(i)$ is the next vertex on this path if i is a vertex on the convex hull that has a successor, or nil otherwise. Axioms **Y0–Y4** express the properties of the convex hull. In these axioms we explicitly write the universal quantifiers.

Y0: $\mathsf{next}(w_1) = \mathsf{nil}$
Y1: $\forall y, i\ (V(y) \wedge \mathsf{left}(u_1, v_1, y) \wedge \mathsf{left}(x_1, w_1, y) \rightarrow i = \mathsf{nil} \vee \mathsf{next}(i) = \mathsf{nil} \vee$
$\qquad y = i \vee y = \mathsf{next}(i) \vee \mathsf{left}(i, \mathsf{next}(i), y) \vee \mathsf{left}(y, \mathsf{next}(i), i))$
Y2: $\forall y, i\ (V(y) \wedge \mathsf{left}(u_1, v_1, y) \wedge \mathsf{left}(x_1, w_1, y) \rightarrow i = \mathsf{nil} \vee \mathsf{next}(i) = \mathsf{nil} \vee$
$\qquad y = i \vee y = \mathsf{next}(i) \vee \mathsf{left}(i, \mathsf{next}(i), y) \vee \mathsf{left}(\mathsf{next}(i), y, i))$
Y3: $\forall y, i\ (V(y) \wedge \mathsf{left}(u_1, v_1, y) \wedge \mathsf{left}(x_1, w_1, y) \wedge \mathsf{left}(\mathsf{next}(i), i, y) \rightarrow$
$\qquad i = \mathsf{nil} \vee \mathsf{next}(i) = \mathsf{nil} \vee y = i \vee y = \mathsf{next}(i) \vee \mathsf{left}(y, \mathsf{next}(i), i))$
Y4: $\forall y, i\ (V(y) \wedge \mathsf{left}(u_1, v_1, y) \wedge \mathsf{left}(x_1, w_1, y) \wedge \mathsf{left}(\mathsf{next}(i), i, y) \rightarrow$
$\qquad i = \mathsf{nil} \vee \mathsf{next}(i) = \mathsf{nil} \vee y = i \vee y = \mathsf{next}(i) \vee \mathsf{left}(\mathsf{next}(i), y, i))$

Axioms **Y11–Y13** express the fact that all the vertices on the path are in $\Delta u_1 q w_1$ and that there is an edge between the vertices of the convex hull.

Y11: $\forall i\ (i = \mathsf{nil} \vee \mathsf{next}(i) = \mathsf{nil} \vee \mathsf{inside}(u_1, v_1, w_1, \mathsf{next}(i)) \vee \mathsf{next}(i) = w_1)$
Y12: $\forall i\ (i = \mathsf{nil} \vee \mathsf{next}(i) = \mathsf{nil} \vee \mathsf{inside}(u_1, x_1, w_1, i) \vee i = u_1)$
Y13: $\forall i\ (E(i, \mathsf{next}(i)) \vee i = \mathsf{nil} \vee \mathsf{next}(i) = \mathsf{nil})$

Any vertex i on the path is located in $\Delta u_1 v_1 \mathsf{next}(i)$ and $\Delta u_1 x_1 \mathsf{next}(i)$ or equal to u_1; $\mathsf{next}(i)$ is located in $\Delta v_1 w_1 i$ and $\Delta x_1 w_1 i$ or equal to w_1.

Y14: $\forall i\ (i = \mathsf{nil} \vee \mathsf{next}(i) = \mathsf{nil} \vee \mathsf{inside}(u_1, v_1, \mathsf{next}(i), i) \vee i = u_1)$
Y15: $\forall i\ (i = \mathsf{nil} \vee \mathsf{next}(i) = \mathsf{nil} \vee \mathsf{inside}(u_1, x_1, \mathsf{next}(i), i) \vee i = u_1)$
Y16: $\forall i\ (i = \mathsf{nil} \vee \mathsf{next}(i) = \mathsf{nil} \vee \mathsf{inside}(v_1, w_1, i, \mathsf{next}(i)) \vee \mathsf{next}(i) = w_1)$
Y17: $\forall i\ (i = \mathsf{nil} \vee \mathsf{next}(i) = \mathsf{nil} \vee \mathsf{inside}(x_1, w_1, i, \mathsf{next}(i)) \vee \mathsf{next}(i) = w_1)$

4   Automated Reasoning

We used the axioms above for automating reasoning about geometric graphs and for proving correctness of results used in the proof of the CP-algorithm.

4.1   Analyzing the Axioms; Decidability

If we analyze the form of the axioms used in the proposed axiomatization (with the exception of the axioms describing the path along the convex hull), we notice that they are all universally quantified formulae over a signature containing only constants and predicate symbols (thus no function symbols with arity ≥ 1), i.e.

all the verification tasks can be formulated as satisfiability tasks for formulae in the Bernays-Schönfinkel class.

Definition 8 (The Bernays-Schönfinkel class). *Let Σ be a signature not containing function symbols with arity ≥ 1. The Bernays-Schönfinkel class consists of all Σ-formulae of the form $\exists x_1 \ldots x_n \forall y_1 \ldots \forall y_m F(x_1, \ldots, x_n, y_1, \ldots, y_m)$, where $n, m \geq 0$ and F is a quantifier-free formula containing only variables $x_1, \ldots, x_n, y_1, \ldots, y_m$.*

Satisfiability of formulae in the Bernays-Schönfinkel class is decidable.

Description of Paths Along the Convex Hull of Points in a Triangle. In [30], McPeak and Necula investigate reasoning in pointer data structures. The language used has sorts p (pointer) and s (scalar). Sets Σ_p and Σ_s of pointer resp. scalar fields are modeled by functions of sort p → p and p → s, respectively. A constant nil of sort p exists. The only predicate of sort p is equality; predicates of sort s can have any arity. The axioms considered in [30] are of the form

$$\forall p \quad \mathcal{E} \vee \mathcal{C} \tag{1}$$

where $\mathcal{E}$ contains disjunctions of pointer equalities and $\mathcal{C}$ contains scalar constraints (sets of both positive and negative literals). It is assumed that for all terms $f_1(f_2(\ldots f_n(p)))$ occurring in the body of an axiom, the axiom also contains the disjunction $p = \text{nil} \vee f_n(p) = \text{nil} \vee \cdots \vee f_2(\ldots f_n(p)) = \text{nil}.$[4]

In [30] a decision procedure for conjunctions of axioms of type (1) is proposed; in [24] a decision procedure based on instantiation of the universally quantified variables of sort p in axioms of type (1) is proposed.

The axioms **Y1–Y17** for the convex hull are all of type (1) and satisfy the nullable subterm condition above. The axioms needed for the verification conditions can therefore be represented as $\mathcal{T}_0 \subseteq \mathcal{T}_1 = \mathcal{T}_0 \cup \{\mathbf{Y1}, \ldots, \mathbf{Y4}, \mathbf{Y11}, \ldots \mathbf{Y17}\}$ where the theory $\mathcal{T}_0$ axiomatized by AxGeom ∪ AxGraphs ∪ $\{\mathbf{R1}, \mathbf{C1}\}$ can be regarded as a theory of scalars. Therefore, satisfiability of ground instances w.r.t. $\mathcal{T}_1$ is decidable, and decision procedures based on instantiation exist.

4.2 Choosing the Proof Systems

For solving the proof tasks, we used SPASS [40] and Z3 [32]. The tests reported here[5] [11] were made with Z3 because the axioms we used were (with the exception of the modelling of the paths along the convex hull) in the Bernays-Schönfinkel class, a class for which Z3 can be used as a decision procedure [36], whereas SPASS might not terminate. In addition, instantiation-based decision procedures exist for the type of list structures used for representing the paths along the convex hulls of points we consider [30]. Since Z3 is an instantiation-based SMT-prover, it performs the necessary instances during the proof process.

---

[4] This nullable subterm property has the role of excluding null pointer errors.

[5] The tests can be found under https://github.com/sofronie/tests-vmcai-2024.git and https://userpages.uni-koblenz.de/~boeltz/CP-Algorithm-Verification/.

We also considered using methods for complete hierarchical reasoning in local theory extensions, proposed in [24,37], since when defining the properties of geometric graphs we proceeded hierarchically – the properties of new predicate symbols used symbols already introduced. This was not possible because the current implementation of the hierarchical reduction in our system H-PILoT [25] can only be used for chains of theory extensions in which for every extension in this chain the free variables occur below extension symbols, which is not the case in the axiomatization we propose here.

4.3 Automated Proof of Properties of Geometric Graphs

In a first step, we analyzed the axiomatization of plane geometry we propose and the axioms of geometric graphs. We automatically proved the following results in plane geometry (the tests can be found in [11] in the folder Axioms):

Dependencies between axioms: We identified axioms of the left predicate which can be derived from the other ones. We proved that some axioms of the inside predicate follow from the others. A discussion can be found in [10, Appendix 6.1].

Pasch's axiom: We proved Pasch's Axiom (stating that if a line intersects one side of a triangle then it intersects exactly one other side) using the axioms **A**, **I** and **T**.

We then automatically proved the following properties of geometric graphs $G = (V, E)$ which are useful in the correctness proof of the CP-algorithm:

For graphs with the coexistence property: We proved that the coexistence property implies the clique property for vertices in the interior of a triangle defined by edges, i.e. the property that for every triangle Δuvw, where u, v, w are vertices of G and $uv, vw, wu \in E$, every two different vertices x, y located in the interior of the triangle according to Definition 2 are connected by the edge $xy \in E$.

For graphs with the redundancy and coexistence property: For three colinear vertices u, v and w, where w is located between u and v, it can be proven that if the edge uv exists, then also the edges uw and wv exist.

4.4 Correctness of the CP-Algorithm: Verifying Main Steps

In order to prove the correctness of the CP-algorithm which is executed on a finite graph $G = (V, E)$, that satisfies the redundancy and the coexistence property, it has to be proven that the resulting graph $G' = (V, F)$ is connected and intersection-free. As mentioned before, the fact that if an edge is added to F then all intersecting edges are removed from the working set W, the graph G' is guaranteed to be intersection-free. Therefore it remains to prove the connectivity of F. We assume that F is not connected after the termination of the algorithm, and we derive a contradiction. For this, the following lemma is proven:

Lemma 1 *For every edge uv that is not in F there exists an edge wx in F that intersects with uv.*

The proof of Lemma 1 is split in two parts:

Part 1: First, it is proven that for an edge $e_0 = u_1v_1$, which is not in F there exists always a sequence of other edges $e_1, ..., e_n$ in E that intersect with e_0. The sequence is constructed in the following way: We start with an edge $d_1 = e_1 = w_1x_1$ that deletes, and therefore also intersects u_1v_1. If $e_1 \in F$ we are done. Assume that we built a sequence $e_1, ..., e_{i-1}$ of edges which are not in F and intersect e_0. Then an edge d_i can be chosen, that deletes e_{i-1}. Let e_i be a redundancy edge of the intersection between e_{i-1} and d_i, such that e_i intersects with e_0.

Part 2: We then show that such a sequence can not be cyclic, by showing that the maximal length of a shortest cycle is 2 and such a cycle can not exist. This proves the statement of Lemma 1.

For the construction of the shortest cycle the following cases have to be considered. If e_2 also deletes u_1v_1, then a sequence of shorter length can be obtained by choosing e_2 instead of e_1. The other situation where a sequence of shorter length can be obtained is if an edge d_a formed by vertices of the edges $d_1, ..., d_i$ deletes an edge e_j for $j < i - 1$. In this case d_a can be chosen instead of d_{j+1} to obtain a sequence of shorter length.

Automated Verification of Part 1. We prove[6] Part 1 in two steps.

Step 1 (see Fig. 6). We prove the following statement: Let u_1v_1 be an edge which is deleted by an edge w_1x_1, i.e. u_1v_1 and w_1x_1 intersect and the edges w_1u_1 and w_1v_1 exist. Assume that w_1x_1 is not in F because an intersecting edge w_2x_2 deletes w_1x_1. Then at least one of the edges w_1w_2 or x_1w_2 exists and intersects with u_1v_1.

For this, we prove that the axioms $\mathsf{AxGeom} \cup \mathsf{AxGraphs} \cup \{\mathbf{R1, C1}\}$ entail:

$$\mathsf{deleting}(u_1, v_1, w_1, x_1) \land \mathsf{deleting}(w_1, x_1, w_2, x_2)$$
$$\rightarrow \mathsf{deleting}(u_1, v_1, w_2, x_1) \lor \mathsf{Intersection}(u_1, v_1, w_1, w_2)$$

The case that x_2w_2 deletes w_1x_1 is analogous, with an intersection of u_1v_1 and x_2w_1 or an intersection of u_1v_1 and x_2x_1. If the coexistence property is dropped, the implication of Step 1 does not hold anymore and the counterexample in Fig. 7 is constructed (cf. [11] folder Counterexamples).

Step 2: In the previous step it was proven that if w_1x_1 deletes u_1v_1 and w_2x_2 deletes w_1x_1 then either w_1w_2 or x_1w_2 intersects with u_1v_1. The case that x_1w_2 (resp. x_1x_2) intersects with u_1v_1 corresponds to Step 1, with w_2 (resp. x_2) instead of w_1. So in Step 2 w_1w_2 is the edge intersecting with u_1v_1.

---

[6] The tests can be found in https://github.com/sofronie/tests-vmcai-2024.git (folder Proof) and https://userpages.uni-koblenz.de/~boeltz/CP-Algorithm-Verification/Proof.

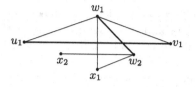
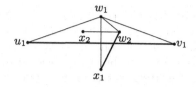

(a) The edges u_1v_1 and w_1w_2 intersect.   (b) The edges u_1v_1 and x_1w_2 intersect.

Fig. 6. One of the edges w_1w_2 or x_1w_2 intersects with u_1v_1.

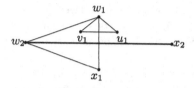
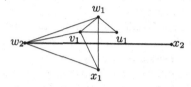

(a) Counterexample for Step 1 without coexistence property.   (b) How the graph would look like if the coexistence property is satisfied.

Fig. 7. v_1 is located inside the triangle $\triangle w_2x_1w_1$.

Furthermore w_3x_3 deletes w_1w_2. We distinguish the following cases:

Case 2a: Only the edge w_3x_3, but not x_3w_3 deletes the edge w_1w_2.

Case 2b: Both of the edges w_3x_3 and x_3w_3 delete the edge w_1w_2.

In **Step 2a** we prove that in Case 2a another edge deletes w_1x_1 (consider Step 1 with this edge instead of w_2x_2) or one of the edges w_1w_3 or w_2w_3 is intersecting with u_1v_1.

Step 2a: We prove that the axioms $\mathsf{AxGeom} \cup \mathsf{AxGraphs} \cup \{\mathbf{R1}, \mathbf{C1}\}$ entail:

$\text{deleting}(u_1, v_1, w_1, x_1) \wedge \text{deleting}(w_1, x_1, w_2, x_2) \wedge$

$\text{intersection}(u_1, v_1, w_1, w_2) \wedge \text{deleting}(w_1, w_2, w_3, x_3)$

$\rightarrow \text{deleting}(w_1, x_1, w_3, x_2) \vee \text{deleting}(w_1, x_1, w_3, x_3) \vee$

$\text{deleting}(w_1, x_1, x_3, w_3) \vee \text{deleting}(w_1, x_1, x_3, x_2) \vee \text{deleting}(w_1, w_2, x_3, w_3) \vee$

$\text{intersection}(u_1, v_1, w_1, w_3) \vee \text{intersection}(u_1, v_1, w_2, w_3)$

In **Step 2b** we prove that in Case 2b another edge deletes w_1x_1[7] (consider Step 1 with this edge instead of w_2x_2) or one of the edges w_1w_3, w_2w_3, w_1x_3, w_2x_3 is intersecting with u_1v_1[8].

Automated Verification of Part 2. It remains to prove that a sequence of edges which intersect with u_1v_1 exists, which is not cyclic, i.e. all edges in the sequence differ. To prove this, it is enough to show that if a sequence of intersecting edges with uv is greater than 2, then also a shorter sequence exists.

This part of the proof is split in the Steps 3–7.

---

[7] Then $d_2 = w_3x_3$ or $d_2 = x_3w_3$ or $d_2 = w_3x_2$ or $d_2 = x_3x_2$.

[8] Depending on the edge intersecting with u_1v_1, d_3 is either w_3x_3 or x_3w_3.

Step 3: In this step it is proven that if an edge w_1x_1 is deleted by an edge w_2x_2, the edge w_1w_2 is deleted by an edge w_3x_3 and w_1w_3 intersects w_2x_2 then there exists another edge between two of the vertices w_2, x_2, w_3, x_3 which is deleting w_1x_1, but is not deleted by w_2x_2.

For the further steps of the proof we therefore assume that the edges inside the triangle $\Delta w_1x_1w_2$ considered in Step 5 as well as the edge w_1w_3 considered in Step 6, do not intersect with w_2x_2.

Step 4: In this step it is proven that if an edge w_1x_1 is deleted by an edge w_2x_2 and w_1w_2 is deleted by w_3x_3, then there exists another edge between two of the vertices w_2, x_2, w_3, x_3 deleting w_1x_1, one of the vertices w_3 or x_3 is located inside the triangle $\Delta w_1x_1w_2$ resp. $\Delta x_1w_1w_2$ (considered in **Step 5**) or the edge w_2x_2 is in F (considered in **Step 6**). The special case of $w_3 = x_1$ is considered in **Step 4a**.

Step 5: It is then proven that if an edge w_1x_1 is deleted by an edge w_2x_2 and an edge yw_2 (**Step 5a**), yw_1 (**Step 5b**) or yz (**Step 5c**) inside the triangle $\Delta w_1x_1w_2$[9] is deleted by an edge w_3x_3 with both vertices located outside the triangle $\Delta w_1x_1w_2$, then there exists another edge between two of the vertices w_2, x_2, w_3, x_3 deleting w_1x_1 or another edge between two of the vertices x_2, w_3, x_3 deleting w_1w_2. Since the formulae are relatively large, we included them in [10, Appendix 7].

Step 6: As in the previous steps an edge w_1x_1 is deleted by an edge w_2x_2 and w_1w_2 is deleted by w_3x_3. In addition in the tests of this step the edge w_1w_3 (**Step 6a** and **Step 6b**) resp. w_2w_3 (**Step 6c**, **Step 6d** and **Step 6e**) is deleted by an edge w_4x_4 with both vertices outside the triangle $\Delta w_1w_2w_3$. It is proven that then there exists another edge between two of the vertices $w_2, x_2, w_3, x_3, w_4, x_4$ deleting w_1x_1 or another edge between two of the vertices x_2, w_3, x_3, w_4, x_4 deleting w_1w_2.

In the special case of $w_4 = w_1$ (**Step 6e**) it is proven that there exists another edge deleting w_1x_1. Analogously in the special case of $w_4 = x_1$(**Step 6b** resp. **Step 6d**) it is proven that there exists another edge deleting w_1w_2.

In the test (**Step 6g**) it is proven that if w_4 is located inside the triangle $\Delta w_1x_1w_2$ and w_1w_4[10] is deleted by w_5x_5, then there exists another edge deleting w_1w_2 or w_1w_3, one of the edges w_1w_3, w_1w_4 or w_1w_5 intersects with w_2x_2 or w_5 or x_5 are located inside a triangle formed by w_1, w_3 and w_4. Since the formulae are relatively large, we included them in [10, Appendix 7].

Step 7: Finally it is proven that if w_1x_1 is deleted by w_2x_2 and x_2w_2, w_1w_2 is deleted by x_1w_3 and w_1x_2 is deleted by x_1w_4 then a contradiction is obtained. For this we prove that AxGeom $\cup$ AxGraph $\cup$ {**R1, C1**} entails:

$$\text{deleting}(w_1, x_1, w_2, x_2) \wedge \text{deleting}(w_1, x_1, x_2, w_2)$$
$$\wedge \text{deleting}(w_1, w_2, x_1, w_3) \wedge \text{deleting}(w_1, x_2, x_1, w_4) \rightarrow \perp$$

---

[9] In **Step5d-f** it is proven that the considered edges do not intersect with w_2x_2.

[10] A similar result is proven in tests **Step 6h-j** for the edges w_2w_4 and w_3w_4.

4.5 Connection Between the Components X and Y

In this section the following Lemma is proven.

Lemma 2 *There exists an edge uv in E with $u \in X$ and $v \in Y$, such that all edges in F intersecting with uv are located in Y.*

We consider the resulting graph $G' = (V, F)$ after the termination of the algorithm with the components X and $Y = G' \setminus X$. In the original graph $G = (V, E)$ there was an edge $u_1 v_1$ with $u_1 \in X$ and $v_1 \in Y$. Now a sequence of edges connecting the components X and Y is constructed. The following tests show that a sequence of edges intersecting with edges connecting the components X and Y can not be cyclic (see Fig. 3a).

For all tests[11] the edge $u_1 v_1$ is deleted by an edge $w_1 x_1$ in F and the redundancy edge for this intersection $v_1 w_1$, is deleted by the edge $w_2 x_2$ in F.

Step 8: If $w_2 x_2$ intersects $v_1 w_1$, then w_2 or x_2 are located inside one of the triangles formed by u_1, v_1 and w_1 ($\Delta u_1 v_1 w_1$ or $\Delta v_1 u_1 w_1$) (considered in Step 9) or $u_1 v_1$ and $w_2 x_2$ intersect (**Step 8a**). In case that $u_1 v_1$ and $w_2 x_2$ intersect, the edge $w_2 x_2$ can be considered instead of the edge $w_1 x_1$. The special case of $w_2 = u_1$ is considered in **Step 8b** and proves that no cycle of length 2 can occur.

Step 9: If w_2 is inside the triangle $\Delta u_1 v_1 w_1$ and the edge in F $w_3 x_3$ deletes $v_1 w_2$, then w_3 or x_3 are located inside the triangle $\Delta v_1 w_1 w_2$ or the edge $w_3 x_3$ is intersecting with the edge $u_1 v_1$ or $v_1 w_1$ (**Step 9a**). If the edge $w_3 x_3$ intersects with $u_1 v_1$, then $w_3 x_3$ can be chosen instead of $w_1 x_1$. If the edge $w_3 x_3$ intersects with $v_1 w_1$, then $w_3 x_3$ can be chosen instead of $w_2 x_2$. The special case of $w_3 = u_1$ is considered in **Step 9b** and proves that no cycle of length 3 can occur.

4.6 Proof of Properties for the Convex Hull

After proving Lemma 1 and 2, the following Lemma is proven.

Lemma 3 *For an edge $u_1 v_1$ and an intersecting edge $w_1 x_1 \in F$, with intersection point q closest to u_1 among all intersecting edges in F there exists a path P on the convex hull of the triangle $\Delta u_1 q w_1$ with all edges in F connecting u_1 and w_1 (see Fig. 3b).*

The concept used for the proof of Lemma 3 is the construction of paths on convex hulls. We consider *the finest path* along the convex hull: If vertices u, v, w are on the convex hull, v is located on the segment uw and we have edges uv and $vw \in F$, then u, v, w are all on the finest path of the convex hull. We prove that a path on the convex hull of a triangle $\Delta u_1 q w_1$ formed by an edge $u_1 v_1$ and an intersecting edge $w_1 x_1 \in F$ with intersection point q closest to u_1 is not

---

[11] The tests can be found in https://github.com/sofronie/tests-vmcai-2024.git (folder Proof) and https://userpages.uni-koblenz.de/~boeltz/CP-Algorithm-Verification/Proof.

intersected by edges of F. For this, we prove that for an edge u_1v_1 and an edge w_1x_1 with intersection point q closest to u_1, no edge $ab \in F$ can intersect with an edge of the convex hull in Δu_1qw_1. From Lemma 1 it now follows that the edges of the convex hull are in F. Therefore there exists a path with all edges in F that connects u_1 with w_1. This path connects the components X and Y and therefore the following Theorem holds.

Theorem 1 *The edges of F form a connected plane spanning subgraph of G after the termination of the CP-algorithm.*

4.7   Constructing a Counterexample Without Coexistence Property

We analyzed whether the proof still can be carried out if the coexistence property is dropped, and obtained a counterexample, i.e. an example of a geometric graph satisfying axiom **R1** but in which axiom **C1** does not hold, for which no connected-intersection free subgraph can be constructed.

The problem here is that connectivity is not first order definable and connectivity in a given number of steps is expressed by $\forall\exists$ sentences, which Z3 cannot handle well. However, we can obtain a counterexample for the case of a redundancy cycle of length 3 (Fig. 8). The cycle can be constructed by using the predicate deleting. An edge u_1y is deleted by an edge u_2v_2. The redundancy edge u_2y exists, but u_2y is deleted by u_3v_3. This leads to the existence of

Fig. 8. Example of a redundancy cycle.

the redundancy edge u_3y. But u_3y is deleted by u_1v_1. This means that the edges u_1y, u_2y and u_3y form a redundancy cycle and the graph containing the vertices u_1, u_2, u_3, v_1, v_2, v_3 and y is connected. The test is in [11], folder Counterexample; a counterexample for a redundancy cycle of length 4 is also included there. Further tests show that also a weak version of the coexistence property, where only one or two coexistence edges have to exist, does not guarantee the connectedness of the resulting subgraph (cf. [10]).

5   Conclusion

We defined an axiom system for geometric graphs, and extended it with possibilities of reasoning about paths along certain convex hulls. We used this axiom system to check the correctness of the CP-algorithm for graphs satisfying the redundancy and coexistence property.

Another verification possibility we would like to investigate is proving that (with the notation in Alg. 1) the connectedness of $W \cup F$ is an inductive invariant. However, the form of the redundancy property changes when edges are removed. We would like to investigate if ideas from [34] could be used in this context for invariant strengthening. Also as future work, we plan to use our axiom system for the verification of other algorithms and transformations in geometric graphs and extend it to a version that covers also the 3-dimensional space.

Acknowledgment. We thank the reviewers for their helpful comments.

References

1. Abdulaziz, M., Mehlhorn, K., Nipkow, T.: Trustworthy graph algorithms (invited talk). In: Rossmanith, P., Heggernes, P., Katoen, J. (eds.) 44th International Symposium on Mathematical Foundations of Computer Science, MFCS 2019, August 26-30, 2019, Aachen, Germany. LIPIcs, vol. 138, pp. 1:1–1:22. Schloss Dagstuhl - Leibniz-Zentrum für Informatik (2019). https://doi.org/10.4230/LIPIcs.MFCS.2019.1

2. Barriére, L., Fraigniaud, P., Narayanan, L.: Robust position-based routing in wireless ad hoc networks with unstable transmission ranges. In: Proceedings of the 5th ACM International Workshop on Discrete Algorithms and Methods for Mobile Computing and Communications (DIAL M 01), New York, New York, USA, pp. 19–27. ACM Press (2001). https://doi.org/10.1145/381448.381451

3. Beeson, M., Wos, L.: Finding proofs in Tarskian geometry. J. Autom. Reason. **58**(1), 181–207 (2017). https://doi.org/10.1007/s10817-016-9392-2

4. Bettstetter, C., Hartmann, C.: Connectivity of wireless multihop networks in a shadow fading environment. Wireless Netw. **11**(5), 571–579 (2005). https://doi.org/10.1007/s11276-005-3513-x

5. Bjørner, N., de Moura, L., Nachmanson, L., Wintersteiger, C.M.: Programming Z3. In: Bowen, J.P., Liu, Z., Zhang, Z. (eds.) SETSS 2018. LNCS, vol. 11430, pp. 148–201. Springer, Cham (2019). https://doi.org/10.1007/978-3-030-17601-3_4

6. Bjørner, N., Nachmanson, L.: Navigating the universe of Z3 theory solvers. In: Carvalho, G., Stolz, V. (eds.) SBMF 2020. LNCS, vol. 12475, pp. 8–24. Springer, Cham (2020). https://doi.org/10.1007/978-3-030-63882-5_2

7. Böhmer, S., Schneider, D., Frey, H.: Stochastic modeling and simulation for redundancy and coexistence in graphs resulting from log-normal shadowing. In: Proceedings of the 22nd International ACM Conference on Modeling, Analysis and Simulation of Wireless and Mobile Systems - MSWIM 2019, New York, New York, USA, pp. 173–182. ACM Press (2019). https://doi.org/10.1145/3345768.3355933

8. Böltz, L., Becker, B., Frey, H.: Local construction of connected plane subgraphs in graphs satisfying redundancy and coexistence. In: XI Latin and American Algorithms, Graphs and Optimization Symposium (LAGOS), pp. 1–10 (2021)

9. Böltz, L., Frey, H.: Existence of connected intersection-free subgraphs in graphs with redundancy and coexistence property. In: Dressler, F., Scheideler, C. (eds.) ALGOSENSORS 2019. LNCS, vol. 11931, pp. 63–78. Springer, Cham (2019). https://doi.org/10.1007/978-3-030-34405-4_4

10. Böltz, L., Sofronie-Stokkermans, V., Frey, H.: On the verification of the correctness of a subgraph construction algorithm (extended version), ArXiv, https://doi.org/10.48550/arXiv.2311.17860 (2023)

11. Böltz, L., Sofronie-Stokkermans, V., Frey, H.: Tests for the verification of the CP-algorithm (2023). https://github.com/sofronie/tests-vmcai-2024.git

12. Chen, T., Chen, J., Wu, C.: Distributed object tracking using moving trajectories in wireless sensor networks. Wirel. Networks **22**(7), 2415–2437 (2016)

13. Courcelle, B.: On the expression of monadic second-order graph properties without quantifications over sets of edges (extended abstract). In: Proceedings of the Fifth Annual Symposium on Logic in Computer Science (LICS 1990), Philadelphia, Pennsylvania, USA, June 4–7, 1990, pp. 190–196. IEEE Computer Society (1990). https://doi.org/10.1109/LICS.1990.113745

14. Courcelle, B.: The monadic second-order logic of graphs VI: on several representations of graphs by relational structures. Discret. Appl. Math. **63**(2), 199–200 (1995). https://doi.org/10.1016/0166-218X(95)00006-D

15. Courcelle, B.: The monadic second-order logic of graphs XVI: canonical graph decompositions. Log. Methods Comput. Sci. **2**(2) (2006). https://doi.org/10.2168/LMCS-2(2:2)2006

16. Courcelle, B.: Monadic second-order logic for graphs: algorithmic and language theoretical applications. In: Dediu, A.H., Ionescu, A.M., Martín-Vide, C. (eds.) LATA 2009. LNCS, vol. 5457, pp. 19–22. Springer, Heidelberg (2009). https://doi.org/10.1007/978-3-642-00982-2_2

17. Cox, D.A., Little, J., O'Shea, D.: Ideals, varieties, and algorithms - an introduction to computational algebraic geometry and commutative algebra (2. ed.). Undergraduate texts in mathematics, Springer (1997)

18. Doczkal, C., Pous, D.: Graph theory in Coq: minors, treewidth, and isomorphisms. J. Autom. Reason. **64**, 795–825 (2020)

19. Dolzmann, A., Sturm, T., Weispfenning, V.: A new approach for automatic theorem proving in real geometry. J. Autom. Reason. **21**(3), 357–380 (1998). https://doi.org/10.1023/A:1006031329384

20. Frey, H., Simplot-Ryl, D.: Localized topology control algorithms for ad hoc and sensor networks. In: Nayak, A., Stojmenovic, I. (eds.) Handbook of Applied Algorithms, chap. 15, pp. 439–464. Wiley, Hoboken (2007)

21. Frey, H., Stojmenovic, I.: On delivery guarantees and worst-case forwarding bounds of elementary face routing components in ad hoc and sensor networks. IEEE Trans. Comput. **59**(9), 1224–1238 (2010). https://doi.org/10.1109/TC.2010.107

22. Heckel, R., Lambers, L., Saadat, M.G.: Analysis of graph transformation systems: Native vs translation-based techniques. In: Electronic Proceedings in Theoretical Computer Science, vol. 309, pp. 1–22 (2019). https://doi.org/10.4204/eptcs.309.1

23. Hilbert, D.: The Foundations of Geometry. 2nd ed. Chicago: Open Court. (1980 (1899))

24. Ihlemann, C., Jacobs, S., Sofronie-Stokkermans, V.: On local reasoning in verification. In: Ramakrishnan, C.R., Rehof, J. (eds.) TACAS 2008. LNCS, vol. 4963, pp. 265–281. Springer, Heidelberg (2008). https://doi.org/10.1007/978-3-540-78800-3_19

25. Ihlemann, C., Sofronie-Stokkermans, V.: System description: H-PILoT. In: Schmidt, R.A. (ed.) CADE 2009. LNCS (LNAI), vol. 5663, pp. 131–139. Springer, Heidelberg (2009). https://doi.org/10.1007/978-3-642-02959-2_9

26. Karamete, B.K., Adhami, L., Glaser, E.: A fixed storage distributed graph database hybrid with at-scale OLAP expression and I/O support of a relational DB: kineticagraph. CoRR **abs/2201.02136** (2022). https://arxiv.org/abs/2201.02136

27. Kuhn, F., Wattenhofer, R., Zollinger, A.: Ad-hoc networks beyond unit disk graphs. In: ACM DIALM-POMC Joint Workshop on Foundations of Mobile Computing, pp. 69–78. San Diego (2003)

28. Mathews, E.: Planarization of geographic cluster-based overlay graphs in realistic wireless networks. In: 2012 Ninth International Conference on Information Technology - New Generations, pp. 95–101. IEEE (2012)

29. Mathews, E., Frey, H.: A localized link removal and addition based planarization algorithm. In: Bononi, L., Datta, A.K., Devismes, S., Misra, A. (eds.) ICDCN 2012. LNCS, vol. 7129, pp. 337–350. Springer, Heidelberg (2012). https://doi.org/10.1007/978-3-642-25959-3_25

30. McPeak, S., Necula, G.C.: Data structure specifications via local equality axioms. In: Etessami, K., Rajamani, S.K. (eds.) CAV 2005. LNCS, vol. 3576, pp. 476–490. Springer, Heidelberg (2005). https://doi.org/10.1007/11513988_47

31. Mirzaie, M., Mazinani, S.M.: MCFL: an energy efficient multi-clustering algorithm using fuzzy logic in wireless sensor network. Wirel. Networks **24**(6), 2251–2266 (2018)

32. de Moura, L., Bjørner, N.: Z3: an efficient SMT solver. In: Ramakrishnan, C.R., Rehof, J. (eds.) TACAS 2008. LNCS, vol. 4963, pp. 337–340. Springer, Heidelberg (2008). https://doi.org/10.1007/978-3-540-78800-3_24

33. Neumann, F., Vivas Estevao, D., Ockenfeld, F., Radak, J., Frey, H.: Short paper: structural network properties for local planarization of wireless sensor networks. In: Mitton, N., Loscri, V., Mouradian, A. (eds.) ADHOC-NOW 2016. LNCS, vol. 9724, pp. 229–233. Springer, Cham (2016). https://doi.org/10.1007/978-3-319-40509-4_16

34. Peuter, D., Sofronie-Stokkermans, V.: On invariant synthesis for parametric systems. In: Fontaine, P. (ed.) CADE 2019. LNCS (LNAI), vol. 11716, pp. 385–405. Springer, Cham (2019). https://doi.org/10.1007/978-3-030-29436-6_23

35. Philip, S.J., Ghosh, J., Ngo, H.Q., Qiao, C.: Routing on overlay graphs in mobile ad hoc networks. In: Proceedings of the IEEE Global Communications Conference, Exhibition & Industry Forum (GLOBECOM'06) (2006)

36. Piskac, R., de Moura, L.M., Bjørner, N.S.: Deciding effectively propositional logic using DPLL and substitution sets. J. Autom. Reason. **44**(4), 401–424 (2010). https://doi.org/10.1007/s10817-009-9161-6

37. Sofronie-Stokkermans, V.: Hierarchic reasoning in local theory extensions. In: Nieuwenhuis, R. (ed.) CADE 2005. LNCS (LNAI), vol. 3632, pp. 219–234. Springer, Heidelberg (2005). https://doi.org/10.1007/11532231_16

38. Sturm, T., Weispfenning, V.: Computational geometry problems in REDLOG. In: Wang, D. (ed.) ADG 1996. LNCS, vol. 1360, pp. 58–86. Springer, Heidelberg (1997). https://doi.org/10.1007/BFb0022720

39. Tarski, A., Givant, S.: Tarski's system of geometry. Bull. Symb. Log. **5**(2), 175–214 (1999). https://doi.org/10.2307/421089

40. Weidenbach, C., Dimova, D., Fietzke, A., Kumar, R., Suda, M., Wischnewski, P.: SPASS Version 3.5. In: Schmidt, R.A. (ed.) CADE 2009. LNCS (LNAI), vol. 5663, pp. 140–145. Springer, Heidelberg (2009). https://doi.org/10.1007/978-3-642-02959-2_10

Efficient Local Search for Nonlinear Real Arithmetic

Zhonghan Wang[1,2], Bohua Zhan[1,2(✉)], Bohan Li[1,2], and Shaowei Cai[1,2]

[1] State Key Laboratory of Computer Science, Institute of Software, Chinese Academy of Sciences, Beijing, China
{wangzh,bzhan,libh,caisw}@ios.ac.cn
[2] University of Chinese Academy of Sciences, Beijing, China

Abstract. Local search has recently been applied to SMT problems over various arithmetic theories. Among these, nonlinear real arithmetic poses special challenges due to its uncountable solution space and potential need to solve higher-degree polynomials. As a consequence, existing work on local search only considered fragments of the theory. In this work, we analyze the difficulties and propose ways to address them, resulting in an efficient search algorithm that covers the full theory of nonlinear real arithmetic. In particular, we present two algorithmic improvements: incremental computation of variable scores and temporary relaxation of equality constraints. We also discuss choice of candidate moves and a look-ahead mechanism in case when no critical moves are available. The resulting implementation is competitive on satisfiable problem instances against complete methods such as MCSAT in existing SMT solvers.

Keywords: Local search · Nonlinear arithmetic · SMT

1 Introduction

Satisfiability Modulo Theories (SMT) is the problem of determining the satisfiability of a formula containing both logical operators and functions interpreted in one or more custom theories [6]. Commonly considered theories include equality, arithmetic, bit-vectors, arrays, and strings. After nearly two decades of development, SMT has gained widespread applications in program verification, model checking, planning, and many other areas.

The arithmetic theories can be divided according to the type of numbers involved into integer and real theories, and according to the operations allowed into difference logic, linear, and nonlinear theories. The case of (quantifier-free) nonlinear real arithmetic (NRA) considers satisfiability of equalities and inequalities involving polynomials of degree greater than one, and where the arithmetic variables take on real values. It has applications in the analysis of nonlinear hybrid automata [13], generating ranking functions for termination analysis [22,28], constraint answer set programming [41,42], and even analysis of biological networks [3]. Problem instances from many of these applications are collected in the SMT-LIB benchmarks [5].

© The Author(s), under exclusive license to Springer Nature Switzerland AG 2024
R. Dimitrova et al. (Eds.): VMCAI 2024, LNCS 14499, pp. 326–349, 2024.
https://doi.org/10.1007/978-3-031-50524-9_15

Similar to the SAT case, methods for solving SMT problems can be roughly divided into *complete* and *incomplete* methods. Complete methods are usually based on DPLL(T) or close variants. They are able to both find solutions and prove unsatisfiability. Incomplete methods, such as those based on local search [24], explore the solution space heuristically, usually by changing the assignment of one variable at a time, in an attempt to find a satisfying solution. Local search methods are not able to prove unsatisfiability, but can have an advantage over complete methods on some satisfiable instances.

Complete methods for the theory of nonlinear real arithmetic make crucial use of cylindrical algebraic decomposition (CAD) [10], which permits deciding the satisfiability of a conjunction of polynomial (in)equalities over real numbers. This can then serve as the theory solver in DPLL(T) [39]. An innovation over DPLL(T) for arithmetic theories is the MCSAT algorithm [25,34], which constructs models involving both boolean and arithmetic variables at the same time. A nice overview of DPLL(T) and MCSAT for nonlinear real arithmetic can be found in Kremer's thesis [27].

Exploration of applying local search to solve SMT problems over arithmetic theories began only recently. The work [7] applied local search to the theory of linear integer arithmetic. It introduced the concept of *critical moves*, a change in one arithmetic variable that satisfies a previously unsatisfied clause. The algorithm iteratively applies critical moves that most improves the score, a weighted count of unsatisfied clauses. The presence of both boolean and arithmetic variables are dealt with by alternately working in the *integer mode* and the *boolean mode*, when assignments to only integer or boolean variables are changed respectively. Switching between modes are performed after the number of non-improving steps reaches a certain threshold.

Compared to linear integer arithmetic, the problem of nonlinear real arithmetic held additional challenges for local search methods. Unlike integer theories, there is an uncountable number of possible assignments to choose from, including infinite number of choices in any finite interval. Unlike the linear case, there is potential need to solve polynomials of degree greater than one, which is both costly in time and may result in variable assignments that are irrational (i.e. algebraic) numbers, causing a slow-down of the ensuing search process. It is also possible that a nonlinear constraint cannot be satisfied by changing the value of one variable alone, resulting in the lack of critical moves, so that other heuristics are needed in such scenarios.

There has already been some work exploring local search for nonlinear real arithmetic. However, largely due to the challenges listed above, none of the existing work covers the entirety of the theory. The work [29] considers the multilinear case, where each variable appears with degree at most one in each polynomial constraint. The work [31] considers problems where all equality constraints contain at least one variable that is linear. In both of these works, the problem of variable assignments to irrational values is avoided by either assuming linearity in each variable, or by limiting higher-degree constraints to strict inequalities only.

1.1 Contributions

In this paper, we propose several improvements to the local search procedure, aimed at addressing the challenges posed by nonlinear real arithmetic. This results in an efficient local search algorithm for the entire NRA theory.

First, we present efficient data structures for caching and updating variable scores used to determine the next critical move. Existing work on local search for arithmetic theories can be thought of as an extension of the GSAT algorithm [40] with adaptive weighting. It is well-known that efficient implementations of GSAT involve caching and updating of variable scores [24, Section 6.2]. For arithmetic theories, this is complicated by the fact that each variable is associated not with one score, but with different scores for changing its assignment to values in different intervals. Hence, current implementations of local search for arithmetic theories recompute the score information for each variable at every iteration, resulting in potential repeated computations. This is especially serious for nonlinear arithmetic, where computations may involve costly root-finding for higher-degree polynomials. We describe data structures maintaining *boundaries* of score changes for each pair of clause and variable appearing in the clause, which only need to be updated on an as-needed basis, and from which the full score information can be quickly recovered.

Second, we address the problem posed by nonlinear equality constraints between variables, which may force assignment to variables that are irrational numbers. Rather than making such assignments directly, we propose to temporarily *relax* such equalities into inequalities, e.g. changing the constraint $p = 0$ into $p > -\epsilon \wedge p < \epsilon$, and continue the local search process. If an (approximate) solution is found that satisfies the relaxed version of these constraints, we restore the equalities to their original form, and try to find an exact solution near the approximate solution. For this step, we try two different heuristic methods, respectively based on analyzing the structure of equations and local search itself. While neither method is guaranteed theoretically (or in practice) to find an exact nearby solution every time, the use of relaxation means local search mainly works with rational assignments, significantly improving its efficiency in those problem instances where irrational assignments may otherwise be needed.

Finally, we present alternative ways to deal with situations where no critical move is available to satisfy a certain literal (that is, when the literal is *stuck*). We first pick heuristically a variable appearing in the literal, and a set of candidate moves for that variable. For each candidate move, we look ahead to see whether that move will lead to the literal having critical moves on the next step. Such moves are then preferred over the others.

The above ideas are implemented as a local search algorithm on top of the Z3 prover [33]. The implementation relies on the existing library of polynomials and algebraic numbers in Z3, but is otherwise independent from its implementation of MCSAT for nonlinear real arithmetic. We perform thorough experiments on the SMT-LIB benchmarks, showing the effect of incremental computation of variable scores and relaxation of equalities, and that the resulting local search algorithm is competitive against complete algorithms in existing SMT solvers on the satisfiable instances. Especially, there is a large amount of complementarity between the problems solved by local search and by the complete algorithms, indicating major improvements can be obtained in a portfolio setting.

1.2 Related Work

Our work builds upon existing work applying local search to SMT over arithmetic theories [7,29,31]. They will be reviewed in more detail in Sect. 2. Besides arithmetic theories, there have also been earlier work applying local search to the theory of bit-vectors [16,21,37]. In particular, the work [16] generalized the scoring function to consider operators on bit-vectors, but the moves remain single-bit flips. Later works [37,38] introduced propagation-based move selection and essential inputs to prune the search.

The most commonly used framework for complete algorithms for nonlinear real arithmetic is MCSAT, first proposed by Jovanovic and de Moura [25,34]. It improves upon the use of CAD within the DPLL(T) framework by assigning to both arithmetic and boolean variables during the search process. Recent innovations in SMT solving for nonlinear arithmetic include variants to the application of CAD [1], and alternative heuristics for choosing variable ordering [30]. Besides complete methods based on CAD, alternative methods based on linearization [11], interval constraint propagation [26,44], and subtropical methods [15,35] are also explored. While these approaches alone do not achieve the same level of overall performance as CAD and MCSAT, they may have advantages in specific classes of problems, making them useful in a portfolio setting.

Also closely-related to our work are various methods for determining whether a given region contains a solution to a set of (in)equality constraints, and their applications to SMT solving. The work of Cimatti et al. [12] first uses global optimization methods to find an initial solution, then uses topological methods to determine whether an exact solution exists nearby. The work of Ni et al. [36] also uses optimization to find a candidate solution, followed by general methods for solving equations [32] to isolate an exact solution.

The work of Gao et al. [17,18] introduced the framework of δ-complete decision procedures, implemented in the dReal tool for solving SMT problems over nonlinear formulas [19]. It can handle polynomials as well as trigonometric and exponential functions. The δ-complete framework allows algorithms to either return δ-sat or unsat, where the δ-sat case returns a solution for a δ-weakening of the input formulas. This permits efficient numerical algorithms to be used, as well as showing decidability for a wide range of problems. The concept of relaxation of constraints in our work is similar to δ-weakening, and we also use it to increase efficiency of our algorithm. However, we still aim to return an exact answer, by restoring the constraints to their original form and try to find an exact solution near any approximate solution that is found.

1.3 Structure of the Paper

We begin by defining the SMT problem over nonlinear real arithmetic, and reviewing existing local search algorithms in Sect. 2. Section 3 presents incremental computation of variable scores. Section 4 presents relaxation and restoring of equality constraints. Section 5 discusses implementation choices, including heuristic move selection and look-ahead mechanism for stuck literals. Section 6 compares with existing SMT solvers, and performs ablation study on each of the proposed improvements. Finally, we conclude in Sect. 7 with a discussion of potential future directions.

2 Preliminaries

In this section, we formally define SMT problems over nonlinear real arithmetic, followed by a review of existing local search algorithms for arithmetic theories.

The syntax of a general SMT formula over nonlinear real arithmetic is as follows:

$$p := x \mid c \mid p + p \mid p \cdot p \qquad\qquad \text{(polynomials)}$$
$$a := b \mid p \ge 0 \mid p \le 0 \mid p = 0 \qquad\qquad \text{(atoms)}$$
$$f := a \mid \neg f \mid f \wedge f \mid f \vee f \qquad\qquad \text{(formulas)}$$

Here x is an arithmetic variable, c is a constant rational value, b is a boolean variable. A *literal* is either an atom or its negation. A *clause* is a disjunction of literals. In other words, we consider (in)equalities on polynomials with rational coefficients[1]. In practice, we assume that input problem instances are given in conjunctive normal form (CNF), that is as a collection of clauses to be satisfied. Note that strict inequalities $p \ne 0$, $p < 0$ and $p > 0$ can be represented as $\neg(p = 0)$, $\neg(p \ge 0)$ and $\neg(p \le 0)$, respectively. We allow problem instances to contain boolean and arithmetic variables at the same time. We define *boolean literal* and *arithmetic literal* to mean literal whose atom is a boolean variable and a polynomial inequality, respectively.

A polynomial p is *linear* in some variable x if all terms of the polynomial have degree at most one in x. Alternatively, p can be written in the form $p_1 \cdot x + p_2$, where p_1, p_2 do not contain x. A polynomial is *multilinear* if it is linear in each of its variables.

Given a problem instance containing boolean variables b_i $(1 \le i \le m)$ and arithmetic variables x_j $(1 \le j \le n)$, a *complete assignment* is a mapping from each b_i to $\{\top, \bot\}$ and each x_j to $\mathbb{R}$. We will only deal with complete assignments in this paper, and hence sometimes call it *assignment* for short. A formula is

---

[1] Our methods can be extended to handle polynomials with coefficients that are algebraic numbers. Alternatively, a coefficient c_i that is an algebraic number can be encoded as a variable x_i satisfying some polynomial $p(x_i) = 0$ together with interval constraints. Hence we limit the discussion to rational coefficients in this paper.

satisfied under an assignment if it evaluates to true under the standard interpretation of boolean and arithmetic operators. An assignment is a solution to a problem instance if it satisfies all its clauses. The SMT problem for nonlinear real arithmetic is to determine whether a given problem instance is satisfiable by some assignment.

Local search algorithms attempt to determine satisfiability of a problem instance by searching in the space of complete assignments, usually by changing the value of one variable at a time. In the SAT case, each move in local search flips the assignment of one boolean variable. Which flip to make is determined by factors such as the number of clauses that become satisfied/unsatisfied as result of the flip, weight of the clauses, which variables are flipped recently, and so on. For SMT over arithmetic theories, the analogous move changes the value of one arithmetic variable, usually in order to make some clause become satisfied. Such moves, called *critical moves*, are introduced in [7] for linear integer arithmetic.

For nonlinear real arithmetic, the basic procedure used to determine possible moves is *root isolation* for polynomials. Given a polynomial p in a single variable x (where here we allow the coefficients of p to be algebraic numbers), the procedure computes the roots of the equation $p(x) = 0$, together with the sign of the polynomial in each interval separated by the roots. Algebraic numbers in the coefficients of p and in the output of root isolation are represented by their minimal polynomials, together with intervals with rational endpoints that bracket a root of that polynomial.

Given a complete assignment, and an arithmetic literal l involving variable x, we can use root isolation to compute the set of values that assignment of x may be moved to in order for l to be satisfied. This is done by substituting in assignments to other variables in l, resulting in a polynomial containing only variable x, then perform root isolation and check the value of l in each resulting interval. The answer is given in terms of a set of intervals (which may contain $\pm\infty$ as one of the endpoints, and may be either open or closed at each endpoint). We state the definitions precisely as follows.

Definition 1. *Given a complete assignment, a literal l and a variable x, the feasible set (resp. infeasible set) is the collection of intervals that the value of x can be moved to in order for l to be satisfied (resp. unsatisfied). Likewise, we define the feasible set (resp. infeasible set) of a clause with respect to a variable. This is computed by taking the union (resp. intersection) of the feasible set (resp. infeasible set) for each literal in the clause.*

A critical move is defined to be a change in the assignment of some variable x to a value that satisfies some previously unsatisfied clause. The basic local search algorithm then performs critical moves at each iteration, using various scoring metrics to determine which move is chosen next. Such moves can also be interpreted as jumping between CAD cells as in [31]. An innovation in [31] is that when no critical moves are available for some literal, moves that change multiple variables at once along some straight line are also explored.

Scoring of critical moves is usually based on a weighted count of unsatisfied clauses. *Adaptive weighting schemes* assign a weight to each clause, reflecting its importance during the current search. Existing work on local search for arithmetic theories mostly use a probabilistic version of the PAWS weighting scheme [43]. This scheme is parameterized by a smoothing probability sp. Whenever there is no moves available that improves the score, with probability $1 - sp$ the weight of each unsatisfied clause is increased by 1, and with probability sp the weight of each satisfied clause with weight greater than 1 is decreased by 1. Then, the *make-break score* of each critical move equals the total weight of clauses that become satisfied by the move, minus the total weight of clauses that become unsatisfied by the move. A move is *improving* if its score is greater than zero.

One key contribution of [29] is the introduction of *make-break intervals*. The idea is that instead of considering only the (in)feasible intervals of a variable x with respect to some clause, we combine the (in)feasible interval information of x with respect to all clauses. This results in a partition of the real line into intervals, with each interval associated to the make-break score for moving the value of x into that interval. We illustrate this idea with the following example.

Example 1. Consider the set of clauses $x^2 + y^2 \leq 1$, $x + y < 1$ and $x + z > 0$. The current assignment is $x \mapsto 1, y \mapsto 1, z \mapsto 1$, and the current weight of clauses are $1, 3, 2$, respectively. The make-break score for variable x with respect to each of the clauses are:

- $x^2 + y^2 \leq 1$ (unsatisfied): $(-\infty, 0) \mapsto 0$, $[0, 0] \mapsto 1$, $(0, \infty) \mapsto 0$.
- $x + y < 1$ (unsatisfied): $(-\infty, 0) \mapsto 3$, $[0, \infty) \mapsto 0$.
- $x + z > 0$ (satisfied): $(-\infty, -1] \mapsto -2$, $(-1, \infty) \mapsto 0$.

Combining the above information, we obtain the following make-break intervals and scores for x: $(-\infty, -1] \mapsto 1$, $(-1, 0) \mapsto 3$, $[0, 0] \mapsto 1$, $(0, \infty) \mapsto 0$. A preferred move would be to change the value of x into the interval $(-1, 0)$, satisfying the clause $x + y < 1$ and leaving the status of the other clauses unchanged, with a make-break score of 3.

If boolean variables are present, the make-break score of flipping each boolean variable is defined in a similar way, as the total weight of clauses that become satisfied by the flip, minus the total weight of clauses that become unsatisfied.

Algorithm 1 shows the structure of the basic local search procedure. Begin by initializing the assignments to all variables (line 1). At each iteration, first try to find a move with the largest make-break score. If the score is greater than 0 (the variable necessarily comes from an unsatisfied clause), then perform the corresponding move (line 9). If no move has score greater than 0, it indicates that we reached a local minimum. Update the clause weights according to the PAWS scheme (line 11), and then try to make a move that makes a randomly chosen clause satisfied (line 16). If that is also not possible after several tries, randomly change the assignment of some variable in some unsatisfied clause according to some heuristic (line 19). This continues until all clauses are satisfied (line 4) or the time or step limit is reached (line 6).

There are possible variations in the choice between boolean and arithmetic variables on line 7. In [7,29], the search is separated into modes where only

---

Algorithm 1: Basic local search algorithm

    Input   : A set of clauses F
    Output: An assignment of variables that satisfy F, or failure

1 Initialize assignment to variables;
2 **while** $\top$ **do**
3 | **if** *all clauses satisfied* **then**
4 | | **return** *success with assignment;*
5 | **if** *time or step limit reached* **then**
6 | | **return** *failure;*
7 | *var, new_value, score* ← best move according to make-break score;
8 | **if** *score > 0* **then**
9 | | Perform move, assigning *var* to *new_value*;
10 | **else**
11 | | Update clause weight according to PAWS scheme;
12 | | **repeat**
13 | | | *cls* ← random unsatisfied clause;
14 | | | *var, new_value, score* ← critical move making *cls* satisfied;
15 | | | **if** *score* $\neq -\infty$ **then**
16 | | | | Perform move, assigning *var* to *new_value*;
17 | | **until** *3 times;*
18 | | **if** *no move performed in previous loop* **then**
19 | | | Change assignment of some variable in some unsatisfied clause;

---

boolean or arithmetic variables are considered. Alternatively, we can combine the lists of moves and decide between them based purely on make-break scores. We take the latter approach in this paper. There are other aspects of the algorithm that are left unspecified, including how the best move is computed on line 7, and the heuristic choice of moves on line 19. These will be specified in more detail in the next sections.

3   Incremental Computation of Variable Scores

One key step in Algorithm 1 is computing the move with the best make-break score. The computation for boolean variables is standard (and is in any case not the bottleneck here), hence we focus on critical moves for arithmetic variables. The default approach is to loop over all variables in all unsatisfied clauses. For each variable, compute its score with respect to each clause and then combine the results (as demonstrated in Example 1). However, it is clear that this may result in repeated computations across iterations. For example, the feasible set of some variable with respect to a clause may be recomputed, even if none of the variables in that clause are changed in the previous step. Following the idea of caching and updating scores in GSAT [40], we propose data structures for caching and updating score information for arithmetic variables.

We define a *boundary* to be a quadruple $\langle val, is\_open, is\_make, cid \rangle$, where *val* is a real number, *is_open* and *is_make* are boolean values, and *cid* is a clause identifier. The boundary indicates that there is a change in make-break score when moving from less than to greater than *val* due to clause *cid*. If *is_make* is $\top$, the score is increased by the weight of clause *cid*, otherwise it is decreased by the weight. If *is_open* is $\top$, the change is not active at *val*, otherwise it is already active at *val*. There is a natural ordering among boundaries, first order by *val* and then by *is_open* (with $\bot < \top$). The make-break score information of each variable with respect to each clause can be characterized by a starting score (indicating the make-break score of large negative values), together with a set of boundaries. The make-break information of each variable with respect to all clauses is formed by summing the starting score and taking the union of the sets of boundaries. We illustrate the computations in the following example.

Example 2. Continuing from Example 1, the starting score and boundary information of variable x with respect to each clause is as follows (we identify the three clauses as 1, 2, 3, respectively).

- $x^2 + y^2 \leq 1$: starting score 0, boundary set $\{(0, \bot, \top, 1), (0, \top, \bot, 1)\}$, indicating no change for large negative values, *make* at boundary $[0, \cdots,$ followed by *break* at boundary $(0, \cdots.$
- $x + y < 1$: starting score 3, boundary set $\{(0, \bot, \bot, 2)\}$, indicating *make* at large negative values, and *break* at boundary $[0, \ldots.$
- $x + z > 0$: starting score -2, boundary set $\{(-1, \top, \top, 3)\}$, indicating *break* at large negative values, and *make* at boundary $(-1, \ldots.$

The combined make-break score information is: starting score 1, with the following (ordered) set of boundaries: $\{(-1, \top, \top, 3), (0, \bot, \top, 1), (0, \bot, \bot, 2), (0, \top, \bot, 1)\}$. Make-break score information in terms of intervals can be easily recovered from the above, by traversing the boundaries in order, increasing the score by the weight of the clause when encountering a boundary with *is_make* $= \top$, and decreasing the score by the weight otherwise.

During local search, after each move of variable v to a new value, only those variables v' that share a clause with v need to have their make-break score information updated (this is analogous to the concept of dependent variables in the SAT case), and moreover boundary information need to be updated for the shared clauses only. This is summarized in Algorithm 2. The set S collects the set of variables sharing a clause with v. Line 5 recomputes starting score and boundary information. Line 7 recomputes best critical move and score for each updated variable.

Example 3. Continuing from Example 2, suppose the move $y \mapsto -2$ is made, making the clause $x + y < 1$ satisfied. Then the score information for variable z does not need to be updated, as y and z do not share a clause. The score information for variable x need to be updated for the first two clauses. For clause $x^2 + y^2 \leq 1$ there is no longer any boundaries (no assignment of x can

---

Algorithm 2: Incremental computation of make-break scores

Input   : Variable v that is modified
Update: Make-break score for all variables

1 $S \leftarrow \{\}$;   // set of updated variables
2 **for** *clause cls that contains v* **do**
3   **for** *variable v' appearing in cls* **do**
4   add v' to S;
5   recompute starting score and boundary of v' with respect to *cls*;

6 **for** *variable v' in S* **do**
7   recompute best critical move and score in terms of boundary information;

---

make the clause true), and for clause $x + y < 1$ the new starting score and boundary set are 0 and $\{(3, \perp, \perp, 2)\}$, respectively. So the overall starting score and boundary set are -2 and $\{(-1, \top, \top, 3), (3, \perp, \perp, 2)\}$.

Remark 1. Data structures such as arrays, linked lists, or binary trees can be used to maintain set of boundaries. If the total number of boundaries for each variable is small (as is the case for most of the problem instances in SMT-LIB), arrays or linked lists are sufficient. Otherwise the use of binary trees result in better asymptotic performance for the required operations.

Remark 2. A further optimization can be made: it is not necessary to immediately recompute the boundary information for a variable v' that does not appear in any unsatisfied clause, as such variables will never be chosen either on line 7 or line 14 of Algorithm 1. Instead, flags can be used to mark that boundary information for certain clauses need to be updated for v'. When at least one of the clauses containing v' becomes unsatisfied, information for those flagged clauses (as well as other clauses that need to be updated for that step) are updated.

4   Relaxation of Equalities

Equality constraints with degree greater than one pose special difficulty for local search, since it may force assignments of variables to irrational (e.g. algebraic) numbers. For example, for the constraint $x^2 + y^2 + z^2 = 1$, with most rational assignments to x and y, the assignment to z would be forced to be irrational in order for the constraint to be satisfied. While it is possible to represent and compute with algebraic numbers during local search, the time-cost of such computation is significantly increased. Even without considering algebraic numbers, numbers with increasingly large denominators are also problematic for slowing down the search process.

    Both [29, 31] avoid algebraic numbers by either limiting themselves to the multilinear case, or considering only equality constraints with at least one linear variable (and solving only for those linear variables). The work [29] further incorporated comparison of size of denominators (as well as absolute value) of potential assignments into the scoring heuristic, in order to keep the complexity of assigned values as low as possible.

We propose a novel approach to address the problem of assignments to irrational values and values with large denominators, that allow the algorithm to be applied efficiently to the full set of nonlinear arithmetic problem instances. The approach still relies on comparing complexity of assigned values, hence we first define it below.

Definition 2 (Complexity of values). *We define a preorder $\prec_c$ on algebraic numbers as follows. $x \prec_c y$ if x is rational and y is irrational, or if both x and y are rational numbers, and the denominator of x is less than that of y. We write $x \sim_c y$ if neither $x \prec_c y$ nor $y \prec_c x$.*

The relaxation mechanism can be described simply as follows: whenever some equality (or inequality) constraints force an assignment of some variable to a comparatively complex value, those constraints are relaxed before continuing the local search process, so that such assignments never actually occur. The implementation is parameterized by two thresholds. The parameter ϵ_v (for *variable* threshold) specifies the complexity of assigned values (according to Definition 2) beyond which relaxation of constraints should be applied. The parameter ϵ_p (for *polynomial* threshold) specifies the amount of relaxation of polynomial constraints. Both ϵ_v and ϵ_p are chosen to be 10^{-4} in the implementation.

It should be noted that the constraints $p \geq 0$ and $p \leq 0$ together can also force the assignment of variables to irrational values. These constraints may appear as part of clauses with more than one literal, and hence are not equivalent to $p = 0$. This means in general we consider relaxation of non-strict inequalities, although we will still use the slightly imprecise (but more intuitive) description of relaxing equalities throughout the paper.

The detailed method for determining which constraints to relax is as follows. When computing the best make-break score of a variable v, if that score comes from a one-point interval, the set of clause identifiers in the boundaries contributing to that interval are recorded. If the variable v is chosen, and the new value of v is more complex than both ϵ_v and any other previously assigned value (according to Definition 2), all equalities and non-strict inequalities in the recorded clauses that contribute to the boundary are relaxed. The result of relaxation is as follows.

- If the constraint is of the form $p = 0$, it is relaxed into the pair of inequalities $p < \epsilon_p$ and $p > -\epsilon_p$.
- If the constraint is of the form $p \geq 0$, it is relaxed into $p > -\epsilon_p$. Likewise, if the constraint is of the form $p \leq 0$, it is relaxed into $p < \epsilon_p$.

Note that strict inequality constraints cannot force a variable to a particular value. After relaxation, the local search process proceeds as before, but with all evaluation of literals and computation of make-break scores according to the relaxed interpretation of literals.

After local search finds a "solution" under the relaxation of some constraints, it is only an *approximate* solution. In fact, there is no guarantee that there is an exact solution nearby. We currently try two different ways to find an exact solution near the approximate solution.

The first method performs a heuristic analysis of the structure of the relaxed constraints, in an attempt to find an order of solving for the variables that is likely to produce variable assignments that satisfies all of the equations. Essentially, we are trying to find an exact solution to a set of equality constraints, nearby an existing approximate solution. The analysis performs the following steps:

1. If any variable is currently assigned to zero, this is substituted into the constraints. We found this very helpful in practice for eliminating many terms in the constraints.
2. Eliminate any variable x in an equation of the form $p \cdot x + q = 0$, where the valuation of p under the current assignment is not close to zero. This is more general than eliminating variables during preprocessing, which requires p to be constant (see Sect. 5.2).
3. Finally, we iteratively look for a variable that appear only in one of the equations. We associate this variable to the corresponding equation, and then remove the equation from consideration in the ensuing iterations.

If no equations remain at the end of Step 3, we attempt to find an exact solution to the equality constraints by solving for the variables in reverse order of the above process. We begin by considering the association between variables and equations in Step 3 in reverse order, solving for each variable using the associated equation. Then we obtain the values of solved variables in Step 2 in reverse order. The final exact solution to these equations is checked again for satisfaction of all clauses (various numerical inaccuracies may prevent it from being so). If there are unsolved equations remaining after Step 3 of the above analysis, or if the resulting solution fails to satisfy all clauses, we move to the second approach.

The second approach uses a simplified version of local search itself to try to move the approximate solution to an exact solution. First, restore the relaxed constraints to their original forms, and then proceed with local search on the arithmetic variables only, until either an exact solution is found, or no improvement can be made. This is a more limited form of the general local search, as we do not attempt changes to boolean variables, or try random moves in case no improvements can be made. Hence it is likely to terminate quickly with either an exact solution or report of failure.

The above description is summarized as a modification of the overall algorithm, shown in Algorithm 3. The main change to the search process is to relax constraints whenever it forces an assignment to values that are more complex than the variable threshold (line 18). If all clauses are satisfied (indicating an approximate solution is found), we first try finding an exact solution nearby by analyzing the structure of relaxed constraints (line 4). If this fails, we try the limited form of local search described above (line 8-9). If this also fails, the search is restarted with a fresh assignment (line 13).

In practice, we find the two heuristic approaches for finding exact solutions to be useful in different scenarios. The first approach deals nicely with cases where the structure of equality constraints involves solving systems of linear equations, but otherwise poses no difficult choices. The second approach can better handle those cases where there may be choices in which variables to modify, but only some of them is correct in order to satisfy the other inequality constraints. It may also be possible to apply more advanced methods for solving equations or determining existence of solutions in [12,32]. However, the methods we implement easily extends to the non-zero-dimensional case, and returns exact solutions that can be verified independently. We leave the incorporation of more advanced methods for solving equations to future work.

Algorithm 3: Relaxation of equalities

    Input   : A set of clauses F
    Output: An assignment of variables that satisfy F, or failure
1 Initialize assignment to variables;
2 **while** $\top$ **do**
3   **if** *all clauses satisfied* **then**
4   *success* ← find exact solution by analyzing structure of equations;
5   **if** *success* **then**
6   **return** *success with assignment*;
7   **else**
8   Restore relaxed constraints to original form;
9   *success* ← find exact solution by limited local search;
10   **if** *success* **then**
11   **return** *success with assignment*;
12   **else**
13   Perform major restart;
14   **if** *time or step limit reached* **then**
15   **return** *failure*;
16   **if** *no improvement for* T_1 *steps* **then**
17   Perform minor restart;
18   Proceed as in line 7-19 of Algorithm 1, except constraints may be relaxed;

---

5 Implementation

In this section, we describe the implementation in more detail. First, we explain our choice of heuristic move selection when encountering a literal without critical moves. Then, we describe some further details on preprocessing, restart mechanism, and other efficiency improvements.

5.1 Heuristic Moves Selection and Look-Ahead

One major difficulty for local search in nonlinear arithmetic is that it is not always possible to find single-variable moves to satisfy a particular constraint. For example, given constraint $x^2 + y^2 < 1$, and the current assignment $x \mapsto 2, y \mapsto 3$, it is not possible to satisfy the constraint by moving only one of x and y. During local search, this is reflected by the situation that no critical move is available for a clause or literal.

Solving this problem in general would likely require complex algorithms such as CAD or polynomial optimization. Indeed, one category in the SMT-LIB benchmarks, Sturm-MBO, coming from analysis of biological networks [3], consists exclusively of problems that require a very complex polynomial to evaluate to zero, subject to positivity constraints of the variables. When the problem has many variables (is high-dimensional), any approach based on heuristically searching for assignments would have difficulty finding the exact combination of assigned values required to satisfy the constraint.

One approach is given in [31], which involves searching in directions other than those parallel to the coordinate axes to look for solutions. The use of gradient information, as well as scoring based on values of polynomials, increase the chance of finding a solution.

In this paper, we propose another approach that still involves moving only one variable at a time. We say a literal is *stuck* if it is currently unsatisfied and has no critical moves to make it satisfied. Given a literal l that is stuck, we first choose a variable x in l whose coefficient is nonzero (according to the current assignment of the other variables), then heuristically pick a set of candidate values to move the assignment of x to. For each candidate value, we compute whether l is still stuck after making that move. We then prefer those moves that result in l no longer being stuck.

Given the current assignment x_0 and the feasible set I of variable x (see Sect. 5.2), the heuristic move selection include the following:

1. rational numbers and integers close to the boundary inside each interval of I. The rational numbers are chosen to be within 10^{-4} of the boundary.
2. the next integer smaller or larger than x_0.
3. three numbers chosen uniformly in the interval $[\frac{x_0}{2}, x_0)$, and three numbers chosen uniformly in the interval $(x_0, 2x_0]$.

The first class reflects what we know about the constraints on x. The second class attempts basic random walk, and prefers (simple) integer values. The third class is the most general, allowing search over large/small values as well as fractions.

The above ideas are summarized in Algorithm 4. The heuristic choice of candidate values are collected into set S (line 2). Then each value in S is tested in turn. If l has critical move after assigning to any value, that value is returned (line 5). Otherwise a randomly chosen value from S is returned (line 7).

Algorithm 4: Heuristic choice of candidate values and look-ahead for critical moves

---

    Input   : Literal l without critical moves

    Output: Candidate variable x and new value x_1

1 $x \leftarrow$ variable in polynomial of l with nonzero coefficient;

2 $S \leftarrow$ heuristic move selection for variable x;

3 **for** *value x_1 in S* **do**

4   **if** *l has critical move after assigning x to x_1* **then**

5   **return** x, x_1

6 $x_1 \leftarrow$ randomly chosen value in S;

7 **return** x, x_1

5.2 Implementation Details

The algorithm is implemented on top of the Z3 prover, and makes use of its library for polynomials and algebraic numbers, as well as data structures for clauses and literals, but otherwise separate from its implementation of the MCSAT algorithm.

Preprocessing. The following preprocessing steps are used. Eliminate clauses with a single boolean variable and propagate assignments. Combine constraints $p \geq 0$ and $p \leq 0$ into equality $p = 0$ (when they appear as clauses on their own; literals $p \geq 0$ and $p \leq 0$ that are parts of larger clauses cannot be combined). Eliminate variable x in an equation of the form $c \cdot x + q = 0$, where c is a constant and q is a polynomial with degree at most 1 and containing at most 2 variables. The conditions on q are designed so that preprocessing does not significantly increase the complexity of the remaining clauses.

Restart Mechanism. We use a two-level restart mechanism with two parameters T_1 and T_2 (both chosen to be 100 in our implementation). Perform a *minor restart* after T_1 moves without improvements, which randomly changes one of the variables in one of the unsatisfied clauses. After T_2 such minor restarts, a *major restart* is performed that resets the value of all variables.

Shortcut for Linear Equations. Root-isolation is done by calling the existing implementation in Z3, except when the variable to be solved is linear in the polynomial, in which case a direct (and more efficient) solution method is used.

Infeasible Sets of Variables. For each clause involving a single variable, derive infeasible set for that variable implied by the clause. Experience shows that excluding assignments from the infeasible set during local search is beneficial for some problem instances but not others. Hence, we exclude such assignments on alternate turns of minor restarts.

Parameter Settings. Values of tunable parameters used in the implementation are summarized in Table 1.

Table 1. Tunable parameters of the algorithm

Symbol	Explanation	Value
sp	Probability sp for PAWS scheme	0.006
T_1	Number of non-improving steps before minor restart	100
T_2	Number of minor restarts before major restart	100
ϵ_v	Threshold for relaxing equality	10^{-4}
ϵ_p	Amount of relaxation	10^{-4}

6 Evaluation

In this section, we compare the implementation with those of complete procedures in existing SMT solvers Z3 [33], cvc5 [4] and Yices [14], as well as previous work on local search for (fragments of) nonlinear arithmetic. We also perform an ablation study on the two improvements described in Sect. 3 and 4.

The benchmark used in the evaluation comes from SMT-LIB's QF_NRA theory. The benchmark consists mostly of industrial problems from various applications of constraint solving in nonlinear real arithmetic, including analysis of nonlinear hybrid automata (hycomp) [13], and generating ranking functions for termination analysis (LassoRanker) [22,28]. The kissing benchmark contains encoding of the *kissing problem*, which asks how many unit spheres in a given dimension can be placed tangent to a single unit sphere without intersecting each other. Each benchmark is labeled with either sat, unsat or unknown, according to whether it is known to be satisfiable/unsatisfiable at time of submission. Many of the unknown problems are in fact shown to be unsatisfiable by complete algorithms implemented in solvers such as Z3 and cvc5. For the experiments, we choose all problems from the benchmarks that are labeled sat or unknown, but excluding those unknown instances that are found to be unsatisfiable by either Z3 or cvc5. We also note that the SMT-LIB benchmarks contain problems with a wide range of difficulties, but without specified difficulty ratings. For example, many problem instances in the metitarski category, from the MetiTarski tool for proving theorems involving special functions [2], are quite small and do not pose much of a challenge for either local search or complete algorithms.

This yields a total of 6216 instances. The experiments are run on a cluster of machines with Intel Xeon Platinum 8153 processor at 2.00 GHz. Each experiment is run with a time limit of 20 min (as in the SMT competition) and memory limit of 30 GB.

6.1 Overall Result

Results of our implementation are compared against that of other SMT solvers in Table 2. One major advantage of our algorithm is in the Sturm-MBO category, which involves a single complicated polynomial that tripped up other solvers. However, we also showed good result across other categories, and solved most instances overall.

There is significant amount of complementarity between our algorithm and both Z3 and cvc5. As shown in Table 2, there are 148 instances across seven different categories that are solved by local search, but none of Z3, cvc5, and Yices. Moreover, there are 291 instances solved by local search but not Z3, and 378 instances solved by local search but not cvc5. Scatter plots comparing solution times against Z3 and cvc5 are shown in Fig. 1, showing there is significant complementarity in solving times as well.

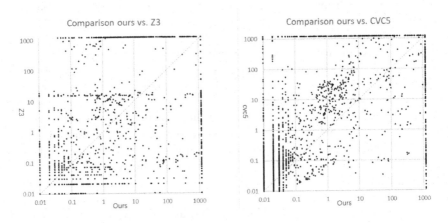

Fig. 1. Scatter plots of running time vs. Z3 and cvc5.

We also note there is a large number of relatively simple problem instances among the SMT-LIB benchmarks for QF_NRA. To put this into quantitative form, we counted the number of instances that are solved within 1 s by all of Z3, cvc5, and our solver. There are 4765 such instances, leaving only 1451 instances that can be considered "challenging" to the solvers. From this view, the overall improvement in the number of solved instances, number of unique solved, and amount of complementarity are quite significant.

Table 2. Comparison with other SMT solvers. Column #inst is the number of instances in the category. Column Z3, cvc5, Yices, and Ours shows number of solved instances by three existing SMT solvers and our implementation. Column Unique is the number of instances solved by our implementation but none of the other three SMT solvers.

Category	#inst	Z3	cvc5	Yices	Ours	Unique
20161105-Sturm-MBO	120	0	0	0	**88**	88
20161105-Sturm-MGC	2	**2**	0	0	0	0
20170501-Heizmann	60	3	1	0	**8**	6
20180501-Economics-Mulligan	93	**93**	89	91	90	0
2019-ezsmt	61	**54**	51	52	19	0
20200911-Pine	237	**235**	201	**235**	224	0
20211101-Geogebra	112	**109**	91	99	101	0
20220314-Uncu	74	73	66	**74**	70	0
LassoRanker	351	155	**304**	122	272	13
UltimateAtomizer	48	**41**	34	39	27	2
hycomp	492	**311**	216	227	304	11
kissing	42	**33**	17	10	**33**	1
meti-tarski	4391	**4391**	4345	4369	4351	0
zankl	133	70	61	58	**100**	27
Total	6216	5570	5476	5376	**5687**	148

6.2 Comparison with Other Work on Local Search

We further compare our results against existing work on local search for fragments of nonlinear real arithmetic. Of the 979 instances that are multilinear considered by [29], our implementation can solve 826 instances, compared to 891 solved instances there. The slightly weaker result is likely due to the more efficient implementation that is possible when only rational numbers need to be considered, and the parameter tuning that is specific to multilinear problems. Of the 2736 instances from SMT-LIB considered by [31], our implementation can solve 2589 instances, compared to 2246 solved instances there. In fact, we solve not only more instances than the local search algorithm given in [31], but also all other SMT solvers used in the comparison. We note that the result given in [31] uses different underlying software (including Maple) and runs on different machines, so this gives only a rough comparison.

6.3 Effect of Incremental Computation of Variable Scores

To show the effect of speedup resulting from incremental computation of variable scores in Sect. 3, we compare three versions of the implementation: with incremental computation (Incremental), without incremental computation (Naive), and without incremental computation, but limiting the number of unsatisfied clauses considered at each turn to 45 (Limit-45). The results are shown in Table 3.

We see that while the difference in total number of problem instance solved is not large, a noticeable effect is still present in the LassoRanker category, whose instances usually require a long time to solve. A closer look at the running time shows that it usually takes 2–10 times longer to solve a particular instance using either (Naive) or (Limit-45) compared to (Incremental), with the exact ratio depending strongly on the specific instance. For a time limit of 20 min the resulting difference in the number of solved problems is not large, but we expect a larger difference with shorter time limits, and especially when local search is incorporated into other methods such as DPLL [8,9].

6.4 Effect of Relaxation of Equalities

We demonstrate the effect of relaxation of constraints by comparing three possible implementations: with relaxation of constraints (Relaxation), without relaxation of constraints, but preferring variable assignments that are less complex than ϵ_v (Threshold), and without relaxation of constraints, with choosing variable assignments without considering complexity order (NoOrder). The results are shown in Table 4.

The results indicate that while taking complexity of assigned values into consideration does have an effect in keeping the search efficient for most categories of problem instances, it is not sufficient for the hycomp category, which involves a large number of nonlinear equalities. In that category using relaxation of constraints have a significant effect, while also performing well in other categories.

Table 3. Comparison showing effect of incremental computation

Category	#inst	Incremental	Naive	Limit-45
20161105-Sturm-MBO	120	88	85	85
20161105-Sturm-MGC	2	0	0	0
20170501-Heizmann	60	8	5	5
20180501-Economics-Mulligan	93	90	89	89
2019-ezsmt	61	19	19	15
20200911-Pine	237	224	222	222
20211101-Geogebra	112	101	101	101
20220314-Uncu	74	70	70	70
LassoRanker	351	272	264	269
UltimateAtomizer	48	27	26	26
hycomp	492	304	298	298
kissing	42	33	32	33
meti-tarski	4391	4351	4352	4352
zankl	133	100	100	100
Total	6216	5687	5663	5665

Table 4. Comparison showing effect of temporary relaxation of constraints

Category	#inst	Relaxation	Threshold	NoOrder
20161105-Sturm-MBO	120	88	100	99
20161105-Sturm-MGC	2	0	0	0
20170501-Heizmann	60	8	9	3
20180501-Economics-Mulligan	93	90	89	86
2019-ezsmt	61	19	19	19
20200911-Pine	237	224	223	222
20211101-Geogebra	112	101	98	92
20220314-Uncu	74	70	70	70
LassoRanker	351	272	277	278
UltimateAtomizer	48	27	26	20
hycomp	492	304	211	164
kissing	42	33	31	27
meti-tarski	4391	4351	4353	4360
zankl	133	100	100	100
Total	6216	5687	5606	5540

6.5 Other Techniques

We also tried other techniques commonly used in works on local search, including tabu search [20], switching between phases for adjusting boolean and arithmetic variables (as applied in [7]), and incorporating random walk (such as variants of WalkSAT [23]). Unlike their applications in earlier work, the use of such methods did not result in noticeable improvements on the current benchmark. However, it remains to investigate whether they will be useful on other types of problems, or in combination with other improvements to the algorithm.

7 Conclusion

In this paper, we presented improvements to the local search algorithm for solving SMT problems in nonlinear real arithmetic. Building upon the basic structure of local search, we presented incremental computation of variable scores and temporary relaxation of constraints. We also described heuristic move selection with look-ahead for dealing with literals without critical moves, and implementation details to improve efficiency. The resulting implementation is competitive against complete algorithms based on DPLL(T) and MCSAT on satisfiable problem instances, as implemented in other SMT solvers. It is the first local search algorithm designed for the entirety of nonlinear real arithmetic, covering a wider range of problems than existing work [29, 31].

While the methods proposed in this paper made progress in addressing challenges of local search for nonlinear real arithmetic, there are remaining problems that represent interesting directions of future work.

- Look-ahead for critical moves presents another way to improve upon random search in cases when no critical move is available. On the other hand, methods based on CAD or polynomial optimization would give a more complete way to determine assignments that satisfy a certain literal. A major challenge is how to incorporate such algorithms into local search in an efficient way.
- In the current work, after an approximate solution is found that satisfies the relaxed version of equalities, we use various heuristic methods to attempt to find an exact solution nearby. Designing or incorporating more general algorithms for finding exact solutions near approximate solutions (or determining that none exist) is an interesting problem that we leave to future work.
- Finally, local search can be incorporated into complete methods such as DPLL, improving its performance even on unsatisfiable instances, as shown by the works [8,9]. It is interesting to investigate this possibility for SMT problems over nonlinear real arithmetic. The improvements in efficiency in our work would be very helpful for such combination, as in those cases local search is only given very short running times.

Acknowledgements. This work was partially supported by the National Natural Science Foundation of China under Grant Nos. 62032024 and 62002351.

References

1. Ábrahám, E., Davenport, J.H., England, M., Kremer, G.: Deciding the consistency of non-linear real arithmetic constraints with a conflict driven search using cylindrical algebraic coverings. J. Log. Algebraic Methods Program. **119**, 100633 (2021). https://doi.org/10.1016/j.jlamp.2020.100633
2. Akbarpour, B., Paulson, L.C.: MetiTarski: an automatic theorem prover for real-valued special functions. J. Autom. Reason. **44**(3), 175–205 (2010). https://doi.org/10.1007/s10817-009-9149-2
3. Akutsu, T., Hayashida, M., Tamura, T.: Algorithms for inference, analysis and control of Boolean networks. In: Horimoto, K., Regensburger, G., Rosenkranz, M., Yoshida, H. (eds.) AB 2008. LNCS, vol. 5147, pp. 1–15. Springer, Heidelberg (2008). https://doi.org/10.1007/978-3-540-85101-1_1
4. Barbosa, H., et al.: cvc5: a versatile and industrial-strength SMT solver. In: TACAS 2022. LNCS, vol. 13243, pp. 415–442. Springer, Cham (2022). https://doi.org/10.1007/978-3-030-99524-9_24
5. Barrett, C., Fontaine, P., Tinelli, C.: The Satisfiability Modulo Theories Library (SMT-LIB) (2016). www.SMT-LIB.org
6. Barrett, C., Tinelli, C.: Satisfiability modulo theories. In: Handbook of Model Checking, pp. 305–343. Springer, Cham (2018). https://doi.org/10.1007/978-3-319-10575-8_11
7. Cai, S., Li, B., Zhang, X.: Local search for SMT on linear integer arithmetic. In: Shoham, S., Vizel, Y. (eds.) Computer Aided Verification - 34th International Conference, CAV 2022, Haifa, Israel, 7–10 August 2022, Proceedings, Part II. LNCS, vol. 13372, pp. 227–248. Springer, Cham (2022). https://doi.org/10.1007/978-3-031-13188-2_12

8. Cai, S., Zhang, X.: Deep cooperation of CDCL and local search for SAT. In: Li, C.-M., Manyà, F. (eds.) SAT 2021. LNCS, vol. 12831, pp. 64–81. Springer, Cham (2021). https://doi.org/10.1007/978-3-030-80223-3_6
9. Cai, S., Zhang, X., Fleury, M., Biere, A.: Better decision heuristics in CDCL through local search and target phases. J. Artif. Intell. Res. **74**, 1515–1563 (2022). https://doi.org/10.1613/jair.1.13666
10. Caviness, B.F., Johnson, J.R.: Quantifier Elimination and Cylindrical Algebraic Decomposition. Texts and Monographs in Symbolic Computation. Springer, Vienna (2004). https://doi.org/10.1007/978-3-7091-9459-1
11. Cimatti, A., Griggio, A., Irfan, A., Roveri, M., Sebastiani, R.: Incremental linearization for satisfiability and verification modulo nonlinear arithmetic and transcendental functions. ACM Trans. Comput. Log. **19**(3), 19:1–19:52 (2018). https://doi.org/10.1145/3230639
12. Cimatti, A., Griggio, A., Lipparini, E., Sebastiani, R.: Handling polynomial and transcendental functions in SMT via unconstrained optimisation and topological degree test. In: Bouajjani, A., Holík, L., Wu, Z. (eds.) Automated Technology for Verification and Analysis - 20th International Symposium, ATVA 2022, Virtual Event, 25–28 October 2022, Proceedings. LNCS, vol. 13505, pp. 137–153. Springer, Cham (2022). https://doi.org/10.1007/978-3-031-19992-9_9
13. Cimatti, A., Mover, S., Tonetta, S.: A quantifier-free SMT encoding of non-linear hybrid automata. In: Cabodi, G., Singh, S. (eds.) Formal Methods in Computer-Aided Design, FMCAD 2012, Cambridge, UK, 22–25 October 2012, pp. 187–195. IEEE (2012). https://ieeexplore.ieee.org/document/6462573/
14. Dutertre, B.: Yices 2.2. In: Biere, A., Bloem, R. (eds.) CAV 2014. LNCS, vol. 8559, pp. 737–744. Springer, Cham (2014). https://doi.org/10.1007/978-3-319-08867-9_49
15. Fontaine, P., Ogawa, M., Sturm, T., Vu, X.T.: Subtropical satisfiability. In: Dixon, C., Finger, M. (eds.) FroCoS 2017. LNCS (LNAI), vol. 10483, pp. 189–206. Springer, Cham (2017). https://doi.org/10.1007/978-3-319-66167-4_11
16. Fröhlich, A., Biere, A., Wintersteiger, C.M., Hamadi, Y.: Stochastic local search for satisfiability modulo theories. In: Bonet, B., Koenig, S. (eds.) Proceedings of the Twenty-Ninth AAAI Conference on Artificial Intelligence, 25–30 January 2015, Austin, Texas, USA, pp. 1136–1143. AAAI Press (2015). http://www.aaai.org/ocs/index.php/AAAI/AAAI15/paper/view/9896
17. Gao, S., Avigad, J., Clarke, E.M.: δ-complete decision procedures for satisfiability over the reals. In: Gramlich, B., Miller, D., Sattler, U. (eds.) IJCAR 2012. LNCS (LNAI), vol. 7364, pp. 286–300. Springer, Heidelberg (2012). https://doi.org/10.1007/978-3-642-31365-3_23
18. Gao, S., Avigad, J., Clarke, E.M.: Delta-decidability over the reals. In: Proceedings of the 27th Annual IEEE Symposium on Logic in Computer Science, LICS 2012, Dubrovnik, Croatia, 25–28 June 2012, pp. 305–314. IEEE Computer Society (2012). https://doi.org/10.1109/LICS.2012.41
19. Gao, S., Kong, S., Clarke, E.M.: dReal: an SMT solver for nonlinear theories over the reals. In: Bonacina, M.P. (ed.) CADE 2013. LNCS (LNAI), vol. 7898, pp. 208–214. Springer, Heidelberg (2013). https://doi.org/10.1007/978-3-642-38574-2_14
20. Glover, F.W., Laguna, M.: Tabu Search. Kluwer (1997). https://doi.org/10.1007/978-1-4615-6089-0
21. Griggio, A., Phan, Q.-S., Sebastiani, R., Tomasi, S.: Stochastic local search for SMT: combining theory solvers with WalkSAT. In: Tinelli, C., Sofronie-Stokkermans, V. (eds.) FroCoS 2011. LNCS (LNAI), vol. 6989, pp. 163–178. Springer, Heidelberg (2011). https://doi.org/10.1007/978-3-642-24364-6_12

22. Heizmann, M., Hoenicke, J., Leike, J., Podelski, A.: Linear ranking for linear lasso programs. In: Van Hung, D., Ogawa, M. (eds.) ATVA 2013. LNCS, vol. 8172, pp. 365–380. Springer, Cham (2013). https://doi.org/10.1007/978-3-319-02444-8_26

23. Hoos, H.H., Stützle, T.: Local search algorithms for SAT: an empirical evaluation. J. Autom. Reason. **24**(4), 421–481 (2000). https://doi.org/10.1023/A:1006350622830

24. Hoos, H.H., Stützle, T.: Stochastic Local Search: Foundations & Applications. Elsevier/Morgan Kaufmann (2004)

25. Jovanović, D., de Moura, L.: Solving non-linear arithmetic. In: Gramlich, B., Miller, D., Sattler, U. (eds.) IJCAR 2012. LNCS (LNAI), vol. 7364, pp. 339–354. Springer, Heidelberg (2012). https://doi.org/10.1007/978-3-642-31365-3_27

26. Khanh, T.V., Ogawa, M.: SMT for polynomial constraints on real numbers. In: Jeannet, B. (ed.) Third Workshop on Tools for Automatic Program Analysis, TAPAS 2012, Deauville, France, 14 September 2012. ENTCS, vol. 289, pp. 27–40. Elsevier (2012). https://doi.org/10.1016/j.entcs.2012.11.004

27. Kremer, G.: Cylindrical algebraic decomposition for nonlinear arithmetic problems. Ph.D. thesis, RWTH Aachen University, Germany (2020). https://publications.rwth-aachen.de/record/792185

28. Leike, J., Heizmann, M.: Ranking templates for linear loops. Log. Methods Comput. Sci. **11**(1) (2015). https://doi.org/10.2168/LMCS-11(1:16)2015

29. Li, B., Cai, S.: Local search for SMT on linear and multilinear real arithmetic. CoRR abs/2303.06676 (2023). https://doi.org/10.48550/arXiv.2303.06676. Accepted for FMCAD

30. Li, H., Xia, B., Zhang, H., Zheng, T.: Choosing better variable orderings for cylindrical algebraic decomposition via exploiting chordal structure. J. Symb. Comput. **116**, 324–344 (2023). https://doi.org/10.1016/j.jsc.2022.10.009

31. Li, H., Xia, B., Zhao, T.: Local search for solving satisfiability of polynomial formulas. In: Enea, C., Lal, A. (eds.) Computer Aided Verification - 35th International Conference, CAV 2023, Paris, France, 17–22 July 2023, Proceedings, Part II. LNCS, vol. 13965, pp. 87–109. Springer, Cham (2023). https://doi.org/10.1007/978-3-031-37703-7_5

32. Li, H., Xia, B., Zhao, T.: Square-free pure triangular decomposition of zero-dimensional polynomial systems. J. Syst. Sci. Complex. (2023)

33. de Moura, L., Bjørner, N.: Z3: an efficient SMT solver. In: Ramakrishnan, C.R., Rehof, J. (eds.) TACAS 2008. LNCS, vol. 4963, pp. 337–340. Springer, Heidelberg (2008). https://doi.org/10.1007/978-3-540-78800-3_24

34. de Moura, L., Jovanović, D.: A model-constructing satisfiability calculus. In: Giacobazzi, R., Berdine, J., Mastroeni, I. (eds.) VMCAI 2013. LNCS, vol. 7737, pp. 1–12. Springer, Heidelberg (2013). https://doi.org/10.1007/978-3-642-35873-9_1

35. Nalbach, J., Ábrahám, E.: Subtropical satisfiability for SMT solving. In: Rozier, K.Y., Chaudhuri, S. (eds.) NASA Formal Methods - 15th International Symposium, NFM 2023, Houston, TX, USA, 16–18 May 2023, Proceedings. LNCS, vol. 13903, pp. 430–446. Springer, Cham (2023). https://doi.org/10.1007/978-3-031-33170-1_26

36. Ni, X., Wu, Y., Xia, B.: Solving SMT over non-linear real arithmetic via numerical sampling and symbolic verification. In: SETTA 2023 (2023)

37. Niemetz, A., Preiner, M., Biere, A.: Precise and complete propagation based local search for satisfiability modulo theories. In: Chaudhuri, S., Farzan, A. (eds.) CAV 2016. LNCS, vol. 9779, pp. 199–217. Springer, Cham (2016). https://doi.org/10.1007/978-3-319-41528-4_11

38. Niemetz, A., Preiner, M., Fröhlich, A., Biere, A.: Improving local search for bit-vector logics in SMT with path propagation (2015)
39. Nieuwenhuis, R., Oliveras, A., Tinelli, C.: Solving SAT and SAT modulo theories: from an abstract Davis–Putnam–Logemann–Loveland procedure to DPLL(T). J. ACM **53**(6), 937–977 (2006). https://doi.org/10.1145/1217856.1217859
40. Selman, B., Levesque, H.J., Mitchell, D.G.: A new method for solving hard satisfiability problems. In: Swartout, W.R. (ed.) Proceedings of the 10th National Conference on Artificial Intelligence, San Jose, CA, USA, 12–16 July 1992, pp. 440–446. AAAI Press/The MIT Press (1992). http://www.aaai.org/Library/AAAI/1992/aaai92-068.php
41. Shen, D., Lierler, Y.: SMT-based constraint answer set solver EZSMT+ for non-tight programs. In: Thielscher, M., Toni, F., Wolter, F. (eds.) Principles of Knowledge Representation and Reasoning: Proceedings of the Sixteenth International Conference, KR 2018, Tempe, Arizona, 30 October–2 November 2018, pp. 67–71. AAAI Press (2018). https://aaai.org/ocs/index.php/KR/KR18/paper/view/18049
42. Susman, B., Lierler, Y.: SMT-based constraint answer set solver EZSMT (system description). In: Carro, M., King, A., Saeedloei, N., Vos, M.D. (eds.) Technical Communications of the 32nd International Conference on Logic Programming, ICLP 2016 TCs, New York City, USA, 16–21 October 2016. OASIcs, vol. 52, pp. 1:1–1:15. Schloss Dagstuhl - Leibniz-Zentrum für Informatik (2016). https://doi.org/10.4230/OASIcs.ICLP.2016.1
43. Thornton, J., Pham, D., Bain, S., Ferreira Jr., V.: Additive versus multiplicative clause weighting for SAT. In: McGuinness, D.L., Ferguson, G. (eds.) Proceedings of the Nineteenth National Conference on Artificial Intelligence, Sixteenth Conference on Innovative Applications of Artificial Intelligence, 25–29 July 2004, San Jose, California, USA, pp. 191–196. AAAI Press/The MIT Press (2004). http://www.aaai.org/Library/AAAI/2004/aaai04-031.php
44. Tung, V.X., Khanh, T.V., Ogawa, M.: raSAT: an SMT solver for polynomial constraints. Formal Methods Syst. Des. **51**(3), 462–499 (2017). https://doi.org/10.1007/s10703-017-0284-9

Author Index

Printed in the United States
by Baker & Taylor Publisher Services

Printed in the United States
by Baker & Taylor Publisher Services